SLEEP MEDICINE

CONTEMPORARY NEUROLOGY SERIES AVAILABLE:

19 THE DIAGNOSIS OF STUPOR AND COMA, Edition 3
    Fred Plum, M.D., and Jerome B. Posner, M.D.

26 PRINCIPLES OF BEHAVIORAL NEUROLOGY
    M-Marsel Mesulam, M.D., *Editor*

32 CLINICAL NEUROPHYSIOLOGY OF THE VESTIBULAR SYSTEM, Edition 2
    Robert W. Baloh, M.D., and Vincente Honrubia, M.D.

36 DISORDERS OF PERIPHERAL NERVES, Edition 2
    Herbert H. Schaumburg, M.D., Alan R. Berger, M.D., and P.K. Thomas, C.B.E., M.D., D.Sc., F.R.C.P., F.R.C.Path.

38 PRINCIPLES OF GERIATRIC NEUROLOGY
    Robert Katzman, M.D., and John W. Rowe, M.D., *Editors*

42 MIGRAINE: MANIFESTATIONS, PATHOGENESIS, AND MANAGEMENT
    Robert A. Davidoff, M.D.

43 NEUROLOGY OF CRITICAL ILLNESS
    Eelco F. M. Wijdicks, M.D., Ph.D., F.A.C.P.

44 EVALUATION AND TREATMENT OF MYOPATHIES
    Robert C. Griggs, M.D., Jerry R. Mendell, M.D., and Robert G. Miller, M.D.

45 NEUROLOGIC COMPLICATIONS OF CANCER
    Jerome B. Posner, M.D.

46 CLINICAL NEUROPHYSIOLOGY
    Jasper R. Daube, M.D., *Editor*

47 NEUROLOGIC REHABILITATION
    Bruce H. Dobkin, M.D.

48 PAIN MANAGEMENT: THEORY AND PRACTICE
    Russell K. Portenoy, M.D., and Ronald M. Kanner, M.D., *Editors*

49 AMYOTROPHIC LATERAL SCLEROSIS
    Hiroshi Mitsumoto, M.D., D.Sc., David A. Chad, M.D., F.R.C.P., and Eric P. Pioro, M.D., D.Phil., F.R.C.P.

50 MULTIPLE SCLEROSIS
    Donald W. Paty, M.D., F.R.C.P.C., and George C. Ebers, M.D., F.R.C.P.C.

51 NEUROLOGY AND THE LAW: PRIVATE LITIGATION AND PUBLIC POLICY
    H. Richard Beresford, M.D., J.D.

52 SUBARACHNOID HEMORRHAGE: CAUSES AND CURES
    Bryce Weir, M.D.

54 BRAIN TUMORS
    Harry S. Greenberg, M.D., William F. Chandler, M.D., and
    Howard M. Sandler, M.D.

55 THE NEUROLOGY OF EYE MOVEMENTS, Edition 3
    R. John Leigh, M.D., and David S. Zee, M.D.
    (book and CD-ROM versions available)

SLEEP MEDICINE

MICHAEL S. ALDRICH, M.D.

Professor of Neurology
Director, Sleep Disorders Center
University of Michigan Medical Center
Ann Arbor, Michigan

New York   Oxford
OXFORD UNIVERSITY PRESS
1999

Oxford University Press

Oxford New York
Athens Auckland Bangkok Bogotá Buenos Aires Calcutta
Cape Town Chennai Dar es Salaam Delhi Florence Hong Kong Istanbul
Karachi Kuala Lumpur Madrid Melbourne Mexico City Mumbai
Nairobi Paris São Paulo Singapore Taipei Tokyo Toronto Warsaw

and associated companies in
Berlin Ibadan

Copyright © 1999 by Oxford University Press, Inc.

Published by Oxford University Press, Inc.,
198 Madison Avenue, New York, New York 10016

Oxford is a registered trademark of Oxford University Press.

All right reserved. No part of this publication may be reproduced,
stored in a retrieval system, or transmitted, in any form or by any means,
electronic, mechanical, photocopying, recording, or otherwise,
without the prior permission of Oxford University Press.

Library of Congress Cataloging-in-Publication Data
Aldrich, Michael S.
  Sleep medicine
  Michael S. Aldrich
     p.   cm. — (Contemporary neurology series ; 53)
  Includes bibliographical references and index.
  ISBN 0-19-512957-1
    1. Sleep disorders. 2. Sleep disorders—Pathophysiology.
  3. Sleep—Physiological aspects.
    I. Title. II. Series.
    [DNLM: 1. Sleep Disorders. 2. Sleep—physiology. 3. Sleep Apnea Syndromes.
  WM 188A365s 1999] RC547.A53 1999
    616.8'498—dc21   DNLM/DLC   for Library of Congress   98-30554

The science of medicine is a rapidly changing field. As new research and clinical
experience broaden our knowledge, changes in treatment and drug therapy do occur.
The author and the publisher of this work have checked with sources believed to be
reliable in their efforts to provide information that is accurate and complete, and in
accordance with the standards accepted at the time of publication. However, in light of
the possibility of human error or changes in the practice of medicine, neither the
author, nor the publisher, nor any other party who has been involved in the preparation
or publication of this work warrants that the information contained herein is in every
respect accurate or complete. Readers are encouraged to confirm the information contained herein with other reliable sources, and are strongly advised to check the product
information sheet provided by the pharmaceutical company for each drug they plan to
administer.

9 8 7 6 5 4 3 2 1

Printed in the United States of America
on acid-free paper.

FOREWORD

When I first met Michael Aldrich, sleep medicine as an organized discipline did not exist. It has been a long, happy journey for the three of us–Mike, me, and sleep medicine.

In 1957, a year out of graduate school, I was hired as a clinical psychologist in the University of Chicago Department of Psychiatry by its chairman, Dr. C. Knight Aldrich–Mike's father. (Knight Aldrich was distinguished by his sober view of Psychiatry as an integral part of medical practice rather than a new religion–and by his failure to hide a very big heart behind a businesslike demeanor.) Four years earlier, Eugene Aserinsky and his mentor Nathaniel Kleitman, also at the University of Chicago, had published their historic report on the association of REM sleep and dreams, but I didn't know anything about that at the time. Aserinsky had left the University a few years before I arrived, William Dement, another student of Dr. Kleitman, had left just before I arrived, and Dr. Kleitman was preparing for his retirement. I learned about REM sleep through an incidental contact with Edward Wolpert, now a prominent Chicago psychiatrist, who had been using Dr. Kleitman's lab for his doctoral research on the relationship between electromyographic potentials and dream content. I fell in love with the research; it seemed to be a wonderful lever for studying the mind-body problem. Wolpert taught me how to run subjects, which I did intermittently in Kleitman's lab whenever I could squeeze it between clinical duties and getting a bit of sleep myself. My only contacts with Dr. Kleitman were in the morning when he arrived as I was finishing the night's work, but he gave me sound advice: "Make sure you clean up before you leave." I passed that good advice along to my graduate students for the next 40 years.

A year or so after I arrived I was faced with near crisis. Dr. Kleitman was closing his lab and moving to California. How to continue? Knight Aldrich came to the rescue. He raised the money for my first EEG machine, and I started my own lab. Over the years, marvelous graduate students and colleagues passed through the lab, including Ruth Benca, Bernard Bergmann, Donald Bliwise, Doris Chernik, Charmane Eastman, Carol Everson, William Flanigan, David Foulkes, Christopher Frederickson, Marcia Gilliland, Kristyna Hartse, Peter Hauri, Clete Kushida, Ralph Mistlberger, Lawrence Monroe, June Pilcher, Richard Rosenberg, Paul Shaw, Gerald Vogel, Robert Watson, Harold Zepelin, and William Zimmerman. One of the younger recruits to the lab was Mike Aldrich, who started working with us during the summer of 1964 while he was between high school semesters. Mike returned to us for a summer fellowship in 1968 and then for another short stint in 1975, just before he started medical school. Obviously, we liked Mike and his work, or we would not have had him back at every opportunity. Even as an adolescent, he showed complete responsibility, a fine mind, and a realistic appraisal of data. One of his tours in the lab resulted in Mike's first professional publication—on a series of four experiments which were among the first to challenge the prevailing model of sleep regulation by serotonin: Rechtschaffen A, Lovell RA, Freedman DX, Whitehead WE, Aldrich M. The effect of parachlorophenylalanine on sleep in the rat: Some implications for the serotonin-sleep hypothesis. In: Barchas J, Usdin E, eds. *Serotonin and Behavior.* New York: Academic Press, 1973: 401–18.

Since those early years, Mike has developed into a sleep authority in his own right. He is now Professor of Neurology at the University of Michigan Medical Center where he directs a busy Sleep Disorders Center. He is also the author of many scientific and clinical papers, with special emphasis on narcolepsy and other sleep disorders associated with neurological pathology. I am very proud of him. My own career has also matured; I expect to retire at the end of this year. And during the period of Mike's growth and my "maturation," *sleep medicine* was born and has flourished

When Mike and I started to study sleep, very little was known about sleep disorders. Insomnias were often casually attributed to hyperthyroidism (almost never the case), hypersomnias were attributed to hypothyroidism (also almost never the case), and sleep apnea was unknown. During the 1960s, like other basic sleep researchers, we began to explore sleep disorders. When I suggested to Lawrence Monroe, one of my first graduate students, that a laboratory study of insomnia might be an interesting dissertation topic, he worried that it might not pass a dissertation committee because nothing like it had ever been done. Nevertheless, Larry undertook the research, finished it in 1965, got his Ph.D., and the study "Psychological and Physiological Differences between Good and Poor Sleepers" became a classic—the first systematic, laboratory-based examination of problems of falling asleep and staying asleep. The following year Peter Hauri earned his Ph.D. with his study of the effects of presleep activity on subsequent sleep—a precursor to his longstanding interest in sleep hygiene. Responding to one of our psychiatrist's interests in the psychological causes of narcolepsy, I administered Rorschach tests to narcoleptic patients. Fortunately we also recorded them electrophysiologically in the laboratory, where their distinctive sleep-onset REM periods became readily apparent. As information about sleep disorders began to accumulate in various basic sleep research laboratories, it became apparent that systematic study of these disorders could be used to diagnose and treat them more rationally. Pioneers like William Dement, Anthony Kales, and Elliot Weitzman in this country and Elio Lugaresi in Italy started sleep clinics. Today the American Sleep Disorders Association lists more than 250 accredited Centers and over 3000 individual members. The Association of Polysomnographic Technologists lists over 1,500 members. Several major centers have training programs for sleep disorder specialists. Over 100 journal articles on sleep disorders are published every month. Obviously, sleep medicine has come of age.

This book is a solid contribution to the field. Because it is accurate, well organized, and compact, it can serve as an entry text for the new sleep clinician, an overview of the field for practitioners in other specialties, and a handy reference for the working sleep clinician. I was most pleased to see the inclusion of introductory chapters on basic sleep phenomena that show the roots from which sleep medicine arose, and the extensive references, which can serve as a springboard for further, more detailed study.

Chicago, Illinois   Allan Rechtschaffen

PREFACE

Sleep medicine is a relatively new medical specialty. Although sleep problems have always been part of medical practice, scientists viewed sleep as a period of quiescence of brain and body until well into the twentieth century, and physicians generally considered sleep disturbance to be a consequence of other medical or psychiatric problems. These views began to change following several discoveries in the latter half of the twentieth century. First, the discovery of REM sleep led inescapably to the conclusion that sleep is an active process, not simply a reduction of waking activity. Second, the discovery that narcolepsy is associated with abnormalities of REM sleep indicated that disturbances of brain mechanisms underlying sleep can lead to daytime symptoms. Third, the identification of sleep apnea and the discovery of periodic limb movements indicated the need for objective study of sleep and led to the development of sleep centers that combined clinical evaluations with sleep laboratory investigations. Fourth, a series of epidemiologic and clinical studies demonstrated that sleep disorders are common, that they can lead to serious consequences, and that they can be treated successfully in many patients.

As a result of these discoveries, the importance of a rational approach to the evaluation of sleep problems has increased. For example, insomnia, a problem for about 15% of adults, has many causes, and treatment directed at the underlying causes is often successful. Sleep apnea, symptomatic in 1% to 3% of the adult population, usually can be treated effectively; if untreated, it is associated with increased risk of cardiovascular morbidity. The sleepiness caused by chronic partial sleep deprivation, a major contributor to industrial accidents and motor vehicle fatalities, can be eliminated in most persons by increased amounts of sleep.

Nonetheless, physicians often ignore sleep complaints, misdiagnose the underlying disorders, or initiate inappropriate treatments, in part because they receive little training in sleep medicine. Most medical schools have little formal curriculum devoted to sleep, and the quality and quantity of sleep medicine education during and after residency is variable. The limited and uneven education of physicians about sleep medicine is unfortunate not only because these disorders are serious but also because of recent major advances in sleep medicine and sleep research. For example, publications in the 1990s have clarified the molecular basis of the circadian timing system; the effects of light on the sleep–wake cycle; the prevalence of sleep apnea, restless legs, and partial sleep deprivation; the types of syndromes related to narcolepsy, sleepwalking, and obstructive sleep apnea; and the morbidity and mortality associated with sleep disorders.

This book meets the need for accurate, concise information concerning the evaluation and treatment of patients with sleep problems. Because management of sleep disorders requires an understanding of sleep itself, the first portion concerns aspects of normal sleep. For the remainder of the book, the approach is based on the clinical method generally used in medicine, in which the chief complaint, the symptoms and signs, and the psychosocial and medical background are used, in conjunction with an understanding of the neurobiological and psychological basis of sleep disorders, to formulate a clinical assessment. With this approach, in which clinical evaluation is primary, the results of polysomnographic studies and other laboratory tests can supplement and refine the clinical information, narrow the differential diagnosis, and assist with consideration of treatment options.

I could not have written this book without the expert training I received in sleep research and sleep medicine. I was first exposed to sleep research at the age of 9 when my father, then chair of Psychiatry at the University of Chicago, arranged for me to be a subject in a study by Ed Wolpert on physiologic correlates of dream mentation.[1] During my night in the sleep laboratory, just 5 years after the first publication on REM sleep, I was awakened several times, reported my dreams, and in the morning received a subject fee of $1. Al Rechtschaffen first taught me about sleep: I worked in his laboratory as an animal caretaker in high school, as an undergraduate research fellow during one summer in college, and as a research assistant just before medical school. Al taught me about the scientific method, about the importance of basing conclusions on data, and about the value of asking the right questions. In his laboratory, I first became fascinated by the wavy EEG tracings emerging from a polygraph. In medical school, I spent much of my psychiatry rotation in Edinburgh, Scotland, with Ian Oswald, gaining an international perspective as well as insights into the psychiatric aspects of sleep.

Yet, despite this exposure to sleep research, I had little knowledge of sleep disorders. When Sid Gilman offered me the opportunity to begin a sleep disorders program in Neurology at the University of Michigan, I realized that I needed some training. Al Rechtschaffen advised me to study with Bill Dement, the leader in the field; with Christian Guilleminault, who, Al told me, knew more about sleep disorders than anyone in the world; and with Tom Roth. These individuals graciously accepted me as a trainee. During my time at Stanford, I gained an appreciation of the extraordinary diversity of sleep pathologies and of the potential to apply basic sleep neurobiology to human sleep problems. At Henry Ford Hospital, I learned about the insufficient sleep syndrome and insomnia, among other sleep problems, and picked up many of the practical aspects of running a sleep center.

Some time after I joined the faculty at Michigan, Sid Gilman, the editor of this series, asked me if I would be interested in developing a book on sleep medicine. I declined, believing that I lacked the expertise necessary for the task. I suggested some other names, but for one reason or another the book did not materialize. When Sid asked me again a few years later, I accepted, believing that now I had something worthwhile to say and that I could write it in a way that would be useful to others.

---

1. Wolpert EA: Studies in psychophysiology of dreams. II. An electromyographic study of dreaming. AMA Arch Gen Psychiatry 2:231–41, 1960.

I am honored that Al Rechtschaffen contributed a foreword to this book. His dedication to sleep research, his scientific rigor, and his tenacity have inspired me for the last three decades. I am also deeply grateful to many colleagues, friends, and family members who contributed to making this book a reality. My colleagues Claudio Bassetti, Ron Chervin, Ivo Drury, Alan Eiser, Doug Gelb, Beth Malow, Sarah Nath, Dan Rifkin, and Linda Selwa reviewed portions of the manuscript and offered many useful comments. Stimulating discussions with these individuals and with Michel Billiard, Don Bliwise, Richard Ferber, Christian Guilleminault, Peter Hauri, Mark Mahowald, Emmanuel Mignot, Wolfgang Schmidt-Nowara, John Shepard, Art Spielman, and Frank Zorick have shaped my views on sleep medicine. Sid Gilman, my mentor in neurology for the past 17 years, provided patient encouragement and advice. My sister, Carol Barkin, a professional author and editor, reviewed and critiqued much of the manuscript. My brother Tom Aldrich, a pulmonologist, reviewed the chapters related to respiratory medicine. As he has with most of my writings over the past 15 years, my father, Knight Aldrich, reviewed the entire manuscript, parts of it more than once, and offered innumerable helpful suggestions for improvements of style and content. His advice concerning psychiatric aspects of sleep medicine has been invaluable, not just for this project but throughout my career. My mother, Julie Aldrich, cheerfully tolerated the hours my father and I spent together poring over the drafts. Most of all, I am grateful to my wife, Leslie, to whom this book is dedicated. She put up with my many evenings in front of the laptop and encouraged me to continue when the task appeared never-ending. Without her love and support, this book would not exist.

Ann Arbor, Michigan M. S. A.
June 1998

CONTENTS

Part 1. Sleep

1. NORMAL HUMAN SLEEP   3

HISTORY   3

Phylogeny   3
Views Concerning the Nature and Function of Sleep   4
Brain Electricity   5
Two States of Sleep   6
Sleep—An Active Process   6
Monoamines and Sleep   7

SLEEP RECORDINGS   7

SLEEP STAGES   9

Wakefulness and Drowsiness   9
Stage 1 NREM Sleep   10
Stage 2 NREM Sleep   11
Stage 3-4 NREM Sleep   12
The transition from NREM Sleep to REM Sleep   13
REM Sleep   14
Arousals and Awakenings   14

CLINICAL FEATURES OF SLEEP   15

The Onset of Sleep   15
Sleep and Consciousness   15

PHYSIOLOGIC CHANGES DURING SLEEP   17

Autonomic Nervous System Activity   17
Cardiovascular Function   17
Respiratory Function   17
Gastrointestinal Function   18
Renal Function   18
Thermal Regulation   18
Endocrine Function   18
Genital Changes   18
Immune Function   19
Cerebral Blood Flow and Metabolism   19
Memory Formation   19

SLEEP HOMEOSTASIS   20

FUNCTIONS OF SLEEP   20

Why Do We Sleep?   20

Functions of REM Sleep	22
How Much Sleep Is Enough?	23

2. NEUROBIOLOGY OF SLEEP — 27

SLEEP-WAKE REGULATION — 27

Wakefulness and the Brainstem Reticular Formation	27
Noradrenaline	29
Serotonin	29
Dopamine	30
Histamine	30
Acetylcholine	30

NREM SLEEP — 30

Sleep Onset	31
Sleep Spindles and Delta Waves	31

REM SLEEP — 32

Postural Atonia	33
Pontine-Geniculate-Occipital Spikes	34
Cortical Activation	34
Rapid Eye Movements	34
Muscle Twitches	34

SLEEP FACTORS — 34

Cytokines	35
Prostaglandins	35
Adenosine	36
Peptides	36
Melatonin	36
Benzodiazepines and Endogenous Benzodiazepine-Receptor Agonists	36

3. BREATHING DURING WAKEFULNESS AND SLEEP — 39

THE RESPIRATORY CONTROL SYSTEM — 39

The Ventilatory Pump	40
The Upper Airway	41
Chemoreceptors and Other Afferents	45
Central Nervous System Controller	46
Metabolic Regulation of Breathing	48
Voluntary Ventilatory Control System	49
Reflex Responses to Altered Respiratory Load	50
The Wakefulness Stimulus	50

SLEEP-RELATED CHANGES IN BREATHING — 50

Changes in Breathing Mechanics	50
Increased Airway Resistance	51
Central Nervous System Changes During Sleep	51
Metabolic Regulation of Breathing During Sleep	52

Changes in Reflex Responses to Respiratory Loads During Sleep	52
Loss of the Wakefulness Stimulus	52
Decreased Ventilation During Sleep	53
State Changes and Arousal Responses	53

4. CHRONOBIOLOGY — 56

PHYSIOLOGIC FUNCTIONS WITH CIRCADIAN RHYTHMICITY — 57

Circadian Distribution of Sleep and Sleepiness — 57

NEUROANATOMY OF THE CIRCADIAN TIMING SYSTEM — 58

GENETIC BASIS OF CLOCK FUNCTION — 59

FACTORS AFFECTING CLOCK FUNCTION — 59

Effects of Light on Clock Function — 59
Effects of Other Exogenous Stimuli on Clock Function — 60
Endogenous Influences on Clock Function — 61

ZEITGEBERS, MASKING STIMULI, AND RHYTHM INTERACTIONS — 61

EFFECTS OF THE CIRCADIAN PACEMAKER ON SLEEP AND WAKEFULNESS — 62

Sleep-Wake Rhythms Under Constant Environmental Conditions — 62
Sleep-Wake Rhythms Under Non–24-Hour Sleep-Wake Schedules — 65

5. ONTOGENY OF SLEEP — 70

DIFFERENTIATION OF STATES OF SLEEP AND WAKEFULNESS — 70

27 to 30 Weeks Conceptional Age — 71
30 to 33 Weeks Conceptional Age — 71
33 to 37 Weeks Conceptional Age — 72
37 to 40 Weeks Conceptional Age — 72

EVOLUTION OF SLEEP-WAKE CYCLES AND ELECTROENCEPHALOGRAPHIC PATTERNS AFTER BIRTH — 72

Term to 12 Months — 72
The First Decade — 74
Adolescence — 76
Adulthood — 76
Late Adulthood — 76

AGE-RELATED CHANGES IN RESPIRATION — 77

Term and Preterm Infants — 77
Childhood — 79
Adulthood — 79

MOVEMENTS DURING SLEEP OVER THE LIFE SPAN — 80

6. DREAMS AND OTHER MENTAL ACTIVITY DURING SLEEP — 82

DREAMS THROUGHOUT HISTORY — 82
DREAMS COMPARED TO OTHER MENTAL ACTIVITY OF SLEEP — 83
PSYCHOPHYSIOLOGY OF REM SLEEP AND DREAMING — 85
DREAM CONTENT — 86

Age and Dream Content — 86
Stage of Sleep and Time of Night — 87
Daytime Events — 87
Emotions — 87

FACTORS INFLUENCING DREAM RECALL — 87
PSYCHOLOGICAL ASPECTS OF DREAMS — 88

Freud's Dream Theories — 88
Postfreudian Views of Dream Psychology — 89

Part 2. Sleep Disorders

7. APPROACH TO THE PATIENT — 95

THE DEVELOPMENT OF SLEEP MEDICINE — 95
CLASSIFICATION OF SLEEP DISORDERS — 96
EVALUATION OF THE PATIENT — 98

Chief Complaint and History — 99
Past Medical History and Review of Systems — 102
Family History of Sleep Problems — 102
Social History — 102
Examination — 103
Formulation and Differential Diagnosis — 103

DIAGNOSTIC PROCEDURES — 104

Questionnaires and Sleep Logs — 104
Monitoring of Sleep — 105
Multiple Sleep Latency Test and Maintenance of Wakefulness Test — 106
Scoring of Sleep Studies — 106
Ancillary Tests — 108

MANAGEMENT — 108

8. SLEEPINESS AND SLEEP DEPRIVATION — 111

SLEEPINESS — 112

Clinical Features of Sleepiness — 112
Causes of Sleepiness — 113
Central Nervous System Basis of Sleepiness — 116
Measures of Sleepiness — 116

TOTAL SLEEP DEPRIVATION — 118

Human Studies	118
Animal Studies	119
INSUFFICIENT SLEEP SYNDROME	**120**
Clinical Features	121
Psychobiologic Basis	121
Diagnosis	122
Management	123

9. INSOMNIA — 127

CLINICAL FEATURES	**128**
Difficulty Falling Asleep	128
Repeated Awakenings and Difficulty Returning to Sleep	128
Early-Morning Awakening	128
Daytime Symptoms	129
Course	129
EPIDEMIOLOGY	**129**
PSYCHOBIOLOGIC BASIS	**130**
Predisposing Conditions	130
Precipitants	131
Perpetuating Circumstances	132
OVERVIEW OF THE MANAGEMENT OF INSOMNIA	**133**
HYPNOTIC MEDICATIONS	**133**
Benzodiazepines and Benzodiazepine-Receptor Agonists	134
Pharmacologic Alternatives to Benzodiazepines and Benzodiazepine-Receptor Agonists	136
DIFFERENTIAL DIAGNOSIS	**136**
PSYCHOPHYSIOLOGIC INSOMNIA	**139**
Behavioral Treatment of Insomnia	140
IDIOPATHIC INSOMNIA	**142**
SLEEP-STATE MISPERCEPTION (PSEUDOINSOMNIA)	**144**
EXTRINSIC SLEEP DISORDERS THAT CAUSE INSOMNIA	**144**
Inadequate Sleep Hygiene	144
Environmental Sleep Disorder	145
Adjustment Sleep Disorder	146
Hypnotic-Dependent Sleep Disorder	146
Sleep-Onset Association Disorder	147
Limit-Setting Sleep Disorder	148
Nocturnal Eating/Drinking Syndrome	148
Food-Allergy Insomnia	149

10. NARCOLEPSY AND RELATED DISORDERS — 152

CLINICAL FEATURES — 153

Sleepiness and Sleep Attacks — 154
Cataplexy — 155
Sleep Paralysis — 155
Hypnagogic and Hypnopompic Hallucinations — 156
Nocturnal Sleep Disruption — 156
Automatic Behavior — 156
Memory and Visual Disturbances — 157
Psychiatric Symptoms and Psychosocial Consequences — 157
Course — 157

BIOLOGIC BASIS — 157

Genetic and Familial Basis — 157
Pathophysiology — 159
Canine Narcolepsy — 160

CLINICAL VARIANTS AND RELATED DISORDERS — 161

Monosymptomatic Narcolepsy — 161
Narcolepsy Associated with Brain Lesions — 162
Idiopathic Hypersomnia — 163
Post-traumatic Hypersomnia — 166

DIFFERENTIAL DIAGNOSIS — 166

Sleepiness — 166
Cataplexy — 166
Sleep Paralysis and Hallucinations — 167

DIAGNOSTIC EVALUATION — 167

MANAGEMENT — 169

Treatment of Sleepiness — 169
Treatment of Other Symptoms — 171

11. RESTLESS LEGS SYNDROME AND PERIODIC LIMB MOVEMENT DISORDER — 175

RESTLESS LEGS SYNDROME — 176

Clinical Features — 176
Biologic Basis — 176
Differential Diagnosis — 178
Diagnostic Evaluation — 179
Management — 179

PERIODIC LIMB MOVEMENT DISORDER — 181

Clinical Features — 181
Diagnosis — 181
Management — 181

12. CHRONOBIOLOGIC DISORDERS — 186

JET LAG — 186

Clinical Features — 186
Biologic Basis — 187
Management — 188

SHIFT-WORK SLEEP DISORDER — 189

Clinical Features — 189
Biologic Basis — 190
Diagnosis — 190
Management — 190

DELAYED SLEEP PHASE SYNDROME — 192

Clinical Features — 192
Psychobiologic Basis — 193
Diagnosis — 194
Management — 194

ADVANCED SLEEP PHASE SYNDROME — 195

IRREGULAR SLEEP-WAKE PATTERN — 195

Clinical Features — 195
Biologic Basis — 196
Diagnosis — 196
Management — 196

NON–24-HOUR SLEEP-WAKE SYNDROME — 197

Clinical Features — 197
Biologic Basis — 197
Diagnosis — 198
Management — 198

13. OBSTRUCTIVE SLEEP APNEA SYNDROME — 202

CLINICAL FEATURES — 203

Snoring, Snorting, Gasping, and Choking Sounds During Sleep — 204
Restless Sleep — 204
Sleepiness and Tiredness — 205
Other Symptoms — 205
Symptoms in Infants and Children — 205
Examination — 206

EPIDEMIOLOGY — 206

Comorbid Conditions — 206

BIOLOGIC BASIS — 207

Narrowing of the Upper Airway — 207
Causes of Airway Narrowing — 212
Snoring — 215

Causes of Arousals from Apneas and Hypopneas	215
Effects of Alcohol and Sedative Medications	216
Causes of Sleepiness	216

SYSTEMIC EFFECTS OF OBSTRUCTIVE SLEEP APNEA — 217

Cognitive and Psychosocial Function	217
Hypoxemia	217
Sympathetic Activity	217
Systemic and Pulmonary Arterial Pressure	217
Cardiac Function	217
Intracranial Hemodynamics and Cerebral Perfusion	219
Renal Function	220
Endocrine Effects	220

COMPLICATIONS — 220

Accidents	220
Vascular Morbidity and Mortality	220
Anesthetic Complications	221
Complications in Infants and Children	221

DIFFERENTIAL DIAGNOSIS — 222

DIAGNOSTIC EVALUATION — 222

MANAGEMENT — 225

Continuous Positive Airway Pressure	225
Surgery	228
Dental Appliances	229
Weight Loss	230
Medications	230
Oxygen	230

14. CENTRAL SLEEP APNEA AND HYPOVENTILATION DURING SLEEP — 237

CENTRAL SLEEP APNEA — 238

Clinical Features	238
Biologic Basis	238
Differential Diagnosis	242
Diagnostic Evaluation	244
Management	247

SUDDEN INFANT DEATH SYNDROME — 249

Clinical Features	249
Biologic Basis	249
Diagnosis	250
Prevention	250

CENTRAL ALVEOLAR HYPOVENTILATION SYNDROME — 251

Clinical Features	251
Biologic Basis	251

Diagnosis	251
Management	251
OBESITY-HYPOVENTILATION SYNDROME	**252**
Clinical Features	252
Biologic Basis	252
Diagnosis	253
Management	253
HYPOVENTILATION ASSOCIATED WITH NEUROMUSCULAR DISORDERS	**253**
Clinical Features	254
Biologic Basis	254
Diagnosis	255
Management	256

15. PARASOMNIAS — 260

AROUSAL DISORDERS	**261**
Clinical Features	261
Biologic Basis	263
Differential Diagnosis	264
Diagnostic Evaluation	265
Management	267
SLEEP-WAKE TRANSITION DISORDERS	**267**
Rhythmic Movement Disorder	267
Sleep Starts	268
Sleeptalking	269
Nocturnal Leg Cramps	269
PARASOMNIAS RELATED TO REM SLEEP	**270**
Sleep Paralysis	270
Rem Sleep Behavior Disorder	271
Rem Sleep–Related Sinus Arrest	274
Impaired Nocturnal Penile Tumescence	274
Sleep-Related Painful Erections	277
Nightmares	277
OTHER PARASOMNIAS	**279**
Sleep Bruxism	279
Sleep Enuresis	281
Nocturnal Dissociative Disorder	283
Sudden Unexplained Nocturnal Death	283

16. PSYCHIATRIC DISORDERS AND SLEEP — 288

SLEEP DISORDERS ASSOCIATED WITH SCHIZOPHRENIA	**288**
Clinical Features	289
Psychobiologic Basis	289

Diagnosis 289
Management 290

SLEEP DISTURBANCE ASSOCIATED WITH MOOD DISORDERS 290

Major Depressive Episodes 291
Manic and Hypomanic Episodes 294
Dysthymic Disorder 295
Seasonal Affective Disorder 295
Bereavement 296

SLEEP DISORDERS ASSOCIATED WITH ANXIETY DISORDERS 297

Generalized Anxiety Disorder 297
Panic Disorder 298
Post-Traumatic Stress Disorder 298
Obsessive-Compulsive Disorder 299

SLEEP DISTURBANCE ASSOCIATED WITH EATING DISORDERS 299

SLEEP DISTURBANCE ASSOCIATED WITH ALCOHOL USE AND ALCOHOLISM 300

Effects of Alcohol on Alertness 300
Effects of Alcohol on Sleep 300
Alcohol-Dependent Sleep Disorder 301
Alcoholism 301

SLEEP DISTURBANCE ASSOCIATED WITH OTHER SUBSTANCE-RELATED DISORDERS 303

17. MEDICAL CAUSES OF DISORDERED SLEEP 307

SLEEP DISTURBANCE IN ISCHEMIC HEART DISEASE 307

Clinical Features 307
Biologic Basis 308
Diagnosis 309
Management 311

SLEEP DISTURBANCE IN CHRONIC LUNG DISEASES 312

Clinical Features 312
Biologic Basis 313
Diagnosis 313
Management 314

SLEEP DISTURBANCE AND GASTROINTESTINAL DISEASE 314

Gastroesophageal Reflux Disease 315
Peptic Ulcer Disease 316
Intestinal Disorders 316

SLEEP DISTURBANCE AND RHEUMATOLOGIC DISORDERS	317
Arthritis	317
Fibromyalgia	318
SLEEP DISTURBANCE AND CHRONIC RENAL FAILURE	320
Clinical Features	320
Biologic Basis	320
Diagnosis	320
Management	321

18. SLEEP DISORDERS IN DEMENTIAS AND RELATED DEGENERATIVE DISEASES — 325

ALZHEIMER'S DISEASE AND SLEEP DISTURBANCE	325
Clinical Features	325
Biologic Basis	327
Diagnosis	328
Management	328
PARKINSON'S DISEASE AND SLEEP DISTURBANCE	329
Clinical Features	329
Biologic Basis	330
Diagnosis	331
Management	331
PROGRESSIVE SUPRANUCLEAR PALSY AND SLEEP DISTURBANCE	332
Clinical Features	332
Biologic Basis	333
Diagnosis and Management	334
SLEEP DISTURBANCE IN DISORDERS WITH MULTI-SYSTEM DEGENERATION	334
Clinical Features	334
Biologic Basis	334
Diagnosis and Management	334

19. DIENCEPHALIC AND BRAINSTEM SLEEP DISORDERS — 338

ENCEPHALITIS LETHARGICA	338
Clinical Features	339
Biologic Basis	339
Diagnosis and Management	339
SLEEPING SICKNESS	339
Clinical Features	340
Biologic Basis	340
Diagnosis and Management	340

FATAL FAMILIAL INSOMNIA	340
Clinical Features	340
Biologic Basis	341
Diagnosis and Management	341
PRADER-WILLI SYNDROME	341
Clinical Features	341
Biologic Basis	342
Diagnosis and Management	342
KLEINE-LEVIN SYNDROME	342
Clinical Features	342
Biologic Basis	343
Diagnosis	343
Management	343
IDIOPATHIC RECURRING STUPOR	344
OTHER DIENCEPHALIC AND BRAINSTEM DISTURBANCES	345
Thalamic Infarcts	345
Diencephalic Lesions	345
Brainstem Lesions	345
20. SLEEP AND EPILEPSY	**350**
OVERVIEW OF EPILEPSY	350
Epilepsy and the Electroencephalogram	351
SLEEP AND EPILEPTIC MANIFESTATIONS	352
Epileptiform Activity During Sleep	352
Effects of Sleep on Seizure Frequency	353
Effects of Sleep Deprivation on Seizures and Epileptiform Activity	353
Effects of Sleep Disorders on Seizures	354
Effects of Seizures and Antiepileptic Medications on Sleep	354
DIAGNOSIS OF SEIZURES DURING SLEEP	354
Daytime Electroencephalography	355
Video-Electroencephalographic Polysomnography	356
Portable Electroencephalographic Recorders and Inpatient Monitoring	357
The Unexpected Electroencephalographic Finding During Polysomnography	357
GENERALIZED EPILEPSIES ASSOCIATED WITH SEIZURES DURING DROWSINESS, SLEEP, OR UPON AWAKENING	359
Generalized Epilepsy with Tonic-Clonic Seizures During Sleep	360
Epilepsy with Grand Mal Seizures on Awakening	360
Juvenile Myoclonic Epilepsy	361
Childhood Absence Epilepsy	361
Lennox-Gastaut Syndrome	362

PARTIAL EPILEPSIES ASSOCIATED WITH SEIZURES DURING SLEEP — 362

Benign Childhood Epilepsy, with Centrotemporal Spikes (Benign Rolandic Epilepsy) — 362
Temporal Lobe Epilepsy — 364
Frontal Lobe Epilepsy — 366
Nocturnal Paroxysmal Dystonia — 370
Autosomal Dominant Nocturnal Frontal Lobe Epilepsy — 370
Epilepsy with Continuous Spike Waves During Slow-Wave Sleep — 370

INDEX — 375

Part 1

Sleep

Chapter 1

NORMAL HUMAN SLEEP

What probing deep has ever solved the mystery of sleep?

THOMAS ALDRICH, HUMAN IGNORANCE, 1837

HISTORY
Phylogeny
Views Concerning the Nature and Function of Sleep
Brain Electricity
Two States of Sleep
Sleep—An Active Process
Monoamines and Sleep
SLEEP RECORDINGS
SLEEP STAGES
Wakefulness and Drowsiness
Stage 1 NREM Sleep
Stage 2 NREM Sleep
Stage 3–4 NREM Sleep
The Transition from NREM Sleep to REM Sleep
REM Sleep
Arousals and Awakenings
CLINICAL FEATURES OF SLEEP
The Onset of Sleep
Sleep and Consciousness
PHYSIOLOGIC CHANGES DURING SLEEP
Autonomic Nervous System Activity
Cardiovascular Function
Respiratory Function
Gastrointestinal Function
Renal Function
Thermal Regulation
Endocrine Function
Genital Changes
Immune Function
Cerebral Blood Flow and Metabolism
Memory Formation
SLEEP HOMEOSTASIS
FUNCTIONS OF SLEEP
Why Do We Sleep?
Functions of REM Sleep
How Much Sleep Is Enough?

Sleep is a necessary and universal part of human existence. The suspension of consciousness that accompanies sleep, the experience of dreaming, and the inability to function without sleep have fascinated poets and philosophers for thousands of years.

In modern terms, sleep is viewed as a state of the brain and body governed by diencephalic and brainstem neural systems and characterized by periodic, reversible loss of consciousness; reduced sensory and motor functions linking the brain with the environment; internally generated rhythmicity; homeostatic regulation; and a restorative quality that cannot be duplicated by rest without sleep or by any food, drink, or drug. Sleep is as essential as food and water: The physiologic and psychological drive to sleep can overwhelm all other needs.

HISTORY

Phylogeny

Although amphibians, fishes, and arthropods have periods of behavioral quiescence, sleep that is associated with slow waves on the electroencephalogram (EEG) and an elevated threshold for arousal probably developed as a separate state approximately 200 million years ago coincident with the development of endothermy, or metabolic control of body temperature. The current understanding is that all mammals and most or all birds sleep, as do some reptiles.[1,2] Rapid eye movement (REM) sleep occurs in almost all mammals, and a similar state without neck muscle atonia occurs in birds.[1]

Views Concerning the Nature and Function of Sleep

What else is sleep but the image of chill death?

OVID, AMORES II

Before the 20th century, sleep was almost always considered a passive process that occurred as a result of a reduction in some vital force. In *The Iliad*, Homer described sleep and death as twin brothers. Descartes (1596–1650) speculated that a diminution or resting of animal spirits in the pineal gland led first to collapse of the ventricles and then to sleep.

The theories of the English physician Thomas Willis (1621–1675) were an exception to the view of sleep as a passive process. Although he believed that bodily animal spirits rested during sleep, he thought that cerebellar spirits became active and that unrestrained animal spirits produced dreams. His concept of dreams as an expression of disinhibited primitive spirits or instincts antedated Freud's theories by more than 200 years.

Before 1900, there were three major theories concerning the cause of sleep: vascular, chemical, and neural. Of these, the vascular theory has been shown to be wrong; variations of the other two remain viable.

VASCULAR THEORIES OF THE CAUSES OF SLEEP

In early Greek civilization, sleep was considered a consequence of changes related to blood. Empedocles of Sicily (5th century BC) believed that sleep occurred when the blood cooled and fire was separated from water, air, and earth—the other three major elements.[3] On the other hand, Hippocrates (460–370 BC), the most influential of the ancient Greek physicians, believed that sleep was a result of warming of the blood as it passed from the limbs to the inner core of the body. Alcmaeon, born in southern Italy in the 5th century BC, was one of the first to link sleep to the brain as well as to blood. He believed that sleep resulted from congestion of the brain with blood and that waking followed when the blood withdrew from the brain.

Theories based on vascular congestion persisted into the 19th century. Albrecht von Haller of Switzerland believed that sleep was caused by pressure on the brain from increased flow of blood to the head, a view strikingly similar to that of Alcmaeon 2400 years earlier. The concept was more fully elaborated by Johannes Purkinje, who believed that vascular congestion of the basal ganglia compressed the corona radiata and disrupted neural pathways, leading to sleep.

In the 18th and 19th centuries, however, the vascular congestion theory began to lose favor in the face of evidence to the contrary. In 1795, Johann Blumenbach of Gottingen observed a human brain directly during sleep, presumably through a skull defect, and noted that the brain became pale, suggesting to him that anemia, or lack of blood, was the reason for sleep. Alexander Fleming, in 1855, induced a sleeplike state by occluding the carotid arteries, and in 1863 Hughlings Jackson noted pallor of the optic disc during sleep, findings that were more consistent with the anemia theory.

At the end of the 19th century, Leonard Hill effectively laid all vascular theories to rest with his studies of intracranial circulation and his demonstration that intracranial pressure was unchanged during sleep.

CHEMICAL THEORIES OF THE CAUSES OF SLEEP

The chemical theories of the causes of sleep originated with Aristotle (384–322 BC), who believed that sleepiness was brought on by warm vapors that passed during digestion from the stomach via the blood to the brain. He thought that sleep occurred as the vapors cooled and removed heat from the brain and that sleep continued as long as food was being digested. Theories in the second half of the 19th century proposed that sleep was caused by a lack of oxygen or that it was caused by a buildup of lactic acid, carbon dioxide, cholesterol, or other toxic wastes. Hypnotoxins were thought to build up in the blood during wakefulness and to decrease during sleep; the early 20th-century experiments of Legendre and Pieron,[4] in which blood from sleep-deprived dogs induced sleep when transferred to dogs that were not sleep-deprived, helped to popularize this belief. Chemical theories of sleep causation remain prominent in sleep research, and a number of sleep factors have been identified (see Chap. 2).

NEURAL THEORIES OF THE CAUSES OF SLEEP

Some of the first experimental evidence supporting neural theories was provided by Luigi Rolando, who demonstrated in 1809 that ablation of the cerebral hemispheres of birds produced a state of sleepiness.[3] His experiments supported the view of sleep as a passive state that occurred in response to reduced cerebral stimulation, and it was widely believed throughout the 19th century that sleep resulted from a reduction in sensory input, with no major distinction between sleep and other states of reduced consciousness, such as coma, stupor, intoxication, and hibernation. The nineteenth-century Scottish physician Robert MacNish described sleep as intermediate between wakefulness and death.[5]

With the demonstration of nerve cell processes by Golgi and others, some investigators proposed that sleep occurred as a result of reduced communication among nerve cells, perhaps by retraction of neural processes or by inhibition. Brown-Sequard, for example, believed that sleep was the result of an inhibitory reflex, and Pavlov thought that sleep was caused by the spread of inhibition throughout the hemispheres and into the midbrain. Toward the end of the 19th century, however, evidence that focal pathology of the upper brainstem could lead to impaired consciousness suggested that wakefulness and consciousness were mediated by specific brainstem structures and that the abolition of wakefulness by sleep might not be caused by a diffuse cerebral process.[6]

Brain Electricity

The discovery of brain electrical activity in 1875 was a major turning point in the study of sleep. In his study on the exposed brains of dogs and rabbits, Caton,[7] a British physiologist, made the following observation: "In every brain, . . . the galvanometer has indicated the existence of . . . feeble currents of varying direction." The significance of Caton's discovery was not widely appreciated until 1930, when Berger,[8] a German physician, reported that the human brain produced electrical signals that could be recorded on paper. The recordings, which he called *electroencephalograms,* provided the first technique for studying the brain during sleep without disturbing the subject.

Berger noted a prominent 10-Hz rhythm in the surface EEG, which he designated as *alpha,* and faster frequencies, which he called *beta.* His demonstration of changes in the EEG during sleep,[8] along with the observation by Adrian and Yamagiwa[9] that the alpha rhythm disappeared during sleep and the EEG studies during sleep by Loomis and colleagues,[10,11] constituted the first definite evidence that the brain is not entirely inactive during sleep. Studies of encephalitis lethargica by von Economo[12] and of the effects of hypothalamic and thalamic stimulation by Hess[13] produced major advances in the understanding of the neuroanatomic substrate of sleep and wakefulness (see Chap. 2).[12,13]

The discovery of brain electrical activity and the capability of observing and recording such activity led to increasingly sophisticated studies of brain function and brain activity. Experiments by Bremer[14,15] in the 1930s appeared to support the view of sleep as a passive process resulting from the withdrawal of waking activity. After transections at the cervicomedullary junction that preserved input from vestibular, auditory, and visual systems (*encephale isolé*), EEG differences between sleep and wakefulness were still present. However, with transection at the mesencephalic level just caudal to the third nerves (*cerveau isolé*), which reduced auditory and vestibular input, continuous slow waves were present similar to those observed during sleep. In addition to demonstrating the importance of brainstem structures for the occurrence of the waking state, the studies supported the view that cerebral waking tone was maintained by specific sensory impulses relayed via the spinal cord or brainstem. In 1939 Kleitman[16] expressed the prevailing view that ". . . there is not a single fact about sleep that cannot be equally well interpreted as a let down of the waking activity."

With the development of implantable electrodes in the 1940s, it became possible to study localized brain activity in much greater detail. Using the encephale isolé preparation, Moruzzi and Magoun[17] demonstrated that electrical stimulation of the medial reticular formation (RF) and the pontine and midbrain tegmentum abolished slow waves and led to the appearance of a low-amplitude, desynchronized EEG, indicating that these areas, under appropriate conditions, could activate forebrain structures. They also demonstrated that stimulation of the posterior hypothalamus and the subthalamus could elicit EEG desynchronization and that the midbrain RF was the area capable of eliciting cortical activation with the least stimulation. These studies emphasized the critical importance of the ascending reticular activating system for maintenance of wakefulness. Subsequent studies from Magoun's laboratory, which demonstrated that sensory collaterals discharge into the RF, also appeared to support the view of sleep as a passive process resulting from decreased sensory input.[18] After the discovery of REM sleep, however, the scientific evidence supporting this view gradually eroded.

Two States of Sleep

Before the discovery of periods of REM sleep in 1953, some investigators suspected that two states of sleep existed, and others recognized some of the elements of REM sleep. Willis, in the 17th century, linked dreams to the periodic activity of cerebellar animal spirits during sleep, and in 1868 Greisinger noted eye movements during sleep.[3] Freud's suggestion that paralysis during dreams prevented the acting out of dreams anticipated the discovery of muscle atonia during REM sleep. Periods of low-voltage EEG during sleep were recognized in the 1930s, but it was generally assumed that these episodes corresponded to "light" sleep, whereas the more common pattern of slower, higher voltage activity corresponded to "deep" sleep.

Nathaniel Kleitman's laboratory at the University of Chicago was the site of the discovery of REM sleep. Kleitman was interested in infant sleep and assigned Eugene Aserinsky, then a graduate student, to work on the project. William Dement, a medical student at the time, also began work in the laboratory. These studies led to the report by Aserinsky and Kleitman[19] of the existence of "regularly occurring periods of eye motility during sleep" that were associated with dreaming. Dement and Kleitman[20] then described the occurrence of repetitive cycles of REM and non–rapid eye movement (NREM) sleep throughout the night.

After the discovery of REM sleep and the recognition that it was accompanied by a low-voltage, "activated" EEG, Jouvet and coworkers[21-24] proposed the term *paradoxical sleep* to describe REM sleep; this term emphasized the paradox of a low-voltage, activated EEG—similar to the waking EEG in the cat—occurring during sleep. The term *slow-wave sleep* (SWS) was used to describe the high-voltage, slow-wave activity that is present during deep NREM sleep.

Sleep–An Active Process

Although the concept of sleep as a "let-down" of waking activity was the predominant view during the first half of the 20th century, there were indications as early as the 1920s that sleep was not entirely a passive process. Hess[13,25] noted that stimulation of the massa intermedia of the thalamus could induce sleep, and Nauta[26] found that lesions of the anterior hypothalamus led to insomnia in rats. Batini and associates[27,28] observed a marked reduction in sleep after lesions at the midpontine level in cats. Although these findings, which suggested that specific areas of the brain facilitate the onset of sleep, were difficult to reconcile with theories of sleep as a passive process, many investigators continued to believe that sleep was caused primarily by a reduction of waking activity. Even after the discovery of REM sleep, some believed that REM sleep was simply a light stage of sleep that was not qualitatively different from NREM sleep.

The passive theories were finally laid to rest by a series of investigations in the 1950s and 1960s:

- The discovery by Jouvet and colleagues[21–24] of muscle atonia during REM sleep
- The demonstration that basal forebrain stimulation could induce EEG synchronization and sleep
- The finding that destruction of the raphe nuclei could cause insomnia
- The discovery that the arousal threshold was not always higher in NREM sleep than in REM sleep
- The evidence from cerebral blood flow studies that the brain is more actively metabolically during REM sleep than during NREM sleep[29–33]

These results made the concept of REM sleep as "light" sleep untenable and led to a new view of sleep as an active process, with distinctive neurophysiologic substrates underlying the two major sleep states. The importance of specific brain regions for sleep initiation was reinforced by another major discovery when brainstem transection studies showed that signs of REM sleep–like activity occur caudally after transection rostral to the pons and rostrally after transections caudal to the pons, thus demonstrating that REM sleep generators are within the pons.[34]

Monoamines and Sleep

With the development in the 1960s of improved methods for assaying enzymes, neurotransmitters, and receptors, it became apparent that monoamines are major contributors to sleep-wake regulation. For example, inhibition of norepinephrine synthesis in cats eliminates behavioral wakefulness and results in continuous EEG sleep spindles, whereas administration of norepinephrine precursors or inhibition of metabolic inactivation of norepinephrine leads to increased arousal and increased wake time.[35,36] In contrast, inhibition of serotonin synthesis by parachlorphenylalanine results in near-total insomnia in rats and cats.[37]

The combination of these findings with evidence that monoamine oxidase inhibitors suppress REM sleep led Jouvet[37] to propose a monoaminergic theory of sleep. According to this theory, serotonergic raphe neurons generate or maintain SWS and prepare the initiation of REM sleep through interactions with noradrenergic neurons of the locus ceruleus (LC). Support came from animal studies showing that lesions of the serotonergic neurons of the anterior raphe nuclei eliminate SWS and reduce REM sleep, whereas destruction of the posterior raphe abolishes REM sleep and reduces SWS.[38]

Further studies showed, however, that many of the changes induced by serotonin and norepinephrine depletion or by destruction of the raphe nuclei or LC were transient. Interspecies differences are pronounced; in cats, depletion of serotonin by parachlorphenylalanine markedly decreases both REM and NREM sleep, while in humans a similar depletion decreases but does not eliminate REM sleep and has little effect on NREM sleep.[39,40] The monoaminergic theory became untenable in its original form, and more sophisticated theories were developed based on the interaction of the thalamus and basal forebrain with monoaminergic and cholinergic systems. Current understanding of the neural control of sleep and wakefulness is presented in Chapter 2.

SLEEP RECORDINGS

Polygraphic recordings during sleep provide the best available approach to assessing normal and abnormal sleep. To assess sleep stage and wakefulness, the EEG, the electrooculogram (EOG), and the surface electromyogram (EMG) of submental muscles are recorded. In the past, most sleep studies were made with analog amplifiers and recorded on paper; sleep laboratories typically had tons of paper filling every available storage space. Increasingly, however, recordings are made with digital amplifiers and computers and are stored on compact media such as digital tapes and optical disks.

The EEG recorded from the scalp is generated by nearby cortical neurons and reflects extracellular potential changes associated mainly with excitatory and inhibitory postsynaptic potentials (EPSPs and IPSPs, respectively) at dendritic synapses. Action potentials are too brief, about 1 millisecond, to produce much effect on the EEG. Most rhythmic EEG activity recorded from the scalp appears to be caused by synchronized changes in dendritic membrane potentials of cortical neurons, particularly the large pyramidal cells of layer V of the cortex, which are produced by rhythmic activity originating in the thalamus (see Chap. 2). The changes in voltage that occur with IPSPs and EPSPs produce electrical fields: EPSPs produce a negative electrical field in the immediate vicinity and a positive electrical field at other points along the neuron. This reciprocal change can be thought of as a dipole, with a negative pole and a positive pole. The activity of cortical pyramidal cells can be approximated by a dipole oriented perpendicular to the surface, with the negative pole on top (Fig. 1–1).

Polygraphic recordings are made with the use of differential amplifiers. Two inputs are required for differential amplifiers: When both inputs come from scalp electrodes, the recording is called *bipolar*; if one electrode is used as a reference for several scalp electrodes, the recording is called *referential* or *monopolar*. Although the ear electrode is often called a reference electrode, it is electrically active and records activity from the ipsilateral temporal lobe. The source and polarity of an electrical potential recorded by a differential amplifier can usually be determined based on the following rules and conventions: (1) Negative potentials at input 1 of the amplifier produce upward deflections, whereas negative potentials at input 2 produce downward deflections. (2) Electrodes close to a source produce larger deflections than those farther away. (3) Electrodes equidistant from a source do not record a potential change, owing to cancellation of equal voltages in the two inputs of a differential amplifier.

The EOG records the potential difference between the cornea and the retina. The eyeball is, in effect, a dipole with a positive pole at the cornea and a negative pole at the retina. As the globe rotates with an eye movement, the cornea moves with respect to a recording electrode placed near the canthus of the eye. Thus, movement of the eyes to the left produces a positive potential in an electrode placed close to the external canthus of the left eye and a negative potential in an electrode close to the external canthus of the right eye.

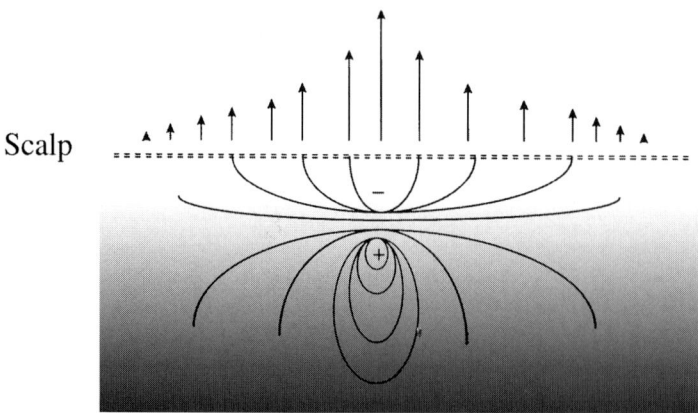

Figure 1–1. Dipoles can be used as models to understand the electrical sources of the EEG recorded at the scalp. The figure shows the electrical field surrounding a vertically oriented dipole located beneath the scalp. Voltage at the surface, indicated by the size of the arrows, is greatest directly over the dipole and declines with increasing distance from the source.

However, a vertical eye movement does not produce potential changes in electrodes placed laterally. Electrodes placed obliquely record both horizontal and vertical eye movements, although an eye movement that is perpendicular to the axis of two electrodes will not be recorded.

Surface EMG activity also results from fluctuations in membrane potentials, in this case from muscle fibers. In contrast to EEG, however, action potentials are responsible for most of the potential changes. Because muscle action potentials are of short duration, the surface activity generated by muscle fibers consists primarily of high frequency potentials.

SLEEP STAGES

Sleep stages are defined by EEG, EOG, and EMG features. After the discovery of REM sleep, the sleep stages described by Loomis and associates[10,11] and Harvey and colleagues[41] were revised by Dement and Kleitman.[20] As sleep research developed in the 1960s, the need for standardized quantitative assessment of sleep to facilitate research and clinical evaluation led to the publication of a manual that remains the standard for assigning sleep stages to epochs of recorded sleep, a process referred to as *sleep scoring*.[42] The manual defines the criteria for identifying REM sleep and the four stages of NREM sleep from stage 1 (the lightest) to stage 4 (the deepest). Many laboratories combine stage 3 and 4 sleep for scoring purposes into a single category called stage 3–4 sleep, because the two stages differ only in the amount of EEG slow-wave activity. The term *polysomnography*, used to describe the simultaneous recording of several physiologic measures during sleep, was introduced in 1974.[43]

Wakefulness and Drowsiness

The EEG recorded at the scalp is classified into four frequency bands: delta, 0 to 3 Hz; theta, 4 to 7 Hz; alpha, 8 to 13 Hz; and beta, greater than 13 Hz. In persons who are awake, alert, and resting quietly with eyes closed, the EEG is dominated by the *alpha rhythm*, a sinusoidal rhythm that is usually 9 to 10 Hz and most apparent over the parietal and occipital scalp (Fig. 1–2). The alpha rhythm is present in 85% to 90% of normal persons; the remainder have a normal variant of the waking EEG characterized by lower voltage, irregular wave forms. With eye opening or with active visual imagery, the alpha rhythm is replaced by a lower-voltage, more irregular pattern with a mixture of frequencies.

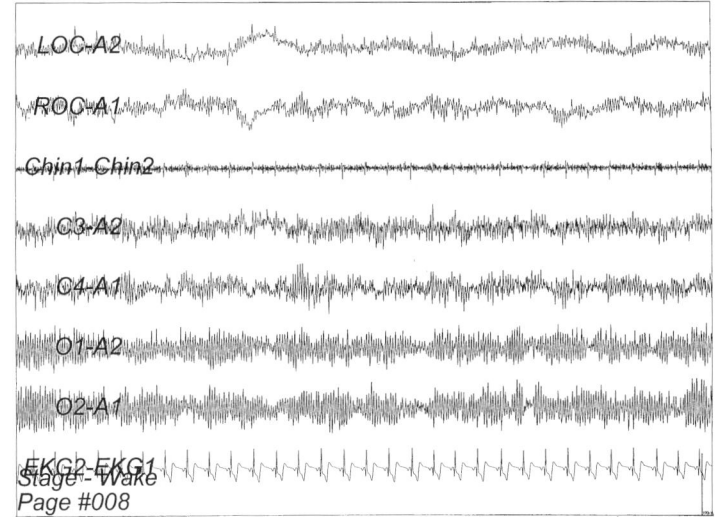

Figure 1–2. Alpha rhythm during a 30-second epoch of relaxed wakefulness in a 35-year-old man. The 9-Hz rhythm is prominent in the occipital derivations and also apparent in the central derivations. The ear electrodes also recorded alpha activity in this patient, accounting for the presence of alpha in channels 1 and 2. The EEG electrodes (A1, A2, C3, C4, O1, O2) were placed in accordance with the International 10-20 system. Other abbreviations: LOC, left outer canthus; ROC, right outer canthus; Chin1-Chin2, submental EMG.

During wakefulness, low-voltage, 14- to 25-Hz beta frequency activity is often intermixed with alpha activity, and is usually most apparent over frontal and central regions. A modest amount of theta activity may be present diffusely, particularly in children. Children may also have a moderate amount of posterior delta activity, referred to as *posterior slow waves of youth*. The *mu rhythm*, which may be present for a few seconds at a time, is a rhythm of 8 to 11 Hz that is most evident over central regions and is attenuated by movement or planned movement of the contralateral hand. The EOG during wakefulness shows REMs and blinks, and the EMG usually demonstrates continuously high levels of muscle activity.

As alertness lessens and drowsiness develops, but before the onset of sleep, the alpha rhythm slows slightly and shows greater waxing and waning in amplitude. There is often an apparent anterior diffusion of alpha activity, sometimes caused by slowing and spreading of the posterior rhythm but more frequently caused by activity of a different generator, possibly in the thalamus.[44] Beta activity becomes more prominent, particularly over the frontal regions, and centrally dominant theta activity becomes more apparent. In elderly persons, low- to-moderate-amplitude delta activity may be seen with a frontocentral predominance. Rhythmic, 4- to 5- Hz theta activity up to 200 μV in amplitude, sometimes called hypnagogic hypersynchrony, is first apparent at about ages 5 to 6 months and remains prominent during drowsiness up to age 6 years. It becomes less prominent in older children and teenagers, but it may persist into early adulthood (see Fig. 5–4). Its voltage maximum is usually over frontal and central regions and less commonly over the parietal-occipital area. Slow, rolling eye movements, sometimes asynchronous, are a prominent feature of drowsiness, and a gradual reduction in chin EMG activity is often evident.

Stage 1 NREM Sleep

The attenuation and drop out of the posterior alpha rhythm and the appearance of centrally predominant theta activity signal the onset of stage 1 sleep. Other EEG features of stage 1 sleep include *vertex sharp waves* and *positive occipital sharp transients* (POSTs).

Vertex sharp waves—sharply contoured waves that are most apparent over the center, or vertex, of the scalp—constitute a distinctive EEG feature of light sleep (Fig. 1–3). They have a prominent negative phase that may exceed 200 μV in amplitude, sometimes followed by a smaller positive phase. These waves, which may appear singly or in repetitive salvos for several seconds, are first apparent at about age 5 months, are most prominent at 2 to 4 years of age, and remain a feature of light sleep throughout life, although their amplitude may decline in aged individuals. They are seen most commonly in stage 1 sleep, less often in deeper stages of

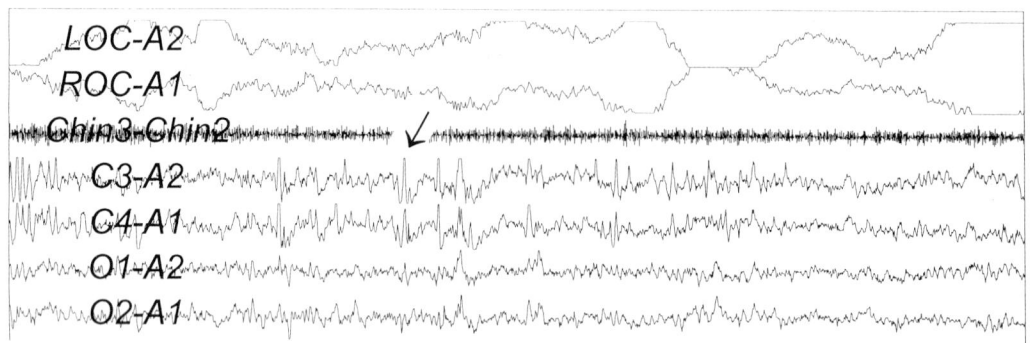

Figure 1–3. Vertex sharp waves during a 30-second epoch of stage 1 sleep. Several sharply contoured voltage-negative transients are apparent from C3-A2 and C4-A1 derivations. The most prominent one is marked by an arrow. For abbreviations, see Figure 1–2.

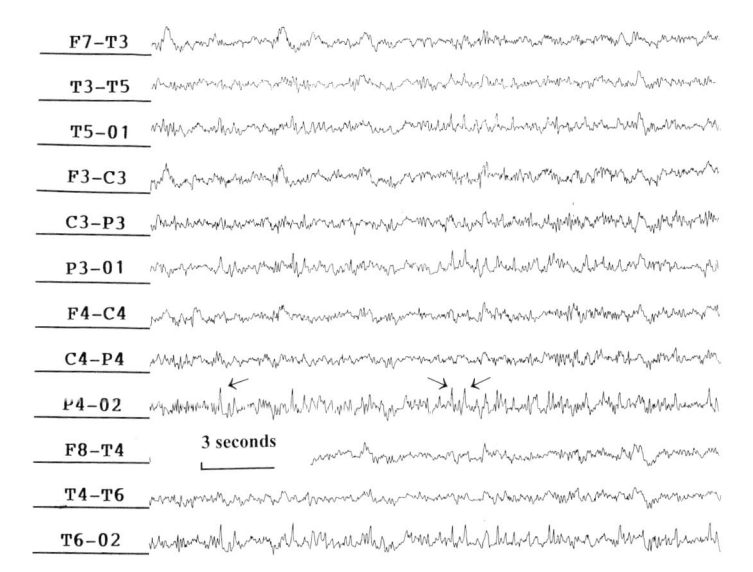

Figure 1–4. Positive occipital sharp transients (POSTs), identified by arrows, in occipital derivations during stage 1 sleep. The EEG electrodes were placed in accordance with the International 10-20 system.

NREM sleep, and not at all in REM sleep. The terms vertex waves, V-waves, vertex sharp transients, and biparietal humps are synonymous with vertex sharp waves.

POSTs, present during stage 1 sleep in 75% to 90% of adults, are diphasic or triphasic waves with a prominent positive component. They are usually 20 to 75 μV in amplitude, occasionally as high as 120 μV, with a duration of 80 to 200 msec (Fig. 1–4). Their voltage maximum is usually in the midline over the occipital lobe.

The EOG during stage 1 sleep shows prominent slow eye movements. Although in some persons the amplitude may be high enough that the polygraphic appearance resembles REMs, the distinction usually can be made based on the duration of the initial pen deflection, which for REMs is less than 280 milliseconds. The submental EMG usually shows a reduction in muscle tone, as well as a reduction in the amount of facial movement, compared to waking.

A number of EEG patterns occur as normal variants during drowsiness and stage 1 sleep, including rhythmic midtemporal theta of drowsiness, formerly called psychomotor variant; wicket spikes; benign epileptiform transients of sleep (BETS), also called small sharp spikes; and 14- and 6-Hz positive bursts or spikes.[45]

Stage 2 NREM Sleep

After a few minutes of stage 1 sleep, the normal sleeping person usually passes into stage 2 sleep, characterized on EEG by sleep spindles and K-complexes on a background of mixed frequencies predominantly in the theta range. Slow eye movements are much less evident during stage 2 sleep than during stage 1 sleep, although they may reappear for short intervals.

The *K-complex*, named by Loomis and colleagues,[11] is one of the most striking features of the EEG during sleep (Fig. 1–5).

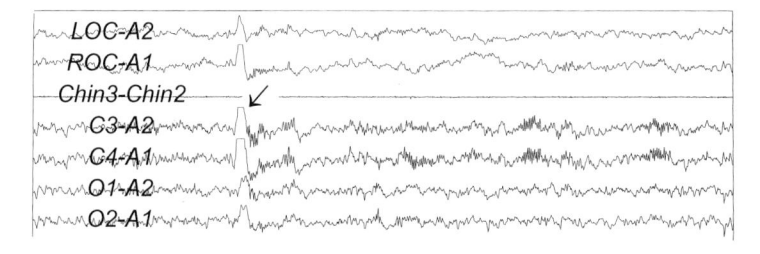

Figure 1–5. K-complex, identified by the arrow, from a 30-second epoch of stage 2 sleep. The peak of the initial negative phase is flattened because the excursion of the recording pen could not accommodate the entire amplitude of the wave. For abbreviations, see Figure 1–2.

The complex, which has a voltage maximum over the vertex or, less commonly, over the midline frontal scalp, has a total duration of at least 0.5 sec and sometimes more than 1 sec. It consists initially of a sharply contoured negative component that rises abruptly out of the background EEG activity and may have an amplitude of several hundred microvolts. This component, which resembles a vertex sharp wave, is followed immediately by a lower amplitude positive slow wave that may have a superimposed sleep spindle.

There are two differing definitions of the K-complex. According to the standard sleep scoring manual,[42] a K-complex must be at least 0.5 seconds in duration and need not be followed by a spindle; other glossaries of EEG terminology do not include a duration criterion but do require the presence of a sleep spindle. The K-complex is first apparent at about age 5 months and is most prominent in older children and younger adolescents. It declines in voltage with advancing age.

The K-complex and the vertex sharp waves are examples of *evoked* responses, a term used to describe brain potentials that are elicited by stimulation. During light stage 2 sleep, the K-complex can be easily elicited by auditory stimuli, such as the tap of a pencil, and by other forms of mild stimulation. It can also occur spontaneously, perhaps in response to internal stimuli.

Sleep spindles are runs of 12- to 14-Hz activity, occasionally up to 16 Hz, lasting 0.5 to several seconds; they are of highest amplitude over the parasagittal regions (Fig. 1–6). Spindles of 14 Hz are most prominent over the central-parietal regions, whereas 12-Hz spindles are most prominent over the frontocentral area.[46] By definition, sleep spindles are not present during stage 1 sleep, although rudimentary forms may be apparent. Sleep spindles are prominent during stage 2 sleep and remain present during stages 3 and 4, although they are more difficult to discern during these latter stages because of the abundant delta frequency activity. They are first apparent at about age 2 months and remain a striking feature of NREM sleep throughout life.

As stage 2 sleep deepens, the amount of intermixed slower frequencies increases. High-amplitude delta waves appear, and as the transition to stage 3 sleep occurs, trains of moderate- to high-amplitude delta activity become more frequent.

Stage 3–4 NREM Sleep

The transition from stage 2 to stage 3 sleep occurs when more than 20% of the recording is dominated by delta activity of 2 Hz or less that has an amplitude of 75 μV or more. Sleep spindles can still be identified, and although K-complexes, POSTs, and vertex sharp waves are more

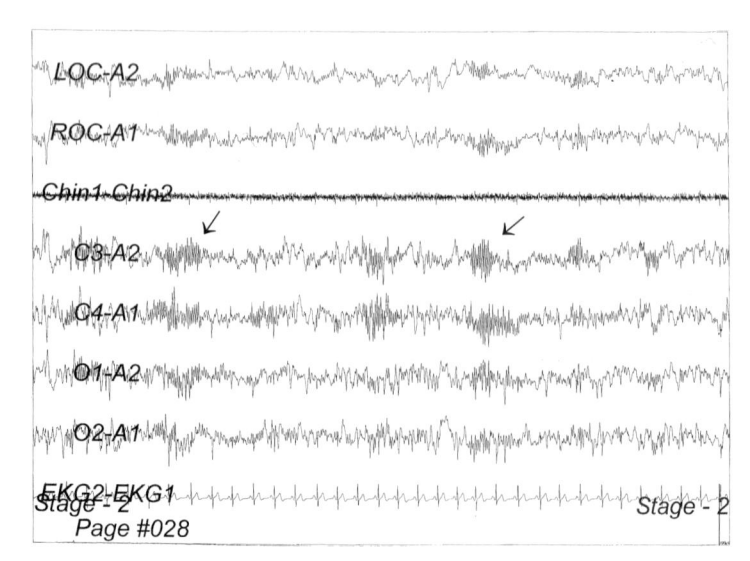

Figure 1–6. Sleep spindles (arrows) during a 30-second epoch of stage 2 sleep. For abbreviations, see Figure 1–2.

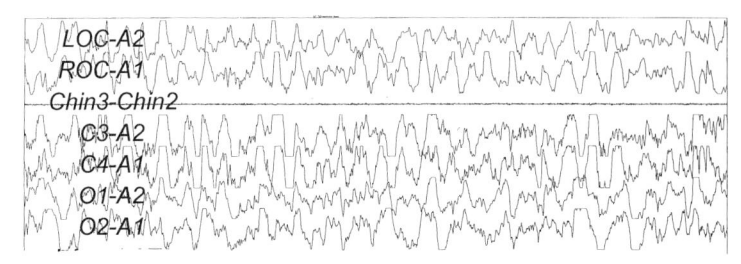

Figure 1–7. Stage 4 sleep in a 17-year-old girl. High-amplitude slow waves, of greatest voltage over the central regions, dominate the 30-second epoch. The slow waves are also apparent in the channels used to record eye movements (top two channels), because these electrodes also record EEG activity from the frontal lobes. The EEG also contains lower amplitude, faster frequencies intermixed with the delta activity. For abbreviations, see Figure 1–2.

difficult to identify by visual analysis, they can be detected with computer analysis. Because of the predominance of delta waves, stage 3–4 sleep is often called slow wave sleep, or *delta sleep* (Fig. 1–7). Slow eye movements are essentially absent during stage 3–4 sleep, and EMG activity usually remains at the same level as in stage 2 sleep.

Some persons have alpha activity during all stages of NREM sleep with the greatest amount during stage 3–4 sleep.[47] The pattern, referred to as *alpha-delta sleep*, is more common in patients diagnosed with fibromyalgia than in control populations (see Chap. 17), but it also occurs in other conditions and in asymptomatic persons, and its clinical significance is uncertain.

Stage 3–4 sleep can be identified as early as 3 months of age, is most prominent in childhood, and declines in proportion to other sleep stages during adulthood. It is most abundant during the first NREM period and constitutes a lower proportion of each successive NREM period (Fig. 1–8).

The Transition from NREM Sleep to REM Sleep

As the transition from NREM sleep to REM sleep approaches, delta activity declines, stage 3–4 sleep ends, and a few minutes of stage 1 or stage 2 NREM occur, often with a brief arousal or awakening accompanied by a change in body position. The transition into REM sleep may occur quickly with the more or less simultaneous occurrence of a change in EEG to a lower voltage, mixed-frequency pattern, a reduction in muscle tone, and the appearance of REMs; or it may develop gradually over several minutes, with initial muscle atonia followed 2 to 3 minutes later by development of a low-amplitude EEG with occasional K-complexes or spindles, followed a few minutes later by the appearance of REMs and the complete disappearance of spindles. The transitional period, which may contain various mixtures of REM and NREM polygraphic components, is sometimes called transitional sleep, indeterminate sleep, or ambiguous sleep.

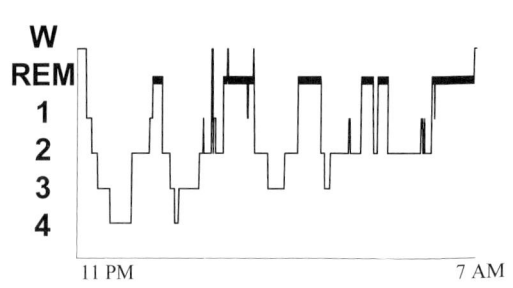

Figure 1–8. Sleep stages during a normal night of sleep in a young adult. Sleep begins after just a few minutes of wakefulness; 5–10 minutes of stage 1 sleep are followed by a rapid descent into stage 4 sleep. After about 80 minutes, sleep lightens and the first REM sleep period begins. Cycles of NREM sleep and REM sleep recur throughout the night with a period of about 100 minutes each. The REM sleep periods tend to become longer as the night progresses, and slow-wave sleep is less prominent with each successive NREM period.

REM Sleep

With the onset of REM sleep, EEG recordings change from a pattern dominated by K-complexes, sleep spindles, and slow waves to a faster frequency, lower amplitude, irregular pattern; this is similar to what occurs during stage 1 sleep, except that during REM sleep vertex sharp waves do not occur (Fig. 1–9). Brief runs of alpha activity may occur, usually 1- to 2-Hz slower than the waking posterior alpha rhythm. Sawtooth waves, the only EEG feature characteristic of REM sleep, are 2- to 6-Hz waves that occur intermittently throughout the REM period, usually in runs lasting 1 to 5 seconds. They frequently accompany bursts of REMs and can also appear just before the onset of the REM period.

During REM sleep, REMs occur singly or in bursts. They usually have a rise time of less than 200 milliseconds and may have amplitudes of up to 300 μV. Brief bursts of facial EMG activity often occur at the same time as bursts of eye movements. These periods of muscle twitching with bursts of eye movements are often referred to as *phasic REM sleep*, in contrast to periods of REM sleep with few or no eye movements and muscle twitches, which are referred to as *tonic REM sleep*. Tonic REM sleep may sometimes last for several minutes.

The muscle atonia of REM sleep, first described in humans by Berger,[48] involves facial muscles, postural muscles of the neck, and accessory muscles of respiration. Limb musculature is less affected.[49] Respiratory muscle activity in REM and NREM sleep is discussed in Chapter 3.

A REM-sleep period may end abruptly with an arousal or awakening or may undergo a gradual transition to stage 1 or 2 sleep followed by a descent into deeper stages of NREM sleep. A normal night of sleep in the adult begins with about 80 to 90 minutes of NREM sleep, followed by a 10- to 15-minute period of REM sleep. The two states then alternate throughout the night with a REM-NREM cycle length of about 100 minutes. The duration of REM sleep periods tends to increase with each successive cycle (see Fig. 1–8).

Arousals and Awakenings

Brief arousals and awakenings are abundant in all sleep stages; they occurred 13 times per hour on average in one study of normal subjects and are usually accompanied by a transient increase in heart rate, an increase in respiratory tidal volume, and an increase in skin conductance.[50] Body movements and shifts in position may also accompany arousals. The EEG also changes, but the type of EEG activity that develops is dependent on the sleep stage from which the arousal occurs.[51] A brief burst of alpha or theta activity is the most common pattern and is typical of most arousals from stage 1 sleep and REM sleep (Fig. 1–10). During stage 2 sleep, arousals may be associated with K-complexes, sometimes followed by brief runs of alpha activity, referred to as K-alpha complexes. Arousals from delta sleep may be associated with runs of synchronous, high-amplitude delta waves.

Arousals may be caused by internal stimuli, such as hypoxia, hypercapnia, bladder or bowel distention, gastroesophageal reflux, pain, and dream-related anxiety, or by external stimulation, such as noise, vibration, or noxious stimuli. Arousals serve other functions than the obvious need to arouse in response to potentially danger-

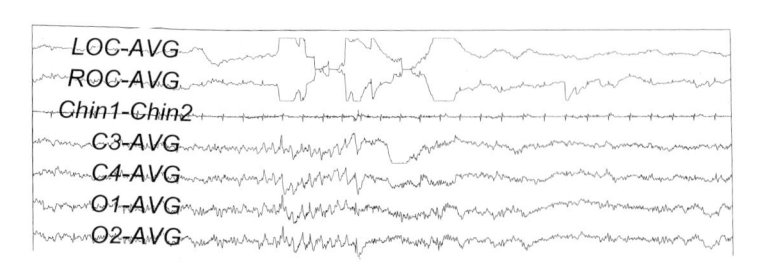

Figure 1–9. A 30-second epoch of REM sleep with a series of sawtooth waves, best defined in the central derivations, occurring simultaneously with rapid eye movements. The chin EMG electrodes demonstrate muscle atonia. For abbreviations, see Figure 1–2.

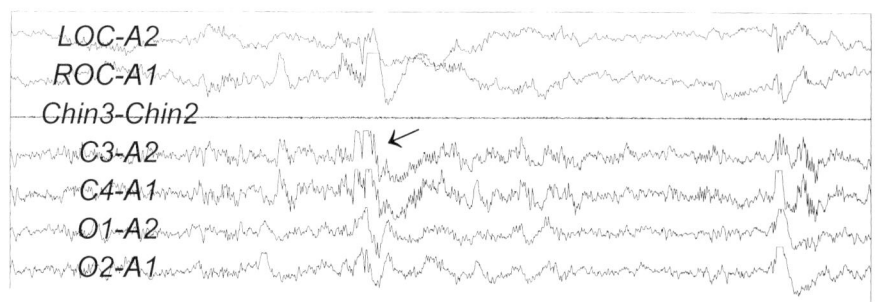

Figure 1–10. A 30-second epoch of sleep interrupted by a brief arousal. The arousal includes a K-complex (arrow) followed by several seconds of alpha- and beta-frequency activity in the central derivations. The faster frequencies are then followed by theta-frequency activity, indicating a resumption of sleep. A K-complex just before the end of the epoch is accompanied by a sleep spindle. For abbreviations, see Figure 1–2.

ous situations. Sighs during arousals reduce atelectasis, the body movements that accompany some arousals relieve pressure points, and postural changes assist with temperature homeostasis.

The strength of the auditory, tactile, or nociceptive stimulation required to induce arousal provides a measure of the depth of sleep. Based on stimulation studies, the depth of sleep is greatest in stage 4 sleep, least in stage 1 sleep, and variable during REM sleep. Sleep depth is greater in children than adults: Auditory stimuli exceeding 120 dB may not produce a full awakening during SWS in young children.[52]

CLINICAL FEATURES OF SLEEP

The Onset of Sleep

Subjective estimates of sleep onset, behavioral features of sleep, and determinations of reduction or loss of consciousness based on responsiveness to meaningful stimuli all have been used to identify the onset of sleep. None of these measures, however, correlates precisely with sleep defined by polygraphic criteria. Subjects who are asked to tap their fingers repetitively usually stop tapping at the transition to electrographically defined stage 1 sleep, although they sometimes continue into stage 1 sleep for several seconds.[53] Similarly, those asked to respond to an auditory or visual stimulus do so less than 20% of the time when the stimulus is presented during stage 1 or 2 sleep.[54,55] Meaningful stimuli produce arousal at lower levels than nonsense stimuli; for example, a nonsense word that has been linked to a noxious stimulus is more likely to induce arousal than one that has not.[56,57]

Clinical signs associated with the onset of sleep include eye closure; muscle relaxation; increased regularity of breathing; and brief, irregular twitches of the distal extremities, called *hypnic jerks*. During NREM sleep the eyes are deviated upward or are dysconjugate, with slow, roving movements; the pupils are small; and if the eyelids are opened, they close smoothly and gradually upon release. Muscle tone is reduced but not flaccid, deep tendon reflexes are present, and plantar responses may be flexor or extensor.

During REM sleep, muscle tone is flaccid, deep tendon reflexes are markedly reduced or absent, and penile erections occur. Breathing is irregular in amplitude and frequency, and the cardiac rhythm is less regular. Conjugate REMs can often be observed through the eyelids, and if the eyes are opened without inducing arousal, the eye movements and fluctuations in pupil size can be observed. Brief clusters of myoclonic jerks of the face and extremities occur at irregular intervals, much more prominently in infants than in adults. The vigorous myoclonic activity is the basis for the term *active sleep,* which is often used to describe infant REM sleep.

Sleep and Consciousness

Perhaps the most striking feature of sleep is the altered consciousness that accompanies it. *Consciousness* is defined as a state of awareness of self and environment that gives meaning or significance to external and internal stimuli. Consciousness has two major components—arousal and cognition—that correlate approximately with two major neuroanatomic systems: (1) the arousal system, including the brainstem reticular activating system and the medial diencephalic structures to which it projects; and (2) the cognitive system, including the cerebral cortex, its associated white matter tracts, and subcortical nuclei (see Chap. 2).

Although consciousness is diminished during sleep, it is not completely absent, and persons may respond to stimuli in meaningful ways, particularly during light sleep. Parents of a new infant may sleep through the sound of an airplane takeoff, yet respond to the much softer cry of the baby. Further, the alteration in consciousness that occurs with sleep is gradual rather than stepwise, and the effects of sleep on self-awareness, memory, sensory awareness, and cognitive continuity may not all occur at the same time. Thus, it is probably impossible to define a precise instant at which wakefulness ends and sleep begins.

Sleep has four distinctive qualities that differentiate it from pathological processes that affect consciousness. First, sleep is a *natural* and necessary event. There is no credible report of a human being who did not need sleep. Second, sleep is *transient*. Despite legends of Rip Van Winkle and other prolonged sleepers, natural sleep episodes rarely exceed 24 hours, and prolonged sleeplike states or behaviors are usually caused by either psychopathology or brain injury. Third, sleep is *periodic*. Although the duration of sleep and the times at which it occurs are functions of a variety of internal processes and social factors, the tendency to sleep at particular phases of the day-night cycle is robust (see Chap. 4). Fourth, sleep is *reversible*. Sleep episodes can be terminated with stimulation.

Coma, in which awareness of self and environment is lost and there is no meaningful response to external stimuli or inner needs, is a pathological state of unconsciousness. Although patients in coma often have a sleeplike appearance and brief observation of comatose persons may lead to the conclusion that they are sleeping, sleep and coma have important physiologic differences as outlined above. In addition, whereas coma is associated with depression of cerebral metabolic activity, overall brain metabolism and oxygen utilization during sleep are altered only slightly.

Two clinical conditions illustrate the distinction between altered consciousness caused by pathologic states and that caused by sleepiness and sleep. Coma associated with severe cortical and subcortical damage (e.g., in cases of anoxia) is frequently followed by the *persistent vegetative state*. In this state, autonomic and vegetative functions are preserved and sleep-wake cycles are readily identifiable: Yawns, chewing, grimacing, and other reflex motor behaviors occur during periods corresponding to physiologic wakefulness, but there is no evidence of awareness or cognition, and cerebral metabolism is markedly reduced. *Akinetic mutism*, a related condition usually caused by large bilateral frontal lobe lesions or diffuse cortical injury, is a state of silent immobility with preserved sleep-wake cycles. During the phase of physiologic wakefulness, patients maintain an alert appearance, but they do not exhibit signs of awareness or cognition. In both of these conditions, consciousness has been lost but sleep-wake cycles remain.

In clinical practice, the distinction between sleep and coma is usually readily made because of the reversibility of sleep with stimulation. Reversal of sleep is sometimes a problem in profoundly sleepy persons, who are difficult to arouse because of severe sleep deprivation or disruption, or in persons in stage 4 sleep who are difficult to arouse with stimulation. In such cases, evaluation of pupillary and oculomotor responses, corneal reflexes, motor responses, and tendon reflexes, along with behavioral responses to stimulation, helps distinguish sleep and sleepiness from stupor or coma and helps determine the cause of pathologically altered consciousness, if present.[58]

PHYSIOLOGIC CHANGES DURING SLEEP

The physiology of every bodily organ is altered during sleep. This section summarizes some of the major changes that accompany NREM sleep and REM sleep.

Autonomic Nervous System Activity

Altered autonomic nervous system activity accounts for many of the physiologic changes that occur during sleep. However, the changes in autonomic activity are not easy to understand because the effects of sleep on sympathetic and parasympathetic activity differ in the various bodily organs and vary across species, and autonomic changes associated with sleep may be attenuated by the effects of compensatory reflexes mediated by chemoreceptors and baroreceptors. For example, arterial blood pressure decreases during NREM sleep in the cat, but not in the rat or rabbit, and heart rate is either decreased or unchanged depending on the species.

In general, parasympathetic tone is augmented during NREM sleep, with a shift toward parasympathetic predominance for organs or structures in which sympathetic and parasympathetic nerves produce opposite effects. For example, the pupil is miotic during NREM sleep, and miosis persists after preganglionic sympathectomy, indicating that the miosis is caused by parasympathetic, rather than sympathetic, activity. Similarly, parasympathetic cardiac tone is greater during sleep than in wakefulness, with its highest levels occurring during SWS. For most organs, parasympathetic activity is probably lower in REM sleep than in NREM sleep.

In contrast to the augmentation of parasympathetic tone, sympathetic tone is reduced during NREM sleep, especially during SWS, with a reduction in circulating levels of epinephrine and norepinephrine.[59,60] The number of sympathetic bursts per minute from sympathetic nerve endings is decreased in human subjects by about 10% to 30% during stage 1–2 NREM sleep and by about 30% to 50% during stage 3–4 sleep compared to wakefulness.[61-63] Sympathetic activity usually increases during arousals, and K-complexes are associated with bursts of sympathetic activity.[61,62,64]

In REM sleep, sympathetic activity is highly variable, although in general it increases compared to NREM sleep, and there are bursts of sympathetic discharge that often correspond with other phasic elements of REM sleep.[62]

Cardiovascular Function

Increased parasympathetic tone and, to a lesser degree, decreased sympathetic tone lead to a decrease in heart rate and cardiac output during NREM sleep.[65,66] In REM sleep, the mean heart rate is similar to the rate in NREM sleep, but it is more variable, particularly during phasic REM sleep, because of fluctuations in sympathetic and parasympathetic activity. Ventricular premature beats are generally reduced during normal sleep, presumably as a result of reductions in sympathetic tone.

In NREM sleep, vasodilatation leads to a reduction in systemic vascular resistance which, combined with the reduction in cardiac output, leads to a 5% to 15% decrease in systemic blood pressure. In contrast, pulmonary artery pressure rises slightly during sleep.[67] In REM sleep, fluctuations in sympathetic activity that affect heart rate and vasomotor tone lead to variability in blood pressure with sudden changes of as much as 40 mm Hg.

Respiratory Function

Increased regularity of breathing is the most striking change in respiration during NREM sleep. Altered breathing affects cardiovascular function: An increase in inspiratory time leads to an increase in venous return, and the decreased interbreath interval variability during NREM sleep leads to greater respiration-related sinus arrhythmia.[68] During REM sleep, inspiratory time decreases and breathing become more irregular. Other respiratory changes associated with sleep are discussed in Chapter 3.

Gastrointestinal Function

The effects of sleep on gastrointestinal function reflect increased parasympathetic activity and a release of intrinsic enteric nervous system activity from central nervous system influences. Functionally, the result is continued but slower digestion. Salivation almost ceases, and swallowing occurs only during arousals. Gastric acid secretion, which rises in the evening, declines in the first hour of sleep, primarily because of increased vagal tone, and remains low for the rest of the night.[69-71] Gastric emptying is slower at night; it may take 50% longer for the stomach to empty after a late evening meal compared to a morning meal.[72,73] The speed of propagation of intestinal contents also is reduced by about 50%.[74]

Renal Function

Urine production decreases during NREM sleep because of reduced renal perfusion, a lower glomerular filtration rate, and an increase in water reabsorption. Urine production decreases even further during REM sleep.[65] Although aldosterone secretion is decreased at night, the day-night difference is largely an effect of the change from an upright to a reclining posture.[75]

Thermal Regulation

Body temperature declines by 1°C to 2°C during sleep at night with about one-half of the decline caused by circadian variation and the other half by a sleep-related change in the setpoint for thermal regulation. The reduced setpoint is accompanied by an increase in heat dissipation at the onset of sleep, with increased sweating and vasodilation, and by decreased heat production. Thermoregulatory responses maintain body temperature at the lower setpoint during NREM sleep.

During REM sleep, however, thermoregulatory responses are attenuated: Sweating, panting, and peripheral vasoconstriction in response to temperature changes are less pronounced, and as a result, body temperature tends to move in the direction of ambient temperature.

Endocrine Function

The secretion of most hormones fluctuates rhythmically during the 24-hour cycle. Some have increased secretion at night, linked either to sleep state or to circadian phase. Cortisol secretion, for example, is a function mainly of circadian phase, as its early-morning peak persists in the absence of sleep. On the other hand, nocturnal secretion of prolactin and growth hormone is dependent on sleep rather than on circadian phase.[76-78] Prolactin secretion starts to rise 1 to 2 hours after sleep onset and peaks in the early morning; during pregnancy, the sleep-related enhancement is even more striking. Growth hormone secretion occurs mainly during SWS, peaking during the first hour of sleep; the circadian contribution is almost nil as the pulse of growth hormone secretion is abolished by SWS deprivation. Parathyroid hormone secretion also is highest in SWS.

Glucose and insulin secretion increase during sleep, probably mainly as a result of the increased levels of growth hormone and cortisol.[79] Testosterone levels rise at sleep onset and continue to rise throughout the night, whereas plasma renin activity fluctuates during sleep, with higher levels occurring during NREM sleep and lower levels during REM sleep.[80] The release of some hormones, such as thyroid-stimulating hormone, is inhibited by sleep.[81]

Genital Changes

Although few genital changes occur during NREM sleep, penile erection is one of the most striking physical features of REM sleep. The erection usually begins within a few minutes of the onset of REM sleep and persists throughout the REM period. Detumescence occurs within a few minutes of the end of the REM period, although the erection may continue during wakefulness, which accounts for the occurrence of morning erections. Sleep-related penile erections occur in healthy boys and men of all ages from infancy to old age and are not substantially affected by the sexual content of dreams, by sexual abstinence, or by pre-sleep sexual fantasy or activity.[82] Similar changes in erectile tissue also occur during REM sleep in women; these changes include increased vaginal pulse pressure and phasic shifts of blood flow within the vaginal blood vessels.[83,84]

Sleep-related penile erections occur as a result of increased blood flow—caused by enhanced parasympathetic activity and local vasodilatation—coupled with constricted vascular outflow and increased cavernosal pressure that is probably a result of increased bulbocavernosus and ischiocavernosus muscle activity. Detumescence is a result of sympathetic activity that promotes venous outflow. The use of recordings of penile circumference changes during sleep as a means of assessing erectile function is discussed in Chapter 15.

Immune Function

The secretion of several cytokines involved in immune regulation is altered by sleep. Interleukin-1 levels increase during sleep, and peak levels occur at the onset of SWS (see Chap. 2). Sleep deprivation leads to a variety of changes in immune function, including reduced antibody response to sheep red blood cells, increased levels of interleukin-1, decreased lymphocyte DNA synthesis, increased numbers of circulating monocytes and granulocytes, and increased natural killer cell activity.[85,86] These changes may affect host defense responses and may contribute to the increased risk of septicemia that occurs with prolonged sleep deprivation in rats.[87]

Cerebral Blood Flow and Metabolism

Both cerebral blood flow and metabolism decline by about 3% to 10% during stages 1 and 2 and by about 25% to 45% during stages 3 and 4.[88–90] The overall reduction in cerebral oxygen consumption of 15% to 20% during NREM sleep is about fivefold higher than the reduction that occurs in other organs, and it appears to be diffuse (i.e., there are small or nonexistent regional differences).

Cerebral blood flow and cerebral metabolism are tightly coupled during the awake state and probably also during sleep, but the coupling may not be as tight during changes in state.[89] During REM sleep, cerebral blood flow and metabolism remain tightly coupled and increase approximately to waking values.[90] The increase compared to NREM sleep is likely to be caused by reduced resistance of cerebral arterioles, as the increase in cerebral blood flow is not accompanied by an increase in cardiac output or in flow to other organs.[91,92] Cerebral autoregulation remains intact during REM sleep.[90] Regional increases in metabolism occur in the visual associative areas, the amygdaloid complexes, and perhaps also in areas of the diencephalon and brainstem.[92–95]

Memory Formation

The change in level of consciousness that occurs with sleep affects cortical processing of the contents of consciousness. In particular, memory is impaired by sleep. For example, words presented less than 4 minutes before sleep onset are usually not remembered if sleep persists for 10 minutes or more.[96] On the other hand, memory is also disrupted by sleep deprivation, which suggests that sleep facilitates memory consolidation. The amount of REM sleep increases after intense learning experiences,[97] and deprivation of REM sleep in humans impairs recall of the previous day's events, although it is possible that stressful aspects of the deprivation lead to the impairment, rather than the deprivation itself. In one study, subjects demonstrated improved performance on a visual discrimination task after a night of normal sleep and after a night with SWS disruption, but did not show improvement after a night of undergoing selective disruption of REM sleep.[98] In rats, hippocampal neurons that fire together during a behavioral task have a greater tendency to fire together during subsequent NREM sleep than during sleep that preceded the task, suggesting that memory traces formed in the hippocampus during wakefulness may be consolidated and perhaps transferred to the cortex during NREM sleep.[99]

SLEEP HOMEOSTASIS

Sleep is regulated by homeostatic mechanisms; as a result, sleep deprivation is followed by sleepiness, increased sleep duration when sleep is allowed to occur, more intense sleep with greater stimulation required for arousal, increased amplitude of slow waves during SWS, and increased density of REMs during REM sleep.

Although sleepiness increases with sleep deprivation, it does not do so monotonically; instead, it maintains a circadian organization (see Chap. 8). Borbely and Achermann[100] proposed a model of sleep and wakefulness based on two components: process S and process C. In this model, process S is a homeostatic process and a measure of sleep need that increases exponentially during wakefulness and decays exponentially during sleep (see Fig. 4–8). The decay of *process S* is closely related to the amount of SWS that occurs. Thus, naps that occur late in the day, when process S has reached a high level, are more likely to contain SWS than naps that occur early in the day.[101] *Process C* is a circadian process with a sinusoidal rhythm that increases sleep propensity during the night and decreases it during the day. Process C thereby helps to maintain sleep as process S declines during the night and to maintain watchfulness as process S increases during the day.

FUNCTIONS OF SLEEP

Sleep, that knits up the ravell'd sleave of care,
The death of each day's life, sore labour's bath,
Balm of hurt minds, great nature's second course,
Chief nourisher in life's feast.

WILLIAM SHAKESPEARE, MACBETH, ACT II

Why Do We Sleep?

If sleep does not serve an absolutely vital function, then it is the biggest mistake the evolutionary process has ever made.[102]

A. RECHTSCHAFFEN

The function of sleep remains its essential mystery. Sleep is ubiquitous among mammals and humans, but despite centuries of speculation by scientists and philosophers, thousands of experiments, and the best efforts of dedicated researchers, its function remains elusive. Sleep is restorative in some basic way: The overwhelming and disabling sleepiness that accompanies prolonged sleep deprivation suggests that sleep is appetitive and fulfills essential needs. Furthermore, both REM and NREM sleep are necessary. Deprivation of either state leads to a rebound increase in that state during the recovery period; in rats, prolonged total deprivation of either REM or NREM sleep leads to death in a matter of weeks (see Chap. 8). There is no reason to believe that it would not be equally lethal in humans.

Knowing that sleep is vital, however, does not answer the question of why this is so. The most inevitable and characteristic consequence of sleep loss is sleepiness, but this observation does not provide an explanation for the function of sleep. To say that the function of sleep is to prevent sleepiness is akin to saying that the function of eating is to prevent hunger; however, although it is evident that food supplies the energy essential to run the biochemical processes of life, a similar unitary concept for sleep need is lacking. Perhaps there is no single function, no parsimonious explanation of all of the deficits caused by sleep deprivation. Instead, it may be more useful to view sleep as a state

Table 1–1. **Proposed Functions of Sleep**

Function	NREM Sleep	REM Sleep
Brain or body restoration, or both	+	+
Replenishment of cerebral glycogen	+	
Tissue synthesis and cell mitosis	+	
Protein synthesis	+	+
Growth hormone release	+	
Thermoregulation	+	
Energy conservation	+	
Regulation of noradrenergic activity		+
Memory consolidation and information processing		+
Brain development		+
Cell maturation		+
Development of oculomotor control		+
Programming of genetically determined behaviors		+
Neural stimulation		+

that serves multiple functions, with some functions more important at one stage of life than at another (Table 1–1). Most current theories concerning the function of sleep focus on restoration and adaptation.

RESTORATION

The refreshed feeling that follows a good night's sleep suggests that sleep serves to restore the brain or body, or both, in some essential manner; the increased sleep that follows illness or surgery also supports the view that sleep has a restorative function. Some investigators believe that body restoration is primarily a function of NREM sleep, whereas brain restoration is primarily a function of REM sleep. A bodily restorative function for NREM sleep is supported by hormonal changes: Levels of anabolic hormones such as growth hormone, testosterone, prolactin, and luteinizing hormone increase, while levels of catabolic hormones, such as cortisol, decrease. A brain-related function of REM sleep is supported by the high proportion of REM sleep in utero and during infancy, when brain development is maximal, and by the finding that REM sleep deprivation in infancy produces impairments of brain development.[103]

Other investigators believe that both NREM and REM sleep are more important for brain restoration than for body restoration on the basis of the following:

1. Quiet resting restores the body but not the brain.
2. Brain activity, as measured by EEG, shows greater differences between sleep and waking than most measures of body activity.
3. Sleep deprivation affects cognitive functions to a greater extent than body functions.
4. EEG changes during recovery from sleep deprivation show some of the most robust homeostatic responses.

The need for alertness and readiness during wakefulness suggests that the brain cannot "relax" during this period; thus, one function of sleep may be to allow the brain to relax and be restored.

The model for sleep proposed by Borbely and Achermann[100] implies that not all sleep is equally restorative: Process S, according to the model, discharges most rapidly during the first few hours of sleep. Horne[104] proposed that there are two kinds of sleep: core sleep and optional sleep. In his view, core sleep, which includes SWS and the first three REM periods in a night, subserves vital needs as reflected by the impaired cerebral functioning that accompanies loss of core sleep. On the other hand, according to his theory, loss of optional sleep, which includes the last two to three REM periods and most of stage 1 and stage 2 sleep, may affect motivation and alertness, but it does not affect vital functions. The preferential recovery of REM sleep and SWS after sleep deprivation, and the failure to recover lost optional sleep after deprivation, support this view.

ADAPTATION

Adaptive theories propose that sleep allows a broader range of adaptation to environmental conditions. Most of these theories posit that sleep plays an essential role in energy conservation or thermal regulation, or both. The autonomic changes that occur during sleep along with the reductions in body temperature and metabolic rate, which maintain homeostasis at a lower level of energy expenditure than occurs in wakefulness, support these theories. The occurrence of sleeplike states, with EEG slowing and increased arousal threshold, in all endothermic species also supports the concept that the major function of sleep is related to cold adaptation and thermoregulation. Furthermore, animals that hibernate enter this state from SWS.

Species differences in sleep also support adaptive theories. In warm-blooded animals, metabolic rate is closely correlated with sleep duration: Small animals with high metabolic rates sleep longer than large animals with low metabolic rates. The metabolic derangements that accompany sleep deprivation also suggest that sleep plays an essential role in metabolic and thermal or energy regulation. Furthermore, the increase in SWS that occurs after 2 to 3 days of starvation is consistent with an energy conservation role for this type of sleep.[105]

SWS at night also increases after afternoon heat exposure caused by exercise, warm baths, or saunas.[106–108] McGinty and Szymusiak[108] proposed that the expression of SWS is dependent on the "heat load" of the organism—that is, the integrated difference between actual hypothalamic temperature and the hypothalamus-mediated temperature setpoint during the preceding period of wakefulness. In their model, the medial preoptic area, the anterior hypothalamus, and the basal forebrain contain heat-sensitive neurons that induce SWS in response to a heat load and when brain temperature, determined by thermosensitive neurons of the anterior hypothalamus and preoptic region, exceeds a threshold level.

Each of these theories has merit, and they are not mutually exclusive; it may be that restoration and the adaptation mechanisms of energy conservation and thermal regulation are all functions of sleep that are of varying importance, depending on the age and physiologic state of the organism.

Functions of REM Sleep

Although REM sleep is present in most mammalian species, its essential functions remain unknown. The need for REM sleep must be critical to allow a state of such vulnerability—with impaired vigilance, muscle atonia, and reduced responsiveness to changes in body temperature and oxygenation—to remain present throughout the life span and across so many species.

One function of REM sleep may relate to brain development and maturation. The amount of REM sleep is highest in the prenatal period and higher in animals that are born relatively immature compared to those born in a more mature state. Neonatal REM sleep deprivation of cats and rats leads to altered visual system development, behavioral changes in adulthood, reduced weight of the cerebral cortex and medulla oblongata, and reduced brain growth in response to environmental stimulation.[103,109] Roffwarg and coworkers[110] suggested that prenatal and neonatal REM sleep may act as a surrogate for wakefulness to provide the brain stimulation required for brain development.

According to other theories, functions of REM sleep include reinforcement of learning and consolidation of memory,[99] elimination of useless memories,[111] and regulation of noradrenergic activity.[112]

Perhaps REM sleep is required for essential maintenance of brain systems that are continually active during wakefulness, such as the noradrenergic system and the systems that maintain posture.

How Much Sleep Is Enough?

The day and night consist of 24 hours. It is sufficient for a person to sleep one third thereof . . .

MAIMONIDES

Sleep need varies from person to person. For a given individual, the amount of sleep required is that which produces optimal alertness throughout the day. Although the mean is 7 to 8 hours for adults, 4 hours is enough for some persons and others require 10 hours. Differences in habitual sleep times reflect, in part, individual differences in tolerance for insufficient sleep: Some persons find mild sleepiness unpleasant and prefer to live life as well rested as possible. Others ignore the sensations associated with sleepiness, do not find sleepiness uncomfortable, or possess the ability to overcome the sensations and effects of sleepiness through motivation or other factors. The varying need for sleep across individuals is yet another mystery of sleep that may not be explained until the functions of sleep are clearly identified. Sleep researchers still have much work to do to explain the basic mysteries of why we sleep.

SUMMARY

Before the 20th century, sleep was viewed as a passive process caused by a reduction in wakefulness. With the discovery of REM sleep and subsequent studies demonstrating vigorous brain activity during REM sleep, distinctive neurophysiologic substrates underlying the two major sleep states, NREM sleep and REM sleep, were identified, and the current view of sleep as an active process evolved. With the use of polygraphic recordings, sleep stages can be identified based on the appearance of the EEG, EOG, and EMG. During a night of sleep, NREM sleep and REM sleep alternate in cycles that average about 100 minutes.

Sleep has four distinctive qualities that differentiate it from pathological processes that affect consciousness: It is natural, transient, periodic, and reversible. The physiologic changes that occur during NREM sleep are caused partly by a shift of the autonomic nervous system to parasympathetic predominance. During REM sleep, however, sympathetic activity increases, and paralysis of skeletal muscles is accompanied by brain activation with increased blood flow and metabolism. Although the functions of sleep are not well defined, sleep is regulated homeostatically and appears to have vital functions. The reasons for its necessity, once unraveled, will solve one of the largest remaining riddles of mammalian neurobiology.

REFERENCES

1. Tobler, I: Evolution and comparative physiology of sleep in animals. In Lydic, R, and Biebuyck, JF (eds): Clinical Physiology of Sleep. American Physiological Society, Bethesda, Md, 1988, pp 21–30.
2. Zepelin, H: Mammalian sleep. In Kryger, MH, et al (eds): Principles and Practice of Sleep Medicine, ed 2. WB Saunders, Philadelphia, 1994, pp 69–80.
3. Thorpy, M: History of sleep and man. In Thorpy, MJ, and Yager, J (eds): The Encyclopedia of Sleep and Sleep Disorders. Facts on File, New York, 1991, pp ix-xxxiii.
4. Legendre, R, and Pieron, H: Le problème des facteurs du sommeil: Résultats d'injections vasculaires et intracérébrales de liquides insomniques. C R Soc Biol 68:1077, 1910.
5. Dement, WC: History of sleep physiology and medicine. In Kryger, MH, et al (eds): Principles and Practice of Sleep Medicine, ed 2. WB Saunders, Philadelphia, 1994, pp 3–15.
6. Gayet, M: Affection encéphalique (encéphalite diffuse probable) localisée aux étages supérieurs des pédoncles cérébraux et aux couches optiques. Arch Physiol 7:341, 1875.
7. Caton, R: The electric currents of the brain. Br Med J 2:278, 1875.
8. Berger, H: Ueber das Elektroenkephalogramm des Menschen. J Psychol Neurol 40:160, 1930.
9. Adrian, ED, and Yamagiwa, K: The origin of the Berger rhythm. Brain 58:323, 1935.
10. Loomis, A, et al: Cerebral states during sleep as studied by human brain potentials. J Exp Psychol 21:127, 1937.
11. Loomis, AL, et al: Distribution of disturbance patterns in the human electroencephalogram,

with special reference to sleep. J Neurophysiol 1:413, 1938.
12. von Economo, C: Sleep as a problem of localization. J Nerv Ment Dis 71:249, 1930.
13. Hess, WR: Das Schlafsyndrom als Folge diencephaler Reizung. Helv Physiol Acta 2:305, 1944.
14. Bremer, F: Cerveau "isolé" et physiologie du sommeil. C R Soc Biol 118:1235, 1935.
15. Bremer, F: L'activité électrique de l'écorce cérébrale at le problème physiologique du sommeil. Boll Soc Ital Biol Sper 13:271, 1938.
16. Kleitman, N: Sleep and Wakefulness. University of Chicago Press, Chicago, 1939.
17. Moruzzi, G, and Magoun, H: Brain stem reticular formation and activation of the EEG. Electroencephalogr Clin Neurophysiol 1:455, 1949.
18. Starzl, TE, et al: Collateral afferent excitation of reticular formation of brain stem. J Neurophysiol 14:479, 1951.
19. Aserinsky, E, and Kleitman, N: Regularly occurring periods of eye motility, and concomitant phenomena, during sleep. Science 118:273, 1953.
20. Dement, W, and Kleitman, N: Cyclic variations in EEG during sleep and their relation to eye movements, body motility, and dreaming. Electroencephalogr Clin Neurophysiol 9:673, 1957.
21. Jouvet, M, and Michel, F: Recherches sur l'activité électrique cérébrale au cours du sommeil. C R Soc Biol 152:1167, 1958.
22. Jouvet, M, et al: Sur un stade d'activité électrique cérébrale rapide au cours du sommeil physiologique. C R Soc Biol 153:1024, 1959.
23. Jouvet, M, et al: L'activité de lesions du rhinencephale au cours du sommeil chez le chat. C R Soc Biol 153:101, 1959.
24. Jouvet, M, and Michel, F: Corrélations electromyographiques du sommeil chez le chat décortiqué et mésencephalique chronique. C R Soc Biol 153:422, 1959.
25. Hess, WR: Hernreizversuche ueber den Mechanismus des Schlafes. Arch Psychiatr Nervenkr 86:287, 1929.
26. Nauta, WJH: Hypothalamic regulation of sleep in rats. J Neurophysiol 9:285, 1946.
27. Batini, C, et al: Persistent patterns of wakefulness in the pretrigeminal midpontine preparation. Science 128:30, 1958.
28. Batini, C, et al: Neural mechanisms underlying the enduring EEG and behavioral activation in the midpontine pretrigeminal cat. Arch Ital Biol 97:13, 1959.
29. Sterman, MB, and Clemente, CD: Forebrain inhibitory mechanisms: Cortical synchronization induced by basal forebrain stimulation. Exp Neurol 6:91, 1962.
30. Hernandez-Peon, R, and Chavez-Ibarra, G: Sleep induced by electrical or chemical stimulation of the forebrain. Electroencephalogr Clin Neurophysiol (suppl)24:188, 1963.
31. Jouvet, M: Biogenic amines and the states of sleep. Science 163:32, 1969.
32. Jouvet, M, and Mounier, D: Effets des lésions de la formation réticulée pontique sur le sommeil du chat. C R Soc Biol 154:2301, 1960.
33. Reivich, M, and Kety, S: Blood flow metabolism couple in brain. In Plum, F (ed): Brain Dysfunction in Metabolic Disorders. Raven Press, New York, 1968, pp 125–140.
34. Moruzzi, G: The sleep-waking cycle. Ergeb Physiol 64:1, 1972.
35. Jones, BE, et al: The effect of lesions of catecholamine-containing neurons upon monoamine content of the brain and EEG and behavioral waking in the cat. Brain Res 58:157, 1973.
36. Jones, BE: The respective involvement of noradrenaline and its deaminated metabolites in waking and paradoxical sleep: A neuropharmacological model. Brain Res 39:121, 1972.
37. Jouvet, M: The role of monoamines and acetylcholine containing neurons in the regulation of the sleep-waking cycle. Ergeb Physiol 64:166, 1972.
38. Monnier, M, and Gaillard, J: Biochemical regulation of sleep. Experientia 36:21, 1980.
39. Wyatt, RJ, et al.: Brain catecholamines and human sleep. Nature 233:63, 1971.
40. Wyatt, RJ, et al: Effects of para-chlorphenylalanine on sleep in man. Electroencephalogr Clin Neurophysiol 27:529, 1969.
41. Harvey, EN, et al: Cerebral states during sleep as studied by human brain potentials. Science 85:443, 1937.
42. Rechtschaffen, A, and Kales, A: A manual of standardized terminology, techniques, and scoring system for sleep stages of human subjects. Brain Information Service/Brain Research Institute, Los Angeles, 1968.
43. Holland, JV, et al: "Polysomnography": A response to a need for improved communication. Presented at the 14th Annual Meeting of the Association for the Psychophysiological Study of Sleep, Jackson Hole, Wyo, June 1974.
44. Broughton, R, et al: Anterior slow alpha of drowsiness: Topography, source dipole analysis. Sleep Res 23:5, 1994.
45. Westmoreland, B: Benign EEG variants and patterns of uncertain clinical significance. In Daly, DD, and Pedley, TA (eds): Current Practice of Clinical Electroencephalography. Raven Press, New York, 1990, pp 243–252.
46. Jobert, M, et al: Topographical analysis of sleep spindle activity. Neuropsychobiology 26:210, 1992.
47. Hauri, P, and Hawkins, DR: Alpha-delta sleep. Electroencephalogr Clin Neurophysiol 34:233, 1973.
48. Berger, R: Tonus of extrinsic laryngeal muscles during sleep and dreaming. Science 134:840, 1961.
49. Jacobson, A, et al: Muscle tonus in human subjects during sleep and dreaming. Exp Neurol 10:418, 1964.
50. Collard, P, et al: Movement arousals and sleep-related disordered breathing in adults. Am J Respir Crit Care Med 154:454, 1996.
51. Atlas Task Force of the American Sleep Disorders Association: EEG Arousals: Scoring rules and examples. Sleep 15:173, 1992.
52. Jobert, M, et al: Topographical analysis of sleep spindle activity. Neuropsychobiology 26:210, 1992.

53. Carskadon, MA, and Dement, WC: Effects of total sleep loss on sleep tendency. Percept Mot Skills 48:495, 1979.
54. Carskadon, MA, and Dement, WC: Normal human sleep: An overview. In Kryger, MH, et al (eds): Principles and Practice of Sleep Medicine, ed 2. WB Saunders, Philadelphia, 1994, pp 16–25.
55. Ogilvie, RD, and Wilkinson, RT: The detection of sleep onset: Behavioral and physiological convergence. Psychophysiology 21:510, 1984.
56. Oswald, I, et al: Discriminative responses to stimulation during human sleep. Brain 83:440, 1960.
57. Williams, HL, et al: Instrumental behavior during sleep. Psychophysiology 2:208, 1966.
58. Plum, F, and Posner, JB: The Diagnosis of Stupor and Coma, ed 3. FA Davis, Philadelphia, 1980.
59. de Leeuw, PW, et al: Effect of sleep on blood pressure and its correlates. Clin Exp Theor Prac A7:179, 1985.
60. Baharav, A, et al.: Fluctuations in autonomic nervous activity during sleep displayed by power spectrum analysis of heart rate variability. Neurology 45:1183, 1995.
61. Hornyak, M, et al: Sympathetic muscle nerve activity during sleep in man. Brain 114:1281, 1991.
62. Okada, H, et al: Changes in muscle sympathetic nerve activity during sleep in humans. Neurology 41:1961, 1991.
63. Takeuchi, S, et al: Sleep-related changes in human muscle and skin sympathetic nerve activities. J Auton Nerv Syst 47:121, 1994.
64. Noll, G, et al: Skin sympathetic nerve activity and effector function during sleep in humans. Acta Physiol Scand 151:319, 1994.
65. Orem, J, and Keeling, J: Appendix : A compendium of physiology in sleep. In Orem, J (ed): Physiology in Sleep. Academic Press, New York, 1980, pp 315–335.
66. Miller, JC, and Horvath, SM: Cardiac output during human sleep. Aviat Space Environ Med 47:1046, 1976.
67. Block, AJ: Respiratory disorders during sleep: Part II. Heart Lung 10:90, 1981.
68. Harper, RM, et al: Cardiac and respiratory interactions maintaining homeostasis during sleep. In Lydic, R, and Biebuyck, JF (eds): Clinical Physiology of Sleep. American Physiological Society, Bethesda, Md, 1988, pp 67–78.
69. Stacher, G, et al: Gastric acid secretion and sleep stages during natural light sleep. Gastroenterology 68:1449, 1975.
70. Moore, JG, and Wolfe, M: The relation of plasma gastrin to the circadian rhythm of gastric acid secretion in man. Digestion 9:97, 1973.
71. Moore, JG: High grade acid secretion after vagotomy and pyloroplasty in man: Evidence for non-vagal mediation. Am J Digest Dis 18:661, 1973.
72. Sheiner, HJ, et al: Gastric motility and emptying in normal and post-vagotomy patients. Gut 21:753, 1980.
73. Goo, RH, et al: Circadian variation in gastric emptying of meals in man. Gastroenterology 93:515, 1987.
74. Kumar, D, et al: Circadian variation in the propagating velocity of the migrating motor complex. Gastroenterology 91:926, 1986.
75. Pratt, JH, et al: Aldosterone excretion rates in children and adults during sleep. Hypertension 8:154, 1986.
76. Sassin, JF, et al: The nocturnal rise of human prolactin is dependent on sleep. J Clin Endocrinol Metab 37:436, 1973.
77. Sassin, JF, et al: Human prolactin: 24-hour pattern with increased release during sleep. Science 177:1205, 1972.
78. Sassin, JF, et al: Human growth hormone release: Relation to slow-wave sleep and sleep-waking cycles. Science 165:513, 1969.
79. Van Cauter, E, et al: Modulation of glucose regulation and insulin secretion by circadian rhythmicity and sleep. J Clin Invest 88:934, 1991.
80. Brandenberger, G: Episodic hormone release in relation to REM sleep. J Sleep Res 2:193, 1993.
81. Parker, DC, et al: Effect of 64-hour sleep deprivation on the circadian waveform of thyrotropin (TSH): Further evidence of sleep-related inhibition of TSH release. J Clin Endocrinol Metab 64:157, 1987.
82. Karacan, I, et al: Erectile mechanisms in man. Science 220:1080, 1983.
83. Abel, GG, et al: Women's vaginal responses during REM sleep. J Sex Marital Ther 5:5, 1979.
84. Rogers, GS, et al: Vaginal pulse amplitude response patterns during erotic conditions and sleep. Arch Sex Behav 14:327, 1985.
85. Krueger, JM, and Majde, JA: Microbial product and cytokines in sleep and fever regulation. Crit Rev Immunol 14:355, 1994.
86. Pollmacher, T, et al: Influence of host defense activation on sleep in humans. Adv Neuroimmunol 5:155, 1995.
87. Everson, CA: Sustained sleep deprivation impairs host defense. Am J Physiol 265:R1148, 1993.
88. Madsen, PL, and Vorstrup, S: Cerebral blood flow and metabolism during sleep. Cerebrovasc Brain Metab Rev 3:281, 1991.
89. Hoshi, Y, et al: Dynamic features of hemodynamic and metabolic changes in the human brain during all-night sleep as revealed by near-infrared spectroscopy. Brain Res 652:257, 1994.
90. Franzini, C, et al: Sleep-dependent changes in regional circulations. News Physiol Sci 11:274, 1996.
91. Zoccoli, G, et al: Brain blood flow and extracerebral carotid circulation during sleep in rat. Brain Res 641:46, 1994.
92. Lenzi, P, et al: Brain circulation during sleep and its relation to extracerebral hemodynamics. Brain Res 415:14, 1987.
93. Maquet, P, et al: Functional neuroanatomy of human rapid-eye-movement sleep and dreaming. Nature 383:163, 1996.
94. Gerashchenko, D, and Matsumura, H: Continuous recordings of brain regional circulation

during sleep/wake state transitions in rats. Am J Physiol 270:R855, 1996.
95. Abrams, RM, et al: Local cerebral glucose utilization non-selectively elevated in rapid eye movement sleep of the fetus. Dev Brain Res 40:65, 1988.
96. Wyatt, JK, et al: Does sleep onset produce retrograde amnesia? Sleep Res 21:113, 1992.
97. De Koninck J: Intensive learning, REM sleep and REM sleep mentation. SRS Bulletin 1:39, 1995.
98. Karni, A, et al: Dependence on REM sleep of overnight improvement of a perceptual skill. Science 265:679, 1994.
99. Wilson, MA, and McNaughton, BL: Reactivation of hippocampal ensemble memories during sleep. Science 265:676, 1994.
100. Borbely, AA, and Achermann, P: Concepts and models of sleep regulation: An overview. J Sleep Res 1:63, 1992.
101. Tobler, I, and Borbely, AA: Sleep EEG in the rat as a function of prior waking. Electroencephalogr Clin Neurophysiol 64:74, 1986.
102. Rechtschaffen, A: The control of sleep. Proceedings of the Symposium on Human Behavior and its Control, Meeting of the American Association for the Advancement of Science, Chicago, Ill, December 1970.
103. Marks, GA, et al: A functional role for REM sleep in brain maturation. Behav Brain Res 69:1, 1995.
104. Horne, J: Why we sleep. Oxford University Press, New York, 1988.
105. MacFayden, H, et al: Starvation and human slow wave sleep. J Appl Physiol 35:391, 1973.
106. Horne, JA, and Shackell, BS: Slow wave sleep elevations after body heating: Proximity to sleep and effects of aspirin. Sleep 10:383, 1987.
107. Horne, JA, and Moore, VJ: Sleep EEG effects of exercise with and without additional body cooling. Electroencephalogr Clin Neurophysiol 60:33, 1985.
108. McGinty, D, and Szymusiak, R: Keeping cool: A hypothesis about the mechanisms and functions of slow-wave sleep. Trends Neurosci 13:480, 1990.
109. Mirmiran, M, and Van Someren, E: The importance of REM sleep for brain maturation. J Sleep Res 2:188, 1993.
110. Roffwarg, HF, et al: Ontogenetic development of the human sleep-wakefulness cycle. Science 152:604, 1966.
111. Crick, F, and Mitchison, G: The function of dream sleep. Nature 304:111, 1983.
112. Siegel, JM, and Rogawski, MA: A function for REM sleep: Regulation of noradrenergic receptor sensitivity. Brain Res Rev 13:213, 1988.

Chapter 2

NEUROBIOLOGY OF SLEEP

SLEEP-WAKE REGULATION
Wakefulness and the Brainstem
    Reticular Formation
Noradrenaline
Serotonin
Dopamine
Histamine
Acetylcholine
NREM SLEEP
Sleep Onset
Sleep Spindles and Delta Waves
REM SLEEP
Postural Atonia
Pontine-Geniculate-Occipital Spikes
Cortical Activation
Rapid Eye Movements
Muscle Twitches
SLEEP FACTORS
Cytokines
Prostaglandins
Adenosine
Peptides
Melatonin
Benzodiazepines and Endogenous Benzo-
    diazepine-Receptor Agonists

SLEEP-WAKE REGULATION

Sleep cannot be localized.
JOSEPH J. DEJERINE, 1914

Although the brain's involvement in the regulation of sleep and wakefulness was suspected at least as early as the 5th century BC (see Chap. 1), the neurobiologic basis for sleep was a mystery until the 20th century. Research studies and clinical studies over the past several decades have shown that sleep is not simply the result of diminished function of waking systems. Rather, it is the consequence of an active process requiring appropriate interactions of a number of brainstem and cerebral systems. In general terms, the waking state is associated with high activity of monoaminergic and cholinergic projection systems; the non–rapid eye movement (NREM) sleep state is associated with low activity of monoaminergic and cholinergic systems; and the rapid eye movement (REM) sleep state is associated with low monoaminergic activity and high cholinergic activity.

Wakefulness and the Brainstem Reticular Formation

The importance for arousal and wakefulness of ascending pathways originating in the reticular formation (RF), often described as *nonspecific* ascending pathways because they do not directly convey sensory or motor information, has been clear since the studies of Moruzzi and Magoun.[1] With the development of techniques for retrograde labeling and intracellular recordings, the pathways involved in arousal and electroencephalographic (EEG) desynchronization have been identified more precisely (Table 2–1). The neurotransmitters involved in these pathways tend to produce changes in neural activity that are slow in onset and long in duration. They are therefore viewed as neuromodulators and probably act via second messengers.

Activity of the pontomesencephalic neurons that mediate arousal is modulated by afferents from sensory pathways, the cor-

Table 2–1. **Ascending Pathways Mediating Arousal**

	Site of Origin	Major Cortical Projection Sites
Norepinephrine	Locus coeruleus and lateral tegmental area	Diffuse with relatively greater innervation of structures involved in visuomotor responses and spatial analysis
Dopamine	Ventral tegmental area	Primary motor cortex and prefrontal cortex; sensory association areas
Serotonin	Dorsal and medial raphe nuclei	Diffuse with some laminar and topographic specificity
Acetylcholine	Basal forebrain and brainstem (pedunculopontine tegmental and laterodorsal tegmental nuclei)	Diffuse
Histamine	Posterior hypothalamus	Diffuse

tex, and the hypothalamus. Some of the neurons with ascending projections increase their firing rates just before awakening and appear to play a key role in initiating arousal. Two ascending pathways exist: One originates from monoaminergic, cholinergic, and glutaminergic neurons of the brainstem and projects to the thalamus; the other, which also originates from cholinergic and monoaminergic groups, passes more ventrally via the medial forebrain bundle to the hypothalamus, basal forebrain, and cortex (Fig. 2–1).[2] In addition to their cortical projections, the neurons of the extrathalamic pathways modulate activity in thalamocortical projection neurons.

Electrolytic lesions of the posterior hypothalamus and the mesencephalic RF that destroy cell bodies and fiber tracts of the ascending system cause long-lasting coma, whereas neurotoxic lesions of these regions that destroy only cell bodies produce only transient loss of wakefulness lasting a few days.[3] Thus, arousal and wakefulness appear to be mediated by cholinergic, noradrenergic, dopaminergic, serotonergic, and histaminergic projection

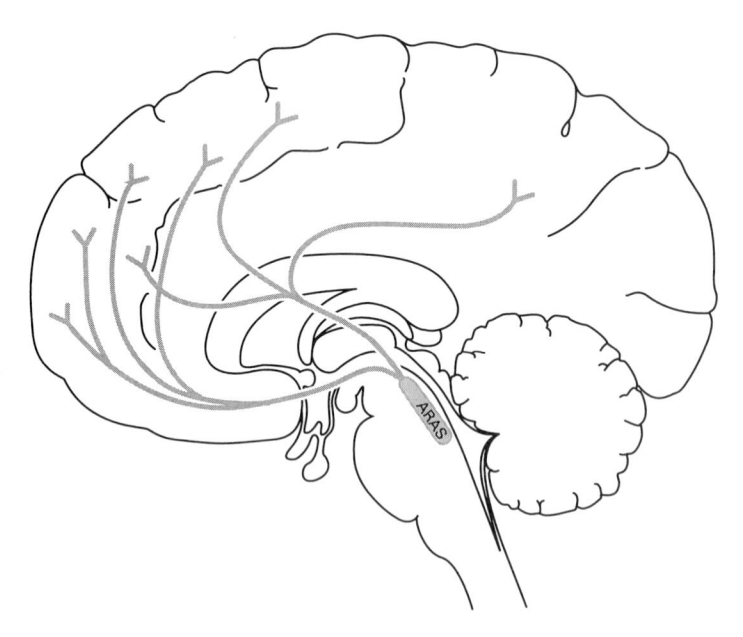

Figure 2–1. The ascending reticular activating system (ARAS), originating in the pons and mesencephalon. The ventral pathway projects to the hypothalamus, the subthalamus, and the basal forebrain, and then diffusely to the cortex. The dorsal pathway projects to the thalamus, and then diffusely to the cortex. (Adapted from Bassetti and Aldrich,[2] p 370, with permission.)

systems that function in parallel with the potential for compensatory mechanisms following damage to one or more of the involved pathways.[4,5] Although not all of the neurotransmitters used by the ascending activating rostral mesencephalic pathways have been conclusively identified, glutamate and other excitatory amino acids are also involved.[4,5]

The content of waking consciousness is mediated by the cortex, which maintains awareness of the environment and integrates perceptual information from sensory modalities. The thalamus provides a mechanism for selective attention by enhancing or attenuating responses to incoming stimuli.[4–6] Sensory processing by the cortex is also facilitated by basal forebrain cholinergic activity.

Electrical stimulation of portions of the tegmentum during sleep leads to suppression of spindles, delta waves, and slow cellular rhythms and to the appearance of a low-frequency, desynchronized EEG similar to a waking pattern. This EEG desynchronization is a consequence of depolarization of thalamic neurons (Fig. 2–2). When these neurons are hyperpolarized with a resting membrane potential less than -65 or -70 μV, they fire in bursts that are associated with the appearance of synchronized rhythmic cortical activity. Ascending pathways from the ascending reticular activating system (ARAS) depolarize the thalamic neurons to a more positive potential than -65 μV, which converts their firing to single spikes instead of bursts and facilitates the appearance of lower voltage, desynchronized EEG activity (see Fig. 2–2).[4–6]

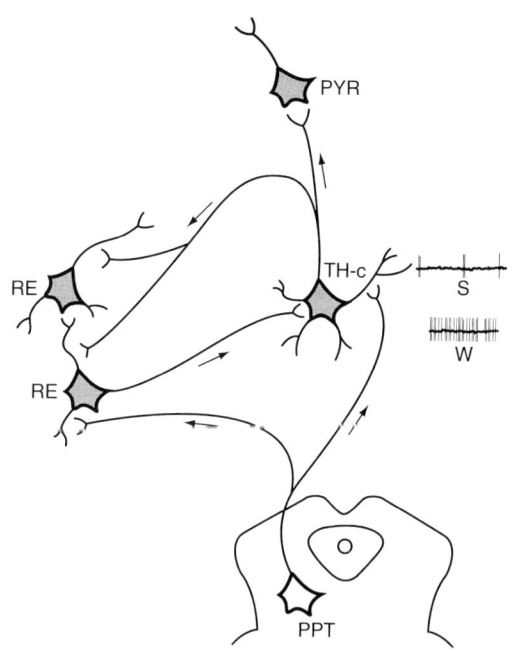

Figure 2–2. Interactions of thalamic reticular neurons (RE), thalamocortical projection neurons (TH-c), pyramidal cells of the cortex (PYR), and cholinergic neurons of the brainstem ascending reticular activating system originating in the pedunculopontine tegmental (PPT) nucleus. During NREM sleep, networks of reticular cells fire rhythmic bursts of action potentials that inhibit thalamocortical projection neurons and lead to sleep spindles. During wakefulness, depolarization of thalamic projection neurons induced by activity of PPT neurons and other ascending afferents disrupts spindle production and leads to EEG desynchronization. (Adapted from Steriade et al.,[5] pp 482, 501, with permission.)

Noradrenaline

Noradrenergic neurons of the locus ceruleus (LC) are almost continually active during wakefulness, become less active during NREM sleep, and are virtually silent during REM sleep. They have widespread efferent connections with two main ascending pathways: the dorsal pathway, which innervates the entire cortex, the hippocampus, and the amygdala and also gives off collaterals to thalamic relay neurons; and the ventral pathway, which projects to the hypothalamus. These extensive projections and the long-lasting effects of norepinephrine on cellular activity suggest that these neurons are ideally suited to playing a role in arousal by modulating cortical and thalamic activity.

Serotonin

Serotonergic neurons in the raphe nuclei of the brainstem have ascending projections and influence the control of the sleep and wake states, but their precise role remains elusive. Serotonergic dorsal raphe neurons fire most rapidly during the waking state, decrease their firing rate during NREM sleep, and become silent during

REM sleep.[7] Furthermore, some dorsal raphe neurons have pacemaker properties and change their firing rates in advance of behavioral state changes. For example, at the end of a REM sleep period, dorsal raphe neurons resume firing before the increase in muscle tone and the cessation of pontine-geniculate-occipital spikes signals the end of the REM period.[7] Thus, dorsal raphe neurons may play a key role in the control and timing of state changes, particularly those between NREM and REM sleep.[7]

Dopamine

The mesencephalic dopamine system includes two major pathways: the nigrostriatal system, which projects from the substantia nigra to the corpus striatum; and the mesocorticolimbic system, which projects from the ventral tegmentum to the nucleus accumbens, the septal nuclei, and the frontal cortex. Dopamine neurons arising in the ventral tegmentum appear to be involved in arousal and maintenance of wakefulness, whereas the nigrostriatal system is involved mainly in the control of the motor system and generally does not play a direct role in the regulation of sleep and wakefulness. The alerting effects of amphetamines appear to be mediated largely through enhancement of release and inhibition of reuptake of dopamine.

Five dopamine receptor subtypes have been identified, and selective ligands exist for at least three: D1, D2, and D3. EEG desynchronization induced by dopamine can be blocked by D1 antagonists, suggesting that these receptors mediate activating effects on the cortex. Dopamine D2 autoreceptors probably play a role in the mediation of sleep through autoinhibition of ventral tegmental dopaminergic neurons.[8]

Histamine

Ascending projections of histamine-containing neurons of the posterior lateral hypothalamus appear to play a role in the maintenance of wakefulness, and the drowsiness induced by antihistamines may be mediated partly by effects on these neurons.[9]

Acetylcholine

Brainstem cholinergic projections originate from the peribrachial area of the pedunculopontine tegmental (PPT) nucleus and from the lateral dorsal tegmental (LDT) nucleus (Fig. 2–3).[10] These neurons project bilaterally to the pontine and bulbar RF, the cranial nerve nuclei, the thalamus, and the basal forebrain. The excitatory (glutamatergic) thalamic relay cells that receive input from the mesopontine cholinergic neurons then send projections to cortical neurons.

A second cholinergic system consists of neurons with cell bodies in the basal forebrain, including the nucleus basalis, the substantia innominata, the diagonal band nuclei, and the septum, which project widely to the cortex.[11] Afferents to the basal forebrain cholinergic system include the mesopontine cholinergic nuclei.

Muscarinic and nicotinic cholinergic agonists can induce cortical activation and wakefulness, indicating involvement of cholinergic systems in the control of wakefulness. The majority of brainstem cholinergic neurons are more active in wakefulness than in NREM sleep, but lesions of these neurons produce at most a limited, transient alteration of consciousness, indicating that these neurons, although involved in the regulation of wakefulness, are not essential.[11] Within the basal forebrain regions, where cholinergic neurons are located, most neurons are most active during wakefulness, although a few are more active during sleep. The role of cholinergic neurons in the initiation and maintenance of REM sleep is discussed below.

NREM SLEEP

Structures involved in the modulation and generation of NREM sleep include the thalamus, the anterior hypothalamus and preoptic area (basal forebrain), afferents

Sleep Onset

Although the neurophysiologic events associated with sleep onset are not fully established, changes in the activity of the thalamus, basal forebrain, and ARAS are key features. Neurons of the preoptic area and basal forebrain are critical for the initiation of sleep: Chemical or electrical stimulation of these regions induces sleep, whereas lesions produce long-lasting reductions in the amounts of deep slow-wave sleep (SWS) and REM sleep.[12] Ventrolateral preoptic neurons that are activated with sleep project to the tuberomamillary nucleus in the posterior hypothalamus; inhibitory effects of these neurons on the posterior hypothalamus and on other neurons of the ARAS may be one mechanism by which sleep is initiated or maintained.[13] Projections from the anterior hypothalamus and basal forebrain to the thalamus may mediate the dampening of sensory input that occurs with sleep onset.[4]

Changes at sleep onset in thalamic activity, perhaps caused by increased transmission from basal forebrain afferents, influence cortical activity during sleep (Fig. 2–4). During light NREM sleep, thalamic reticular neurons show slow fluctuations in resting membrane potential, with superimposed, rhythmic 7- to 14-Hz bursts of spikes generated by calcium influx (Ca^{2+} spikes). These rhythmic bursts inhibit thalamic projection neurons and produce inhibitory postsynaptic potentials, some of which are followed by a rebound Ca^{2+} spike and an associated burst of action potentials. The bursts of activity in thalamocortical cells feed back to the reticular neurons to enhance rhythmicity and also to produce excitatory postsynaptic potentials on cortical neurons, leading to the appearance of surface sleep spindles (see Fig. 2–2).

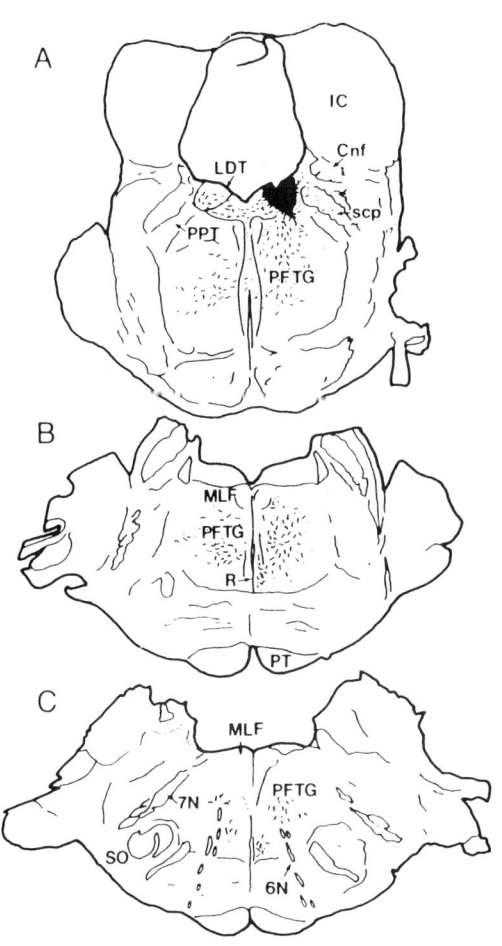

Figure 2–3. Frontal section at the level of the pons-midbrain transition, showing the location of cholinergic neurons of the lateral dorsal tegmental (LDT) nucleus and the pedunculopontine tegmental (PPT) nucleus. Other abbreviations: IC = inferior colliculus; Cnf = cuneiform nucleus; scp = superior cerebellar peduncle; PFTG = giant cell portion of the pontine reticular formation; MLF = medical longitudinal fasciculus; R = raphe nuclei; PT = pyramidal tracts; 7N = 7th nerve nucleus; SO = superior olive; 6N = 6th nerve nucleus. (Adapted from Mitani et al.,[10] p 399, with permission.)

to the thalamus from the RF, and projections from the thalamus to the cortex (thalamocortical projections).[6] Other areas that show increased firing during NREM sleep include the dorsal medullary RF and the nucleus tractus solitarius; however, the mechanisms and pathways by which these structures contribute to initiation and maintenance of sleep are not well understood.

Sleep Spindles and Delta Waves

Sleep spindles, hallmarks of light NREM sleep, are generated by interactions of inhibitory (γ-aminobutyric acid (GABA)-

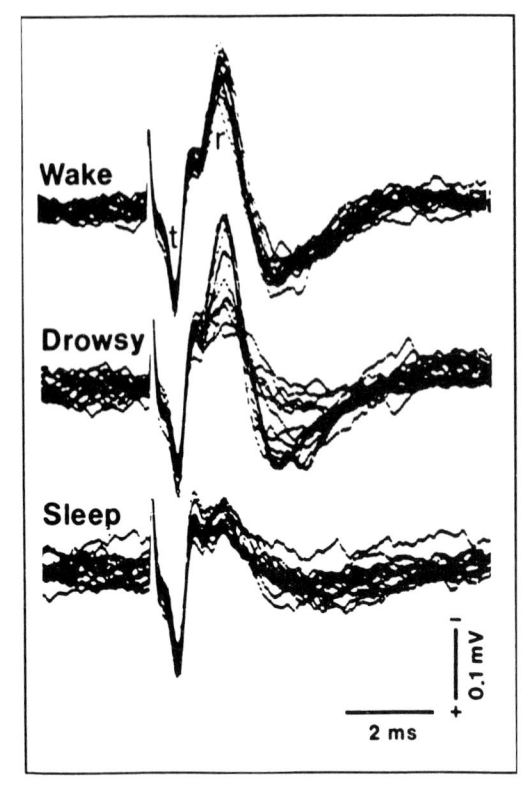

Figure 2–4. Reduced synaptic transmission in the ventrolateral thalamus at the onset of sleep. The afferent volley (t) was produced by stimulation of cerebellothalamic axons. At the onset of sleep, the amplitude of the potential relayed to the cortex (r) is reduced. (From Steriade,[4] p 13, with permission.)

ergic thalamic reticular neurons, thalamocortical projection neurons, and cortical pyramidal neurons. Although isolated networks of reticular cells are capable of generating spindles in the absence of the cortex, feedback via corticothalamic projections contributes to the widespread coherence of spindle rhythms (see Fig. 2–2).[14,15]

As NREM sleep deepens, there is progressive hyperpolarization of the thalamic reticular neurons associated with a reduction in spindles and an increase in delta wave activity. Although spindles require a minimum of a network of thalamic reticular cells, delta frequency rhythms can be generated in single cells by the interaction between a hyperpolarization-activated cation current (I_h) and a transient low-threshold Ca^{2+} current (I_t), similar to the ionic currents that generate activity in sinoatrial cardiac cells. The expression of this rhythm, however, requires hyperpolarization of the thalamocortical cells. An additional slow rhythm of approximately 0.3 Hz that can be recorded from neocortical cells persists after extensive thalamic lesions and appears to be generated within the neocortex, although it is probably modulated by activity in reticular thalamic neurons.

When the thalamic projection neurons and reticular thalamic neurons are depolarized by activating pathways, the low-threshold Ca^{2+} current is inactivated, burst firing by the thalamic neurons is inhibited, and the delta rhythm is abolished. As with spindles, ascending brainstem noradrenergic and serotonergic projections, as well as brainstem and forebrain cholinergic projections, inhibit delta activity, probably by reducing the hyperpolarization needed to generate the delta waves. Ascending glutamatergic and histaminergic projections may also contribute to delta wave suppression during wakefulness and REM sleep. Thus, the delta rhythm can be viewed as a result of loss of excitatory afferent input; in this respect it may be similar to the delta rhythms seen on the EEG recordings of subjects with focal structural lesions.

REM SLEEP

Transection experiments by Jouvet and others demonstrated that REM sleep is generated in the pons, and in the 1960s Hernandez-Peon and colleagues[16–18] and Mazzuchelli-O'Flaherty and associates[19] showed that acetylcholine crystals placed directly on the pons could induce sleep. These and other experiments demonstrated that mesencephalic and pontine cholinergic neurons are essential for the generation of REM sleep. Cholinergic cells of the median and dorsolateral pons increase their firing rates during REM sleep.[20–22] Neurons of the anterodorsal pontine tegmentum and the area just rostral to the LC, which fire selectively during REM sleep and are essential for its expression, are almost certainly cholinergic.[23–25] The precise locus of the cholinoceptive trigger zone for REM sleep within the anterodorsal pontine tegmentum has not been determined, but muscarinic agonists injected into selected areas of the pontine tegmentum facilitate the appearance of REM sleep and can induce partial expression of the REM sleep state by the production of isolated muscle atonia or isolated REMs.[26–28]

Brainstem monoaminergic neurons, particularly the serotonergic neurons of the raphe nuclei and noradrenergic neurons of the LC, become inactive during REM sleep and appear to play a permissive role through modulation of cholinergic activity.[23,29] Neurons that are inactive during REM sleep are sometimes referred to as REM-off neurons, in contrast to REM-on neurons that are active during REM sleep. In addition to the serotonergic neurons of the raphe and the noradrenergic neurons of the LC, histamine-containing neurons of the posterior hypothalamus also appear to be REM-off neurons.[9] Dopamine appears to be involved in the regulation of REM sleep through effects mediated by D1 and D2 receptors. A model of REM sleep generation and modulation based on the interactions between cholinergic REM-on neurons and monoaminergic REM-off neurons was presented by McCarley and Hobson[30] and subsequently revised by McCarley.[28]

Postural Atonia

The muscle atonia that accompanies REM sleep, discovered by Jouvet and Michel,[31] is controlled by neurons that are presumed to be cholinergic and are located just outside the LC in regions referred to by Sakai[24,32] as LC-α (ventromedial to LC), locus subceruleus (ventrolateral to LC-α), and peri-LC-α (medial to LC-α). These areas are sometimes referred to collectively as the dorsolateral small cell reticular group.[28]

The pontine neurons that initiate postural atonia project to relay neurons of the medial medullary RF.[33] These medullary neurons project via the ventrolateral reticulospinal tract to inhibitory glycinergic spinal interneurons, which then produce postsynaptic inhibition and membrane hyperpolarization of spinal alpha motoneurons (Fig. 2–5).[34,35] Glutamatergic neurons also may be involved, as muscle atonia can be induced by injections of non-N-methyl-D-aspartate (non-NMDA) agonists into the regions of the peri-LC-α and the nucleus magnocellularis,[36] but the locations of the cell bodies of such neurons are unknown.

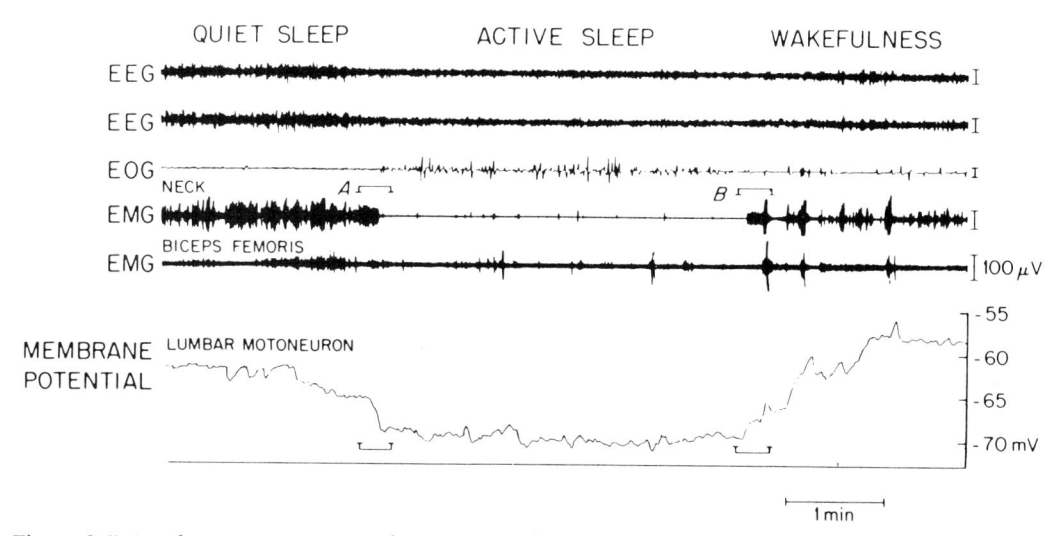

Figure 2–5. Lumbar motoneuron membrane potential during NREM sleep (quiet sleep), REM sleep (active sleep), and wakefulness. The lumbar motoneuron is hyperpolarized during REM sleep because of the inhibitory effects of medullary neurons; the period of hyperpolarization corresponds to the period of neck muscle atonia. (From Morales and Chase,[35] p 825, with permission.)

Lesions in the region of the dorsolateral small cell group inhibit the postural atonia of REM sleep and may lead to orienting behavior and apparent acting out of dreams with attack behavior. Although small lesions in this region can inhibit muscle atonia, more extensive lesions are required to produce complex behaviors. To cause attack behavior, lesions must extend into the midbrain; orienting behavior is associated with dorsolateral pontine lesions.[37]

Pontine-Geniculate-Occipital Spikes

A striking neurophysiologic feature of REM sleep is the occurrence of bursts of rapid firing of collections of neurons in the pons, the lateral geniculate nucleus, and the occipital cortex; the firing patterns associated with these bursts are referred to as *pontine-geniculate-occipital (PGO) spikes*.[38,39] The spikes, which are frequent during REM sleep and show a rebound increase after REM sleep deprivation, appear to be generated by PPT and LDT neurons.[32,40] Noradrenergic neurons of the LC and serotonergic neurons of the raphe become silent during PGO spiking and appear to play a permissive role.[32]

Cortical Activation

Cholinergic neurons of the LDT-PPT that project to thalamic nuclei appear to mediate EEG desynchronization during REM sleep. Noradrenergic LC neurons, which contribute to cortical activation during wakefulness, are not involved because they are inactive during REM sleep. However, EEG desynchrony occurs after partial lesions of the mesencephalon and rostral pontine RF, after thalamectomy, and with lesions of the subthalamus and hypothalamus, suggesting that the ascending pathways responsible for cortical activation are diffusely distributed at these levels.[32] Thus, other neurons of the brainstem RF that project to the thalamus also may contribute to cortical activation. On the other hand, lesions of the most caudal nonmonoaminergic portion of the ARAS—the nucleus reticularis magnocellularis of the medulla oblongata—abolish cortical desynchrony, suggesting that these neurons play a critical role in EEG desynchrony.[32] They appear to project dorsally to medial and intralaminar thalamic nuclei and ventrally to the posterior hypothalamus, and then are relayed to the cortex.

Rapid Eye Movements

The horizontal REMs that occur during REM sleep appear to be generated by the saccade generators of the paramedian pontine RF. The less frequent vertical REMs are presumably generated by neurons of the mesencephalic RF. Although the pathways that mediate increased excitability of saccade generators remain to be determined, changes in cerebral metabolism accompanying the REMs of REM sleep suggest that some eye movements are saccadic movements to targets in the dream scene.[41]

Muscle Twitches

Muscle twitches during REM sleep are produced by strong excitatory postsynaptic input to alpha motoneurons that is mediated presumably by glutamate, although by a non-NMDA receptor. The excitatory input is superimposed on the tonic inhibition that leads to hyperpolarization and is slightly preceded by an even greater hyperpolarization. This pattern of brief hyperpolarization followed by intense excitation is apparently unique to REM sleep, as it has not been observed in NREM sleep or wakefulness.[34,42,43]

SLEEP FACTORS

At the beginning of the twentieth century, Legendre and Pieron[44,45] in France and

Ishimori[46] in Japan reported that certain substances that accumulated in the cerebrospinal fluid (CSF) during prolonged wakefulness could induce sleep when transferred to other animals. Since then, scientists have identified a number of *sleep factors,* defined as humoral or cerebrospinal substances, that induce physiologic sleep (Table 2–2).[47–51] For example, the CSF of sleep-deprived goats contains muramyl peptides, originally called factor S, that can increase SWS.[52] The effects of muramyl peptides may be mediated by interleukin-1 (see below).

Cytokines

Cytokines are regulatory proteins that are involved in the response to tissue injury, inflammation, and infection.[53] Several cytokines, including interleukin-1 (IL-1), tumor necrosis factor, some of the interferons, and acidic fibroblast growth factor, have somnogenic properties. Current evidence suggests that the sleep-inducing effects of cytokines are mediated largely by IL-1, which is released from macrophages and contributes to T-cell activation. Its production increases with sleep deprivation, and administration of the protein enhances NREM sleep.[54] Inhibitors of IL-1 reduce sleep time, and the increase in sleep induced by IL-1 can be blocked by administration of an IL-1 receptor antagonist. Although the precise mechanisms by which IL-1 promotes sleep have not been determined, its ability to increase growth-hormone release and its effects on anterior hypothalamic neurons probably contribute to the promotion of sleep.[55]

Prostaglandins

Prostaglandins (PG) are 20-carbon, polyunsaturated fatty acids that exhibit a wide variety of biologic effects. Two prostaglandins present in the CSF—PGD_2 and PGE_2—have effects on sleep. PGE_2 inhibits sleep, probably by actions in or near the posterior hypothalamus; endogenous PGE_2 levels, which are low during sleep, start to rise before the onset of wakefulness, suggesting that PGE_2 may be involved in mechanisms that initiate wakefulness.[56]

On the other hand, PGD_2 infused into the CSF promotes sleep, apparently by exciting sleep-active neurons and inhibiting wake-active neurons in the basal forebrain and preoptic area.[57–60] Inhibition of the activity of PGD synthase, which catalyzes production of PGD_2, leads to a marked decrease in SWS and REM sleep, suggesting that PGD synthase plays an important role in the regulation of sleep and wakefulness through its effects on PGD_2 production. The greatest increases in the amounts of SWS and REM sleep occur when PGD_2 is applied to the ventral surface of the basal forebrain, suggesting that PGD_2 may act as a neurohormone, rather than as a classical neurotransmitter, to influence sleep.[58]

Table 2–2. **Sleep Factors**

	Increase NREM Sleep	Increase REM Sleep
cis-9,10-Octadecenamide	+	
Cholecystokinin	+	
Cortistatin	+	
Insulin	+	
Interleukin-1	+	
Interferon alfa-2	+	
Muramyl peptides	+	
Tumor necrosis factor	+	
Growth hormone		+
Prolactin		+
Somatostatin		+
Vasoactive intestinal peptide		+
Adenosine	+	+
Arginine vasotocin	+	+
Delta sleep–inducing peptide	+	+
Growth hormone–releasing hormone	+	+
Prostaglandin D_2	+	+
Uridine	+	+
Serotonin	+	+

Adenosine

Administration of adenosine, adenosine precursors, and adenosine receptor agonists increases REM and NREM sleep, perhaps in part because of their effects on tegmental cholinergic neurons; the sleep-disrupting effects of caffeine may be partly a result of its blockade of adenosine receptors. Adenosine receptors appear to mediate the sleep-promoting effects of PGD_2.[61]

Peptides

Several peptides present in the brain have effects on NREM sleep and REM sleep, including delta sleep–inducing peptide, growth hormone–releasing hormone, arginine-vasotocin, vasoactive intestinal peptide, and cortistatin. *Delta sleep—inducing peptide (DSIP)*, an endogenous nonapeptide, increases SWS when injected intraventricularly or systemically.[62] In humans, it has a circadian rhythm with peak plasma levels in mid afternoon and minimal levels shortly after midnight.[63] Cortistatin, present in the hippocampus and cortex, depresses neuronal activity and facilitates slow-wave sleep, perhaps via its antagonistic effects on acetylcholine.[64] Growth hormone–releasing hormone (GHRH) promotes sleep, especially NREM sleep, while inhibition of GHRH activity suppresses sleep. Furthermore, the rebound increase in sleep following sleep deprivation is abolished by administration of a competitive antagonist of GHRH.[55] The effects of GHRH on sleep are probably not mediated by GH.[55]

Two peptides, vasoactive intestinal peptide (VIP) and arginine-vasotocin, may contribute to REM sleep initiation and modulation. Intraventricular injections of VIP enhance REM sleep, and VIP antiserum inhibits REM sleep; furthermore, VIP concentration increases with sleep deprivation, and extraction of this compound from CSF reduces subsequent REM sleep rebound.[65] Arginine-vasotocin may be involved in REM sleep modulation, as intraventricular injections block REM sleep and injections of its antiserum promote REM sleep. Enhanced REM sleep may also occur with administration of somatostatin and prolactin.

Melatonin

Melatonin is secreted by the pineal gland and appears to have sleep-promoting effects. Because of its intimate relation to circadian rhythmicity, melatonin and its relation to sleep are discussed in Chapter 4.

Benzodiazepines and Endogenous Benzodiazepine-Receptor Agonists

Benzodiazepines, which have muscle relaxant, anticonvulsant, and anxiolytic properties, are also potent hypnotics. Although conclusive proof that endogenous benzodiazepine receptor ligands are sleep factors is lacking, such ligands exist and contribute to the syndrome of idiopathic recurring stupor (Chap. 19). The hypnotic effects appear to be mediated by the benzodiazepine receptor complex, which is associated with GABA receptors and chloride channels. Activation of benzodiazepine receptors facilitates GABAergic transmission, increases chloride ion flow, and leads to membrane hyperpolarization. The hypnotic effects of benzodiazepines appear to be the result of the inhibitory effects of increased GABAergic transmission on neurons of the ARAS.

SUMMARY

Sleep is the consequence of an active process involving cortical, diencephalic, and brainstem structures. Wakefulness is mediated by the ARAS, which includes cholinergic and monoaminergic pathways that project to the thalamus and diffusely to the cortex. These pathways function in parallel, so that lesions involving only one neurotransmitter system generally produce transient, rather than permanent, alterations of consciousness.

Neurons of the basal forebrain play a critical role in the initiation and maintenance of NREM sleep. The cortical changes that occur with NREM sleep are mediated by networks of thalamic reticular neurons, which become hyperpolarized during sleep and generate the rhythmic activity that leads to sleep spindles and to the delta waves of deep NREM sleep. REM sleep and its associated postural muscle atonia are generated by pontine cholinergic neurons of the pedunculopontine nuclei and the lateral dorsal tegmentum. The brainstem activity associated with REM sleep depolarizes thalamic neurons and inhibits the production of sleep spindles and delta waves.

Sleep factors, endogenous substances that induce sleep, also play a role in sleep onset and maintenance, although their precise physiologic contributions have not been determined.

REFERENCES

1. Moruzzi, G, and Magoun, H: Brain stem reticular formation and activation of the EEG. Electroencephalogr Clin Neurophysiol 1:455, 1949.
2. Bassetti, C, and Aldrich, MS: Consciousness, delirium, and coma. In Albin, MS (ed): Textbook of Neuroanesthesia with Neurosurgical and Neuroscience Perspectives. McGraw-Hill, New York, 1996, pp 369–408.
3. Denoyer, M, et al: Neurotoxic lesion of the mesencephalic reticular formation and/or the posterior hypothalamus does not alter waking in the cat. Brain Res 539:287, 1991.
4. Steriade, M: Basic mechanisms of sleep generation. Neurology (7 suppl 6)42:9, 1992.
5. Steriade, M, et al: Report of IFCN Committee on Basic Mechanisms: Basic mechanisms of cerebral rhythmic activities. Electroencephalogr Clin Neurophysiol 76:481, 1990.
6. Steriade, M, et al: Thalamocortical oscillations in the sleeping and aroused brain. Science 262:679, 1993.
7. Lydic, R: Central regulation of sleep and autonomic physiology. In Lydic, R, and Biebuyck, JF (eds): Clinical Physiology of Sleep. American Physiological Society, Bethesda, Md, 1988, pp 1–19.
8. Bagetta, G, et al: Ventral tegmental area: site through which dopamine D2-receptor agonists evoke behavioural and electrocortical sleep in rats. Br J Pharmacol 95:860, 1988.
9. Lin, JS, et al.: Involvement of histaminergic neurons in arousal mechanisms demonstrated with H3-receptor ligands in the cat. Brain Res 523:325, 1990.
10. Mitani, A, et al: Cholinergic projections from the laterodorsal and pedunculopontine tegmental nuclei to the pontine gigantocellular tegmental field in the cat. Brain Res 451:397, 1988.
11. Jones, BE: Basic mechanisms of sleep-wake states. In Kryger, MH, et al (eds): Principles and Practice of Sleep Medicine, ed 2. WB Saunders, Philadelphia, 1994, pp 145–162.
12. Sallanon, M, et al: Long-lasting insomnia induced by preoptic neuron lesions and its transient reversal by muscimol injection into the posterior hypothalamus in the cat. Neuroscience 32:669, 1989.
13. Sherin, JE, et al: Activation of ventrolateral preoptic neurons during sleep. Science 271:216, 1996.
14. Contreras, D, et al: Spatiotemporal patterns of spindle oscillations in cortex and thalamus. J Neurosci 17:1179, 1997.
15. Contreras, D, et al: Control of spatiotemporal coherence of a thalamic oscillation by corticothalamic feedback. Science 274:771, 1996.
16. Hernandez-Peon, R, and Chavez-Ibarra, G: Sleep induced by electrical or chemical stimulation of the forebrain. Electroencephalogr Clin Neurophysiol (suppl)24:188, 1963.
17. Hernandez-Peon, R, et al: Limbic cholinergic pathways involved in sleep and emotional behavior. Exp Neurol 8:93, 1963.
18. Hernandez-Peon, R, et al: Sleep and other behavioral effects induced by acetylcholinergic stimulation of basal temporal cortex and striate structures. Brain Res 4:243, 1967.
19. Mazzuchelli-O'Flaherty, AL, et al: Sleep and other behavioral responses induced by acetylcholinergic stimulation of frontal and medial cortex. Brain Res 4:268, 1967.
20. Hobson, JA, et al: Time course of discharge rate changes by cat pontine brain stem neurons during sleep cycle. J Neurophysiol 37:1297, 1974.
21. Hobson, JA, et al: Selective firing by cat pontine brain stem neurons in desynchronized sleep. J Neurophysiol 37:497, 1974.
22. Hobson, JA, et al: Sleep cycle oscillation: Reciprocal discharge by two brainstem neuronal groups. Science 189:55, 1975.
23. Siegel, JM: Brainstem mechanisms generating REM sleep. In Kryger, MH, et al: Principles and Practice of Sleep Medicine, ed 2. WB Saunders, Philadelphia, 1994, pp 125–144.
24. Sakai, K: Executive mechanisms of paradoxical sleep. Arch Ital Biol 126:239, 1988.
25. Webster, HH, and Jones, BE: Neurotoxic lesions of the dorsolateral pontomesencephalic tegmentum-cholinergic area in the cat: II. Effects upon sleep-waking states. Brain Res 458:285, 1988.
26. Yamamoto, K, et al: A cholinoceptive desynchronized sleep induction zone in the anterodorsal pontine tegmentum—locus of the sensitive region. Neuroscience 39:273, 1990.
27. Baghdoyan, HA, et al: Site-specific enhancement and suppression of desynchronized sleep signs following cholinergic stimulation of three brainstem regions. Brain Res 306:39, 1984.
28. McCarley, RW: Neurophysiology of sleep: Basic mechanisms underlying control of wakefulness and sleep. In Chokroverty, S (ed): Sleep Disor-

ders Medicine: Basic Science, Technical Considerations, and Clinical Aspects. Butterworth-Heinemann, Boston, 1994, pp 17–36.
29. Jones, BE: Paradoxical sleep and its chemical/structural substrates in the brain. Neuroscience 40:637, 1991.
30. McCarley, RW, and Hobson, JA: Neuronal excitability modulation over the sleep cycle: A structural and mathematical model. Science 189:58, 1975.
31. Jouvet, M, and Michel, F: Corrélations electromyographiques du sommeil chez le chat décortiqué et mésencephalique chronique. C R Soc Biol 153:422, 1959.
32. Sakai, K: Central mechanisms of paradoxical sleep. Brain Dev 8:402, 1986.
33. Holmes, CJ, and Jones, BE: Importance of cholinergic, GABAergic, serotonergic and other neurons in the medial medullary reticular formation for sleep-wake states studied by cytotoxic lesions in the cat. Neuroscience 62:1179, 1994.
34. Chase, MH, and Morales, FR: The atonia and myoclonia of active (REM) sleep. Ann Rev Psychol 41:557, 1990.
35. Morales, FR, and Chase, MH: Intracellular recording of lumbar motoneuron membrane potential during sleep and wakefulness. Exp Neurol 62:821, 1978.
36. Lai, YY, and Siegel, JM: Medullary regions mediating atonia. J Neurosci 8:4790, 1988.
37. Hendricks, LC, et al: Different behaviors during paradoxical sleep without atonia depend on pontine site lesion. Brain Res 239:81, 1982.
38. Jouvet, M: Biogenic amines and the states of sleep. Science 163:32, 1969.
39. Jouvet, M, et al: Sur un stade d'activité électrique cérébrale rapide au cours du sommeil physiologique. C R Soc Biol 153:1024, 1959.
40. Steriade, M, et al: Different cellular types in mesopontine cholinergic nuclei related to pontogeniculo-occipital waves. J Neurosci 10:2560, 1990.
41. Hong, CC, et al: Localized and lateralized cerebral glucose metabolism associated with eye movements during REM sleep and wakefulness: A positron emission tomography (PET) study. Sleep 18:570, 1995.
42. Chase, MH, and Morales, FR: Subthreshold excitatory activity and motoneuron discharge during REM periods of active sleep. Science 221:1195, 1983.
43. Steriade, M, and McCarley, RW: Brainstem Control of Wakefulness and Sleep. Plenum, New York, 1990.
44. Legendre, R, and Pieron, H: Le problème des facteurs du sommeil: Résultats d'injections vasculaires et intracérébrales de liquides insomniques. C R Soc Biol 68:1077, 1910.
45. Legendre, R, and Pieron, H: Recherches sur le besoin de sommeil consécutif à une veille prolongée. Z Allg Physiol 14:235, 1912.
46. Ishimori, K: True cause of sleep: A hypnogenic substance as evidenced in the brain of sleep-deprived animals. Tokyo Igakkai Zasshi 23:429, 1909.
47. Konoda, Y, et al: SPS-B, a physiological sleep regulator, from the brainstems of sleep-deprived rats, identified as oxidized glutathione. Chem Pharm Bull 38:2057, 1990.
48. Borbely, AA, and Tobler, I: Endogenous sleep-promoting substances and sleep regulation. Physiol Rev 69:605, 1989.
49. Cravatt, BF, et al: Chemical characterization of a family of brain lipids that induce sleep. Science 268:1506, 1995.
50. Honda, K, et al: Oxidized glutathione regulates physiological sleep in unrestrained rats. Brain Res 636:253, 1994.
51. Inoue, S, and Borbely, AA: Endogenous Sleep Substances and Sleep Regulation. Japan Scientific Societies Press, Tokyo, 1985.
52. Pappenheimer, JR, et al: Sleep-promoting effects of cerebrospinal fluid from sleep-deprived goats. Proc Natl Acad Sci USA 58:513, 1967.
53. Opp, MR, and Krueger, JM: Interleukin-1 is involved in responses to sleep deprivation in the rabbit. Brain Res 639:57, 1994.
54. Krueger, JM, and Majde, JA: Microbial product and cytokines in sleep and fever regulation. Crit Rev Immunol 14:355, 1994.
55. Krueger, JM, and Obal, F: Growth hormone-releasing hormone and interleukin-1 in sleep regulation. FASEB J 7:645, 1993.
56. Gerozissis, K, et al: Changes in hypothalamic prostaglandin E2 may predict the occurrence of sleep or wakefulness as assessed by parallel EEG and microdialysis in the rat. Brain Res 689:239, 1995.
57. Hayaishi, O: Sleep-wake regulation by prostaglandins D2 and E2. J Biol Chem 263: 14593, 1988.
58. Hayaishi, O, and Matsumura, H: Prostaglandins and sleep. Adv Neuroimmunol 5:211, 1995.
59. Hayaishi, O: Molecular mechanisms of sleep-wake regulation: Roles of prostaglandins D2 and E2. FASEB J 5:2575, 1991.
60. Osaka, T, and Hayaishi, O: Prostaglandin D2 modulates sleep-related and noradrenaline-induced activity of preoptic and basal forebrain neurons in the rat. Neurosci Res 23:257, 1995.
61. Satoh, S, et al: Promotion of sleep mediated by the A2a-adenosine receptor and possible involvement of this receptor in the sleep induced by prostaglandin D2 in rats. Proc Natl Acad Sci USA 93:5980, 1996.
62. Monnier, M, and Hosli, L: Dialysis of sleep and waking factors in blood of rabbit. Science 146:796, 1964.
63. Friedman, TC, et al: Diurnal rhythm of plasma delta-sleep–inducing peptide in humans: Evidence for positive correlation with body temperature and negative correlation with rapid eye movement and slow wave sleep. J Clin Endocrinol Metab 78:1085, 1994.
64. De Lecea L, et al.: A cortical neuropeptide with neuronal depressant and sleep-modulating properties. Nature 381:242, 1996.
65. Jimenez-Anguiano, A, et al: Cerebrospinal fluid (CSF) extracted immediately after REM sleep deprivation prevents REM rebound and contains vasoactive intestinal peptide (VIP). Brain Res 631:345, 1993.

Chapter 3

BREATHING DURING WAKEFULNESS AND SLEEP

THE RESPIRATORY CONTROL SYSTEM
The Ventilatory Pump
The Upper Airway
Chemoreceptors and Other Afferents
Central Nervous System Controller
Metabolic Regulation of Breathing
Voluntary Ventilatory Control System
Reflex Responses to Altered
 Respiratory Load
The Wakefulness Stimulus
**SLEEP-RELATED CHANGES
 IN BREATHING**
Changes in Breathing Mechanics
Increased Airway Resistance
Central Nervous System Changes
 During Sleep
Metabolic Regulation of Breathing
 During Sleep
Changes in Reflex Responses to Respiratory
 Loads During Sleep
Loss of the Wakefulness Stimulus
Decreased Ventilation During Sleep
**STATE CHANGES AND AROUSAL
 RESPONSES**

The study of breathing during sleep is important for several reasons. First, of all the organ functions of the body, breathing is unique in that it is under voluntary as well as automatic control. Although little volitional control can be exerted over cardiac, renal, gastrointestinal, and endocrinologic activity, breathing can easily be increased, decreased, or modulated voluntarily, which permits singing, talking, whispering, and breath holding. Voluntary systems, however, are operative only during wakefulness; with the onset of sleep, control of breathing becomes entirely involuntary. Second, breathing is more vulnerable than other major organ functions to the physiologic changes that occur with sleep. Third, disorders associated with disturbed breathing during sleep are both frequent and severe, and they often have serious complications. Sleep disorders associated with breathing disturbances are discussed in Chapters 13 and 14.

Understanding of central nervous system (CNS) control of breathing has advanced dramatically in recent years. Before the 1980s, most views on control of breathing were based on the influential work of Lumsden,[1,2] who proposed that pneumotaxic and apneustic centers in the pons and inspiratory and expiratory centers in the medulla were responsible for regulation of respiration. Newer concepts of breathing control have moved away from models based on respiratory centers and instead view control of breathing in terms of integrated activity of diffusely distributed networks of neurons.

THE RESPIRATORY CONTROL SYSTEM

The respiratory system may be viewed as consisting of five components:
 1. The lungs, which perform gas exchange

2. The chest wall and the inspiratory and expiratory muscles, which act as a pump to fill and empty the lungs
3. The upper airway, a compliant tube extending from the nose and mouth to the trachea, which, along with the bronchial tree, conducts respired gases to and from the lungs
4. The sensory elements of the CNS and the peripheral nervous system, which provide information about the status of the respiratory system
5. The CNS neurons that integrate information about respiration and control the activity of the muscles of the pump and tube

This chapter focuses on components 2, 3, 4, and 5.

The Ventilatory Pump

INSPIRATION

To achieve inspiration, forces generated by the inspiratory muscles (Table 3–1) must overcome the elastic properties of the chest wall and lungs, air flow resistance, and tissue viscosity and inertia. Of the inspiratory muscles, the diaphragm is clearly the most important. It can be thought of as two muscles: the costal diaphragm, which originates from the lower six ribs and the sternum; and the crural diaphragm, which originates from the upper lumbar vertebrae. Both converge on the aponeurotic central tendon.

Lung inflation occurs as a result of rib cage expansion, downward movement of the diaphragm, or both. Diaphragm contraction has an inspiratory effect via three mechanisms:

1. Shortening of diaphragm fibers draws the central tendon in a caudal direction, enlarging the chest with a pistonlike action.
2. Positive abdominal pressure produced by descent of the diaphragm and compression of abdominal contents pushes outward on the lower rib cage against which the diaphragm is apposed (*zone of apposition*).
3. To the extent that abdominal contents resist descent of the diaphragm, shortening of the costal fibers elevates the lower ribs onto which they insert.

Table 3–1. **The Ventilatory Pump**

Muscles	Nerves	Roots	Actions
INSPIRATORY MUSCLES			
Diaphragm	Phrenic nerves	C3–5	Elevates lower ribs; expands lower rib cage; increases length of chest
External and parasternal intercostals	Intercostal nerves	T1–12	Elevate ribs
Scaleni	Cervical nerves	C2–5	Elevate and fix upper ribs
Sternocleidomastoids	Spinal accessory and 2nd cervical nerves	CN XI, C2	Elevate sternum
EXPIRATORY MUSCLES			
Abdominals	Intercostal, subcostal, iliohypogastric, and ilioinguinal nerves	T7–11, L1	Depress lower ribs; compress abdominal cavity; elevate diaphragm
Internal intercostals	Intercostal nerves	T1–12	Depress ribs

CN = cranial.
C = cervical.

Although the diaphragm does most of the work of inspiration, optimal diaphragm function requires appropriate action of the other respiratory muscles. For example, during exercise or with other causes of increased ventilation, the action of abdominal muscles during expiration raises the diaphragm to its most efficient position for inspiration. Intercostal muscle activity helps to stabilize the rib cage during inspiration and prevent its inward movement. Contraction of the costal and crural diaphragm provides the abdominal component of inspiration, whereas rib cage expansion is produced by the action of the external intercostal muscles, the accessory muscles that elevate the first rib, and the costal diaphragm. During quiet breathing in the sitting position, the rib cage component is greater than the abdominal component; in the supine position, the abdominal component is greater. Increases in tidal volume above resting levels are usually the result of increased diaphragmatic activity.

EXPIRATION

With quiet breathing, expiration is passive because of the elastic recoil of the lungs and chest wall. The expiratory muscles perform a small amount of work at higher levels of ventilation or when airway resistance is increased. They also assist with forced expiration and with explosive expiration associated with coughing and sneezing.

THE ENERGY COST OF BREATHING

The energy cost of breathing refers to the amount of energy required to maintain adequate ventilation. The energy required for each breath is approximately proportional to the work of breathing, which can be measured as the integrated product of pleural pressure and airflow or approximated as the product of tidal volume and mean transpleural pressure. As breathing efficiency is about 10%,[3] the caloric requirements are approximately ten times greater than the pressure-volume product. In normal persons at rest, the energy cost of breathing requires about 0.25 mL of oxygen uptake by respiratory muscles per liter of ventilation and accounts for approximately 5% of total oxygen uptake. With increasing ventilation, and especially with increased respiratory loads, breathing efficiency declines and the oxygen requirement per liter of ventilation increases.

Respiratory loads, which determine the relationship between change in volume and change in pressure, are determined by elastic recoil of the chest wall, elastic recoil of the lungs, frictional resistance to displacement of lung tissue (tissue resistance), and airway resistance.

Airway resistance accounts for about 80% to 90% of total nonelastic resistance. Although the upper airway is responsible for only about 20% to 30% of airway resistance during quiet breathing, this proportion increases to 50% or more when minute ventilation is increased. During nasal breathing, nasal resistance—which varies with posture, mechanical and chemical irritation, exercise, and air temperature—may account for as much as 50% of airway resistance.

According to Poiseuille's law, airway resistance is a function of the fourth power of the radius of the airway for laminar air flow. Thus, reduction of airway radius by half leads to a 16-fold increase in airway resistance. Airway resistance is also affected by lung volume: At higher lung volumes, small airways expand and airway resistance falls. The degree of expansion is a function of the compliance of the tissue. The trachea, with its hard, cartilaginous rings, has lower compliance than the bronchioles, which lack cartilaginous support. Of the upper airway structures, only the pharynx lacks cartilaginous support, and it is therefore the most compliant structure of the upper airway.

The Upper Airway

Efficient breathing requires appropriately coordinated contractions of upper airway muscles and ventilatory pump muscles. The upper airway is a complex structure that runs from the nose and mouth to the

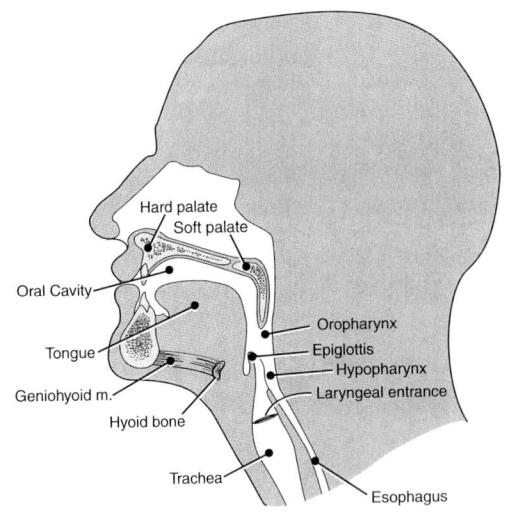

Figure 3–1. Lateral view of the upper airway.

trachea (Fig. 3–1).[4,5] More than 20 muscles are involved in its function, including muscles of the nose, soft palate, oropharynx, hyoid and hypopharynx, and larynx (Fig. 3–2). Although many of these muscles have specialized functions that subserve activities such as talking and swallowing, they also are major determinants of airway resistance.

The nasal group of upper airway muscles influences nasal resistance by altering the position of the nasal cartilage (Table 3–2). The inspiratory activity of nasal dilators reduces nasal resistance in response to hypoxia, hypercapnia, or resistive loading.[5] Activity of nasal muscles also may contribute to regulation of breathing rhythmicity.

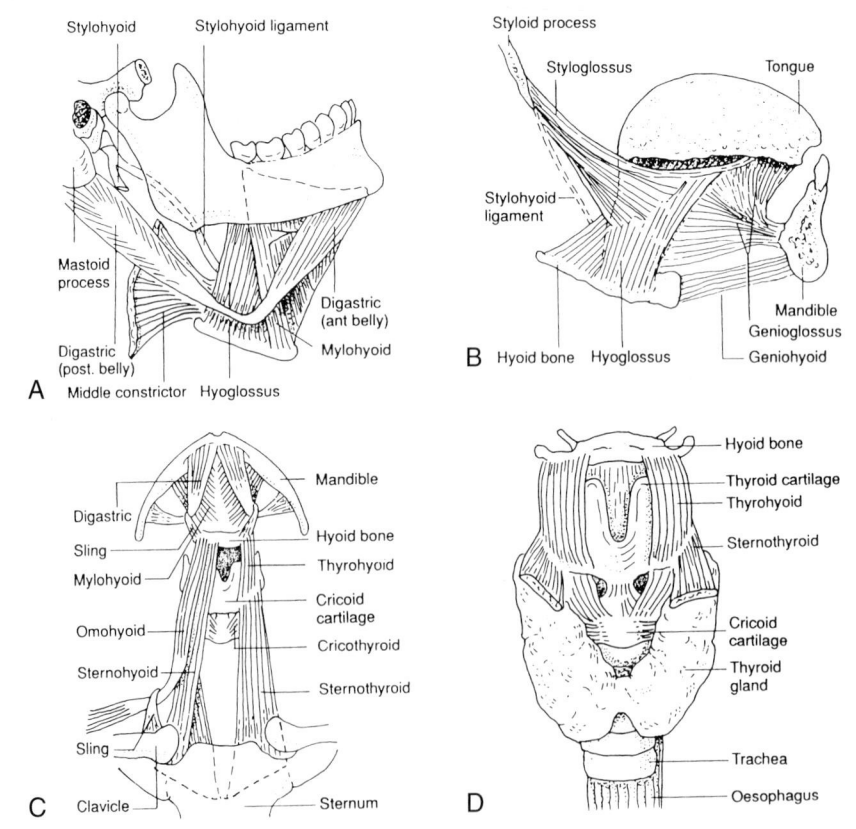

Figure 3–2. Anatomy of upper airway muscles. (A) Lateral view of muscle attachments to the mandible, styloid process, mastoid process, and hyoid bone. (B) Lateral view of the muscles of the tongue. (C) Anterior view of the muscles of the hyoid bone. (D) Anterior view of the deeper muscles that attach to the hyoid bone, thyroid cartilage, and cricoid cartilage. (From Stradling,[4] p 25, with permission.)

Table 3–2. **Nasal, Palatal, and Oropharyngeal Upper Airway Muscles**[5]

Muscles	Nerves	Actions
NASAL GROUP		
Compressor naris	CN VII	Compresses nasal cartilage and constricts nasal passages
Dilator naris, Levator alae nasi	CN VII	Dilate nasal passages during inspiration and reduce nasal resistance
SOFT PALATE		
Levator veli palatini	CN IX, X	Elevate and stiffen soft palate; promote oral airflow
Tensor veli palatini	CN V	Active during inspiration; elevate and stiffen soft palate
Musculus uvulae	CN IX, X	Elevates uvula
Palatoglossus, palatopharyngeus	CN IX, X	Move palatoglossus arches together, narrow oropharyngeal isthmus, and promote nasal airflow
OROPHARYNGEAL		
Genioglossus	CN XII	Protrudes tongue; increases oropharyngeal and hypopharyngeal airway diameter
Masseters	CN V	Close jaw and control mandible position; indirectly affect position of hyoid to increase airway diameter
Pterygoids	CN V	Open jaw
Superior pharyngeal constrictors	CN IX, X	Constrict pharynx

Source: Adapted from Lydic et al.,[5] pp 116–117, with permission.

The muscles regulating the position of the soft palate alter the balance between nasal and oral airflow. Movement of the soft palate away from the posterior wall of the oropharynx promotes nasal airflow, whereas elevation of the palate narrows the nasopharyngeal isthmus and promotes oral airflow. Stiffening of the soft palate, which occurs with activation of the tensor veli palatini and accompanies elevation of the soft palate, helps to stabilize the nasopharyngeal isthmus and prevent collapse.

The oropharyngeal muscles affect the diameter of the oropharyngeal airway by their effect on the position of the tongue and mandible. Tongue protrusion by the genioglossus and jaw closure by the masseters increase oropharyngeal airway diameter.

The hyoid bone contributes rigidity to the hyoid region of the pharynx and serves as the point of insertion for several muscles involved in the control of head movement, swallowing, and speech. The hyoid muscles also alter the position of the hyoid bone and influence the size of the hypopharynx. Anatomically, the hyoid muscles are divided into two groups: the suprahyoid muscles, which include the geniohyoid, mylohyoid, hyoglossus, stylohyoid, and digastric; and the infrahyoid muscles, which include the sternohyoid, omohyoid, and sternothyroid.

Functionally, the hyoid muscles comprise three suspensions that provide three vectors of force acting on the hyoid bone (Table 3–3) a ventral-rostral vector produced by the mandibular suspension, a ventral-caudal vector produced by the styloid suspension, and a dorsal vector produced by the vertebral suspension.[5] The effects of the various hyoid muscles on air-

Table 3–3. **Hyoid and Hypopharyngeal Upper Airway Muscles**[5]

Muscles	Nerves	Actions
Mylohyoid and anterior digastric (hyoid mandibular suspension)	CN V	Move hyoid anteriorly and superiorly during inspiration; increase hypopharyngeal diameter
Stylohyoid and posterior digastric (hyoid styloid suspension)	CN VII	Move hyoid anteriorly and inferiorly
Middle pharyngeal constrictor (hyoid vertebral suspension)	CN IX, X	Active during inspiration; moves hyoid posteriorly
Geniohyoid and hyoglossus (hyoid mandibular suspension)	C1–2	Active during inspiration; dilate hypopharynx
Sternohyoid and sternothyroid	C1–2	Active during inspiration; dilate and stiffen hypopharynx; stabilize rib cage
Inferior pharyngeal constrictor		Dilates hypopharynx
Medial pharyngeal constrictor		Constricts hypopharynx

Source: Adapted from Lydic et al.,[5] pp 116–117, with permission.

way compliance and diameter are complex because the positions of the sites of attachment for the hyoid muscles on the hyoid, mandible, and sternum vary during different patterns of breathing.

Laryngeal muscles control the position of the vocal cords and the opening and closure of the glottis (Table 3–4). Activity of the posterior cricoarytenoids is a major determinant of laryngeal resistance to airflow, as these muscles are the only abductors of the vocal cords.

UPPER AIRWAY MUSCLE ACTIVITY DURING INSPIRATION

Upper airway size and patency are determined by the anatomy of the bones and soft tissues of the neck and face, by the compliance of the pharyngeal walls, and by the pressures acting on those walls.[6] During wakefulness, contraction of pharyngeal muscles just before the onset of inspiration increases outward pressure on the pharyngeal wall, which facilitates energy-efficient inspiration and reduces the likelihood of collapse from the negative intrathoracic pressure generated by the diaphragm.[7] Nasal dilators expand the nasal passages, the tensor veli palatini stiffens the soft palate, the genioglossus muscle pulls the tongue forward, the mylohyoid and digastric muscles move the hyoid bone anteriorly and superiorly, and the posterior cricoarytenoids abduct the vocal cords and stiffen the larynx. The net result is a sharp decrease in airway resistance that occurs a fraction of a second before the onset of inspiration and continues throughout inspiration.

Table 3–4. **Laryngeal Upper Airway Muscles**[5]

Muscles	Nerves	Actions
Cricothyroid	CN X	Lengthen and tense vocal cords; active during inspiration
Posterior cricoarytenoid	CN X	Abduct vocal cords and open glottis; active during inspiration
Lateral cricoarytenoid	CN X	Close glottis
Interarytenoid	CN X	Close glottis; move arytenoid cartilage together
Thyroarytenoid	CN X	Expiratory: adduct cords, close glottis, and relax vocal cords

Source: Adapted from Lydic et al.,[5] pp 116–117, with permission.

The actions of the genioglossus and tensor veli palatini during inspiration are critical. The inspiratory activity of the genioglossus increases in the supine position to overcome the effects of gravity, and genioglossal activation plays a major role in reopening the pharyngeal airway after occlusion, which contributes to termination of obstructive apneas (see Chap. 13). Genioglossal activity is depressed by diazepam, by alcohol, and after sleep deprivation.

Activity of upper airway dilator muscles is affected by arterial blood gas concentration; by proprioceptive input to respiratory neurons from the jaw, mouth, and thorax; and by pressure sensors that respond to changes in upper airway transmural pressure. For example, increased negative upper airway pressure leads to increased activity of the genioglossus, which pulls the tongue forward and tends to open the airway.

Although airway muscle activity during inspiration appears to decrease airway compliance, other factors also affect airway shape and compliance, including flexion and extension of the neck, the vascularity and perfusion pressure of the upper airway tissues, and the elasticity and amount of connective tissue.[8]

EXPIRATION

The actions of the upper airway muscles are less important during expiration because the positive airway pressure generated by the elastic recoil of the lungs and chest wall helps to maintain airway patency. Airway resistance, however, tends to increase as the airway dilator muscles relax. At end-expiration, upper airway diameter reaches its minimum and airway resistance becomes maximal because most of the positive expiratory pressure has dissipated and inspiratory dilator activity has not yet resumed.

Increased expiratory upper airway resistance is not necessarily harmful. In obstructive airways diseases, impaired elastic support of large and medium-sized airways allows them to be compressed by increased expiratory pleural pressure, further worsening the expiratory airway obstruction. If laryngeal resistance increases, the higher positive pressure in the large airways tends to oppose such compression.

Chemoreceptors and Other Afferents

Chemoreceptors and afferents from the lungs, chest wall, and upper airway provide information about the state of the respiratory system to the network of CNS controller neurons (see below).

CHEMORECEPTORS

Chemoreceptors respond to changes in the plasma concentrations of oxygen (O_2), carbon dioxide (CO_2), and hydrogen ions (pH). Peripheral chemoreceptors of the aortic arch are sensitive only to large changes in CO_2 tension (P_{CO_2}), O_2 tension (P_{O_2}), and pH; consequently, they probably have little role in breathing control under most circumstances. Peripheral chemoreceptors of the carotid body are sensitive to small changes in arterial P_{O_2} and P_{CO_2} (Pa_{O_2} and Pa_{CO_2}, respectively): Their discharge rate increases as Pa_{O_2} falls or Pa_{CO_2} increases. Information from the carotid body is transmitted via the glossopharyngeal nerve and from the aortic arch via the vagus nerve.

Central chemoreceptors, located just beneath the surface of the ventrolateral medulla, are sensitive to the pH of extracellular brain fluid, which is in equilibrium with the cerebrospinal fluid (CSF). The blood-brain barrier is permeable to CO_2, but not to bicarbonate or to the hydrogen ion; as a result, the pH of CSF changes slowly in response to changes in Pa_{CO_2}. An increase in Pa_{CO_2} leads to a gradual increase in CSF P_{CO_2}, which is buffered by CSF bicarbonate. Thus, high levels of CSF bicarbonate, such as occurs with chronic hypercapnia or metabolic alkalosis caused by diuretic use, buffers changes in CSF pH and tends to reduce the ventilatory response to hypercapnia. The central medullary chemoreceptors are responsible for approximately 80% of the increase in ventilation that occurs with hypercapnia. The peripheral chemoreceptors account for the remaining 20% and are not essential for maintenance of life-sustaining levels of ventilation, although they are required for precise modulation.

UPPER AIRWAY, LUNG, AND CHEST WALL AFFERENTS

Upper airway, lung, and chest wall afferents also assist in the regulation of breathing. Upper airway neurons sensitive to changes in intraluminal hypopharyngeal pressure project via the glossopharyngeal nerve to respiratory centers. Pressure changes can thereby cause alterations in genioglossal tone; this effect is abolished by topical upper airway anesthesia.[9,10] Laryngeal receptors respond to laryngeal muscle contraction, airflow, and changes in transmural laryngeal pressure, and their activation may alter the pattern of breathing.[11] Intrapulmonary receptors that respond to stretch associated with lung inflation project via the vagus nerve to inspiratory cells of the nucleus tractus solitarius, and the activity of these vagal neurons contributes to termination of respiration. Parenchymal lung receptors respond to congestion and may contribute to the increase in ventilatory effort associated with congestion. Spindles, pacinian corpuscles, and Golgi tendon end-organs are numerous in intercostal muscles and provide a system for monitoring tidal volume and load.

Central Nervous System Controller

The CNS controller integrates afferent information from peripheral and central sources to maintain adequate alveolar ventilation, ensure appropriate oxygenation, and assist with the regulation of acid-base balance. In addition, it coordinates the neuronal activity required for appropriate timing of respiratory muscle activation and relaxation during inspiration and expiration. The controller achieves these functions by modulating the depth and frequency of breathing in response to information received from the central and peripheral chemoreceptors.

Although the specific neuroanatomic and physiologic interactions required for this control are not entirely elucidated, two major groups of respiratory neurons, defined as neurons that fire in phase with inspiration or expiration, are concentrated in the lower medulla (Fig. 3–3). The first group is called the dorsal respiratory group (DRG); the second group includes the ventral respiratory group (VRG), the Bötzinger complex (BOT), and the pre-Bötzinger complex (pre-BOT).[12]

Respiratory neurons of the DRG, VRG, and BOT can be subdivided into neurons that fire during inspiration and those that fire during expiration. Further subdivisions can be made based on the firing patterns of the neurons; for example, neurons that increase their firing during inspiration are called *inspiratory augmenting neurons*; those that decrease their firing rates during inspiration are called *inspiratory decrementing neurons*. The DRG neurons are almost exclusively inspiratory augmenting neurons that project to motoneurons and interneurons of the spinal cord; a few also have collateral axons that project to inspiratory neurons of the VRG.

Functionally, the VRG neurons are made up of two types: inspiratory neurons, which are located in the more rostral part of the VRG in the area of the nucleus ambiguus; and expiratory neurons, which are located more caudally in the region of the nucleus retroambigualis. Most inspiratory neurons of the rostral VRG are augmenting neurons that project to the spinal cord, mainly contralaterally, and synapse with neurons of the phrenic nerve nucleus and of the thoracic ventral horn. A smaller number of inspiratory neurons of the VRG are decrementing neurons that project to wide areas of the VRG and BOT. Most expiratory neurons of the caudal VRG are augmenting neurons that project contralaterally to the spinal cord; they do not have medullary collaterals and thus appear to function as respiratory upper motoneurons.[12]

The BOT contains mostly expiratory neurons—consisting mainly of *expiratory augmenting neurons*—which project bilater-

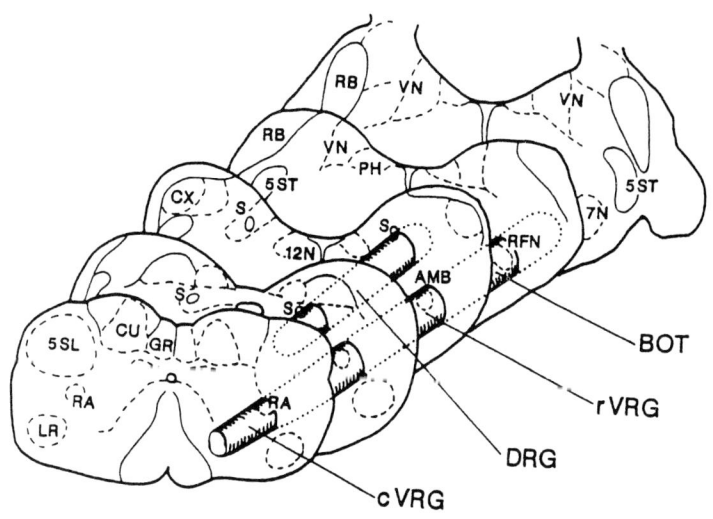

Figure 3–3. Location of the dorsal respiratory group (DRG), the rostral ventral respiratory group (rVRG), the caudal ventral respiratory group (cVRG), and the Bötzinger complex (BOT) in the medulla of the cat. The DRG, located in the region of the solitary tract (S) and the surrounding reticular formation, has a more restricted extent than the VRG, which is ventrolateral to the DRG and includes the regions of the nucleus ambiguus (AMB) and the nucleus retroambigualis (RA). The BOT, which is in continuity with the VRG, is in the vicinity of the retrofacial nucleus (RFN). The pre-Bötzinger complex, not shown in this diagram, is just caudal to the BOT. Other abbreviations: CU, cuneate nucleus; CX, external cuneate nucleus; DX, dorsal motor nucleus of the vagus; GR, nucleus gracilis; LR, lateral reticular nucleus; PH, nucleus prepositus hypoglossi; RB, restiform body; VN, vestibular nucleus; 5SL, laminar trigeminal nucleus; 5ST, spinal trigeminal tract; 7N, facial nucleus; 12N, hypoglossal nucleus. (From Ezure,[12] p 431, with permission.)

ally to the DRG and VRG as well as to the BOT itself. *Expiratory decrementing neurons* are also present in the BOT and project to the DRG, VRG, and BOT. The pre-BOT contains a mixture of inspiratory and expiratory neurons as well as neurons whose firing patterns span inspiration and expiration.[13]

It is known that interactions among DRG, VRG, and BOT neurons form the basis of respiratory rhythm generation,[12] but the precise mechanisms have not yet been fully established (Table 3–5). Cells with rhythmic bursting properties are present in the pre-BOT, and lesions in this area eliminate the respiratory rhythm.[14,15] Thus, some neurons of the pre-BOT may be pacemaker neurons responsible for generation of the basic respiratory rhythm. The rhythm is made up of four phases:

1. Inspiratory phase
2. Inspiratory-to-expiratory transition
3. Expiratory phase
4. Expiratory-to-inspiratory transition

During the inspiratory phase, phrenic nerve activity increases monotonically to reach a maximum at the end of inspiration. The increasing phrenic nerve activity is produced by the activity of augmenting inspiratory neurons of the DRG and rostral VRG. Self-reexcitation and positive feedback appear to be the principal mechanisms for producing the augmentation of activity during inspiration. Toward the end of inspiration, reduced firing of inspiratory decrementing neurons of the VRG and BOT appears to release expiratory decrementing neurons of the BOT from inhibition.

During the inspiratory-to-expiratory transition, there is some overlap between the activity of inspiratory neurons and that of expiratory neurons. The decrementing expiratory neurons of the BOT begin to fire near the end of inspiration and appear to inhibit the inspiratory neurons of the DRG and VRG. In addition, some inspiratory neurons of the DRG and VRG are active only at the end of inspiration and may contribute to inhibition of augmenting inspiratory neurons.[16] As the decrementing expiratory neurons become active, augmenting expiratory neurons of the BOT also begin to fire and lead to in-

Table 3–5. **Summary of Respiratory Rhythm Generation**[12-16]

	Augmenting Neurons	Decrementing Neurons
Inspiratory phase	Self-reexcitation and positive feedback of augmenting inspiratory neurons of the DRG and rostral VRG produce synchronous burst firing of inspiratory neurons.	Gradual reduction of firing by decrementing inspiratory neurons of the VRG and BOT releases expiratory neurons from inhibition.
Inspiratory-to-expiratory transition		Decrementing expiratory neurons of the BOT that have been released from from inhibition begin to fire and inhibit inspiratory neurons.
Expiratory phase	Augmenting expiratory neurons of the BOT become active and produce complete inhibition of inspiratory neurons.	
Expiratory-to-inspiratory transition		Subsiding activity of decrementing expiratory neurons releases inspiratory neurons from inhibition.

BOT = Bötzinger complex.
DRG = dorsal respiratory group.
VRG = ventral respiratory group.

hibition of the augmenting inspiratory neurons of the DRG and VRG, thus completing the transition to expiration.

The expiratory-to-inspiratory transition is more sharply delineated than the inspiratory-to-expiratory transition, with virtually no overlap of activity between expiratory and inspiratory neurons. At the transition, the inspiratory neurons abruptly start firing, probably because they are released from inhibition by the subsiding activity of the decrementing expiratory neurons of the BOT. Once these neurons begin firing, self-reexcitation networks produce synchronous burst firing of inspiratory neurons. The burst firing of decrementing inspiratory neurons inhibits the augmenting expiratory neurons of the BOT and caudal VRG.

CNS inputs to brainstem neurons involved in respiratory control come from the following sources: central autonomic neurons, brainstem neurons in the Kolliker-Fuse nuclei (the pneumotaxic center described by Lumsden), the dorsolateral caudal pons (the apneustic center described by Lumsden),[1,2] brainstem neurons involved in sleep-wake regulation, diencephalic structures, and a number of cortical regions. These inputs permit modulation of the basic respiratory rhythm to facilitate changing ventilatory needs as well as nonventilatory functions, such as speech.

Output from the respiratory controller occurs via two main groups of premotor respiratory neurons, mainly from the BOT and VRG. The first group consists of vagal and glossopharyngeal motoneurons that innervate upper airway muscles. The second group includes bulbospinal neurons that decussate in the lower medulla and upper cervical cord, traverse the reticulospinal tracts in the ventrolateral spinal cord, and synapse with lower motoneurons that innervate respiratory muscles.

Functionally, at least three systems are involved in breathing control: the metabolic system, the voluntary system, and the system mediating reflex responses to changes in the respiratory load. There is also a hypothesized fourth component, referred to as the wakefulness stimulus. These four components are discussed below.

Metabolic Regulation of Breathing

The metabolic system, also called the automatic system, regulates ventilation to maintain near-constant PaO_2, $PaCO_2$, and cerebrospinal fluid (CSF) PCO_2, thereby matching minute ventilation to bodily needs. For example, increased bodily CO_2 production leads to increased ventilation, which increases CO_2 dissipation and returns $PaCO_2$ a normal level. The ventilatory response to changing $PaCO_2$ is linear (Fig. 3–4),[17] and as bodily CO_2 stores are substantial, large increases in ventilation are required to affect arterial and CSF CO_2.

In contrast, bodily stores of oxygen are low, and the relation between oxygenation and ventilation is hyperbolic, with relatively little change in ventilation over a wide range of PaO_2 and an increase in ventilation most apparent when PaO_2 falls to less than 60 mm Hg. (Fig. 3–5).[18]

The ventilatory response to hypoxia or hypercapnia is a function of the magnitude of the stimulus and the "gain" of the negative feedback system. Experimental studies indicate that increased gain of the feedback loop leads to increased instability of ventilation, and spontaneous oscillations in healthy subjects may be converted to periodic waxing and waning of breathing by such manipulation.[19] Subjects with high ventilatory responses to hypoxia and hypercapnia are most likely to convert to periodic breathing. Individual variations in ventilatory responsiveness contribute to the development of periodic breathing in patients with disorders associated with this type of breathing, such as congestive heart failure (see Chap. 17), and help to explain why some patients exhibit periodic breathing of the Cheyne-Stokes type and others do not.

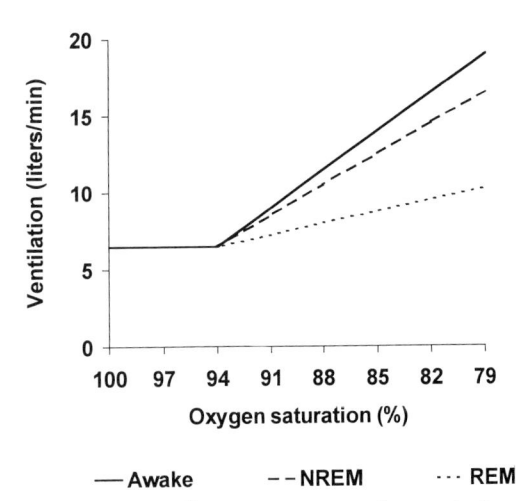

Figure 3–5. Ventilatory response to changes in PO_2. The ventilatory response is relatively flat at high levels of oxygenation and rises rapidly with progressive hypoxia. As with the ventilatory response to PCO_2, the slope of the response is greatest in wakefulness and least in REM sleep. (Adapted from Douglas,[18] p 564, with permission.)

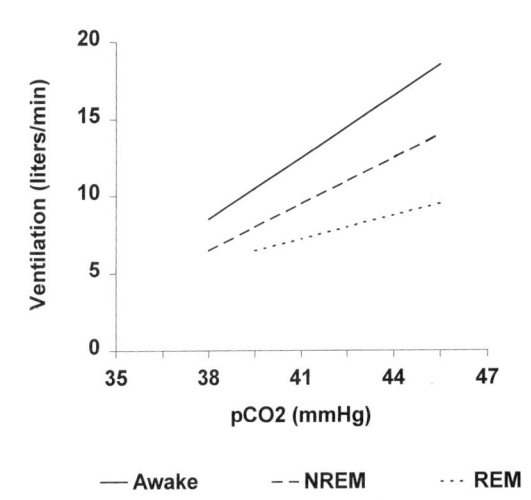

Figure 3–4. Ventilatory response to changes in PCO_2. Although the response is linear in wakefulness and all stages of sleep, the slope of the response is greatest during wakefulness and least in REM sleep. (Adapted from Douglas et al.,[17] p 759, with permission.)

Voluntary Ventilatory Control System

Regulation of breathing is also partly under voluntary control. Some neurons involved in voluntary ventilatory control originate in the cerebral cortex and descend to integrate with medullary neurons involved in automatic control of breathing. Other cortical neurons descend independently with corticobulbar tracts to the reticular formation and with corticospinal tracts to spinal respiratory motoneurons. Voluntary changes in the depth and frequency of breathing as well as changes in respiration associated with talking, eating, and swallowing are mediated by these pathways.

Reflex Responses to Altered Respiratory Load

Changes in airway resistance or in the respiratory load lead to changes in respiration-related neural activity. With increased inspiratory resistance during wakefulness, there is an increase in tidal volume and a decrease in respiratory frequency with little or no overall change in minute ventilation. These changes occur too rapidly to be mediated by changes in oxygen or carbon dioxide tension. For example, by the second breath after application of a load, both motor output to respiratory muscles and inspiratory duration increase, leading to preserved minute ventilation. This increased motor output is probably mediated at least in part by upper airway reflexes, but the precise neuroanatomic substrate for these reflexes is unknown.

The Wakefulness Stimulus

The wakefulness stimulus is a hypothetical tonic stimulus to breathe that is posited to account for the increase in ventilation during wakefulness compared to sleep. The wakefulness stimulus is presumed to be the reason that posthyperventilation apnea is more prominent in sleep than in wakefulness. In addition, the loss of the wakefulness stimulus at the onset of sleep is thought to be responsible for central apneas that occur in some persons upon the return to sleep after brief arousals. Although the pathways mediating the state-related shift in the setpoint of ventilation are poorly understood, they appear to involve reticular activating system inputs to brainstem respiratory neurons as well as descending pathways from more rostral structures.

SLEEP-RELATED CHANGES IN BREATHING

A number of changes in breathing occur during sleep. During non–rapid eye movement (NREM) sleep, the most notable include an increase in airway resistance, withdrawal of the wakefulness stimulus, changes in the rate and rhythm of breathing, a change in breathing mechanics that increases the contribution of the rib cage to breathing, and a reduction in load compensation. The net result is a decrease in ventilation. During rapid eye movement (REM) sleep, changes are even more marked because of the tonic inhibition of respiratory muscle activity, wide fluctuations in respiratory muscle activity that accompany phasic events of REM sleep, and reduced homeostatic responses.

Changes in Breathing Mechanics

Respiratory muscle activity is altered during NREM sleep (Table 3–6).[20-22] Under isocapnic conditions, diaphragm activity during NREM sleep is reduced compared to wakefulness.[23] However, the increase in $Paco_2$ that occurs during NREM sleep leads to an increase in phasic inspiratory activity of the diaphragm above waking levels. There is also an increase in inspiratory intercostal activity and a reduction in postural muscle activity. The results of these changes are an increased lateral expansion of the rib cage, a relative increase in the contribution of the rib cage to tidal volume, and an increase in transdiaphragmatic pressure.

Ventilatory mechanics change to a greater extent during REM sleep, with the loss of tonic activity of the diaphragm, the intercostals and other accessory inspiratory muscles, and most of the upper airway dilator muscles. Phasic inspiratory activity of the genioglossus is also reduced, but the diaphragm remains phasically active in REM sleep, which allows breathing to continue. As a result of these changes, stiffening of the upper airway and rib cage is reduced. The reduced contribution to breathing of the intercostal and accessory breathing muscles means that the diaphragm does most of the work of breathing. As a result, there is less ventilatory reserve. Loss of laryngeal tone and reduced chest wall stiffness during expiration contribute to a reduction in end-expiratory volume and a reduction in functional

Table 3-6. **Respiratory Muscle Activity During Sleep**[20-22]

	NREM SLEEP COMPARED TO WAKEFULNESS		REM SLEEP COMPARED TO NREM SLEEP	
	Inspiratory Phasic Activity	Tonic Activity	Inspiratory Phasic Activity	Tonic Activity
VENTILATORY PUMP MUSCLES				
Diaphragm	↑		Similar or ↑	↓
Accessory inspiratory muscles	↑		↓ or absent	↓
Expiratory muscles	↑			
UPPER AIRWAY MUSCLES				
Levator alae nasi	↓			
Tensor veli palatini	↓	↓		
Genioglossus	↑	Slight ↓	↓	Marked ↓
Sternohyoid, geniohyoid		Same		Same
Sternothyroid				Marked ↓
Cricothyroid			↑	
Posterior cricoarytenoid	↓	↓	↑	↓
Lateral cricoarytenoid		Absent		
Interarytenoid		Absent		
Thyroarytenoid		Absent		Bursts

residual capacity. As a consequence of the reduced functional residual capacity, oxygenation reserve is reduced, causing a more rapid onset of hypoxemia during apneas or hypopneas.

Increased Airway Resistance

Nasal, pharyngeal, and laryngeal airway resistance increases during sleep, mainly because of the decreased activity of dilator muscles at these levels.[24] Reduction in lung volume may also contribute to increased upper airway resistance,[25] but there is little change in the elastic and flow-resistive properties of the lung.[26] Among normal subjects, the increase in resistance varies considerably—anywhere from 1.5 to 7 times the waking value.[27] The decreased tone of the tensor palatini appears to be a major contributor to increased airway resistance during NREM sleep, although decreased tone of the genioglossus and geniohyoid muscles also contributes.

On the other hand, laryngeal inspiratory and expiratory muscle activity increases during hypoxia and during hypercapnia, leading to a reduction in upper airway resistance.[28] Furthermore, the increase in upper airway muscle activity that occurs in response to increased negative pharyngeal pressure is maintained in NREM sleep.[9,29] The hypotonia of pharyngeal and laryngeal dilators is more pronounced in REM sleep, and the inspiratory activity of the genioglossus is markedly inhibited during phasic REM sleep.[30,31]

Central Nervous System Changes During Sleep

Discharge patterns of some respiratory neurons in the brainstem change during sleep. For example, some neurons that fire predominantly during inspiration switch to a pattern of firing predominantly during expiration,[32] and some brainstem neurons fire synchronously with respiration during wakefulness but not during sleep.[33] A marked reduction in the spontaneous activity of some ventral medullary neurons may contribute to the reduced tonic activity of airway dilator muscles.[34]

Although sleep reduces tonic output to respiratory muscles, it does not increase the CO_2 threshold for generation of rhythmic respiration and appears not to change the output from the central controller to the ventilatory pump muscles.[35] Thus, central respiratory drive does not change to any substantial degree in either NREM or REM sleep.[36] In REM sleep, therefore, the reduction in respiratory muscle activity is caused by active inhibition of respiratory motoneurons (i.e., hyperpolarization), not by reduced output from medullary areas involved in respiratory control.

Metabolic Regulation of Breathing During Sleep

During NREM sleep, the voluntary system of breathing control is not functional; as a result, the metabolic control of breathing becomes predominant. The responsiveness of chemoreceptors to changes in gas tensions during sleep appears unchanged in sleep compared to wakefulness. Respiration is more rhythmic during NREM sleep than during wakefulness, whereas rate changes may be more pronounced because of the absence of voluntary control. Because of its predominant role during sleep, metabolic system impairment has more severe consequences during sleep than during wakefulness. For example, vagal blockade during sleep produces marked slowing of respiratory rate to 3 to 5 breaths per minute, and hypercapnia is more pronounced during sleep than during wakefulness in animals with carotid body denervation.[37]

Although the metabolic system is predominant during sleep, ventilatory responses to changes in PaO_2 and $PaCO_2$ are attenuated. The attenuation is greater in men than in women during NREM sleep, but it is similar in both sexes during REM sleep. The reduced responsiveness probably reflects two factors: decreased sensitivity to changes in gas tensions and increased upper airway resistance to breathing.

Changes in Reflex Responses to Respiratory Loads During Sleep

The increase in upper airway resistance that accompanies sleep produces a respiratory load, and during NREM sleep, the magnitude of the reflex compensation for increased loads on respiratory muscles is reduced, perhaps in part because of the lack of conscious perception of the increased resistive load. The lack of immediate load compensation leads to reduced tidal volume and transient hypoventilation during NREM sleep that lasts for about 2 to 5 minutes.[38,39] The variability in the time required for ventilation to return to near-normal levels is probably a reflection of the variability in ventilatory responses to hypercapnia. Although load compensation during REM sleep has not been well studied, it is probably reduced compared to that occurring during NREM sleep.

Loss of the Wakefulness Stimulus

The homeostatically regulated setpoint for $PaCO_2$ in humans is about 1 to 2 mm Hg less during wakefulness than during NREM sleep. Consequently, if $PaCO_2$ is reduced during NREM sleep by mechanical ventilation to the waking $PaCO_2$ value, and then mechanical ventilation is stopped, a period of apnea ensues.[40] Although this change in setpoint is usually attributed to the loss of the wakefulness stimulus for breathing, medullary neuronal activity appears to be reduced only slightly or not at all during NREM sleep. The change, therefore, may be caused mainly or entirely by an increase in cerebral blood flow during sleep that leads to a reduction in the cerebral arterial-venous PCO_2 difference. The lower cerebral venous effluent PCO_2 means that the PCO_2 at medullary chemoreceptors may be reduced during sleep for a given systemic $PaCO_2$.

Decreased Ventilation During Sleep

In humans, minute ventilation is about 7 L/min during quiet wakefulness, and it decreases by about 6% to 15% during NREM sleep.[22,41,42] Part of the reduction is the result of a lower metabolic rate, with a 10% to 30% reduction in tissue CO_2 production.[43] However, ventilation decreases to a greater extent than body metabolism, and the result is a rise in $PaCO_2$ of about 2 to 8 mm Hg and a decrease in SaO_2 of 1% to 2%.

The rise in $PaCO_2$ appears to arise from two factors: altered setpoint, as described above, and increased airway resistance. As the change in setpoint appears to be no more than 1 to 2 mm Hg, the remaining increase in $PaCO_2$ is probably mainly caused by increased upper airway resistance.[44] Because there is little or no change in chemoreceptor response during sleep, the increase in $PaCO_2$ leads to increased inspiratory (and sometimes expiratory) phasic activity in pump muscles. Although the increase in upper airway resistance may contribute to the reduction in ventilation during sleep, it does not appear to be the sole cause because continuous positive airway pressure (CPAP) applied during stage 4 sleep in normal nonsnoring subjects to normalize resistance does not lead to an increase in ventilation to waking levels.[42]

Minute ventilation is similar between REM sleep and NREM sleep, but the respiratory rate is more irregular during REM sleep. Phasic REM sleep is associated with rapid, shallow breathing, which may produce transient ventilation decreases of 10% to 25%.[41,45]

STATE CHANGES AND AROUSAL RESPONSES

The onset of sleep is associated with a change in breathing patterns. About 50% to 75% of normal subjects have periodic breathing with one or more cycles during which breathing is initially reduced or absent, then increased during an arousal, and followed again by a period of reduced ventilation.[22] Periodic breathing is facilitated by increased gain in the feedback system of respiratory control: As the ventilatory setpoint falls at sleep onset, an apnea occurs; the apnea is followed by hyperventilation during a subsequent arousal, which drives $PaCO_2$ below normal, which in turn facilitates the next apnea. Periodic breathing is abolished in normal subjects with the onset of stage 2 sleep, and breathing becomes regular with little variability in rate or amplitude.

Changes in respiratory function (e.g., hypoxia, hypercapnia, increased airway resistance to breathing, and airway irritation) can cause arousal from either NREM or REM sleep. These arousal responses are important protective mechanisms, allowing correction of the ventilatory problem. The likelihood for each of these changes to cause arousal is a function of the stage of sleep and the severity of the respiratory aberration.

Isocapnic hypoxia is a relatively weak stimulus for arousal from either NREM or REM sleep. Hypercapnia is a stronger stimulus, and progressive hypercapnia usually induces an arousal before end-tidal PCO_2 has risen 15 mm Hg above baseline. Although hypercapnia accompanied by hypoxia is an even more potent arousal stimulus than hypercapnia alone, the combined stimulus has less effect in REM sleep than in NREM sleep.

Increased resistance to breathing or occlusion of the airway at the onset of inspiration tends to produce arousal, and the intensity of breathing effort correlates with the likelihood of arousal.[46,47] The effect is similar in light NREM sleep and REM sleep, whereas stage 3–4 sleep is most resistant to the arousing effect of increased resistance. The effect is mediated partly by airway receptors, as upper airway anesthesia impairs the arousal response to airway occlusion.[48]

Bronchial and upper airway irritation can also produce arousal. Arousal responses to nasal irritants are similar in REM and NREM sleep, whereas the cough response to laryngeal stimulation is decreased in REM sleep compared to NREM sleep. Pharyngeal stimulation sometimes induces a brief postarousal central apnea.

SUMMARY

The respiratory control system is made up of the lungs, the chest wall, the inspiratory and expiratory muscles, the upper airway, the respiratory afferents, and the CNS neurons that integrate respiratory information and control the respiratory muscles. Efficient breathing requires appropriately coordinated activation of ventilatory pump muscles and the more than 20 muscles involved in upper airway function. Contraction of pharyngeal muscles, particularly the genioglossus and the tensor palatini, just before the onset of inspiration reduces the likelihood of pharyngeal collapse from the negative intrathoracic pressure generated by the diaphragm and facilitates energy-efficient inspiration. Regulation of breathing is dependent on chemoreceptors and other afferents, which provide information about the state of the respiratory system to two sets of respiratory control neurons located in the medulla: (1) the dorsal respiratory group; and (2) a more ventrolateral collection of neurons that includes the ventral respiratory group, the Bötzinger complex, and the pre-Bötzinger complex. These neurons, which interact to generate the basic respiratory rhythm, respond to changes in PaO_2, $PaCO_2$, and respiratory load.

With the onset of NREM sleep, airway resistance increases, and the responses to hypercapnia, hypoxia, and respiratory loads are reduced. The net result is a decrease in ventilation. During REM sleep, changes in breathing are even more marked as a result of tonic inhibition of respiratory muscle activity, wide fluctuations in respiratory muscle activity that accompany phasic events of REM sleep, and a further reduction in homeostatic responses. As a result of these changes, upper airway narrowing and periodic breathing are more likely to develop in sleep than in wakefulness.

REFERENCES

1. Lumsden, T: Observations on the respiratory centres in the cat. J Physiol 57:153, 1922.
2. Lumsden, T: The regulation of respiration. J Physiol 58:111, 1924.
3. Campbell, EJM, et al: The Respiratory Muscles: Mechanics and Neural Control. Lloyd-Luke, London, 1970.
4. Stradling, JR: Handbook of Sleep-Related Breathing Disorders. Oxford University Press, Oxford, 1993, pp 24–26.
5. Lydic, R, et al: Sleep-dependent changes in upper airway muscle function. In Lydic, R, and Biebuyck, JF (eds): Clinical Physiology of Sleep. American Physiological Society, Bethesda, Md, 1988, pp 97–123.
6. Remmers, JE, et al: Pathogenesis of upper airway occlusion during sleep. J Appl Physiol 44:931, 1978.
7. Isono, S, and Remmers, JE: Anatomy and physiology of upper airway obstruction. In Kryger, MH, et al (eds): Principles and Practice of Sleep Medicine, ed 2. WB Saunders, Philadelphia, 1994, pp 642–656.
8. Strohl, KP, and Olson, LG: Concerning the importance of pharyngeal muscles in the maintenance of upper airway patency during sleep: An opinion. Chest 92:918–920, 1987.
9. Mathew, OP, et al: Influence of upper airway pressure changes on genioglossus muscle respiratory activity. J Appl Physiol 52:438, 1982.
10. Mathew, OP, et al: Genioglossus muscle responses to upper airway pressure changes: Afferent pathways. J Appl Physiol 52:445, 1982.
11. Sant'Ambrogio, FB, et al: Laryngeal influences on breathing pattern and posterior cricoarytenoid muscle activity. J Appl Physiol 58:1298, 1985.
12. Ezure, K: Synaptic connections between medullary respiratory neurons and considerations on the genesis of respiratory rhythm. Prog Neurobiol 35:429, 1990.
13. Connelly, CA, et al: Pre-Botzinger complex in cats: Respiratory neuronal discharge patterns. Brain Res 590:337, 1992.
14. Johnson, SM, et al: Pacemaker behavior of respiratory neurons in medullary slices from neonatal rat. J Neurophysiol 72:2598, 1994.
15. Smith, JC, et al: Pre-Botzinger complex: A brainstem region that may generate respiratory rhythm in mammals. Science 254:726, 1991.
16. Oku, Y, et al: Possible inspiratory off-switch neurones in the ventrolateral medulla of the cat. Neuroreport 3:933, 1992.
17. Douglas, NJ, et al: Hypercapnic ventilatory response in sleeping adults. Am Rev Respir Dis 126:758, 1982.
18. Douglas, NJ: Control of ventilation during sleep. Clin Chest Med 6:563, 1985.
19. Chapman, KR, et al: Possible mechanisms of periodic breathing during sleep. J Appl Physiol 64:1000, 1988.
20. Harding, R, et al: Respiratory activity of laryngeal muscles in awake and sleeping dogs. Respir Physiol 66:315, 1986.
21. Kuna, ST, et al: Thyroarytenoid muscle activity during wakefulness and sleep in normal adults. J Appl Physiol 65:1332, 1988.
22. Krieger, J: Breathing during sleep in normal subjects. In Kryger, MH, et al (eds): Principles

and Practice of Sleep Medicine, ed 2. WB Saunders, Philadelphia, 1994, pp 212–223.
23. Rist, KE, et al: Effects of non-REM sleep upon respiratory drive and the respiratory pump in humans. Respir Physiol 63:241, 1986.
24. Wheatley, JR, et al: Influence of sleep on alae nasi EMG and nasal resistance in normal men. J Appl Physiol 75:626, 1993.
25. Begle, RL, et al: Effect of lung inflation on pulmonary resistance during NREM sleep. Am Rev Respir Dis 141:854, 1990.
26. Hudgel, DW, et al.: Mechanics of respiratory system and breathing pattern during sleep in normal humans. J Appl Physiol 52:607, 1984.
27. Wiegand, L, et al: Collapsibility of the human upper airway during normal sleep. J Appl Physiol 66:1800, 1989.
28. England, SJ, et al: Laryngeal muscle activities during progressive hypercapnia and hypoxia in awake and sleeping dogs. Respir Physiol 66:327, 1986.
29. Mathew, OP: Upper airway negative-pressure effects on respiratory activity of upper airway muscles. J Appl Physiol 56:500, 1984.
30. Orem, J, and Lydic, R: Upper airway function during sleep and wakefulness: Experimental studies on normal and anesthetized cats. Sleep 1:49, 1978.
31. Parisi, RA, et al: Correlation between genioglossal and diaphragmatic responses to hypercapnia during sleep. Am Rev Respir Dis 135:378, 1987.
32. Harper, RM, et al: Cardiac and respiratory interactions maintaining homeostasis during sleep. In Lydic, R, and Biebuyck, JF (eds): Clinical Physiology of Sleep. American Physiological Society, Bethesda, Md, 1988, pp 67–78.
33. Orem, J, et al: Changes in the activity of respiratory neurons during sleep. Brain Res 82:309, 1974.
34. Orem, J, et al: Activity of respiratory neurons during NREM sleep. J Neurophysiol 54:1144, 1985.
35. Horner, RL, et al: Effects of sleep on the tonic drive to respiratory muscle and the threshold for rhythm generation in the dog. J Physiol 474:525, 1994.
36. White, DP: Occlusion pressure and ventilation during sleep in normal humans. J Appl Physiol 61:1279, 1986.
37. Phillipson, EA, and Sullivan, CE: Respiratory control mechanisms during NREM and REM sleep. In Guilleminault, C, and Dement, WC (eds): Sleep Apnea Syndromes. AR Liss, New York, 1978, pp 47–64.
38. Hudgel, DW, et al: Neuromuscular and mechanical responses to inspiratory resistive loading during sleep. J Appl Physiol 63:603, 1987.
39. Wiegand, L, et al: Sleep and the ventilatory response to resistive loading in normal men. J Appl Physiol 64:1186, 1988.
40. Henke, KG, et al: Inhibition of inspiratory muscle activity during sleep. Am Rev Respir Dis 138:8, 1988.
41. Gould, GA, et al: Breathing pattern and eye movement density during REM sleep in humans. Am Rev Respir Dis 138:874, 1988.
42. Morrell, MJ, et al: Changes in total pulmonary resistance and pCO_2 between wakefulness and sleep in normal human subjects. J Appl Physiol 78:1339, 1995.
43. Dempsey, JA, et al: Effects of sleep on the regulation of breathing and respiratory muscle function. In Crystal, RG, et al (eds): The Lung: Scientific Foundations. Raven Press, New York, 1991, pp 1615–1629.
44. Henke, KG, et al: Effects of sleep-induced increases in upper airway resistance on ventilation. J Appl Physiol 69:617, 1990.
45. Millman, RP, et al: Changes in compartmental ventilation in association with eye movements during REM sleep. J Appl Physiol 65:1196, 1988.
46. Gleeson, K, et al: Arousal from sleep in response to ventilatory stimuli occurs at a similar degree of ventilatory effort irrespective of the stimulus. Am Rev Respir Dis 142:295, 1989.
47. Gleeson, K, et al: The influence of increasing ventilatory effort on arousal from sleep. Am Rev Respir Dis 142:295, 1990.
48. Berry, RB, et al: Effect of upper airway anesthesia on obstructive sleep apnea. Am J Respir Crit Care Med 151:1857, 1995.

Chapter 4

CHRONOBIOLOGY

PHYSIOLOGIC FUNCTIONS WITH CIRCADIAN RHYTHMICITY
Circadian Distribution of Sleep and Sleepiness
NEUROANATOMY OF THE CIRCADIAN TIMING SYSTEM
GENETIC BASIS OF CLOCK FUNCTION
FACTORS AFFECTING CLOCK FUNCTION
Effects of Light on Clock Function
Effects of Other Exogenous Stimuli on Clock Function
Endogenous Influences on Clock Function
ZEITGEBERS, MASKING STIMULI, AND RHYTHM INTERACTIONS
EFFECTS OF THE CIRCADIAN PACEMAKER ON SLEEP AND WAKEFULNESS
Sleep-Wake Rhythms Under Constant Environmental Conditions
Sleep-Wake Rhythms Under Non–24-Hour Sleep-Wake Schedules

According to Babylonian legend, a heliotropic plant, which opened its leaves in the day and closed them at night, symbolized the unrequited love of Clytie, the daughter of the Babylonian king, for Helios, the sun god.[1]

The ability to regulate physiologic functions rhythmically is a phylogenetically ancient adaptation to the change in environment that accompanies the earth's rotation, and circadian rhythms have been identified in almost all organisms from unicellular algae to primates. Although circadian rhythms of physiologic functions have been recognized since ancient times, for much of human history it was thought that this rhythmicity represented a passive response to environmental changes induced primarily by the rising and setting of the sun. In 1729, de Mairan[2] was probably the first to demonstrate that rhythmicity has an endogenous basis. He noted that *Mimosa pudica* opened and closed its leaves in a 24-hour cycle, even when placed in a dark room. Although De Candolle in 1832 showed that the rhythms of plants raised in constant conditions did not exactly follow a 24-hour cycle, indicating that external stimuli regulate the phase of the internal clock, more than a century passed before investigators demonstrated that the phase of endogenous circadian rhythms in humans could be shifted with appropriately timed exposure to light.[3]

The anatomic basis for the control of endogenous rhythms was unknown before the 1960s, when Richter[4] showed that they were eliminated by lesions of the ventromedial hypothalamus. Subsequent studies demonstrated that the suprachiasmatic nucleus (SCN) of the hypothalamus is the major source of circadian rhythmicity and that the retinohypothalamic pathway to the SCN mediates the effects of light on circadian rhythms.[5] Although early studies of human circadian rhythms suggested that light had only a minor effect on circadian rhythms, studies performed in the 1970s provided conclusive evidence that light is the most potent stimulus affecting the internal clock.

Disturbances of circadian rhythms may lead to disorders of sleep and wakefulness (see Chap. 12).

PHYSIOLOGIC FUNCTIONS WITH CIRCADIAN RHYTHMICITY

Hundreds of human physiologic functions show circadian rhythmicity, including basic intracellular processes, enzyme activities, hormonal secretions, and more complexly regulated functions such as core body temperature, sleep and wakefulness, and rest and activity. Depending on the function, the wave form of the rhythm may be sinusoidal, square-waved, pulsatile, or asymmetric, with substantial changes across the 24-hour day. For example, the toxicity of ionizing radiation as well as that of some anticancer chemotherapeutic agents is highly dependent on the time of day of administration; in mice, the mortality caused by a dose of *Escherichia coli* endotoxin is almost 50% greater when the injection is administered at the midpoint of the day compared to the midpoint of the night.[6]

Each physiologic rhythm has a typical phase relationship to the day-night cycle and to other rhythms; one function of the circadian timekeeping system is to maintain appropriate phase relationships of the various rhythms. However, phase relationships to the output of the circadian pacemaker are maintained more tightly for some rhythms than for others. A tightly coupled rhythm, such as core temperature, provides a means of approximating pacemaker activity, whereas physiologic functions that are more loosely coupled to the pacemaker, such as the sleep-wake rhythm, may shift in their phase relationship to the pacemaker.

Circadian Distribution of Sleep and Sleepiness

Under usual conditions, with a sleep period from 11 pm to 7 am, sleepiness and the propensity to fall asleep have a circadian rhythm with a bimodal distribution.[7-11] Sleepiness is greatest during the early morning hours, when body temperature is at a minimum; a second peak of sleepiness occurs during the afternoon, when body temperature is high (Fig. 4–1). Similarly, alertness is high in the morning, from about 8 to 11 am, and again in the evening, from about 8 to 10 pm. Sleep deprivation studies also indicate a striking correlation between body temperature and alertness (Fig. 4–2).

Under most circumstances, the rhythm of sleep and wakefulness remains coupled to the endogenous pacemaker and to rhythms strongly linked to it, such as the temperature rhythm. Under certain circumstances, however, the sleep-wake

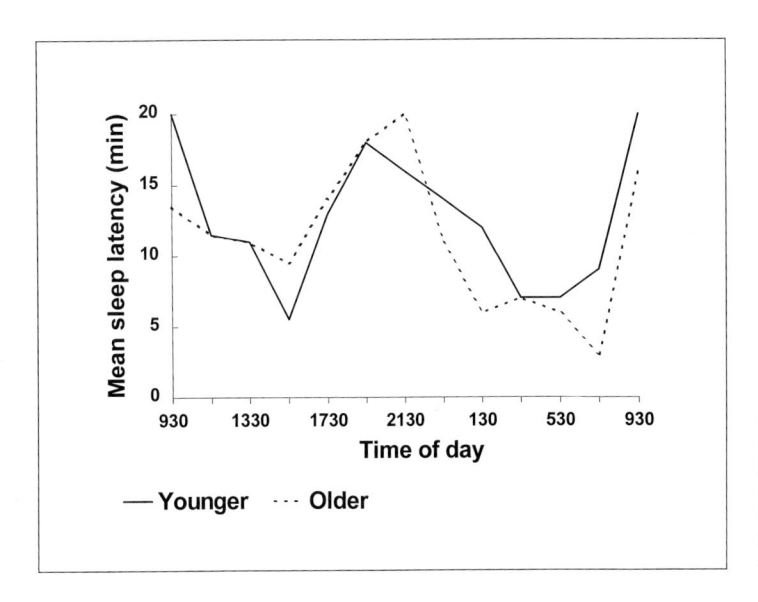

Figure 4–1. A bimodal distribution of sleepiness, with a major peak in the middle of the usual sleep period and a secondary peak in the mid afternoon. (Adapted from Carskadon and Dement,[8] p 312, with permission.)

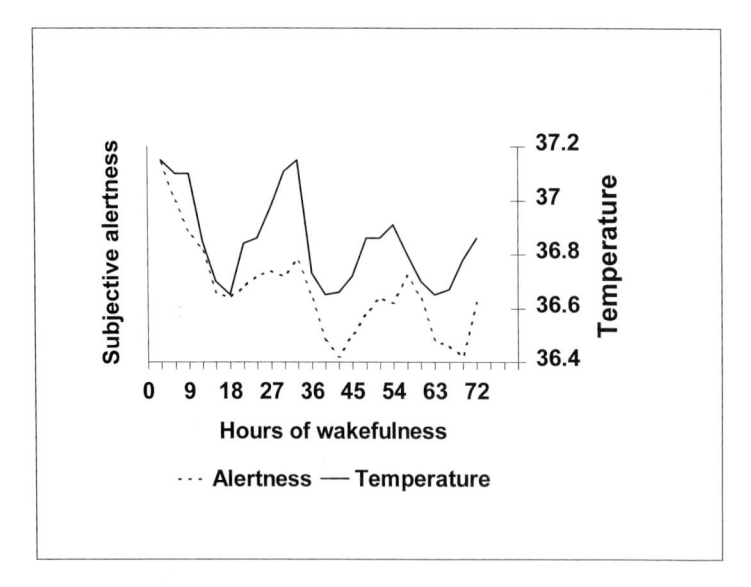

Figure 4–2. Subjective alertness and its relation to body temperature during 72 hours of sleep deprivation. (Adapted from Fröberg,[11] pp 126, 129, with permission.)

rhythm becomes uncoupled from the major oscillator: that is, its rhythm is separate from the temperature rhythm and from other tightly coupled rhythms. The condition in which the sleep-wake rhythm becomes uncoupled from the temperature rhythm is called *internal desynchronization*.

NEUROANATOMY OF THE CIRCADIAN TIMING SYSTEM

The SCN, located in the anterior hypothalamus just above the optic chasm, is the primary locus of the circadian pacemaker. Neurons of this nucleus fire in a circadian pattern under constant environmental conditions and after isolation of the nucleus from external inputs. The nucleus is tiny, with a volume of about 0.1 mm^3, and is made up of about 20,000 neurons. It contains two major subdivisions: the ventrolateral portion, with neurons containing vasoactive intestinal peptide and neuropeptide Y; and the dorsomedial region, with neurons containing vasopressin. Other neurotransmitters and peptides present in the SCN include acetylcholine, serotonin, γ-aminobutyric acid (GABA), somatostatin, nerve growth factor, and cholecystokinin.[12]

Isolated cells of the SCN demonstrate circadian rhythmicity, indicating that the biologic clock functions at the cellular level, and when SCN tissue from mutant rats with a genetically based 22-hour period of the rest-activity cycle is transplanted into rats who have undergone SCN ablation, the rats who receive the transplant adopt the 22-hour period of the transplanted tissue.[13]

The primary afferent tract to the SCN is the retinohypothalamic tract, made up of axons from specialized retinal ganglion cells with wide receptive fields. Retinal information is also conveyed to the SCN via retinal pathways to the intergeniculate leaflets of the lateral geniculate nuclei and from there via the geniculohypothalamic tract to the SCN. Another major input to the SCN comes from serotonergic neurons originating in the raphe nuclei. Although in mammals the effects of light on circadian rhythms are mediated almost exclusively via the retina (enucleated and congenitally blind mammals do not respond to light), certain birds and reptiles have encephalic photoreceptors that permit entrainment.[14]

Efferents from the SCN project to the paraventricular nuclei and other areas of the hypothalamus, to the thalamus, and to the basal forebrain. The SCN exerts effects on pineal secretion of melatonin via a multisynaptic pathway (see below).

Although the SCN is the major circadian pacemaker, it may not be the sole source of circadian rhythmicity. Some circadian rhythms, such as feeding in rats, persist after SCN ablation, suggesting that other areas of the nervous system are involved in control of such behavior. Areas that might be involved include the retina and the pineal body, as both contain neurons that exhibit circadian rhythmicity in vitro. The *clock* gene, which is highly expressed in the SCN (see below), is also expressed in the retina, gut, and kidneys.[15-17] According to one model, the SCN is a dominant pacemaker that controls the output of a number of other pacemakers; after ablation of the SCN, the subordinate pacemakers may contribute to persisting circadian or ultradian rhythmic behaviors.[18]

GENETIC BASIS OF CLOCK FUNCTION

Three genes identified in the 1990s play major roles in the biologic clock. In the fruit fly, the *per* (period) gene is involved in circadian rhythmicity, and mutations of this gene can cause fast, slow, or absent circadian rhythms. Messenger ribonucleic acid (mRNA) produced by the *per* gene has a circadian rhythm with peak production at or shortly after dusk; the protein produced by this mRNA, called PER protein, reaches a peak level about 6 to 8 hours after the peak in mRNA production and appears to form a feedback loop that negatively regulates *per* mRNA production.[19] This feedback loop appears to be a major component of the intracellular clock, and direct injection of protein synthesis inhibitors into the SCN produces a phase shift in SCN output that is a function of the time of injection.[20]

The *tim* (timeless) gene is also involved in clock function in the fruit fly; its protein, TIM, dimerizes with the PER protein to facilitate entry of the PER protein into the cell's nucleus.[19,21] The production of TIM is influenced by light, suggesting that altered interactions between the PER and TIM proteins may provide a mechanism by which light affects *per* mRNA production.

The third gene, *clock*, is present in mice; it appears to drive transcription of an oscillator gene that was unidentified at this writing.[15] In mice with a point mutation of one copy of the *clock* gene, the clock runs with a day-length of 24.2 hours, compared to 23.5 hours for normal mice. If both copies are mutants, the clock period slows to 27.4 hours and homozygous mutants confined to darkness have no circadian rhythm at all, indicating that in the absence of light, normal *clock* gene function is required for circadian rhythmicity.

FACTORS AFFECTING CLOCK FUNCTION

To maintain appropriate synchronization between the internal clock and the seasonal changes in day length and time of dawn and dusk, the clock must be able to respond to environmental changes. In theory, a stimulus that affects the clock could induce four major types of changes: a change in the length of the cycle, a change in phase, a change in the amplitude of the output of the clock, or a change in the shape of the circadian output. Because direct measurements of SCN activity are difficult, changes in amplitude and shape of the output are rarely determined, and the response of the clock to stimuli is usually assessed by determination of phase-shifting effects and changes in period length.

Effects of Light on Clock Function

Light is the most potent of the stimuli that can alter the phase of the biologic clock[22,23]; its effect on the phase is determined by its intensity, duration, and timing. Although effects are most striking with bright light, even dim light has phase-shifting effects; the magnitude of the phase shift is approximately proportional to the cube root of the light intensity. Light exposure just before or during the first half of the dark phase simulates later sunsets and produces phase delays (backward resetting of the clock), whereas light during the second half of the dark phase or just after the end of the dark phase simulates earlier sunrises and produces phase advances (forward resetting of the clock) (Fig. 4–3).[24,25]

The timing of light exposure has a profound impact on the extent of the shift in phase of the pacemaker. A single exposure to bright light of several hours' duration, timed to occur just before or just after the minimum of the temperature curve, can produce shifts of the temperature rhythm and melatonin rhythm in humans by as much as 3.5 hours; in contrast, administration of similar light pulses during most of the subjective daylight period produces little shift.[26,27] Therefore, based on a maximum phase shift per day of 3.5 hours, the longest sustainable cycle length, in theory, is about 27.5 hours, and the minimum is 20.5 hours. Such cycles, however, would require exposure to intense light that was precisely timed in relation to the output of the clock.

In practice, phase shifts of this magnitude cannot be achieved, and under most circumstances, humans can maintain synchronized sleep-wake rhythm cycles of only about 23 to 26 hours. In one study, evening bright light pulses allowed 74% of subjects to entrain to a 26-hour sleep-wake schedule.[28] With shorter or longer cycle lengths, the biologic clock does not shift its phase sufficiently, and the sleep-wake pattern becomes disordered. With rapid phase shifts, such as those that accompany intercontinental jet travel, the various hormonal and other physiologic rhythms become desynchronized, and it may take several days or weeks for them to resume their normal phase relationships.

Effects of Other Exogenous Stimuli on Clock Function

Other stimuli that can have phase-shifting effects include social cues, food, activity, and ambient temperature. For example, the 24-hour activity rhythm of songbirds deprived of all other time cues can be entrained to the sound of the song of birds of the same species.[29] Rats given food only at a particular time each day develop premeal increases in body temperature, motor activity, and synthesis of digestive enzymes[30]; if they are then placed in constant light, the premeal activity occurs every 24 hours while the endogenous rest-activity adopts a free-running cycle. Ablation of the SCN eliminates the endogenous rest-activity rhythm, but the premeal activity cycle is preserved, indicating that there is an anatomically distinct pacemaker for this rhythm.[30,31]

Activity or arousal can induce phase shifts, at least in rodents, but it has a less

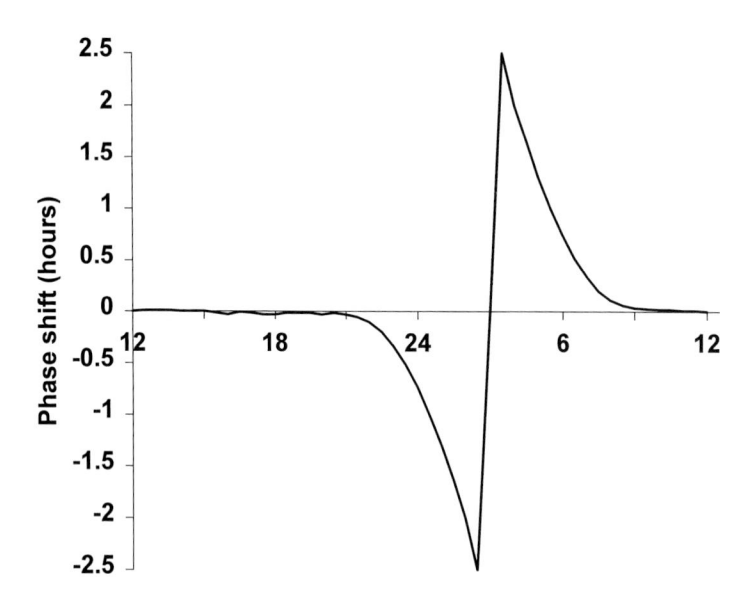

Figure 4–3. Phase-shifting effects of light. The shift of phase of the circadian pacemaker induced by light is strongly dependent on the timing of light exposure. Light exposure just before the nadir of the endogenous component of the temperature rhythm, about 3 AM, produces the greatest phase advance; light exposure just after the nadir produces the greatest phase delay. The magnitude of the shift is also dependent on the intensity and duration of light exposure. (Adapted from Takahashi and Zatz,[25] p 1106, with permission.)

potent effect than light.[31–33] Changes in environmental temperature can produce phase shifts in poikilothermic animals, but such manipulations have little effect on the circadian system in warm-blooded animals. Although social stimuli may alter circadian rhythms in birds and rodents, the ability of these stimuli to produce phase shifts directly in humans has not been demonstrated.

Endogenous Influences on Clock Function

Several endogenous factors appear to affect clock function.[34] For example, estradiol and testosterone influence activity rhythms in animals, and hormones involved in thyroid function may also affect clock function. However, the hormone with the most pronounced effects on circadian rhythmicity is melatonin.

Melatonin, an indoleamine, is the primary hormone secreted by the pineal gland. In humans, most secretion occurs at night in response to the absence of photic stimulation; nocturnal release of melatonin can be suppressed by light in a dose-dependent manner, with some attenuation even by dim light (300 lux). However, circadian rhythmicity of melatonin secretion persists even in total darkness.

The effects of light on melatonin secretion are mediated by the retinohypothalamic tract and the SCN. Efferents from the SCN synapse in the hypothalamic paraventricular nuclei; from there, axons descend via the medial forebrain bundle to the intermediolateral cell column of the spinal cord. Preganglionic sympathetic fibers then pass to the superior cervical ganglion, and postganglionic sympathetic fibers from the superior cervical ganglion travel with pineal blood vessels to the pineal gland. Melatonin release is therefore a function of sympathetic input to the pineal gland: Its secretion is increased by noradrenergic reuptake inhibitors and decreased by β-adrenergic receptor blockers, such as propranolol.[35] A major function of melatonin is to provide a mechanism that allows body changes in response to seasonal changes in the length of day and night. These seasonal changes are particularly important for reproductive physiology and behavior.

Although the effects of endogenous melatonin on the circadian system are not well defined, melatonin can induce phase shifts of SCN electrical activity.[36] When administered exogenously, it can produce phase shifts that are opposite to the effects of light: that is, evening administration of melatonin produces phase advances, whereas morning administration produces phase delays.[37,38]

ZEITGEBERS, MASKING STIMULI, AND RHYTHM INTERACTIONS

Periodic stimuli that alter the phase or the period of the circadian timing system may entrain a rhythm so that the phase and period of the rhythm are linked to the specific stimulus. When entrainment occurs, the stimulus is termed a *zeitgeber* (time giver) or a *synchronizer*. Alternatively, a stimulus may impose a rhythm without affecting the underlying oscillator. For example, a regular stressor may affect the shape of the cortisol secretion curve over the 24-hour cycle without affecting the clock that governs basal cortisol secretion. Similarly, regular meals affect the shape of the 24-hour cycle of gastric acid secretion, but they do not alter the underlying basal rhythm. The effects of such stimuli tend to *mask* the phase of the underlying oscillator.

The interrelation among various rhythmic functions may also affect the pattern of their rhythmicity. For example, although growth hormone secretion has a diurnal variation, with peak secretion occurring during the first part of the night, the pattern is largely dependent on the occurrence of slow-wave sleep (SWS). With sleep deprivation, the pulsatile secretion is markedly attenuated.

Some rhythms reflect the combined effects of the endogenous pacemaker and other rhythms. For example, the core body temperature rhythm, which has

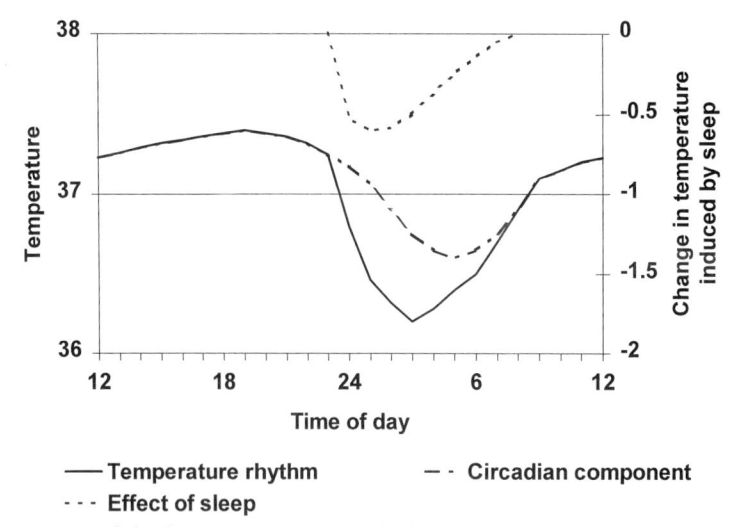

Figure 4–4. Components of the human temperature rhythm. Under usual conditions of regular nighttime sleep and daytime wakefulness, the core body temperature rhythm has a 24-hour cycle with a nadir at about 4–6 AM and a peak during the mid afternoon. This rhythm, which has about a 1.2°C peak-to-nadir amplitude, is the summation of the rhythm generated by the circadian pacemaker and the temperature change induced by sleep. Under most circumstances, these two rhythms are in phase, and each contributes about 0.6°C to the total rhythm amplitude; however, the temperature minimum of the combined rhythm occurs 1–3 hours earlier than the minimum of the circadian component alone.

about a 1.2°C peak-to-nadir amplitude, is the summation of the rhythm generated by the pacemaker and the temperature change associated with sleep (Fig. 4–4).

EFFECTS OF THE CIRCADIAN PACEMAKER ON SLEEP AND WAKEFULNESS

Although it has been clear for decades that the circadian pacemaker has a major impact on the timing of sleep and wakefulness, the specifics of its effects have been difficult to quantify because of the interactions among rhythms, masking effects, and the impracticality of measuring SCN output directly. Furthermore, sleep itself has a strong effect on subsequent sleep patterns. To separate the contribution of the circadian pacemaker from the effects of prior sleep and other periodic influences, two techniques can be used: (1) constant environmental conditions without time cues, and (2) sleep-wake schedules that are outside the range of entrainment. In addition, mathematical techniques applied to data obtained in other ways also can be used to assess the endogenous component.[39]

Sleep-Wake Rhythms Under Constant Environmental Conditions

To study circadian rhythms and sleep-wake patterns under constant environmental conditions, time cues and periodic variations in external stimuli must be absent. Kleitman performed such studies in Mammoth Cave, Kentucky (described later in this chapter), and Aschoff[29] developed the first experimental facility for time-isolation studies, a set of two underground apartments shielded from sound and light. Newer facilities provide no temporal cues concerning light, sound, temperature, or social activities, and permit physiologic monitoring for periods of 6 months or longer. In initial studies conducted in temporal isolation, most subjects had a temperature rhythm of 25 to 25.5 hours and adopted a similar sleep-wake schedule. As a result, the sleep-wake cycle and the temperature rhythm remained in a fairly constant phase relationship, although the phase relationship differed from that observed under normal conditions: for example, the nadir of the temperature rhythm tended to occur at sleep onset rather than at the midpoint of the sleep period.

In subjects living in such facilities for several weeks, the sleep-wake rhythm sometimes becomes dissociated from the temperature rhythm for a portion of the experiment—a phenomenon referred to as *spontaneous internal desynchronization*. During the period of desynchronization, sleep-wake cycles become elongated to between 30 and 50 hours, while the temperature rhythm continues with a rhythm close to 24 hours (Fig. 4–5).[40] As a result, the phase relation between the temperature rhythm and the sleep-wake cycle changes continuously.

This spontaneous internal desynchronization, which leads to variability in the phase relationship between sleep and the output of the circadian pacemaker, provides researchers the opportunity to assess the effect of the pacemaker on sleep (see Fig. 4–5). For example, the duration of the sleep episode tends to be longest with sleep initiated at the peak of the temperature cycle and shortest with sleep initiated at its nadir (Fig. 4–6).[41] In addition, subjective alertness parallels the temperature curve, and subjects rarely choose to begin sleep when the core temperature is greater than 98°F (see Fig. 4–2). The end of sleep is also determined in part by circadian phase: Subjects tend to end sleep while core temperature is rising.[42,43] On the other hand, during spontaneous internal desynchronization, subjects often choose to sleep when body temperature has reached its plateau, which corresponds to the late afternoon under usual conditions. Sleep occurring at this phase of the temperature curve may consist of a nap, or subjects may sleep for more than 12 hours, not awakening until body temperature is rising again.[44] These findings indicate that decisions about when to go to sleep, as well as sleep duration, are influenced by circadian phase.

The basis for spontaneous internal desynchronization, which occurs in only a minority of subjects in temporal isolation studies, is uncertain. One explanation is based on the presumed existence of a second pacemaker with a period of 29 to 30 hours.[45] In the model proposed by Aschoff[29] and elaborated by Kronauer and colleagues,[40,45] the pacemaker controlling the temperature rhythm is strong, whereas a second oscillator is weak and labile. The two can become dissociated under certain circumstances, such as extended free-running conditions. According to the model, the initial 25- to 25.5-hour sleep-wake rhythm observed in temporal isolation reflects the combined influence of the two systems; when they become dissociated, the temperature rhythm assumes a 24.5-hour cycle while the sleep-wake rhythm has a mean cycle length of 29 to 30 hours.

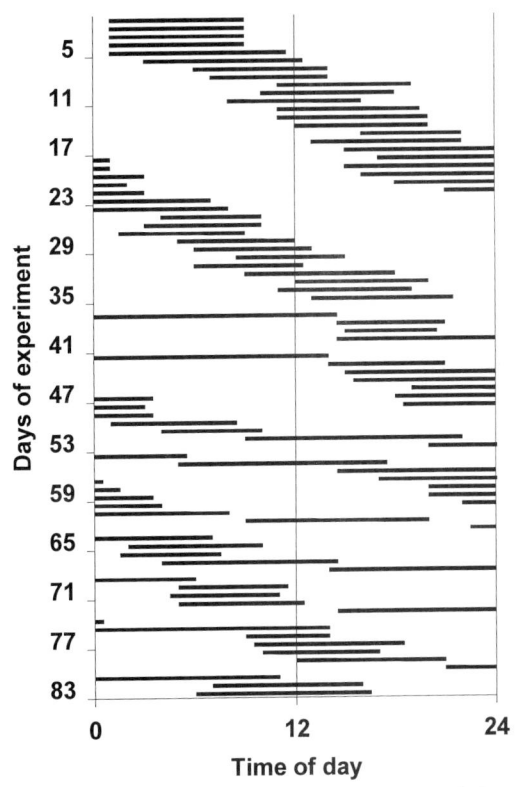

Figure 4–5. Sleep-wake rhythm in a subject living under constant environmental conditions, without time cues, for 78 days. Sleep periods are shown in black. During the first 4 days, time cues were provided and the subject maintained a 24-hour cycle. On the fifth day, time cues were removed, and the cycle length of the sleep-wake rhythm increased to approximately 25 hours. Spontaneous internal desynchronization of the sleep-wake rhythm from the temperature rhythm occurred on day 36; subsequently, the sleep-wake rhythm had an average period of about 29 hours, while the temperature rhythm had a period of about 24.5 hours. (Adapted from Kronauer et al.,[40] p 175, with permission.)

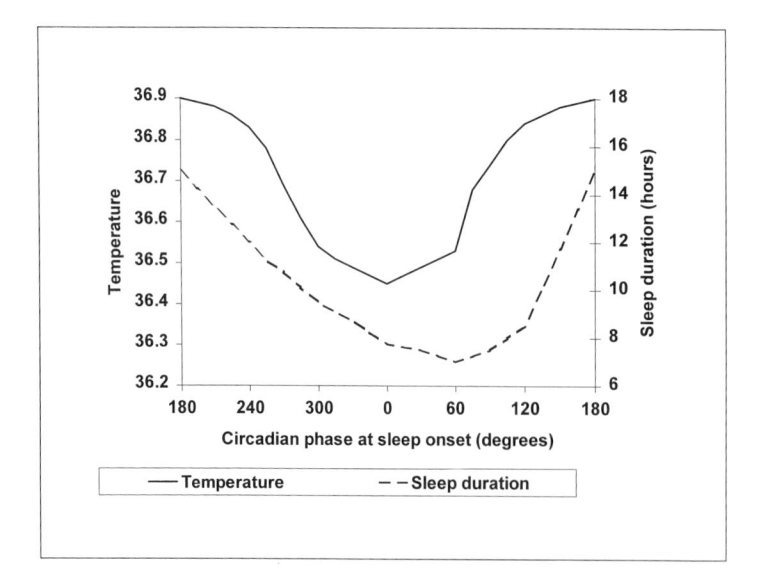

Figure 4–6. The relation of sleep duration to circadian phase at the onset of sleep during spontaneous internal desynchronization. The nadir of the temperature curve corresponds to a phase angle of 0°. Sleep episodes that begin at the peak of the temperature curve tend to be long, leading to preferential awakening during the rising phase of the temperature curve. Sleep episodes that begin at or just after the nadir of the temperature curve tend to be short, also leading to preferential awakening during the rising phase of the temperature curve. (Adapted from Czeisler and Jewett,[41] p 130, with permission.)

Although the two-pacemaker model can account for many of the findings associated with spontaneous internal desynchronization, an anatomic locus for the posited second pacemaker has not been identified. In addition, subsequent studies suggested that both the 25- to 25.5-hour day length adopted by many subjects in temporal isolation and the tendency to develop spontaneous internal desynchronization were due at least in part to instructions not to nap. When subjects placed in such a facility are instructed to sleep as desired, sleep is often biphasic: There is a major sleep period associated with the temperature nadir and a second 1-to 2-hour sleep period at the time of the temperature maximum. In such cases, the endogenous period is between 24 and 24.5 hours, and spontaneous internal desynchronization does not occur.[46,47] At present, it appears unlikely that a second endogenous pacemaker with a cycle length of 29 to 30 hours exists. Instead, current evidence suggests that timing of retinal exposure to light, the effects of processes underlying sleep homeostasis, and the choice of when to go to sleep create the appearance of a second pacemaker.[48] The interactions of sleep homeostasis with the circadian variation in alertness and sleep propensity are discussed below.

Although temporal isolation facilities provide a powerful setting for the study of circadian processes, the length and cost of experiments in such facilities limit their usefulness. The *extended constant routine*, an experimental paradigm in which stimuli that may influence the endogenous temperature rhythm are minimized or provided at a constant or near-constant level, does not require the days and weeks of recording needed for temporal isolation experiments. With one version of this technique, subjects undergo a night of sleep deprivation and then maintain a constant routine for at least one circadian cycle, during which they receive snacks at fixed intervals, remain awake but inactive in a semirecumbent position, and are exposed to a constant level of dim illumination.[49,50] The suppression of masking components induced by light, meals, sleep, and activity allows assessment of the endogenous component of the temperature rhythm, thus permitting estimation of the phase and period of the circadian pacemaker output.

Although the extended constant routine is a powerful technique, the sleep deprivation that accompanies it prevents any assessment of interactions of the circadian pacemaker with homeostatic sleep processes. In addition, the effort to stay awake varies during the course of the experiment, which may in itself induce masking effects. A variation of the extended constant routine is the ultra-short sleep-wake schedule. In one version of this protocol, 13 minutes of enforced wakefulness alternates with a 7-minute period during which sleep is allowed. Using this technique, Lavie and associates[51–53] demonstrated that the 8 to 10 pm period is associated with a substantially reduced ability to fall asleep, and they referred to this interval as the "forbidden zone" for sleep.

Sleep-Wake Rhythms Under Non–24-Hour Sleep-Wake Schedules

Although time-isolation experiments provide valuable information about the links between sleep and circadian phase, it is difficult to separate the effects on sleep of the duration of prior wakefulness from the effects of circadian phase in experiments in which sleep is allowed to occur *ad lib*. Non–24-hour sleep-wake schedules, an alternative approach, have been used experimentally for more than 50 years.

Kleitman,[54] who performed a number of experiments using non–24-hour sleep-wake schedules, attempted without success to establish 12-hour, 18-hour, and 48-hour temperature rhythms using sleep-wake periods of similar length. Kleitman was able, however, to alter temperature rhythms by using a 21-hour day and a 28-hour day. In one of his most innovative studies, Kleitman and a colleague, serving as both investigators and subjects, spent just over 1 month in 1938 on a 28-hour sleep-wake schedule in the near-constant environmental conditions of Mammoth Cave, Kentucky. One subject's temperature rhythm shifted in the direction of the new day length, while the other's remained close to 24 hours, indicating for the first time that the sleep-wake cycle could oscillate with a period separate from that of body temperature.

The importance of light's contribution to phase shifts was not well understood at the time Kleitman performed his experiments; as described above, it is now clear that the ability to entrain the circadian pacemaker and the temperature rhythm to a non–24-hour schedule is highly dependent on the timing and intensity of light exposure. When phase-shifting light pulses are not administered, the 28-hour day and the 21-hour day lead to forced desynchrony between the temperature rhythm and the sleep-wake cycle.

The forced desynchrony that occurs with 28-hour schedules allows investigators to evaluate circadian contributions to sleep and wakefulness while minimizing the contribution of prior wakefulness to sleep episodes. Dijk and Czeisler[55,56] placed eight subjects, who lived in a time-free environment for 33 to 36 days, on a 28-hour schedule of 18.67 hours of enforced wakefulness and 9.33 hours available for sleep. With this schedule, sleep periods occurred at all phases of the circadian cycle, whereas variations in prior wakefulness were minimized. This experiment yielded the surprising result that the circadian drive for sleep was *low* at bedtime and was significantly higher at about the usual time for awakening (Fig. 4–7).

How can this result be understood? The answer lies in *process S*, the hypothetical construct representing the homeostatic drive for sleep (see Chap. 1) and its interaction with the circadian rhythm of propensity for sleep and wakefulness (Fig. 4–8).[57] Process S, which accumulates during wakefulness and is discharged during sleep, causes a monotonic increase in sleep tendency during wakefulness. Under usual conditions, the circadian process, *process C*, produces an increase in propensity for wakefulness during much of the day, which counteracts the increase in sleepiness caused by process S. As a result, process C helps to maintain wakefulness throughout the day. During the night, as process S discharges, the circadian propensity for wakefulness declines, which helps to maintain sleep. These findings lead to the conclusion that a major sleep-related role of the circadian pacemaker is to facilitate consolidation of sleep and wakefulness and that the propensity to discontinue sleep is a function of circadian phase and the amount of sleep obtained.[55]

The experiment by Dijk and Czeisler and other experiments in time-free environments also reveal striking differences

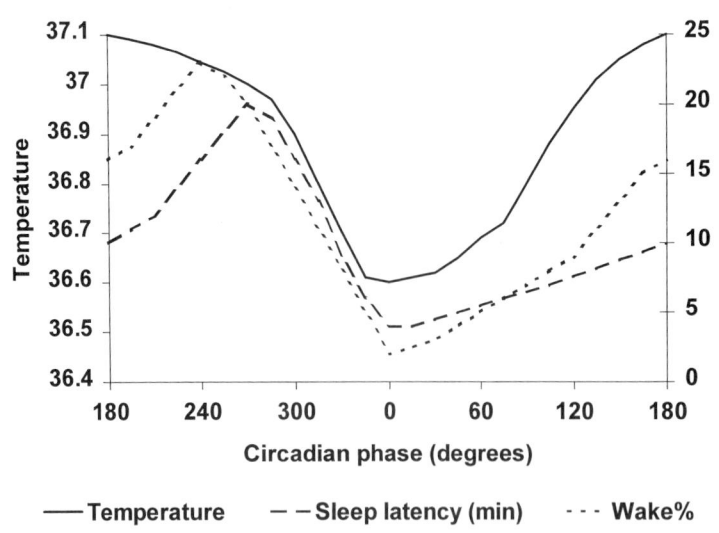

Figure 4–7. The relationship among circadian phase, sleep latency, and the proportion of wakefulness during the sleep episode in subjects on a 28-hour schedule of 18.67 hours of wakefulness and 9.33 hours allowed for sleep. The nadir of the temperature curve corresponds to a phase angle of 0°, or about 3 AM for subjects who usually sleep from 11 PM to 7 AM. In this experiment, the duration of prior wakefulness was relatively constant at all phases of the temperature rhythm. The circadian contribution to sleep latency rises during the rising phase of the temperature curve; at 300° (corresponding to 11 PM under usual conditions), the sleep latency is higher than it is during much of the usual waking period, indicating a tendency of the circadian process to maintain and consolidate wakefulness. Similarly, the propensity for wakefulness is at a low level at a phase angle of 60° (corresponding to 7 AM under usual conditions), indicating that the circadian process helps to maintain and consolidate sleep during the usual sleep period. (Adapted from Dijk and Czeisler,[55] pp 65, 66, with permission.)

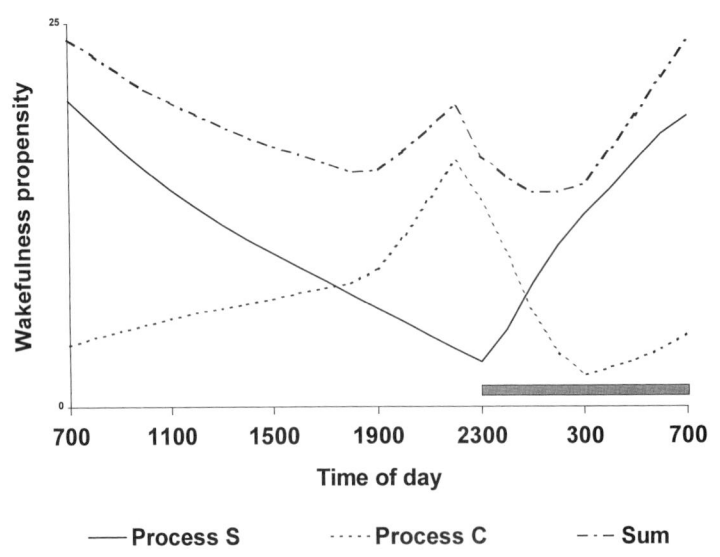

Figure 4–8. Model of process S, the homeostatic sleep process, and process C, the circadian sleep-wake tendency, and their influence on wakefulness propensity. Sleep, represented by the gray bar, occurs from 2300 to 0700 hours. Process S accumulates during wakefulness and is discharged during sleep. Process C produces an increase in the propensity for wakefulness during much of the day, which counteracts the increase in sleepiness caused by process S. During the night, as process S discharges, the circadian propensity for wakefulness declines, which helps to maintain sleep. The sum of the two processes approximates sleep latencies across the day and night, as demonstrated in Figure 4–1. Based in part on data from Dijk and Czeisler[55,56] and on models described by Borbély.[57]

in the relations of REM sleep and SWS to circadian phase. The propensity for REM sleep shows a clear relation to circadian phase, with sleep-onset REM periods much more likely to occur during the rising phase of the temperature curve, whereas SWS propensity shows little correlation with circadian phase after controlling for effects of prior wakefulness. Furthermore, the propensity for REM sleep increases with increasing time asleep, independent of circadian phase. On the other hand, the propensity for SWS increases with increasing time awake and decreases with increasing time asleep, suggesting that the delta waves of SWS correlate closely with the discharge of process S.

The findings from these studies suggest that 8 hours of continuous or nearly continuous sleep can be obtained only within a narrow range of sleep-onset times. If one goes to bed too early, the circadian propensity for wakefulness is at a high level and process S may not have accumulated sufficiently to allow rapid sleep onset. If one goes to bed too late, the circadian propensity for wakefulness rises before 8 hours of sleep are obtained, leading to early awakening and poor sleep thereafter.

In addition to the effects of circadian variation in wake propensity and sleep homeostasis, sleep patterns may be affected by rhythms that have more than one cycle per day, or *ultradian rhythms*. Under conditions of enforced bed rest with encouragement to nap, normal subjects take three or four naps during a 24-hour day, and the onset of sleep has an ultradian rhythm of 4 to 6 hours.[58,59] Using the ultrashort sleep-wake schedule described previously, Lavie and coworkers[52,53] demonstrated a 90-minute rhythmic variation in sleep ability.

SUMMARY

The ability to regulate physiologic functions rhythmically is a phylogenetically ancient adaptation that allows organisms to make physiologic changes in response to environmental changes. Hundreds of human physiologic functions show circadian rhythmicity, including basic intracellular processes, enzyme activities, hormonal secretions, core body temperature, and sleep and wakefulness. The SCN, located in the anterior hypothalamus, is the primary locus of the biologic clock; its projections to other areas of the hypothalamus, to the thalamus, and to the forebrain provide the output that drives circadian fluctuations in physiologic functions. Isolated cells of the SCN demonstrate circadian rhythmicity, indicating that the biologic clock functions at the cellular level. Some rhythms, such as core body temperature, are tightly coupled to the intrinsic circadian pacemaker; others, such as the sleep-wake rhythm, are loosely coupled and may become dissociated from the output of the circadian pacemaker.

The function of the SCN is strongly influenced by light, which can shift the phase of the pacemaker. The effect of light is a function of its intensity and duration as well as its timing: Light pulses administered close to the nadir of the core temperature rhythm, close to 3 am under usual conditions, produce marked phase shifts, whereas light administered at the apex, in the mid afternoon, has virtually no effect on phase. Melatonin appears to have phase-shifting properties that are opposite to those of light.

Experimental techniques that uncouple the sleep-wake rhythm from the output of the circadian pacemaker, such as the use of time-free environments, non–24-hour sleep-wake schedules, and extended constant routines, provide a means to assess the effects of circadian phase on sleep and wakefulness. Studies using these techniques indicate that the propensity for sleep and sleepiness appears to be governed principally by a homeostatic drive for sleep and a circadian variation in sleep propensity. The homeostatic drive for sleep, often called process S, increases during wakefulness and discharges during sleep; the theoretical amplitude correlates closely with the slow-wave activity of NREM sleep. The circadian component of sleep propensity is highest at and just after the nadir of the temperature rhythm, corresponding to the end of the usual sleep period. The circadian propensity for wakefulness is highest at the end of the plateau of the temperature curve, corresponding to the end of the usual period of wakefulness. Thus, the circadian pacemaker facilitates consolidation of sleep and wakefulness, whereas the homeostatic process facilitates changes from sleep to wakefulness and wakefulness to sleep.

REFERENCES

1. Moore-Ede, MC, et al: Circadian timekeeping in health and disease: Part I. Basic properties of circadian pacemakers. N Engl J Med 309:469, 1983.
2. De Mairan, JJD'O: Observation Botanique. Histoire de L'Academie Royale des Sciences, Paris, 1729.
3. Burckard, E, and Kayser, C: L'inversion du rythme nycthéméral de la température chez l'homme. C R Soc Biol 141:1265, 1947.
4. Richter, CP: Sleep and activity: Their relation to the 24 hour clock. In Kety, S, et al (eds): Sleep and Altered States of Consciousness. Williams and Wilkins, Baltimore, 1967, pp 8–28.
5. Moore, RY, and Lenn, NJ: A retinohypothalamic projection in the rat. J Comp Neurol 146:1, 1972.
6. Halberg, F: Implications of biologic rhythms for clinical practice. Hosp Pract 12:139, 1977.
7. Roth, T, et al: Daytime sleepiness and alertness. In Kryger, MH, et al (eds): Principles and Practice of Sleep Medicine, ed 2. WB Saunders, Philadelphia, 1994, pp 40–49.
8. Carskadon, MA, and Dement, WC: Daytime sleepiness: Quantification of a behavioral state. Neurosci Biobehav Rev 11:307, 1987.
9. Carskadon, MA, and Dement, WC: Multiple sleep latency tests during the constant routine. Sleep 15:396, 1992.
10. Monk, TH: Circadian rhythms in subjective activation, mood, and performance efficiency. In Kryger, MH, et al (eds): Principles and Practice of Sleep Medicine, ed 2. WB Saunders, Philadelphia, 1994, pp 321–330.
11. Fröberg, JE: Twenty-four-hour patterns in human performance, subjective and physiological variables and differences between morning and evening active subjects. Biol Psychol 5:119, 1977.
12. Rusak, B, and Bina, KG: Neurotransmitters in the mammalian circadian system. Ann Rev Neurosci 13:387, 1990
13. Ralph, MS, et al: Transplanted suprachiasmatic nucleus determines circadian period. Science 247:975, 1990.
14. Rusak, B, and Zucker, I: Neural regulation of circadian rhythms. Physiol Rev 59:449, 1979.
15. King, DP, et al: Positional cloning of the mouse circadian clock gene. Cell 89:641, 1997.
16. Cahill, GM, and Besharse, JC: Circadian rhythmicity in vertebrate retinas: regulation by a photoreceptor oscillator. Prog Retinal Eye Res 14: 267, 1995.
17. Tosini, G, and Menaker, M: Circadian rhythms in cultured mammalian retina. Science 272:419, 1996.
18. Harrington, ME, et al: Anatomy and physiology of the mammalian circadian system. In Kryger, MH, et al (eds): Principles and Practice of Sleep Medicine, ed 2. WB Saunders, Philadelphia, 1994, pp 286–300.
19. Schwartz, WJ: Understanding circadian clocks: From c-Fos to fly balls. Ann Neurol 41:289, 1997.
20. Inouye, ST, et al: Inhibitor of protein synthesis phase shifts a circadian pacemaker in mammalian SCN. Am J Physiol 255:R1055, 1988.
21. Sehgal, A: Genetic dissection of the circadian clock: A timeless story. Semin Neurosci 7:27, 1995.
22. Czeisler, CA, et al: Bright light induction of strong (type 0) resetting of the human circadian pacemaker. Science 244:1328, 1989.
23. Minors, D, et al: Sleep and circadian rhythms of temperature and urinary excretion on a 22.8 hr "day." Chronobiol Int 5:65, 1988.
24. Drennan, M, et al: Bright light can delay human temperature rhythm independent of sleep. Am J Physiol 257:R136, 1989.
25. Takahashi, JS, and Zatz, M: Regulation of circadian rhythmicity. Science 217:1104, 1982.
26. Buresova, M, et al: Early morning bright light phase advances the human circadian pacemaker within one day. Neurosci Lett 121:47, 1991.
27. Van Cauter, E, et al: Preliminary studies on the immediate phase-shifting effects of light and exercise on the human circadian clock. J Biol Rhythms 8:S99, 1993.
28. Eastman, CI, and Miescke, K-J: Entrainment of circadian rhythms with 26-h bright light and sleep-wake schedules. Am J Physiol 259:R1189, 1990.
29. Aschoff, J: Circadian systems in man and their implications. Hosp Pract 11:51, 1976.
30. Boulos, Z, et al: Feeding schedules and the circadian organization of behavior in the rat. Behav Brain Res 1:39, 1980.
31. Mistlberger, RE, and Rusak, B: Circadian rhythms in mammals: Formal properties and environmental influences. In Kryger, MH, et al (eds): Principles and Practice of Sleep Medicine, ed 2. WB Saunders, Philadelphia, 1994, pp 286–300.
32. Edgar, DM, et al: Triazolam fails to induce sleep in suprachiasmatic nucleus-lesioned rats. Neurosci Lett 125:125, 1991.
33. Edgar, DM, et al: Influence of running wheel activity on free-running sleep/wake and drinking circadian rhythms in mice. Physiol Behav 50: 373, 1991.
34. Wever, RA: The Circadian System in Man: Results of Experiments Under Temporal Isolation. Springer-Verlag, New York, 1979.
35. Brown, GM: Melatonin in psychiatric and sleep disorders: Therapeutic implications. CNS Drugs 3:209, 1995.
36. McArthur, AJ, et al: Melatonin directly resets the rat suprachiasmatic circadian clock in vitro. Brain Res 565:158, 1991.
37. Arendt, J: Clinical perspectives for melatonin and its agonists. Biol Psychiatry 35:1, 1994.

38. Lewy, A, et al: Melatonin shifts human circadian rhythms according to a phase-response curve. Chronobiol Internat 9:380, 1992.
39. Minors, DS, and Waterhouse, JM: Investigating the endogenous component of human circadian rhythms: A review of some simple alternatives to constant routines. Chronobiol Int 9:55, 1992.
40. Kronauer, RE, et al: Mathematical representation of the human circadian system: Two interacting oscillators which affect sleep. In Chase, MH, and Weitzman, ED (eds): Sleep Disorders: Basic and Clinical Research. Advances in Sleep Research, Vol. 8. Spectrum, New York, 1983, pp 173–194.
41. Czeisler, CA, and Jewett, ME: Human circadian physiology: Interaction of the behavioral rest-activity cycle with the output of the endogenous circadian pacemaker. In Thorpy, MJ (ed): Handbook of Sleep Disorders. Marcel Dekker, New York, 1990, pp 117–137.
42. Strogatz, SH, et al: Circadian regulation dominates homeostatic control of sleep length and prior wake length in humans. Sleep 9:353, 1986.
43. Strogatz, SH, et al: Circadian pacemaker interferes with sleep onset at specific times each day: Role in insomnia. Am J Physiol 253:R172, 1987.
44. Campbell, SS, and Zulley, J: Napping in time-free environments. In Dinges, DF, and Broughton, RJ (eds): Sleep and Alertness: Chronobiological, Behavioral, and Medical Aspects of Napping. Raven Press, New York, 1989, pp 121–138.
45. Kronauer, RE, et al: Mathematical model of the human circadian system with two interacting oscillators. Am J Physiol 242:R3, 1982.
46. Murphy, PJ, and Campbell, SS: Physiology of the circadian system in animals and humans. J Clin Neurophysiol 13:2, 1996.
47. Campbell, SS, et al: When the human circadian system is caught napping: Evidence for endogenous rhythms close to 24 hours. Sleep 16:638, 1993.
48. Borbély, AA, and Achermann, P: Concepts and models of sleep regulation: An overview. J Sleep Res 1:63, 1992.
49. Mills, JN, et al: Adaptation to abrupt time zone shifts of the oscillators controlling human circadian rhythms. J Physiol (London) 285:455, 1978.
50. Czeisler, C, et al: A clinical method to assess the endogenous circadian phase (ECP) of the deep circadian oscillator in man. Sleep Res 14:295, 1985.
51. Lavie, P: Ultrashort sleep-waking schedule: III. "Gates" and "forbidden zones" for sleep. Electroencephalogr Clin Neurophysiol 63:414, 1986.
52. Lavie, P, and Scherson, A: Ultrashort sleep-waking schedule: I. Evidence of ultradian rhythmicity in "sleepability." Electroencephalogr Clin Neurophysiol 52:163, 1981.
53. Lavie, P, and Zomer, J: Ultrashort sleep-waking schedule: II. Relationship between ultradian rhythms in sleepability and the REM-non-REM cycles and effects of the circadian phase. Electroencephalogr Clin Neurophysiol 57:35, 1984.
54. Kleitman, N: Sleep and Wakefulness, ed 2. University of Chicago Press, Chicago, 1963, pp 172–184.
55. Dijk, DJ, and Czeisler, CA: Paradoxical timing of the circadian rhythm of sleep propensity serves to consolidate sleep and wakefulness in humans. Neurosci Lett 166:63, 1994.
56. Dijk, DJ, and Czeisler, CA: Contribution of the circadian pacemaker and the sleep homeostat to sleep propensity, sleep structure, electroencephalographic slow waves, and sleep spindle activity in humans. J Neurosci 15:3526, 1995.
57. Borbély, AA: Sleep homeostasis and models of sleep regulation. In Kryger, MH, et al (eds): Principles and Practice of Sleep Medicine, ed 2. WB Saunders, Philadelphia, 1994, pp 309–320.
58. Campbell, SS: Duration and placement of sleep in a "disentrained" environment. Psychophysiology 21:106, 1984.
59. Zulley, J, and Campbell, SS: Napping behavior during "spontaneous internal desynchronization": Sleep remains in synchrony with body temperature. Hum Neurobiol 4:123, 1985.

Chapter 5

ONTOGENY OF SLEEP

DIFFERENTIATION OF STATES OF
  SLEEP AND WAKEFULNESS
27 to 30 Weeks Conceptional Age
30 to 33 Weeks Conceptional Age
33 to 37 Weeks Conceptional Age
37 to 40 Weeks Conceptional Age
EVOLUTION OF SLEEP-WAKE CYCLES
  AND ELECTROENCEPHALOGRAPHIC
  PATTERNS AFTER BIRTH
Term to 12 Months
The First Decade
Adolescence
Adulthood
Late Adulthood
AGE-RELATED CHANGES
  IN RESPIRATION
Term and Preterm Infants
Childhood
Adulthood
MOVEMENTS DURING SLEEP
  OVER THE LIFE SPAN

Although sleep researchers have been interested for centuries in the development of the sleep state and its differentiation from wakefulness, little information about sleep patterns early in life could be obtained until the mid 20th century. Before that time, few tools were available to study prenatal sleep, and infants born 2 or 3 months prematurely rarely survived. After the discovery of the electroencephalogram (EEG), however, Lindsley,[1] using monitors placed on the maternal abdomen, recorded fetal cardiac and cerebral electrical activity in 1942, and in 1951 Hughes and colleagues[2] performed systematic EEG studies in infants born prematurely.

After the discovery of rapid eye movement (REM) sleep, several investigators noted that REM and non–rapid eye movement (NREM) sleep (at this age, also called active sleep and quiet sleep, respectively) could be differentiated by 30 weeks conceptional age (CA).[3] In these early studies, the inability to differentiate sleep states before 30 weeks CA was probably a result of the hypoxic brain injury associated with prematurity. With assisted ventilation and other improvements in the care of premature infants, infants of 26 to 30 weeks CA are now more likely to survive with little or no hypoxic injury, and state differentiation in relatively healthy premature infants can be detected as early as 27 weeks CA.

Patterns of sleep and wakefulness change rapidly during the prenatal period and continue to change during the first few years of postnatal life. Basic sleep patterns then remain relatively stable for several decades before changes again occur in late adulthood.

DIFFERENTIATION OF STATES OF SLEEP AND WAKEFULNESS

In the measures used to determine sleep-wake state—body and eye movements, respiratory patterns, and brain electrical activity—differentiating features develop at different rates. Although body movements and brainstem electrical activity are detectable as early as 10 weeks CA, and hemispheric brain electrical activity can be identified at 17 weeks gestational age, no significant rhythmicity is present before 20 weeks.

Rhythmic cycling of body movements begins at 20 to 24 weeks with intermittent periods of activity separated by quiet periods.[4] The period of the rest-activity cycle at this age is about 40 to 60 minutes. At 26 weeks CA, the EEG consists of bursts of high-voltage slow waves and runs of low-voltage (8- to 14-Hz) activity separated by 20- to 30-second intervals of nearly isoelectric background. Activity often occurs independently over the two hemispheres, a pattern referred to as *hemispheric asynchrony*. The discontinuous pattern—sometimes called *tracé discontinu*—is accompanied by irregular respirations and occasional eye movements.

EEG patterns and eye movements at this age permits differentiation of *quiet sleep* (QS) from *active sleep* (AS) during portions of recordings.[5,6] The EEG is discontinuous, and eye movements are rare during QS, whereas during AS the EEG shows continuous delta or theta-delta activity. Ultrasonographic monitoring studies with analysis of body movements, breathing movements, and REMs of 28- to 30-week-old fetuses in utero also suggest that there are stable periods of QS and AS at this age.[7] The terms *indeterminate sleep* and *transitional sleep* are used to describe periods of sleep that cannot be classified as either AS or QS.

27 to 30 Weeks Conceptional Age

Between 27 and 30 weeks CA, hemispheric asynchrony remains common and the EEG is usually discontinuous, although with progressively shorter interburst intervals. Temporal sharp waves are prominent EEG features at this age, and delta waves with superimposed 14- to 24-Hz activity—called *delta brushes*—begin to appear with a posterior predominance.

Although the states of sleep and wakefulness are not fully differentiated at 27 weeks CA, analysis of the combination of

30 to 33 Weeks Conceptional Age

At 30 to 33 weeks CA, the EEG is nearly continuous during AS, with a low-voltage, mixed-frequency pattern; however, during QS, the EEG remains discontinuous. Delta brushes and frontal and temporal sharp waves remain prominent EEG features.

As QS and AS become more distinct, the amount of indeterminate sleep decreases over the next several weeks (Fig. 5–1).[5] With the reduction in indeterminate sleep, the NREM-REM cycle becomes

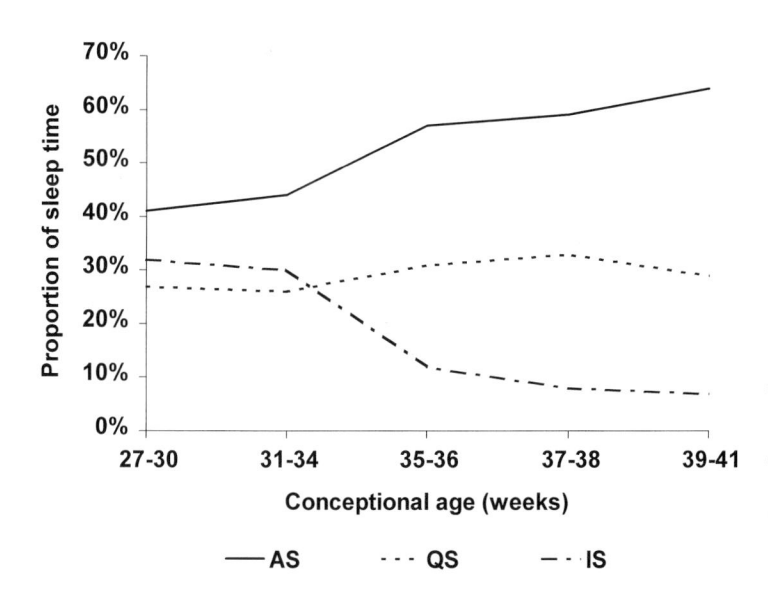

Figure 5–1. Differentiation of sleep stages during prenatal life. The amount of active sleep (AS) increases while the amount of indeterminate sleep (IS) decreases. The amount of quiet sleep (QS) remains fairly constant. (Adapted from Curzi-Dascalova et al.,[5] and Curzi-Dascalova[6], p. 154, with permission.)

more evident; its duration is about 45 minutes at 34 weeks CA and increases to 60 to 70 minutes at term. State differences in cardiac and respiratory rhythms are also more apparent, with more regular rhythms during QS than during AS. At about 33 to 34 weeks, a difference in muscle tone between AS and QS is detectable, with a decrease in muscle tone occurring during AS.[8]

33 to 37 Weeks Conceptional Age

Between 33 and 37 weeks CA, additional features of AS are evident, including REMs, smiles, grimaces, and other facial and body twitches. In addition, a third state begins to emerge in prematurely born infants; it is similar to AS, except that the eyes are open and muscle atonia is not consistently present. This third state, which develops eventually into the waking state, has EEG features similar to AS and is most easily differentiated from the sleep states by the open eyes. It is uncertain whether a similar "waking" state occurs in a fetus of similar CA.

37 to 40 Weeks Conceptional Age

By 37 weeks CA, AS is well developed, having the following features: low-to-moderate-voltage, continuous EEG; REMs; irregular breathing patterns; muscle atonia; and phasic twitches of the face and extremities. During QS, respirations are regular, eye movements are sparse or absent, and body movements are few; the discontinuous EEG pattern seen at earlier ages evolves to an alternating pattern (*tracé alternant*) in which 1- to 10-second bursts of moderate-to-high-voltage, mixed-frequency activity alternate with 6- to 10-second periods of low-voltage, mixed-frequency activity. Delta brushes become less frequent and gradually disappear between 37 and 42 weeks. In addition, periods of more continuous slow-wave activity begin to appear during QS at about 37 weeks CA and become more prevalent with increasing age.

EVOLUTION OF SLEEP-WAKE CYCLES AND ELECTROENCEPHALOGRAPHIC PATTERNS AFTER BIRTH

Term to 12 Months

Although the term infant spends about one-third of its existence in each of the three states—wakefulness, NREM sleep, and REM sleep—the proportions change rapidly during the first year of life. Total sleep time declines to 13 hours at 12 months, almost entirely as a result of a decline in REM sleep from 7 to 8 hours at birth, to 6 hours at 6 months, and then to 4 to 5 hours at 12 months.

The decline in the amount of REM sleep during the first year of life is accompanied by a change in its timing. Sleep-onset REM periods, which occur with 65% of sleep episodes at age 3 weeks, become less frequent with age, occurring with 20% of sleep episodes at 6 months and only occasionally after the age of 12 months.[6,9] As the frequency of sleep-onset REM periods declines, NREM sleep becomes more prominent in the first half of the night, while REM sleep begins to predominate in the second half. The REM-NREM cycle length remains at about 60 to 70 minutes during the first year.

DEVELOPMENT OF CIRCADIAN SLEEP-WAKE RHYTHMS

Although a 24-hour rhythm of sleep and wakefulness is essentially nonexistent in utero, premature infants of 37 weeks CA tend to spend more time awake during the day than at night.[10] Such rhythms, however, are probably a result of timing of light exposure, feeding schedules, and other social cues, rather than an indication of a functioning internal clock.[11] In most infants at term, although more sleep occurs between 8 PM and 8 AM than between 8 AM and 8 PM,[12] the 24-hour sleep-wake rhythm is weak, and a term neonate may spend 4 hours awake during the day and 2 to 3 hours awake at night.

Kleitman and Engelmann[13] studied the sleep-wake patterns of an infant "[whose] parents were sufficiently indulging to permit her to set her own sleep-wakefulness patterns and feeding schedule." Data from this infant showed a circadian periodicity that began to emerge at 4 to 6 weeks of age, suggesting that clock function is essentially nonexistent before this age and that periodicity in the neonate is largely a result of external factors. During the first 3 months of life, the internal clock probably begins to function in tandem with the development of the core body temperature rhythm, and a circadian sleep-wake rhythm becomes more clearly established.[12,14] Fortunately for their parents, by the time most infants are 3 months of age the major sleep period occurs at night.

NAPS AND SLEEP DURATION

As the circadian rhythm of wakefulness and sleep becomes stronger, the average duration of the major sleep period increases from about 3½ hours at age 3 weeks to 6 hours at age 6 months.[15] However, considerable inter-individual variability exists, partly related to parenting practices, and some infants sleep 6 to 7 hours at night in the first week or two of life.

The initial inability of infants to sleep for 8 hours continuously is of particular significance for parents, who may have to get up at night to feed or care for the infant. However, parental reports of nighttime awakenings indicate the number of times that the parents' sleep is disturbed rather than the number of actual awakenings. Parental observations are not always accurate, as infants sometimes may lie awake quietly in their cribs without awakening the parents, and many infants return to sleep on their own. The degree of parental sleep disruption is a function of cultural practices, parenting styles, and individual differences in infants. In Western countries, nighttime awakenings requiring parental intervention still occur once or twice per night in the majority of infants, but by 6 months of age, about half of all infants sleep without disturbing their parents.[16] Nighttime awakenings become somewhat more frequent in the second half of the first year, and about 10% of infants awaken three or more times.

During the first few weeks of life, most infants nap several times during the day. As the amount of nighttime sleep increases, the number of daytime naps and the amount of daytime sleep decrease (Fig. 5–2).[13,17,18] By age 7 to 8 months, most infants take two naps, one in the late morning and one in the mid afternoon.[16,18]

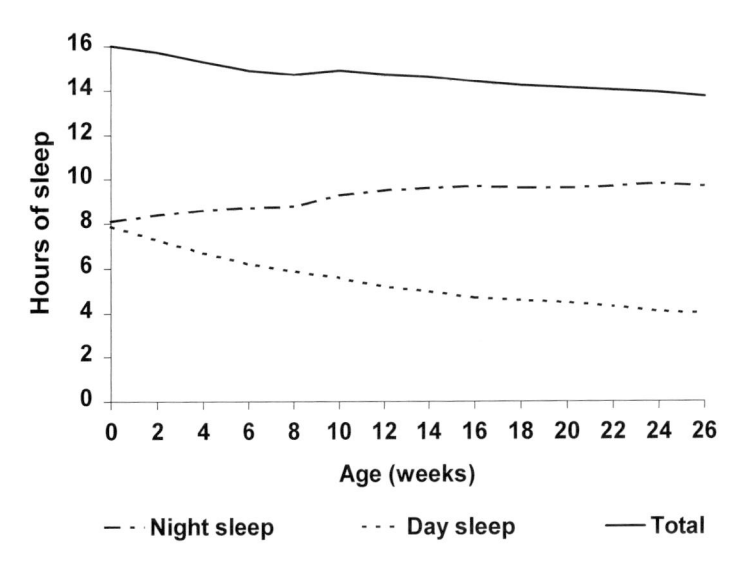

Figure 5–2. Hours of daytime sleep, nighttime sleep, and total sleep during the first 6 months of life. Nighttime sleep increases by about 1.5–2 hours over the 6-month period, while daytime sleep is reduced by half. (Adapted from Webb,[17] p 32, with permission, with data from Kleitman and Engelmann[13] and from Parmalee et al.[16])

ELECTROENCEPHALOGRAPHIC CHANGES

The EEG during wakefulness changes markedly during the first year of life. At birth, the EEG shows mixed frequencies without a clearly dominant posterior rhythm. Over the subsequent weeks, the waking posterior rhythm becomes apparent at 3 to 4 Hz and is well established at 3 months. The frequency, which eventually develops into the alpha rhythm, increases gradually to 5 Hz at 5 months, 6 Hz at 1 year, 7 Hz at 2 years, 8 Hz at 3 years, and 9 to 10 Hz at 6 to 10 years.

In the weeks after birth, the EEG also changes during NREM sleep. At birth, as described above, periods of continuous slow activity and periods of *tracé alternant* are apparent. During the first month of life, the *tracé alternant* pattern becomes less common, usually disappearing by age 1 month; it is replaced entirely by continuous slow-wave activity, which eventually develops into the slow-wave sleep (SWS) pattern characteristic of stage 3–4 sleep. Sleep spindles become apparent at about 6 to 8 weeks, although harbingers of this activity may be seen as early as term. By age 3 to 6 months, sleep spindles appear in well-developed trains lasting for several seconds. Spindles are often asynchronous during the first year, but by age 1 year about 70% of spindles are synchronous over the two hemispheres, and by age 2 virtually all spindles are synchronous.

Vertex waves and K-complexes also become evident during NREM sleep between 3 and 6 months of age; as they begin to appear, stage 1 and stage 2 sleep can be differentiated from stage 3–4 sleep. By 6 months of age, the distinction is clear. The amplitude of the slow waves also increases, and as a result, the amount of SWS increases during the second half of the first year.[9] Unlike the EEG during NREM sleep and wakefulness, the EEG during REM sleep changes little during the first year, and the REM-NREM cycle duration remains about 60 to 70 minutes.

One additional prominent EEG pattern that develops in the first year is a rhythmic, moderate-to-high-amplitude (4- to 5-Hz) activity called *hypnagogic hypersynchrony*. The rhythm, first apparent at about age 5 to 6 months, is present mainly during drowsiness and during brief arousals. By age 1 year, runs of this activity may almost entirely replace other rhythms for 5 to 20 seconds at a time (Fig. 5–3).

The First Decade

By the end of the first year of life, the major EEG patterns of NREM sleep—spin-

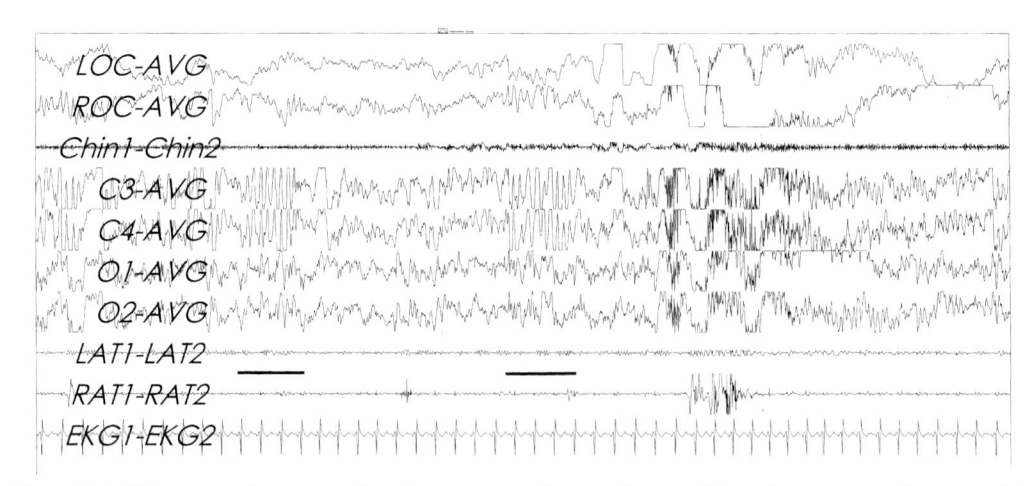

Figure 5–3. Thirty-second segment of a polysomnogram from an 8-year-old boy demonstrating hypnagogic hypersynchrony. Two prominent runs of high-amplitude, synchronous, 4- to 5-Hz activity are underlined. The EEG electrodes (A1, A2, C3, C4, O1, O2) were placed in accordance with the International 10-20 system. Abbreviations: LOC, left outer canthus; ROC, right outer canthus; AVG, average reference; Chin1–Chin2, submental EMG; LAT1–LAT2 and RAT1–RAT2, left and right anterior tibialis surface EMG; EKG1–EKG2, electrocardiogram.

dles, K-complexes, and delta waves—are well established, and the waking EEG shows a well-developed, posterior-dominant rhythm with intermixed slower frequencies. The amount of intermixed slower activity decreases gradually throughout the first and second decades, although individual delta waves over the occipital region, referred to as *posterior slow waves of youth*, are prominent throughout the first decade.

Although substantial inter-individual differences exist, the total amount of sleep continues to decline from 13 hours at 1 year to 12 hours at 2 years, 10 to 11 hours at 4 years, and 9 to 10 hours at 10 years. The amount of REM sleep also continues to decline from 4 to 5 hours out of 24 at age 12 months to 2 to 2½ hours at age 5 years, at which time it has reached the adult proportion of 20% to 25% of total sleep time. Between ages 3 and 9, boys on the average sleep more than girls. The NREM-REM cycle time increases to the adult length of 80 to 90 minutes by age 5 years.

SLOW-WAVE SLEEP

One of the most striking features of sleep during the first decade is the evolution of stage 3–4 sleep. After its initial appearance at age 3 to 5 months, stage 3–4 sleep increases to about 50% of total NREM sleep by about 1 year and remains at a high level for the next several years. During SWS, which may consist of virtually continuous high-amplitude slow waves during these years (Fig. 5–4), children are extremely difficult to awaken and may not be aroused with auditory stimuli of up to 123 dB.[19] The amount of delta sleep begins to decline after the first few years of life, and by age 9 years, delta sleep makes up about 22% to 28% of sleep, slightly more in boys than in girls.[20,21]

NAPS

Although two naps each day is the typical pattern of daytime sleep between the ages of about 7 months and 12 months, the number of naps each day during the second year of life may be one or two. In one series, at age 2 about 25% of children took two naps each day, 68% took one nap each day, and about 8% had irregular nap patterns.[22] Between ages 2 and 3, about 25% of children stop taking naps entirely, and another 25% take them only on some days (Fig. 5–5). The proportion of children who take daily naps continues to decline over the next few years of life, so that by age 6, most children have stopped taking regular naps.

After the first year, nighttime sleep patterns gradually become more regular, with fewer awakenings, although 45% of 2-year-olds still awaken at least once each night (see Fig. 5–5). By age 6, most children are excellent sleepers who fall asleep quickly and sleep soundly with few awakenings.[21] In these children, stage 2 sleep usually appears within a few minutes of sleep onset, and SWS appears about 7 to 10 minutes later.[23,24] Continuous nighttime sleep remains the norm for the next several years, and most 7- to 10-year-olds spend about 10 hours per night in bed with a sleep efficiency, defined as the proportion of time in bed that is spent asleep, of about 93% to 95%. During the day, preadolescents are usually highly alert: The mean sleep latency on the Multiple Sleep Latency Test (MSLT) is at least 17 to 18 minutes,[21] and many children between

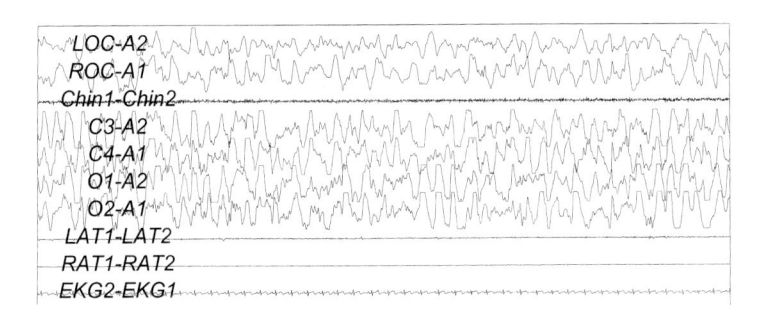

Figure 5–4. Virtually continuous high-amplitude slow waves during slow-wave sleep in a 6-year-old child. For abbreviations, see Figure 1–2.

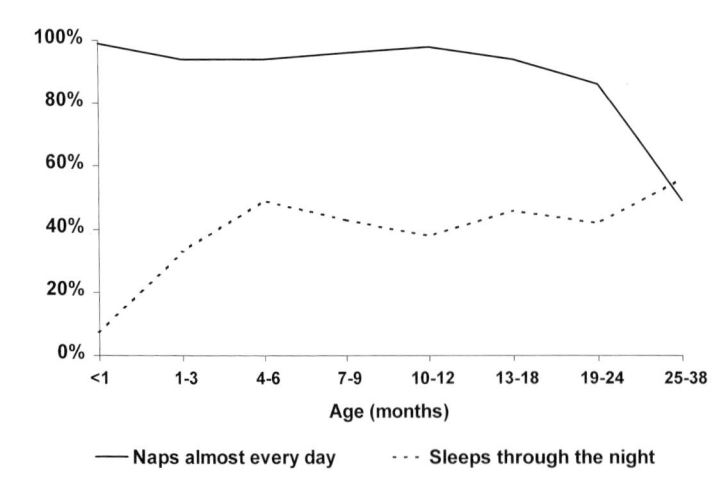

Figure 5–5. Proportion of children who sleep through the night and who take naps almost every day during the first 3 years of life. (Adapted from Armstrong et al.,[18] p. 203, with permission.)

ages 7 and 10 do not fall asleep at all on the MSLT.

Adolescence

During adolescence, the amount of sleep required for optimal alertness appears to be about the same as or slightly less than that required in the preadolescent years. For example, when required to spend 10 hours in bed in a highly structured environment, adolescents sleep slightly more than 9 hours. However, most adolescents sleep less than 9 hours per night. For 13-year-olds in the United States, mean sleep hours are 8½ hours on weeknights and 9½ on weekends. For college students aged 18 to 22 years, mean sleep hours are 7 hours on weeknights and about 8¼ hours on weekends. The discrepancy between the amount of sleep obtained and the amount required accounts for part of the increase in sleepiness that occurs during puberty.

Adulthood

Although the amount of sleep required for maximal alertness remains fairly constant throughout adult life, the quality of sleep begins to decline as early as the fourth decade. Although most middle-aged persons fall asleep without difficulty, the amount of stage 1 sleep, the number of awakenings and brief arousals, and the amount of time awake during the night all begin to increase, and the decrease in delta sleep that began in the preteen years continues. In contrast, there is little or no decline in the amount of REM sleep.

Late Adulthood

After age 65, the decline in sleep quality that begins during middle age accelerates. Although healthy older persons usually fall asleep easily at bedtime, they have frequent awakenings; in the eighth decade, most persons are awake for 10% to 30% of their time in bed at night. Some of this increase is related to an increase in time spent in bed, which rises from an average of between 7½ and 8 hours to about 8½ hours. The amount of stage 1 sleep increases, while the amount of REM sleep remains about the same.

The EEG changes somewhat after age 65. The frequency of the alpha rhythm is often slightly slower than in earlier years, and a small amount of slowing over the temporal regions during wakefulness is usually considered normal. Rhythmic delta slowing during drowsiness is also considered a normal finding at this age. Although delta waves remain present during deep NREM sleep, their amplitude declines, and many older persons no longer have enough high-amplitude delta waves to allow stage 3 and stage 4 sleep to be scored.

The common belief that older persons need less sleep is probably wrong. Daytime drowsiness and nap frequency tend to increase in older adults, and the combination of nighttime and daytime sleep often equals the midlife norm. It appears that the major reason for daytime naps and drowsiness is simply that it is more difficult for older persons to obtain continuous restful sleep at night. Nonetheless, there is marked inter-individual variability, and some older adults sleep well with few interruptions.

Some of the causes of sleep fragmentation in older people are shown in Table 5–1. The decline in the amplitude of the temperature rhythm may be caused by either reduced output of the circadian pacemaker or by diminished responsiveness of the physiological system for control of body temperature.[25] A similar process may affect the circadian variation in sleep tendency.

In most older persons, however, psychosocial factors and medical problems are more important contributors to sleep fragmentation than is neuronal degeneration. Two conditions so common that they are almost a part of normal aging—benign prostatic hypertrophy and osteoarthritis—produce sleep fragmentation by causing, respectively, (1) increased need to urinate at night, and (2) increased difficulty finding a comfortable sleeping position, which in turn causes frequent postural adjustments during the night. When the various psychological, medical, and social factors that can contribute to disrupted sleep are controlled for, the increase in sleep disturbance in older persons is much less striking, or even nonexistent.[26,27]

The circadian sleep tendency in older persons appears to shift forward; thus, sleepiness in the evening is common and falling asleep at bedtime may not be difficult, but early morning awakening is much more common. Reduced evening light exposure resulting from retinal or ocular pathology or because of less time spent outside may contribute to the apparent phase advance. Dampening of circadian processes that contribute to the maintenance of the sleep-wake cycle also may contribute.

AGE-RELATED CHANGES IN RESPIRATION

Respiratory patterns change substantially over the life span. Respiratory movements and the initial development of respiratory control systems begin during the fetal period. Fetal breathing movements, which initially occur during AS but not during QS, increase in response to an increase in Pa_{CO_2} and are altered by a variety of pharmacological agents.[28,29]

Term and Preterm Infants

Human infants born prematurely at 27 to 30 weeks CA have irregular respirations and frequent brief episodes without attempts to breathe. *Respiratory pause*, a term used to describe breathing patterns in term and preterm infants, refers to a period of 3 seconds or longer without an attempt to breathe. Periodic fluctuations in the amplitude of breathing also occur in a pattern referred to as *periodic breathing*. According to one consensus definition, periodic breathing refers to the occurrence of three or more respiratory pauses of at least 3 seconds' duration with less than 20 seconds of normal breathing between pauses.[30] Preterm infants may also have central apneas and periods of waxing and waning of respiratory amplitude or frequency (*Cheyne-Stokes pattern*).

Table 5–1. **Contributors to Sleep Disturbance in Older Adults**

Degenerative central nervous system changes
Reduced amplitude of circadian rhythms
Decreased light exposure
Inactivity and bed rest
Increased daytime sleep
Hypnotic and alcohol use
Medical illness
Psychiatric illness
Sleep apnea
Periodic limb movements
Bereavement
Retirement and social isolation; loss of time cues

Between 31 weeks CA and term, as QS and AS become more clearly differentiated, state-related differences in respiratory patterns become more apparent. Irregular breathing and breathing pauses are confined mainly to AS, whereas breathing is more regular during QS.[5,31] Although preterm infants show some variation in breathing in response to hypoxia and hypercapnia, the feedback control systems are not well developed. For example, hypoxia in preterm infants sometimes leads to ventilatory depression instead of an increased ventilatory response, and it may not induce an arousal even with oxyhemoglobin saturations as low as 70%.

Although respiratory control systems continue to develop during the third trimester, they are not fully developed at term. At that time, the activity of posterior cricoarytenoid muscles, which assist in maintaining upper airway patency, is not fully coordinated with diaphragm activity. As a consequence, the preinspiratory upper airway muscle activity that contributes to stabilization of the airway during inspiration may not occur, leading to brief obstructive apneas, particularly during AS.

At term, state-related differences in breathing patterns are clear: Respiratory rate, minute ventilation, and variability of respiratory rate and amplitude are higher in AS than QS. The greater variability in respiratory cycle length in AS compared to QS is mainly a result of variation in expiratory time and in tidal volume. In AS, tidal volume is often reduced during periods associated with prominent phasic activity. Central apneas and respiratory pauses do occur during sleep at term, more often in AS than in QS, but they occur less frequently in term infants than in preterm infants. Periodic breathing occurs mainly during AS; in normal infants, up to 5% of breathing during sleep may be periodic breathing. Occasional central apneas of up to 15 seconds' duration are probably normal if they are not associated with cyanosis, pallor, or bradycardia. In healthy term infants, periodic breathing gradually disappears during the first few weeks of life, and apneas gradually become less frequent during the first 4 months post term. The respiratory rate gradually slows during the first several years of life (Fig. 5–6).[32,33]

The mechanics of breathing are significantly different in the full-term infant than in the adult. First, the thorax is approximately circular in newborns (versus elliptical in adults); this contributes to their less efficient diaphragm contraction, because diaphragm insertion is horizontal rather than oblique. Second, rib cage compliance is high in the infant; it diminishes throughout childhood, but it does not reach adult levels until adolescence. As a

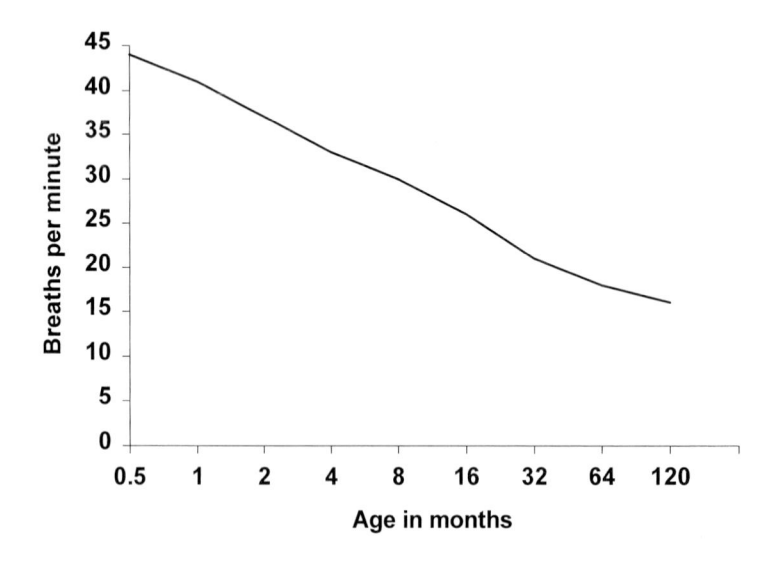

Figure 5–6. Change in respiratory rate during sleep in infancy and childhood (schematized). With age plotted on a logarithmic scale, the decline in respiratory rate is approximately linear. Based in part on data from Rusconi et al.[32] and Litscher et al.[33]

result, the rib cage is easily deformed by diaphragm motion.

During AS, when the phasic inspiratory activity of the intercostal muscles and accessory muscles of respiration is inhibited, the contraction of the diaphragm during inspiration combined with a highly compliant ribcage produces *inward* motion of the rib cage, a phenomenon referred to as *paradoxical breathing*. The inward motion of the rib cage during AS leads to reduced end expiratory volume, reduced oxygenation, and reduced diaphragm efficiency. Paradoxical breathing, which in adults is usually an indication of airway obstruction or abnormal respiratory muscle activity, is normal during AS in the infant, and it may still be seen in REM sleep during the first few years of life, but it becomes gradually less apparent as the rib cage contribution to breathing increases. During QS, paradoxical breathing is much less common because intercostal activity stabilizes the chest wall.

The upper airway configuration of the infant is significantly different from that of the adult. At birth the upper airway is narrow, but the epiglottis is large and covers the soft palate, forming the "low epiglottic sphincter" that favors nasal breathing. As a result, neonates breathe almost entirely through the nose, and partial nasal obstruction reduces ventilation, more so in REM sleep than in NREM sleep. During the first 2 years of life, the epiglottis, larynx, and hyoid bone gradually change position, and the velolingual sphincter develops. These changes facilitate oral breathing and allow speech to develop.

Patterns of neural activity in response to hypoxia and hypercapnia in infancy differ from the patterns that occur in adults. For example, hypoxia in adults leads to sustained hyperventilation; in infants, hypoxia that occurs during QS causes sustained hyperventilation, whereas hypoxia that occurs during AS or during wakefulness may cause an initial period of hyperventilation followed by hypoventilation. Hypoxia may also cause an increase in the amount of periodic breathing during sleep. Responses to airway stimulation also differ. Airway stimulation that produces arousal in adults may not do so in infants and may instead depress respiration or trigger an apnea.

Childhood

By the end of the first year, breathing is more regular and control of breathing is much more fully developed than in infancy. The healthy child rarely snores, and normal preadolescents seldom have more than 1 to 2 apneas per hour of sleep, although brief hypopneas may occur more frequently (Table 5–2).[21,34,35]

Adulthood

Breathing patterns during sleep remain stable during middle age, but they tend to deteriorate in older people. Hypopneas and brief apneas, both obstructive and central, increase after age 65. About one-fourth of all community-dwelling persons over age 65 have more than 5 apneas per hour of sleep, and about 60% have more than 10 respiratory events (apneas plus hypopneas) per hour of sleep; in most older persons, the majority of such respiratory events are obstructive.[36] The clinical significance of the age-related increase in apnea is uncertain (see Chap. 13).

Table 5–2. **Rates of Apneas and Hypopneas in Healthy Older Children**

	Apnea Index	Apnea-Hypopnea Index	Minimum Sao_2
Boys 13 ± 2 years[34]	1.0 ± 1.0	1.3 ± 1.3	93.5 ± 2.7
Girls 14 ± 2 years[34]	0.9 ± 0.7	1.1 ± 0.7	94.3 ± 1.7
Boys and girls 10 ± 5 years[35]	0.1 ± 0.5		96 ± 2

MOVEMENTS DURING SLEEP OVER THE LIFE SPAN

The frequency of body movements, limb movements, and eye movements during sleep changes with development. Body movements are first identifiable at about 8 to 10 weeks CA, and a rhythmic rest-activity cycle of quiet periods and active periods can be detected at about 20 weeks CA. State-related differences in body movements accompany the state-related EEG changes that develop at 27 to 30 weeks CA. Periods of sleep without movement lasting 20 or more seconds occur during QS at about 28 to 30 weeks CA, and the majority of 30-second sleep epochs are quiescent after about 33 weeks CA.

Phasic twitching of the limbs and face is a prominent feature of AS throughout the third trimester and at term. Limb twitches and jerks may be quite vigorous, and the "smiles" that occur in infants are usually caused by phasic contractions of facial muscles during AS. The amplitude of these phasic contractions gradually decreases over time.

Repetitive rhythmic movements, such as body rocking, head banging, and head rolling, are common at the transition between wakefulness and sleep during the first year. They may also occur during light NREM sleep and occasionally during SWS or REM sleep. Such movements become much less frequent after age 4 years (see Chap. 15).

Apart from brief arousals and shifts in body position and the eye movements and phasic twitches of REM sleep, movements during sleep are infrequent between the ages of 4 and about 50. After age 50, the prevalence of periodic leg movements during sleep increases. About 45% of healthy persons over age 65 have five or more periodic leg movements per hour in bed (see Chap. 11).[37] Abnormal movements and behaviors during REM sleep are also more common in older persons (see Chap. 15).

SUMMARY

Differentiation of states of sleep and wakefulness begins during fetal life, and a cycle of active sleep (REM sleep) and quiet sleep (NREM sleep) begins to emerge between 27 and 33 weeks CA. At term, the infant spends about one-third of its existence in each of three states—wakefulness, NREM sleep, and REM sleep.

Electroencephalographic features of NREM sleep—sleep spindles, K-complexes, and high-amplitude delta waves—develop during the first 6 months. During the first year of life, sleep becomes consolidated into nighttime hours, the total amount of sleep and the amount of REM sleep decline, and the amount of SWS increases. The amount of daytime sleep and the number of naps continue to decline during the first few years of life, and by age 6, daytime sleep is uncommon.

During the second decade, because most adolescents decrease their sleep time despite little or no change in sleep need, daytime sleepiness is prevalent. During adulthood, sleep quality gradually deteriorates, with increased time awake at night and increased amounts of stage 1 sleep. This deterioration becomes more pronounced in older people, leading to an increase in daytime drowsiness and nap frequency.

Although respiratory movements are first apparent during fetal life, respiratory control systems are not fully developed at birth; as a result, irregular respirations and brief apneas are common in preterm and term infants. Respiratory mechanics and upper airway anatomy, which are significantly different in the full-term infant than in the adult, contribute to the increase in central and obstructive respiratory events early in life. Breathing becomes much more regular during the first year of life and remains stable during childhood and most of adulthood; however, central and obstructive apneas become more common during late adulthood.

The frequency of body and limb movements during sleep changes with development. Phasic twitching of the limbs and face is a prominent feature of active sleep throughout the third trimester and at term, but it becomes much less prominent thereafter. Apart from brief arousals and shifts in body position and the eye movements and phasic twitches of REM sleep, movements during sleep are infrequent between the ages of 4 and about 50. After age 50, the prevalence of periodic leg movements during sleep increases. Abnormal movements and behaviors during REM sleep are also more common in older persons.

REFERENCES

1. Lindsley, DB: Heart and brain potentials of human fetuses *in utero*. Am J Psychol 55:412, 1942.
2. Hughes, JG, et al: Electroencephalography of the newborn infant: VI. Studies on premature infants. Pediatrics 7:707, 1951.
3. Parmalee, AH, et al: Maturation of EEG activity during sleep in premature infants. Electroencephalogr Clin Neurophysiol 24:319, 1968.
4. Parmalee, AH, et al: Sleep states in premature infants. Dev Med Child Neurol 9:70, 1967.
5. Curzi-Dascalova, L, et al: Sleep state organization in premature infants of less than 35 weeks gestational age. Pediatr Res 34:624, 1993.
6. Curzi-Dascalova, L: Physiological correlates of sleep development in premature and full-term neonates. Neurophysiol Clin 22:151, 1992.
7. Okai, T, et al: A study on the development of sleep-wakefulness cycle in the human fetus. Early Hum Dev 29:391, 1993.
8. Dreyfus-Brisac, C: Ontogenesis of sleep in human premature infants after 32 weeks conceptual age. Dev Psychobiol 3:91, 1970.
9. Coons, S, and Guilleminault, C: Development of sleep-wake patterns and non-rapid eye movement sleep stages during the first six months of life in normal infants. Pediatrics 69:793, 1982.
10. Ardura, J, et al: Development of sleep-wakefulness rhythm in premature babies. Acta Paediatr 84:484, 1995.
11. McMillen, IC, et al: Development of circadian sleep-wake rhythms in preterm and full-term infants. Pediatr Res 29:381, 1991.
12. Ma, G, et al: The development of sleep-wakefulness rhythm in normal infants and young children. Tohuku J Exp Med 171:29, 1993.
13. Kleitman, N, and Engelmann, TG: Sleep characteristics of infants. J Appl Physiol 6:269, 1953.
14. Glotzbach, SF, et al: Biological rhythmicity in normal infants during the first 3 months of life. Pediatrics 94:482, 1994.
15. Coons, S: Development of sleep and wakefulness during the first 6 months of life. In Guilleminault, C (ed): Sleep and Its Disorders in Children. Raven, New York, 1987, pp 17–27.
16. Armstrong, KL, et al: The sleep patterns of normal children. Med J Aust 161:202, 1994.
17. Parmelee, AH, et al: Infant sleep patterns from birth to 16 weeks of age. J Pediatr 65:578, 1964.
18. Webb, WB: Development of human napping. In Dinges, DF, and Broughton, RJ (eds): Sleep and Alertness: Chronobiological, Behavioral, and Medical Aspects of Napping. Raven, New York, 1989, pp 31–52.
19. Busby, K, and Pivik, RT: Failure of high intensity auditory stimuli to affect behavioral arousal in children during the first sleep cycle. Pediatr Res 17:802, 1983.
20. Coble, PA, et al: EEG sleep of normal healthy children: Part I. Findings using standard measurement methods. Sleep 7:289, 1984.
21. Carskadon, MA, et al: Nighttime sleep and daytime sleep tendency in preadolescents. In Guilleminault, C (ed): Sleep and Its Disorders in Children. Raven, New York, 1987, pp 43–52.
22. Moore, T, and Ucko, LE: Night waking in early infancy. Arch Dis Child 32:333, 1957.
23. Carskadon, MA, et al: Pubertal changes in daytime sleepiness. Sleep 2:453, 1980.
24. Ross, JJ, et al: Sleep patterns in pre-adolescent children: An EEG-EOG study. Pediatrics 42:324, 1968.
25. Czeisler, CA, et al: Phase advance and reduction in amplitude of the endogenous circadian oscillator correspond with systematic changes in sleep-wake habits and daytime functioning in the elderly. Sleep Res 15:268, 1986.
26. Gislason, T, and Almqvist, M: Somatic diseases and sleep complaints. Acta Med Scand 221:475, 1987.
27. Ford, DE, and Kamerow, DB: Epidemiologic study of sleep disturbances and psychiatric disorders: An opportunity for prevention? JAMA 262:1479, 1989.
28. Dawes, GS, et al: Respiratory movements and rapid eye movement sleep in the foetal lamb. J Physiol (Lond) 220:119, 1972.
29. Rigatto, H: Control of breathing during sleep in the fetus and neonate. In Ferber, R, and Kryger, M (eds): Principles and Practice of Sleep Medicine in the Child. WB Saunders, Philadelphia, 1995, pp 29–43.
30. National Institutes of Health Consensus Development Conference: Infantile apnea and home monitoring. NIH Pub No 87–2905. US Department of Health and Human Services, Bethesda, Md, 1987.
31. Curzi-Dascalova, L: Physiological correlates of sleep development in premature and full-term neonates. Neurophysiol Clin 22:151, 1992.
32. Rusconi, F, et al: Reference values for respiratory rate in the first three years of life. Pediatrics 94:350, 1994.
33. Litscher, G, et al: Respiration and heart rate variability in normal infants during quiet sleep in the first year of life. Klin Paediatr 205:170, 1993.
34. Acebo, C, et al: Sleep, breathing, and cephalometrics in older children and young adults: Part I. Normative values. Chest 109:664, 1996.
35. Marcus, CL, et al: Normal polysomnographic values for children and adolescents. Am Rev Respir Dis 146:1235, 1992.
36. Ancoli-Israel, S, et al: Sleep-disordered breathing in community-dwelling elderly. Sleep 14:486, 1991.
37. Ancoli-Israel, S, et al: Periodic limb movements in sleep in community-dwelling elderly. Sleep 14:496, 1991.

Chapter 6

DREAMS AND OTHER MENTAL ACTIVITY DURING SLEEP

DREAMS THROUGHOUT HISTORY
DREAMS COMPARED TO OTHER MENTAL ACTIVITY OF SLEEP
PSYCHOPHYSIOLOGY OF REM SLEEP AND DREAMING
DREAM CONTENT
Age and Dream Content
Stage of Sleep and Time of Night
Daytime Events
Emotions
FACTORS INFLUENCING DREAM RECALL
PSYCHOLOGICAL ASPECTS OF DREAMS
Freud's Dream Theories
Postfreudian Views of Dream Psychology

Jacob . . . . lay down . . . to sleep. And he dreamed, and behold a ladder set up on the earth, and the top of it reached to heaven . . . And, behold, the Lord stood above it, and said, I am the Lord God of Abraham thy father, and God of Isaac: the land whereon thou liest, to thee will I give it and to thy seed . . .

GENESIS 28:10–16.

DREAMS THROUGHOUT HISTORY

Dreams have been a source of fascination for all of known history. In many ancient civilizations, dreams were viewed as communications from gods and spirits and were considered of great import because of their presumed potential to reveal the future. As early as 1350 BC, dreams were viewed in ancient Egypt as contrary predictors—a dream of illness, for example, signified the coming of good health.[1]

Because the significance of dreams is often obscure, dream interpreters have played important roles for thousands of years. In the Bible, Pharaoh's dream in which seven thin cows devoured seven fat ones and seven shriveled ears of corn replaced seven healthy ones was interpreted by Joseph to mean that seven years of prosperity and abundance would be followed by seven years of famine. In early Greece, Aristotle was only one of the philosophers who believed that dreams could foretell the future, and dream interpreters accompanied military leaders during their campaigns. According to Plutarch, Alexander the Great considered abandoning his campaign against Tyre until an interpreter of dreams told him that his dream of satyrs dancing in triumph presaged victory. In the 2nd century BC, Artemidorius developed a dream interpretation aid based on the plausibility and continuity of the dream as well as the dreamer's character, mood, and life situation.

Varying and contradictory causes have been ascribed to dreams. Many philosophers believed that dreams were caused by external forces; others thought that dreams reflected inner psychological processes. Socrates held that dreams were an expression of conscience, whereas Plato considered dreams to be a window to an inner, more savage self. In *The Republic*, Plato stated: "The point which I desire to note is that in all of us, even in good men, there is a lawless wild-beast nature, which peers out in sleep." In the Middle Ages, the external forces theory prevailed, and dreams were often thought to be the work of gods, demons, and other spirits. Mohammed's dreams, which he attributed to Allah's influence, provided much of the basis for the Koran.[2] In modern times, the idea that dreams reflect inner psychological processes gained strength and reached its fullest expression in the works of Freud,[3] who viewed dreams as "the royal road to the unconscious," and those of Jung.[4]

In the 18th and 19th centuries, however, the rise of rationalism led to a new theory that dreams were merely meaningless random expressions of physiologic activity during sleep, a view not too different from that expressed in the more recent activation-synthesis hypothesis (see Psychophysiology of REM Sleep and Dreaming below).[5]

The discovery of rapid eye movement (REM) sleep and its association with dreaming led to new approaches to the study of dreams and new theories concerning the functions of dreams and REM sleep. Initially it appeared that dreams occurred only during REM sleep, and therefore it was thought that just as REM sleep is a unique physiologic state, dreams were a unique form of consciousness. Subsequent studies, however, revealed that dreams also occur during non-rapid eye movement (NREM) sleep, and that other forms of mental activity, dissimilar to dreams, can occur during all stages of sleep.

DREAMS COMPARED TO OTHER MENTAL ACTIVITY OF SLEEP

One dreams as one thinks.

N. KLEITMAN.[6]

Although almost everyone has had dreams that are clearly different from waking mental activity (Table 6–1), it is not easy to define what constitutes a dream. At first, the discovery of the association between REM sleep and dreaming appeared to make a definition of dreams simple: Dreams were the cognitive, perceptual, and emotional experiences that accompanied REM sleep. Aserinsky and Kleitman,[7] in their initial description of REM sleep, reported that 74% of awakenings from REM sleep were associated with dreams, compared to 9% of awakenings from NREM sleep. Impressed by the strength of this association, early investigators suspected that dream recall during NREM sleep was a result of memories of dreams from preceding REM periods, rather than of dreaming during NREM sleep.

It was soon discovered, however, that awakenings from NREM sleep are often accompanied by reports of some type of mental activity. Furthermore, reports of mental activity from the first hour of sleep, before any REM sleep has occurred, provided strong evidence that mental activity of sleep is not confined to REM sleep. Foulkes[8] found that 70% of awakenings

Table 6–1. **Dreams Compared to Wakefulness**

More Prominent During Dreams	More Prominent During Wakefulness
Hallucinations, especially visual	Taste and smell
Delusions with lack of insight	Depression, shame, guilt
Emotional intensification: anxiety, fear, surprise, obsessional concerns	Pain
Uncritical belief	Discontinuities of thought
Discontinuities of visual events	Reflection, imagination
Bizarre elements: incongruous, improbable, or impossible events	
Thematic continuity	
Poor recall	

from stage 3–4 sleep, 74% of awakenings from stage 2 sleep, and 81% of awakenings at the time of the onset of sleep spindles were associated with recall of some mental activity. Furthermore, the perceptual and emotional features accompanying mentation at sleep onset and during NREM sleep are sometimes similar to those accompanying mentation in REM sleep.[9]

On the other hand, subjects awakened from NREM sleep are more likely to describe mental activity as "thoughtlike" than as "dreamlike." Physical activity, bizarreness, sudden scene shifts, and intense emotional experiences—most commonly anxiety, surprise, and anger—are common in reports from REM sleep awakenings, whereas the mental activity of NREM sleep tends to be more conceptual and more like waking mental activity (Table 6–2).[10] Undisguised memories of the day's events and recreations of recent events, consistent with the "day residues" described by Freud (see below), are much more common in NREM sleep than in REM sleep.[8] Differences in descriptions of mentation enabled raters in one study to predict successfully the origin of a report as REM sleep or NREM sleep 77% of the time.[11] In addition, they could discriminate between reports from REM sleep awakenings and those from NREM sleep awakenings, even when both were described as dreams by the subjects.

These findings indicate that, although there are clear differences in the mental activity that occurs during REM sleep versus NREM sleep, dreams do occur during NREM sleep, and at times they are indistinguishable from dreams occurring during REM sleep. Furthermore, as dreamlike mentation can occur during the transition from wakefulness to NREM sleep and even during periods of relaxed wakefulness, the state in which mental activity occurs does not differentiate dreaming from other mental activity of sleep.

What, then, distinguishes a dream from other cognitive activity? Hauri and associates,[12] after comparing the mental activity during sleep described by subjects as "dreams" with that described as "thoughts," suggested that "dreaminess" is related to three principal factors: (1) unreality, characterized by impossibilities, absurdities, and distortions; (2) intensity of experience, characterized by clarity, dramatization, and emotion; and (3) the richness and duration of the mental experience, as indicated by the length of the report.

Dreams differ from nondream mentation in other ways as well. Comparing dreams to waking mental activity, Rechtschaffen[13] noted that dreams are characterized by absence of reflection, single-mindedness, thematic continuity, and poor recall. The lack of reflection is apparent in the almost universal failure of the dreamer to consider the impossibilities and absurdities that often accompany dreams and to recognize that the mental activity is a dream. Awareness of dreaming during a dream occurs in less than 5% of dreams.[13]

Table 6–2. **Frequency of Reports of Types of Mental Activity in REM Sleep and NREM Sleep**

Mental Activity	REM Sleep (%)	NREM Sleep (%)
Dreamlike	87	41
Thoughtlike	8	19
Vivid	74	24
Primarily visual	61	40
Primarily conceptual	16	36
Bizarre or implausible	37	6
Pleasant	38	63
Emotional	74	20

Source: Adapted from Rechtschaffen et al.,[10] pp 409–414, with permission.

Dreamers are single-minded in that they tend to think only about the dream events. Furthermore, the mental activity of dreaming seems to be isolated from the rest of consciousness: There is little or no voluntary control over the content of the dream, and the events of the dream are not interrupted by reflection and unrelated thoughts.

Thematic continuity is also a distinguishing feature of dreams; dreamers rarely report that they have spontaneous disconnected thoughts unrelated to the theme of the dream, but such thoughts are common during waking mental activity. Often, on a single night, the same themes occur in reports obtained from REM and NREM sleep, indicating that dreams are related to a larger body of sleep-related mental activity.[14]

Finally, dream recall is poor despite the bizarre and dramatic content that would be memorable if it occurred during wakefulness. Most dreams are forgotten within 8 minutes of the end of the REM period and most persons forget 90% or more of all their dreams (see Factors Influencing Dream Recall below).[15] The lack of reflection and the continuous thought stream of dreams probably contribute to the failure of memory formation.

PSYCHOPHYSIOLOGY OF REM SLEEP AND DREAMING

After the discovery of the link between REM sleep and dreaming, intense interest developed in the psychophysiology of dreaming and the relation between the mental phenomena of dreams and the physical phenomena of REM sleep.

Cortical activation and increased metabolic activity, both of which are characteristic of REM sleep, are probably required for the cognitive and emotional activity that accompany dreams. The visual aspect of dreams is presumably related to the activation of visual systems, while the relative isolation of dreams from perceptual experiences may be related to afferent inhibition. Dreamers often fail to incorporate either external or internal stimuli. For example, thirsty subjects seldom dream about drinking, and despite the penile erections and vaginal engorgement that accompany REM sleep, few dreams contain overt sexual content.[13] Kleitman[6] noted that sounds used to awaken subjects from dreams were incorporated into the dream in only about 10% of dream reports. In one study, water sprayed onto the dreamer's face was incorporated into the dream in only about 40% of cases.[16] On the other hand, stimuli sometimes alter dream content; for example, a subject with a full bladder may dream of searching for a bathroom.

Output blockade—postural atonia with hyperpolarization of alpha motor neurons that inhibits movement during sleep—may protect the dreamer from acting out the dream contents. McCarley[17] proposed that motor system activation during REM sleep causes activation of a corollary discharge system that induces a perception of movement in the absence of actual movement. The perception of movement is then incorporated into the dream. When the control of muscle atonia during REM sleep is defective, persons in REM sleep appear to act out dreams (see Chap. 15). Dream-related autonomic expression also appears to be partially blocked, as heart rate changes during REM sleep do not correlate with dream content.

The eye movements themselves may have some effect on dreams. Periods of increased eye movements are associated with more vivid, active, and emotional dreams,[18,19] and the direction of the eye movement correlates to some degree with the final movement in the dream imagery.[20]

The findings described in the previous paragraphs suggest that the physiologic features of REM sleep affect the nature of dreams, but they do not provide an explanation for the association between REM sleep and dreams, and they do not provide an answer to the question of whether dreaming occurs because of the necessity for REM sleep. According to one theory, called the *activation-synthesis hypothesis*, dreams are necessary mainly to protect sleep from the activation of the cortex that occurs with REM sleep activation: "[The cortex produces] . . . partially coherent dream imagery from the relatively noisy signals sent up to it from the brainstem."[5] Theories such as this one, however, do not provide an explanation for dreams that occur during NREM sleep.

Foulkes[21] suggested that dreaming is the form assumed by consciousness when cortical activation occurs in the relative absence of direction from voluntary self-control and perceptions of the environment. In his view, although REM sleep provides optimal conditions for dreaming, NREM sleep and wakefulness can be associated with dreaming with the appropriate combination of cortical activation dissociated from environmental influences and self-control.

DREAM CONTENT

Dream content is influenced by a number of factors, including age, stage of sleep, time of night, and the activities, thoughts, and emotions that occur during wakefulness. Dreams almost always include a visual component, less often an auditory or vestibular component, and rarely a tactile or gustatory sensation.[22] However, the dreams of congenitally blind persons lack visual imagery, and among persons with acquired blindness, the duration of blindness influences visual content; visual content may continue for 10 to 15 years after the onset of blindness, but it is usually absent in persons who have been blind for 30 or more years.[23]

Age and Dream Content

Although REM sleep is present before birth, it has been impossible to determine whether dreams occur in infancy. Theoretically, one can learn about children's mental activity as soon as the child can communicate meaningfully, probably in the second year of life. In practice, dream reports are rarely obtained before age 3. Foulkes[24] suggested that dreaming is not possible before the development of symbolic thinking at about age 2. If symbolic thinking is needed for dreams, then nonhuman species probably do not dream despite their sometimes vigorous twitching during REM sleep—and despite the old saying, "The dog dreams of his bone, the goose of corn."

At ages 3 and 4, dream reports are infrequent and short, usually 18 or fewer words.[24] Reports of dreams from REM and NREM sleep are similar at this age and often include animals, but they have few human characters and little movement or affect. Children often dream that they are asleep in a different setting, such as on top of a school bus, and the likelihood of obtaining a dream report is higher in children with strong visuospatial aptitudes.[24] The frequency of dream reports increases with age; by age 9, the rate is similar to that of adults (Table 6–3).[24] By age 5, dream reports are obtained more frequently from REM sleep than from NREM sleep, and by late childhood, NREM dream reports are more thought-

Table 6–3. **Frequency of Reports of Mental Activity During Sleep in Children**

Age (years)	Report Rate After Awakenings from REM Sleep (%)	Report Rate After Awakenings from NREM Sleep (%)
3–5	15	0
5–7	31	6
7–9	43	22
9–11	79	33
11–13	79	24
13–15	73	39

Source: Adapted in part from Foulkes,[24] p. 348, with permission.

like, with fewer visual and narrative components.[24]

Stage of Sleep and Time of Night

Mental activity during sleep differs depending on the time of night and the stage of sleep. The initial minutes of stage 1 sleep are often accompanied by vague, fragmentary visual imagery and drifting thoughts, which may be perseverative and are often based on daily activities or concerns. Longer REM periods are associated with longer dreams, and dream reports from awakenings late in a REM period are generally more detailed and more emotional, but not more bizarre, than dream reports from early in the REM period. Dream recall is somewhat higher with an awakening from phasic REM sleep than with one from tonic REM sleep.

Daytime Events

The events of the previous day affect dream content. For example, subjects for dream research often dream about the laboratory setting. Recent events usually have a greater impact on dreams than remote events: Subjects in psychoanalysis tend to dream about material covered during the most recent analytic hour.[25] In persons with post-traumatic stress disorder, however, dream content often relates to traumatic events that may have occurred years or decades previously (see Chapter 16).

Emotions

Emotional state also has an impact on dream content, and the contents of dreams tend to reflect the more emotional experiences of the previous day.[26] In a study that compared presleep viewing of a violent and a nonviolent television episode, REM sleep reports were longer, more imaginative, and more vivid after the violent episode, but they were not more violent, and the specific content of the television episodes was usually not included.[27] Thus, it seems likely that the emotional experience associated with the viewing was incorporated to a greater extent than the actual content.

Dream mentation is also affected in persons with affective disorders. Dreams in persons with major depression tend to be short, past-oriented, masochistic, and repetitive,[28,29] and persons who are in the process of divorce more often have dreams with anxiety and negative affect than persons who are not.[28,30]

FACTORS INFLUENCING DREAM RECALL

Because most dreams are forgotten, it is not surprising that dream recall in everyday life is much different from the dream reports obtained in the sleep laboratory after awakenings. A number of factors influence dream recall; the most important is probably the occurrence of an awakening during or immediately after the dream, which appears to be a necessary but not sufficient condition for later dream recall. Longer, more complete awakenings are more likely to lead to dream recall than brief arousals with a rapid return to sleep. Dreams at the end of the night from longer REM periods with denser eye movements are more likely to be recalled in the morning than dreams from shorter REM periods earlier in the night. Other factors that may be required include an awakening of sufficient length to form lasting memories, probably about 4 minutes, and a dream that has substantial emotional impact.

Dream recall later in the day often varies from dream recall during the night or upon awakening in the morning. In a study that compared reports obtained in the sleep laboratory to reports concerning the same dreams obtained during psychotherapy, the daytime reports were distorted by waking concerns and fantasies.[31] Selection of dreams to report and therapist expectation also contribute to the differences between dream reports obtained in psychotherapy and those obtained in the sleep laboratory.

PSYCHOLOGICAL ASPECTS OF DREAMS

Dreaming . . . discharges the Unconscious excitation, serves . . . as a safety valve and . . . preserves sleep . . . in return for a small expenditure of waking activity.

SIGMUND FREUD, 1913[3]

Although psychological and philosophical theories concerning dreams have existed for many centuries, the views of Sigmund Freud have had the most profound impact on ideas about the psychological function of dreams.[3,32]

Freud's Dream Theories

At the time when he formulated his dream theories, Freud viewed the mind as made up of the unconscious system, the preconscious system, and consciousness. The unconscious system could function independently, but its material was only accessible to consciousness via the preconscious. He considered consciousness a sense organ that used attention to survey and to become aware of the contents of the preconscious, which had executive functions and organizational and synthetic abilities. The preconscious was separated from the unconscious by a barrier, the repression barrier, which required psychic energy for its maintenance.

Freud viewed sleep as a period in which mental activity was inaccessible to consciousness; he believed the wish to sleep was associated with withdrawal of interest in the external world (i.e., the sleeper gave up relations with reality in preparation for sleep). This withdrawal from reality, a form of "narcissistic regression," was a return to a childlike state. In addition, relaxation and reduced cortical activity associated with sleep was considered to reduce the amount of energy available for maintenance of the barrier between the preconscious and the unconscious and thus to make it easier for unconscious materials, specifically unconscious wishes, to find their way into the preconscious system. According to Freud, most of these unconscious wishes were related to sex or hostility, and the dream provided a mechanism for the indirect expression of these unacceptable unconscious impulses.

In Freud's theories, unconscious wishes were able to pass into the preconscious by attaching to a *day residue,* a memory of events of the preceding day. However, these unconscious wishes could not be fulfilled during sleep because of the motor inhibition that accompanies sleep; instead, a memory of a prior gratification or a fantasy occurred. He believed that dream formation arose from these derivatives of primary unconscious motives and that the early aspects of dream organization occurred in the unconscious. During this initial process, the unconscious wish and the prior memory were transformed into a hallucinatory, but seemingly real, perception of wish-fulfillment.

Almost immediately after the initial dream formation, Freud postulated that the dream "attracted" consciousness to itself, either by arousing mechanisms or by an ascent to a lighter stage of sleep. The preconscious, although less energized in sleep than during wakefulness, did not allow the expression of the unacceptable wishes and transformed them by cortical inhibition into perceptions that would be more acceptable to the self. In doing so, the preconscious acted as a censor and also synthesized the dream elements into a coherent, if not always logical, story. According to Freud, the *dreamwork* was the process by which the residues of the day's experiences and prior memories were transformed, first to express the unconscious desires and then to disguise them from conscious awareness. Thus, the dream content was a function of the prior perceptual experiences of the dreamer—particularly recent experiences, such as those of the previous day—that were incorporated into the dream and represented disguised, unconscious motives.

According to Freud, dreams had three major functions. First, they allowed the expression of unconscious wishes. Second, by disguising the wishes and allowing them to be expressed, they provided a safety valve—a means of "discharging" the unconscious and releasing the psychic tension and excitation that resulted from the unconscious wishes. Third, they served as a "guardian" for sleep and allowed it to continue, because the dreamwork had transformed the wish so thoroughly that full arousal and awakening did not occur despite the release of the unconscious excitation.

Postfreudian Views of Dream Psychology

The 19th-century neurophysiologic principles of cortical function, on which Freud's mechanistic model of dreams was based, soon became obsolete. With the discovery of REM sleep and subsequent studies of dream reports from awakenings in the sleep laboratory, it became obvious that some of Freud's views, such as the idea that dreams are formed in an instant, were untenable. On the other hand, his ideas about the unconscious provided a powerful platform for the growth of psychotherapy, psychoanalysis, and dream interpretation.

Some of Freud's students adapted his theories to incorporate additional knowledge about sleep and cognition. Adler,[33] for example, believed that the distinction between primary-process thinking in the dream and reality-oriented thinking in wakefulness was not absolute. He also suggested that only some dreams were motivated by repressed unconscious sexual or hostile wishes; others reflected unresolved problems from conscious experience. He proposed that dreams had an adaptive function that allowed the ego to sort out and rearrange conflicting needs.

Current thinking goes further than Adler went in postulating two functions for dreams. According to these views, most dreams have relatively little concern with unconscious conflict and do not interfere with sleep. Dreams on which unconscious conflicts are superimposed, however, are more likely to be remembered. Thus, emotional conflict does not cause dreaming but instead uses dreams as a vehicle to express conflict indirectly.[34,35]

Aside from their possible role in emotional conflict, the function of dreams is unknown. In some persons, REM-sleep deprivation leads to an increase in dreaming during NREM sleep, which suggests that dreams have a function independent of REM sleep. Attempts to prove or disprove Freud's theories that dreams express unconscious wishes and protect sleep have been inconclusive. The 19th-century ideas of Hughlings Jackson, that irrelevant memories are swept away and significant memories are consolidated during dreams, have been revived recently but remain unproven.[36] Other posited functions of dreams include problem solving, enhancement of creativity, control of affect, and adaptation of the ego.[4,33,37-40]

SUMMARY

Dreams have been a source of fascination for all of known history. Many philosophers believed that dreams were caused by external forces; others believed that dreams reflected inner psychological processes. With the discovery of REM sleep, it was thought initially that dreams occurred only during REM sleep. However, subsequent studies showed that dreams, as well as other forms of mental activity, occurred during both REM and NREM sleep. Nonetheless, differences exist: Mentation during NREM sleep is more often conceptual and thoughtlike, whereas mentation during REM sleep is more often bizarre and associated with intense emotional experiences. Unreality, intensity and duration of the mental experience, and absence of reflection distinguish dreams from other mental activities that occur during sleep and wakefulness.

Dream content is affected by the age of the dreamer, the stage of sleep, and thoughts and emotions during wakefulness. The physiologic features of REM sleep also appear to influence the nature of dreams occurring during this state.

Dream recall is highly selective, and dream recall during the night often differs from recall of the same dream several hours later, when it often becomes distorted by waking concerns and fantasies.

Because dreams appear to represent a form of consciousness relatively free of voluntary control, they provide an opportunity to assess the psychology of the dreamer. Freud believed that dreams provided a mechanism for the indirect expression of unacceptable unconscious impulses, which then were disguised and transformed in a process called the dreamwork. Later psychologists suggested that dreams also may be caused by unresolved problems from conscious experience and that they may have adaptive functions that allow the ego to sort out and rearrange conflicting needs.

REFERENCES

1. Thorpy, M: History of sleep and man. In Thorpy, MJ, and Yager J (eds): The Encyclopedia of Sleep and Sleep Disorders. Facts on File, New York, 1991, pp. ix-xxxiii.
2. Parkes, JD: Sleep and Its Disorders. WB Saunders, London, 1985, p 23.
3. Freud, S: The Interpretation of Dreams. Macmillan, New York, 1913.
4. Jung, C: Dreams. Princeton University Press, Princeton, NJ, 1974.
5. Hobson, JA, and McCarley, R: The brain as a dream-state generator: An activation-synthesis hypothesis of the dream process. Am J Psychiatry 134:1335, 1977.
6. Kleitman, N: Sleep and Wakefulness, ed 2. University of Chicago Press, Chicago, 1963, pp 102–104.
7. Aserinsky, E, and Kleitman, N: Regularly occurring periods of eye motility, and concomitant phenomena, during sleep. Science 118:273, 1953.
8. Foulkes, WD: Dream reports from different stages of sleep. J Abnorm Soc Psychol 65:14, 1962.
9. Vogel, GW, et al: Limited discriminability of REM and sleep onset reports and its psychiatric implications. Arch Gen Psychiatry 26:449, 1972.
10. Rechtschaffen, A, et al: Reports of mental activity during sleep. Can Psychiatr Assoc J 8:409, 1963.
11. Monroe, LJ, et al: Discriminability of REM and NREM reports. J Pers Soc Psychol 2:456, 1965.
12. Hauri, P, et al: Dimensions of dreaming: A factored scale for rating dream reports. J Abnorm Psychol 72:16, 1967.
13. Rechtschaffen, A: The single-mindedness and isolation of dreams. Sleep 1:97, 1978.
14. Rechtschaffen, A, et al: Interrelatedness of mental activity during sleep. Arch Gen Psychiatry 9:536, 1963.
15. Dement, W, and Kleitman, N: The relation of eye movements during sleep to dream activity: An objective method for the study of dreaming. J Exp Psychol 53:339, 1957.
16. Dement, W, and Wolpert, EA: The relation of eye movements, body motility and external stimuli to dream content. J Exp Psychol 55:543, 1958.
17. McCarley, RW: Dreams and the biology of sleep. In Kryger, MH, et al (eds): Principles and Practice of Sleep Medicine, ed 2. WB Saunders, Philadelphia, 1994, pp 373–383.
18. Dement, W, and Kleitman, N: Incidence of eye motility during sleep in relation to varying EEG pattern. Fed Proc 14:37, 1955.
19. Foulkes, D, et al: Effects of awakenings in phasic vs. tonic stage REM on children's dream reports. Sleep Res 1:104, 1972.
20. Roffwarg, HP, et al: Dream imagery: Relationship to rapid eye movements of sleep. Arch Gen Psychiatry 7:235, 1962.
21. Foulkes, D: A contemporary neurobiology of dreaming? Sleep Res Soc Bull 3:2, 1997.
22. McCarley, RW, and Hoffman, EA: REM sleep dreams and the activation-synthesis hypothesis. Am J Psychiatry 138:904, 1981.
23. Berger, OJ, et al: The EEG, eye movements and dreams of the blind. Am J Exp Physiol 14:183, 1966.
24. Foulkes, D: Dream ontogeny and dream psychophysiology. In Chase, MH, and Weitzman, ED (eds): Sleep Disorders: Basic and Clinical Research. Spectrum, New York, 1983, pp 347–362.
25. Greenberg, R, and Pearlman, C: REM sleep and the analytic process. Psychoanal Q 44:392, 1975.
26. Piccione, P, et al: The relationship between daily activities, emotions, and dream content. Sleep Res 6:133, 1977.
27. Foulkes, D, and Rechtschaffen, A: Presleep determinants of dream content: Effects of two films. Percept Mot Skills 19:983, 1964.
28. Cartwright, RD: Dreams and their meaning. In Kryger, MH, et al (eds): Principles and Practice of Sleep Medicine, ed 2. WB Saunders, Philadelphia, 1994, pp 400–406.
29. Hauri, P: Dreams in patients remitted from reactive depression. J Abnorm Psychol 85:1, 1976.
30. Trenholme, I, et al: Dream dimension differences during a life change. Psychiatry Res 12:36, 1984.
31. Whitman, R, et al: Which dream does the patient tell? Arch Gen Psychiatry 8:277, 1963.
32. Trosman, H: Freud's dream theory. In Carskadon, MA (ed): Encyclopedia of Sleep and Dreaming. Macmillan, New York, 1993, pp 251–254.
33. Adler, A: On the interpretation of dreams. Int J Indiv Psychol 1:3, 1936.
34. Kupfer, A: The structure of dream sequences. Cult Med Psychiatry 7:153, 1983.
35. Cartwright, RD: Interpretation of dreams. In Carskadon, MA (ed): Encyclopedia of Sleep and Dreaming. Macmillan, New York, 1993, pp 316–318.

36. Crick, F, and Mitchison, G: The function of dream sleep. Nature 304:111, 1983.
37. French, T, and Fromm, E: Dream interpretation. Basic Books, New York, 1964.
38. Erikson, E: The dream specimen of psychoanalysis. J Am Psychoanal Assoc 2:5, 1954.
39. Jones, R: The New Psychology of Dreaming. Grune and Stratton, New York, 1970.
40. Stone, MD: Creativity in dreams. In Carskadon, MA (ed): Encyclopedia of Sleep and Dreaming. Macmillan, New York, 1993, pp 149–151.

Part 2

Sleep Disorders

Chapter 7

APPROACH TO THE PATIENT

THE DEVELOPMENT OF SLEEP
  MEDICINE
CLASSIFICATION OF SLEEP DISORDERS
EVALUATION OF THE PATIENT
Chief Complaint and History
Past Medical History and Review of Systems
Family History of Sleep Problems
Social History
Examination
Formulation and Differential Diagnosis
DIAGNOSTIC PROCEDURES
Questionnaires and Sleep Logs
Monitoring of Sleep
Multiple Sleep Latency Test and
  Maintenance of Wakefulness Test
Scoring of Sleep Studies
Ancillary Tests
MANAGEMENT

THE DEVELOPMENT OF SLEEP MEDICINE

Sleep medicine has probably been part of medical practice since the time of the first shamans and healers. Democritus (420 BC) in ancient Greece recognized that sleep disturbance is often a sign of ill health or disease. For example, he viewed insomnia as a consequence of a poor diet and daytime sleepiness as a symptom of physical illness. In the Middle Ages, it was recognized that sleep could affect disorders such as asthma and epilepsy, and in the 17th-century, Thomas Willis described insomnia, nightmares, and restless legs in his work, *The Practice of Physick*.

Until well into the 20th-century, however, sleep problems had generally been viewed as secondary to other medical or psychiatric problems rather than as primary diseases or disorders. Narcolepsy, recognized in the 19th-century, was often considered a form of epilepsy or a psychiatric disturbance until the discovery in the 1960s of its association with abnormal REM sleep. This discovery made it the first identified *primary sleep disorder*: that is, a disorder associated primarily with sleep-related symptoms and with abnormalities of the sleep process.

Narcolepsy would probably have remained a curiosity were it not for the development of techniques for monitoring sleep, which led to two discoveries in the 1960s: (1) abnormal breathing patterns during sleep,[1,2] and (2) regular, recurring patterns of leg movements during sleep, now called *periodic limb movements of sleep*.[3] Further studies of breathing during sleep revealed that upper airway narrowing and occlusion are the most important causes of sleep-related breathing disturbance.

The recognition that sleep could facilitate specific disorders and that a variety of sleep pathologies existed led to the development in the 1970s of sleep clinics designed to diagnose and treat sleep disorders and in 1978 to the creation of a journal, *Sleep*, devoted specifically to sleep disorders medicine. The Association of Sleep Disorders Centers, established in 1975 and later renamed the American Sleep Disorders Association (ASDA), provided further impetus for the formation of centers that combined clinical evaluation with laboratories designed to uncover abnormal sleep patterns. These centers, which often combine teaching and research with clinical practice, have facilitated the development of a rational approach to sleep complaints based on clinical assessment, underlying psychobiology, and appropriately applied laboratory studies. They have also provided an opportunity for physicians to obtain special skills in the practice of sleep medicine. The ASDA and later the American Board of Sleep Medicine have been examining and certifying sleep specialists since 1978.

The importance of sleep medicine increased as it became apparent that sleep disorders are common, often serious, and frequently treatable. About one-sixth of adults have a major problem with insomnia over the course of any one year, and 1% to 4% have symptoms of sleep apnea. Chronic partial sleep deprivation in adults is probably more common than it was at the turn of the century, and most adolescents in Western countries do not get enough sleep to maintain full alertness throughout the day. Disturbed sleep patterns are also common in shift workers and night workers.

CLASSIFICATION OF SLEEP DISORDERS

As knowledge of the extent of sleep disorders increased in the 1960s and 1970s, it became important to classify sleep disorders in a rational system. The first detailed classification, published in 1979 by the Association of Sleep Disorders Centers,[4] was based on symptoms. It defined more than 60 syndromes and grouped them into four major classes: (1) disorders of initiating and maintaining sleep (insomnias); (2) disorders of excessive somnolence; (3) disorders of the sleep-wake schedule; and (4) parasomnias.

Although the symptom-based 1979 classification was useful for generating differential diagnoses, it soon became apparent that a nosology based on pathophysiology had greater value than one based principally on symptoms. The current International Classification of Sleep Disorders provides a closer approximation of such an approach.[5] This classification divides

Table 7–1. International Classification of Sleep Disorders—Dyssomnias

Intrinsic Sleep Disorders	*Extrinsic Sleep Disorders*	*Circadian Rhythm Sleep Disorders*
Psychophysiologic insomnia	Inadequate sleep hygiene	Time zone change (jet-lag) syndrome
Sleep state misperception	Environmental sleep disorder	Shift-work sleep disorder
Idiopathic insomnia	Altitude insomnia	Irregular sleep-wake pattern
Narcolepsy	Adjustment sleep disorder	Delayed sleep phase syndrome
Recurrent hypersomnia	Insufficient sleep syndrome	Advanced sleep phase syndrome
Idiopathic hypersomnia	Limit-setting sleep disorder	Non–24-hour sleep-wake disorder
Post-traumatic hypersomnia	Sleep-onset association disorder	Circadian rhythm sleep disorder NOS
Obstructive sleep apnea syndrome	Food allergy insomnia	
Central sleep apnea syndrome	Nocturnal eating (drinking) syndrome	
Central alveolar hypoventilation syndrome	Hypnotic-dependent sleep disorder	
Periodic limb movement disorder	Stimulant-dependent sleep disorder	
Restless legs syndrome	Alcohol-dependent sleep disorder	
Intrinsic sleep disorder (NOS)	Toxin-induced sleep disorder	
	Extrinsic sleep disorder NOS	

NOS = not other specified.
Source: From American Sleep Disorders Association,[5] pp 15–16, with permission.

sleep disorders into four major categories: dyssomnias, parasomnias, sleep disorders associated with medical and psychiatric disorders, and proposed sleep disorders.

Dyssomnias (Table 7–1) are defined as disorders that cause either excessive sleepiness or complaints of insomnia. These disorders can be subdivided as follows:
1. Intrinsic sleep disorders (i.e., originating or developing within the body), such as psychophysiologic insomnia, narcolepsy, and obstructive sleep apnea
2. Extrinsic sleep disorders (i.e., originating or developing from causes outside the body), such as disturbed sleep caused by the noise from a busy airport or excessive sleepiness caused by chronically insufficient amounts of sleep
3. Circadian rhythm sleep disorders (i.e., related to the timing of sleep within the 24-hour day), such as jet lag syndrome and delayed sleep phase syndrome

Parasomnias (Table 7–2) are behaviors, movements, sensations, or other phenomena that occur during sleep but are not associated with primary complaints of insomnia or excessive sleepiness. They are subdivided into four categories: (1) arousal disorders, (2) sleep-wake transition disorders, (3) parasomnias usually associated with REM sleep, and (4) other parasomnias (see Chap. 15).

Sleep disorders associated with medical and psychiatric disorders (Table 7–3) are not primary sleep disorders; they tend to vary in severity depending on the underlying disorder (see Chaps. 16 and 17). Disturbed sleep is a prominent symptom in most patients with major depression, and it is common in a variety of medical diseases, including chronic obstructive pulmonary disease, end-stage renal disease, and congestive heart failure.

Proposed sleep disorders (Table 7–4) are those for which insufficient information is available to confirm their existence as distinct disorders. Some of these, such as "pregnancy-associated sleep disorder," will probably be accepted eventually as valid sleep disorders; others, such as the "subwakefulness syndrome," probably will not.

Table 7–2. **International Classification of Sleep Disorders—Parasomnias**

Arousal Disorders	*Other Parasomnias*
Confusional arousals	Sleep bruxism
Sleepwalking	Sleep enuresis
Sleep Terrors	Sleep-related abnormal swallowing syndrome
	Nocturnal paroxysmal dystonia
Sleep-Wake Transition Disorders	Sudden unexplained nocturnal death syndrome
Rhythmic movement disorder	Primary snoring
Sleep starts	Infant sleep apnea
Sleeptalking	Congenital central hypoventilation syndrome
Nocturnal leg cramps	Sudden infant death syndrome
Parasomnias Usually Associated with REM Sleep	Benign neonatal sleep myoclonus
Nightmares	Other parasomnias NOS
Sleep paralysis	
Impaired sleep–related penile erections	
Sleep-related painful erections	
REM sleep–related sinus arrest	
REM sleep behavior disorder	

NOS = not otherwise specified.
Source: From American Sleep Disorders Association,[5] pp 16–17, with permission.

Table 7–3. **International Classification of Sleep Disorders—Sleep Disorders Associated with Medical and Psychiatric Disorders**

Associated with Mental Disorders	*Associated with Neurological Disorders*	*Associated with Other Medical Disorders*
Psychoses	Cerebral degenerative disorders	Sleeping sickness
Mood disorders	Dementia	Nocturnal cardiac ischemia
Anxiety disorders	Parkinsonism	Chronic obstructive pulmonary disease
Panic disorder	Fatal familial insomnia	Sleep-related asthma
Alcoholism	Sleep-related epilepsy	Sleep-related gastroesophageal reflux
	Electrical status epilepticus of sleep	Peptic ulcer disease
	Sleep-related headaches	Fibrositis syndrome

Source: From American Sleep Disorders Association,[5] p. 17, with permission.

EVALUATION OF THE PATIENT

Although good sleep and its associated sense of restoration are usually taken for granted by those who sleep well, poor sleep can lead not only to fatigue but to irritability, anxiety, depression, and misery. Sleep disorders can affect academic and occupational performance, can lead to accidents at work or while driving, and can contribute to marital problems and social maladjustment. In addition, sleep disorders may cause or contribute to serious medical and psychiatric problems. It behooves the clinician, therefore, to take sleep complaints seriously and to approach their evaluation in a logical, systematic fashion.

The foundations for the diagnosis of sleep disorders are the chief complaint and the history of the sleep problem. Although sleep-related complaints are common, assessment of such complaints may present diagnostic challenges to the clinician because of the many possible causes for some of the most common symptoms. In addition to the patient's history, information from the spouse or bed partner should be obtained, as most patients are unaware of the frequency and severity of snoring and cannot describe leg jerks, twitches, sleep walking, or violent behavior during sleep. The partner may confirm the patient's complaints or may give a somewhat different story, suggesting that the patient is overestimating or underestimating the degree of sleep disturbance.

As with other medical and psychiatric problems, essential information in the diagnosis of sleep disorders is provided by the patient's medical, psychiatric, and family history; the history of medication use; the psychosocial assessment; and the physical, neurological, and mental status examinations.

The clinical evaluation, combined with knowledge about the manifestations of sleep disturbances, provides the clinician with information that generally yields a provisional diagnosis or a set of diagnostic

Table 7–4. **International Classification of Sleep Disorders—Proposed Sleep Disorders**

Short sleeper
Long sleeper
Subwakefulness syndrome
Fragmentary myoclonus
Sleep hyperhidrosis
Menstrual-associated sleep disorder
Pregnancy-associated sleep disorder
Terrifying hypnagogic hallucinations
Sleep-related neurogenic tachypnea
Sleep-related laryngospasm
Sleep choking syndrome

Source: From American Sleep Disorders Association,[5] p. 17, with permission.

possibilities, which can then be confirmed or excluded with diagnostic investigations. Most patients with accurately diagnosed sleep disorders respond well to treatment.

Chief Complaint and History

The chief complaint indicates the patient's concerns. Some patients are more concerned about daytime consequences of sleep problems, such as sleepiness or fatigue; others are more concerned about their inability to sleep or about unpleasant or worrisome events or sensations during the night. Still others are more concerned about the potential consequences of their sleep problem. I usually ask patients whether a specific event motivated them to come for evaluation and what their expectations are regarding evaluation and treatment. For example, some patients who have not previously sought treatment despite being excessively sleepy for years present for evaluation because of an automobile accident or near-accident related to sleepiness.

Some patients have no complaint; they present because their spouses or parents have urged them to do so, because their physicians have referred them, or because of problems at work or school. If the chief complaint is from the spouse or bed partner, it is important to determine whether the patient recognizes the problem, is unaware of it, or minimizes or denies its existence. A patient referred because of loud snoring that is bothersome to the bed partner may be less willing to comply with treatment than a patient who has been passed over for promotion because of falling asleep while at work.

Once the chief complaint is established, details can be obtained concerning the circumstances at its onset, the temporal course and severity, the factors that lead to exacerbation or improvement, and any associated symptoms during sleep or during nocturnal awakenings. The clinician should assess the usual sleep-wake schedule (Table 7–5), its degree of variability, and any changes of the schedule during weekends or days off. Questions concerning the sleeping environment are important: Excessive noise, extreme temperatures, or an uncomfortable bed may all adversely affect sleep continuity. The history should also include the patient's or family's cognitive or emotional response to, as well as interpretation of, the symptoms. Assessment of drug use is critical because of the wide variety of medications that alter sleep and wakefulness (Table 7–6). It is important to ask specifically about the patient's use of over-the-counter and prescription medications.

Chief complaints fall into three major categories: sleepiness and fatigue, insomnia, and sensations or events that occur during sleep or during awakenings. The three categories have significant overlap; thus, many patients with complaints of insomnia also report daytime fatigue. Furthermore, a particular sleep disorder may be associated with more than one of the major categories of symptoms. For example, obstructive sleep apnea may be associated with restless sleep and frequent

Table 7–5. **Significant Aspects of the Sleep-Wake Schedule**

Sleep Schedule	Waking Schedule
Time of lights out	Naps: timing, duration, intentional or unintentional
Time from "lights out" to sleep onset	Meals
Number and times of awakenings	Social activities
Time of final awakening	Work
Method of awakening (spontaneous, with an alarm, or by a family member)	School
Time of getting out of bed	Times and amounts of alcohol, caffeine, and nicotine use
	Times and doses of medication

Table 7–6. Drug Effects on Sleep and Wakefulness

Increased nocturnal time awake: β adrenergic agonists (e.g., bronchodilators, decongestants), stimulants (e.g., appetite suppressants, caffeine), diuretics

Increased daytime sleepiness: Anticonvulsants, antihistamines, antipsychotics, benzodiazepines, sedatives (prescription and over-the-counter), tricyclic antidepressants

Suppression of REM sleep: Amphetamines, antipsychotics, lithium, monoamine oxidase inhibitors, nicotine, opiates, tricyclic antidepressants

Suppression of slow wave sleep: Benzodiazepines

---

awakenings; with choking, gasping, and mumbling during sleep; and with daytime drowsiness.

SLEEPINESS AND FATIGUE

Patients with complaints of sleepiness may describe drowsiness that interferes with daytime activities, irresistible daytime sleep episodes, increased need for sleep, or all three. Some patients complain that daytime drowsiness occurs regardless of how much sleep is obtained at night. Although falling asleep while driving or at other inappropriate or dangerous times often is the impetus that brings the patient to the clinician, many patients have little awareness of their daytime sleep episodes or believe that they represent normal sleep patterns. In such cases, the spouse or companion frequently can give a more accurate account of the extent of sleepiness than the patient.

Some patients complain of fatigue or tiredness rather than sleepiness. For such patients, it is essential to determine the precise nature of the complaint. Some patients use the term "fatigue" to mean lack of energy or motivation, others use it to mean loss of exercise capacity or endurance, and still others use it synonymously with sleepiness. The distinction between fatigue with drowsiness and fatigue without drowsiness is important because the differential diagnosis of excessive sleepiness consists mainly of sleep disorders, whereas the causes of fatigue and tiredness without drowsiness include not only sleep disorders but also neuromuscular problems, psychiatric disorders, cardiac disease, anemia, rheumatologic conditions, endocrinopathies, chronic lung disease, infections, physical deconditioning resulting from prolonged bed rest or reduced mobility, and reduced motivation as a result of secondary gain or other factors.

Some patients with complaints of fatigue also have excessive sleepiness but are unaware of it. In such cases, the occurrence of "tiredness" during sedentary situations and the association of tiredness with unwanted sleep episodes are helpful clinical features. Other patients with complaints of fatigue or tiredness, loss of a sense of well-being, difficulty with attention and concentration, and inability to function at an expected level may believe that their symptoms are a result of abnormal sleep when in fact they are related to psychiatric disorders or to the systemic effects of a medical illness.

Chronic sleepiness is often accompanied by lack of energy, lack of motivation, impairment of attention and memory, and loss of a sense of well-being, any of which may be the chief complaint in a patient with excessive sleepiness. With each of these additional symptoms, although a sleep disturbance may be the cause, other diagnoses must also be considered, including psychiatric and cognitive disorders.

The presence or absence of associated symptoms provides essential information in patients who complain of sleepiness. Loud snoring, gasping, snorting, and episodes of apnea suggest the diagnosis of *obstructive sleep apnea syndrome* (see Chap. 13). A history of episodic muscle weakness associated with laughter or strong emotion suggests cataplexy and a diagnosis of *narcolepsy* (see Chap. 10). Reduced severity of sleepiness during vacations may be an indication of *insufficient sleep syndrome* (see Chap. 8).

INSOMNIA

Patients with insomnia complain that their nocturnal sleep is inadequate in some way. They may describe difficulty falling asleep, difficulty staying asleep, or awakening too early, and they often complain that sleep does not provide the usual sense of restoration. Some patients complain of inability to function when adequate sleep is not obtained.

Questions about mood, cognition, and behavior during the evening and during sleepless intervals at night may help elucidate the diagnosis. Depressed daytime mood, crying spells, reduced appetite, and early morning awakening suggest that disturbed sleep may be caused by depression. Insomniacs who become aroused and anxious when preparing for sleep and who worry about their insomnia and its effects on performance at work or at home the following day may have *psychophysiologic insomnia*. Patients who have an irregular bedtime schedule, exercise late at night, watch television in bed, consume caffeinated or alcoholic beverages in the evening, or go from work directly to bed may have *inadequate sleep hygiene*. Patients who report "resting" for up to 8 hours at night without sleeping or leaving the bed may have *sleep-state misperception*.

Medical, psychologic, and social factors all may contribute to insomnia. Pain, dyspepsia, nocturia, immobility, and orthopnea are common medical symptoms that can cause or exacerbate insomnia. Depression, anxiety, and loneliness are typical psychological factors. Social factors include social isolation, loss of scheduled activities, and frequent napping. Stressful life events are common precipitants of insomnia, but the nature of the stress may not be immediately apparent in some patients; for example, positive life events, such as a promotion at work or a recent marriage, may not be recognized as stressful by the patient or the clinician.

SENSATIONS OR EVENTS DURING SLEEP OR AROUSALS FROM SLEEP

Patients or their bed partners may report a variety of nocturnal sensations, movements, or other bodily events (Table 7–7). Some of the more distressing symptoms include nocturnal urinary incontinence, shouting or screaming, and violent behavior, and patients may be particularly concerned about the possibility that serious medical illnesses are responsible for nighttime episodes of chest pain, gasping or choking, or heart palpitations.

Some patients are entirely unaware of the nighttime activity, and information about it can be obtained only from the bed partner or parents. In such cases, the bed partner should be asked to describe the events, to relate episodes to sleep onset

Table 7–7. **Nocturnal Movements, Sensations, and Other Phenomena**

Movements	Sensations	Other
Twitching or kicking of the extremities	Headaches	Seizures
	Dyspnea, wheezing	Urination
Chewing movements	Heart palpitations	Sweating
Sleepwalking	Heartburn	Apnea
Talking in sleep	Leg cramps	Death or near-death
Screams	Numbness, tingling, paralysis	
Violent behavior		

and time of night, and to note the degree of the patient's responsiveness during the episode.

Past Medical History and Review of Systems

The history of current or past medical, surgical, and psychiatric illnesses and the review of systems may uncover symptoms of medical or psychiatric illnesses that can cause or contribute to sleep disorders (Table 7–8). Mood disturbances increase the likelihood of a psychiatric etiology, whereas recent weight gain and an increase in collar size increase the likelihood of obstructive sleep apnea syndrome. Nocturnal angina, orthopnea, dyspnea, and wheezing may indicate that sleep disturbance is caused by cardiac or pulmonary disease. Leg cramps, neuropathic or arthritic pain, and nocturia are common causes of disturbed sleep, particularly in older persons.

The sexual history should also be explored. Sexual dysfunction may be caused by sleep disturbance, as in some patients with obstructive sleep apnea, or it may be the cause of sleep disturbance, as in some patients with insomnia. Elements of the sexual history include the sexual orientation, the type and frequency of sexual activity, the number of sexual partners and their relationship to the patient, the degree of satisfaction with sexual functioning, and the relation of sleep complaints and sleep disturbance to sexual concerns and activities.[6]

Family History of Sleep Problems

Patients who present with disordered sleep should be asked whether other family members have a sleep disorder. Table 7–9 provides a list of sleep disorders that tend to run in families. The occurrence of a sleep disorder in more than one family member may be a result of genetic susceptibility, cultural practices, or psychosocial factors.

Social History

Assessment of occupational and academic functioning as well as marital satisfaction can yield valuable information about the psychosocial impact of disordered sleep and about potential causes of disturbed sleep, such as marital conflict. When the chief complaint comes from the spouse, the patient may have one of the many

Table 7–8. **Medical, Neurological, and Psychiatric Disorders That Cause or Contribute to Sleep Disturbance**

Medical	Neurological	Psychiatric
Acromegaly	Dementing illnesses	Acute psychosis
Allergic rhinitis	Diencephalic and brainstem disorders	Alcoholism
Anemia		Anxiety disorders
Arthritis	Epilepsy	Dissociative disorders
Chronic lung diseases	Headache disorders	Mood disorders
Chronic renal failure	Painful neuropathies	
Cushing's syndrome	Parkinsonism	
Cystitis	Stroke	
Gastroesophageal reflux	Traumatic brain injury	
Hypothyroidism		
Ischemic heart disease		
Obstructive uropathy		

Table 7–9. **Sleep Disorders with Increased Familial Incidence**

Narcolepsy
Obstructive sleep apnea
Restless legs syndrome
Insomnia
Arousal disorders
Enuresis
Sleep terrors
Fatal familial insomnia

sleep disorders whose symptoms may not be noticed by the patient, but the possibility of marital difficulties should also be considered and explored. In patients complaining of excessive sleepiness, potential occupational hazards should be assessed. The ability of the work environment to provide support for the patient should be evaluated. For children and adolescents, school performance should be determined.

Alcohol, caffeine, nicotine, and illicit drug use should be determined, including frequency of use, time of the day or night that the drug is ingested, and the amount used each day. Caffeine may contribute to insomnia, and nicotine dependence may lead to nocturnal awakenings. Alcohol use or abuse may intensify snoring and obstructive sleep apnea, may be a contributor to insomnia, and in alcoholics may produce long-lasting changes in sleep patterns (see Chap. 16). Cocaine, amphetamines, barbiturates, and opiates can cause major disruptions of sleep architecture.

Examination

Patients with obesity, elevated blood pressure, craniofacial abnormalities, nasal obstruction, or excessive oropharyngeal tissue are at increased risk for obstructive sleep apnea. On abdominal examination, hepatomegaly may suggest that alcohol abuse is contributing to sleep disturbance or, in conjunction with other findings, that congestive heart failure is a factor. Examination of the extremities may reveal evidence of painful musculoskeletal conditions that contribute to disordered sleep. However, although the physical examination often provides clues to the diagnosis, many patients with sleep disorders have no abnormalities on examination.

Findings on mental status testing and neurological examination may indicate substance abuse or the presence of a psychiatric or neurological disease that may be causing or contributing to disturbed sleep. Reduced alertness, slurred speech, and nystagmus may be signs of hypnotic or sedative abuse. Impairment of short-term memory and abstract reasoning suggests the presence of a dementing illness that may be causing insomnia or nocturnal confusion. Psychomotor slowing, downcast eyes, blunted affect, or sad expression may suggest major depression and its associated sleep disturbance. Assessment of mood may suggest the presence of depression or mania, either of which may be associated with insomnia. Delusional thoughts and agitation may indicate that acute psychosis is the cause of insomnia. Impaired sensation and reduced or absent tendon reflexes may indicate peripheral neuropathy, which is sometimes accompanied by sleep disruption caused by nocturnal dysesthesias.

Formulation and Differential Diagnosis

Once the initial evaluation is complete, the clinician is in a position to formulate the clinical problem. For some patients, a diagnosis can be established on the basis of the history and physical examination alone, without laboratory testing or additional evaluation. In others, although a diagnosis may be suspected, laboratory tests are needed to determine either the severity of the disorder or the response to treatment. For example, obstructive sleep apnea may be strongly suspected based on the history and physical examination, but treatment recommendations may depend on the frequency of apneas, the severity of associated hypoxemia and sleepiness, and the occurrence of cardiac arrhythmias. As in other medical conditions, when the diagnosis is uncertain, a differential diagnosis is formulated; based on the relative likelihood of the diagnostic possibilities, appropriate diagnostic studies are obtained.

In patients with excessive sleepiness, the nature and severity of sleepiness alone are rarely helpful in making a diagnosis. Instead, the answers to four important questions are needed:

1. Does the patient snore? If so, the cause of sleepiness may be *obstructive sleep apnea syndrome* (see Chap. 13).
2. Are there episodes of muscular weakness? If so, the patient may have *narcolepsy* (see Chap. 10).
3. Is the patient taking medications? If so, the cause of sleepiness may be one of the several hundred medications that have drowsiness as a side effect.
4. What is the patient's usual sleep schedule? Persons who habitually sleep less than 6 hours, sleep longer on weekends, and feel more rested after longer sleep periods may have *insufficient sleep syndrome* (see Chap. 8).

For patients with complaints of insomnia, the description of sleep disturbance and its course may help to determine etiology. For example, if insomnia is related to psychological stress, such as the recent death of a family member, the diagnosis may be *adjustment sleep disorder* (see Chap. 9) or insomnia secondary to grief or depression (see Chap. 16). In children, the lack of a bedtime routine may suggest *limit-setting sleep disorder* (see Chap. 9). Patients who complain of difficulty falling asleep in the evening followed by normal sleep and difficulty waking up in the morning may have *delayed sleep phase syndrome* (see Chap. 12).

In patients with sleep disturbance caused by medical or psychiatric diseases, the severity of sleep problems usually varies according to the underlying illness. For example, in patients with congestive heart failure, the degree of insomnia may correlate with the cardiac ejection fraction. Similar relationships may be found in patients with sleep disturbance related to depression, asthma, gastroesophageal reflux, and arthritis.

DIAGNOSTIC PROCEDURES

Questionnaires and Sleep Logs

Many clinicians use sleep questionnaires as a source of information. These may be brief questionnaires designed to assess specific symptoms, such as the Epworth Sleepiness Scale,[7] or they may be more detailed instruments that provide medical, psychosocial, and demographic information. The Sleep Disorders Questionnaire includes psychometric scales that have predictive power for diagnoses of sleep apnea, periodic limb movement disorder, narcolepsy, and sleep disturbance related to psychiatric illness.[8]

The sleep log (Fig. 7–1) is particularly useful in patients with insomnia or sus-

	PM						AM											PM						
	6	7	8	9	10	11	12	1	2	3	4	5	6	7	8	9	10	11	12	1	2	3	4	5	
Su					↓									↑	c	c									M
M					↓									↑	c										Tu
Tu					↓									↑	c										W
W					↓									↑	c										Th
Th					↓									↑	c										F
F		A	c			↓									↑	c									Sa
Sa		A	A		A		↓								↑	c									Su

Figure 7–1. Sample of a 1-week sleep log. Abbreviations can be used to indicate events of possible significance to sleep disturbance. In this patient, the sleep log suggests that difficulty falling asleep Friday night may have been due to coffee intake on Friday evening. A sleep-phase delay on the subsequent Sunday evening may have been due to the nap taken on Sunday afternoon. Sleep disruption on Saturday night may have been related to alcohol intake the prior evening. Abbreviations: c, coffee; A, alcohol. Symbols: horizontal line, sleep time; down arrow, time of going to bed; up arrow, time of arising. (Adapted from Spielman et al.,[9] p 532.)

pected circadian rhythm disturbances.[9] At a minimum, the log should include time of going to bed, time of sleep onset, amount of time spent asleep, final awakening time, and nap times. Other events can be recorded as indicated. The sleep log may provide a new perspective on the sleep problem for both the clinician and the patient; some patients are surprised and relieved to note that their sleep is better than they had believed.

Monitoring of Sleep

Monitoring of physiologic variables assists the clinician in determining the presence and severity of sleep disorders. Monitoring can be done at home, in the sleep laboratory, or in the hospital; it may be limited to a single channel of monitoring, such as an activity monitor or oximetry, or it may encompass dozens of channels that assess a variety of physiologic functions. The miniaturization of microprocessors and data storage devices has greatly increased the options for monitoring.

POLYSOMNOGRAPHY

The most widely used monitoring technique for assessing suspected sleep disorders is a polysomnogram performed in a sleep laboratory (Table 7–10).[10,11] By definition, polysomnography includes the three measures used to assess state and to determine sleep stage: electroencephalogram (EEG), submental electromyogram (EMG), and electro-oculogram (EOG). During a typical polysomnogram, respiratory effort, airflow at the nose and mouth, oxyhemoglobin saturation, electrocardiogram (ECG), and leg movements also are assessed with continuous recording throughout the night. Additional sensors are sometimes used to assess expired CO_2 concentration, esophageal pH, body position, snoring sound, and intrathoracic pressure. For suspected parasomnias, electrodes can be applied to all four limbs to monitor movement, and additional electrodes can be applied to the scalp to provide more extensive EEG monitoring. Simultaneous audio and video monitoring adds to the polygraphic information.

Polysomnographic paper recordings are usually made with polygraphs that display the data on paper at a recording speed of 10 or 15 mm/s. Polysomnographic data can be also displayed on a video terminal, stored on optical or magnetic media, and analyzed and scored without a paper recording. The accuracy of the display and the ease of its use depend on the method used to transfer data to the screen, the sampling rate, the screen resolution, and the accompanying software. Laboratory-based polysomnography is indicated in a number of clinical settings.[10,11] Although its routine use in the evaluation of insomnia is not indicated, it is helpful for some patients with refractory insomnia and for those suspected to have sleep-state misperception.

Polysomnography can also be performed at home. With digital equipment,

Table 7–10. **Indications for Polysomnography**[10,11]

1. Diagnosis of any of the following suspected disorders:
    Sleep-related breathing disorders
    Narcolepsy or idiopathic hypersomnia
    Nocturnal seizures, when clinical evaluation and daytime EEG are inconclusive
    Violent or potentially violent parasomnias
    Periodic limb movement disorder
2. Titration of continuous positive airway pressure in patients with sleep-related breathing disorders
3. Follow-up after treatment of sleep-related breathing disorders with surgery, oral appliances, and weight loss.

it is now possible to record 18 or more channels with lightweight portable devices that transmit data via telephone or store it for subsequent review at the laboratory. The major advantages of home polysomnography include the clinician's ability to record the patient in the usual sleep setting and the reduced cost compared to laboratory polysomnography. The disadvantages include the clinician's inability to observe the patient during the procedure, correct any technical problems, and make therapeutic interventions.

Devices that record only a few physiologic variables are also used. Many of these devices do not include EEG and therefore are not considered polysomnography, but they may still provide considerable information in specific clinical settings. Activity monitoring devices that record body movement can be used for several weeks at a time; the activity level allows the physician to obtain an approximation of times of sleep and wakefulness. Devices that record the ECG or esophageal pH also are available.

The interpretation of nocturnal monitoring studies requires skill and experience, in part because for many sleep variables, the age-related normal range is unknown, arbitrary, or controversial. For example, the normal range (mean ± 3 SD) for periodic limb movements per hour of sleep in 60-year-olds is unknown. For hypopneas, the normal range depends on which definition of hypopnea is used and the type of equipment used to monitor breathing. Although individual laboratory norms are of great value, they are probably impractical to establish for most laboratories.

In the assessment of the results of the studies, the sleep specialist must consider the possibility of night-to-night variability. For example, apneas and periodic limb movements may be absent on one night and present the next, or they may differ substantially in frequency from one night to the next. Although laboratory studies can be interpreted in the absence of clinical information, their interpretation is more helpful when correlated with the clinical evaluation.

Multiple Sleep Latency Test and Maintenance of Wakefulness Test

Daytime sleep laboratory testing is also commonly used, particularly the Multiple Sleep Latency Test (MSLT). During the MSLT, which is usually performed the day after a polysomnogram, the patient is asked to take four or five brief naps at 2-hour intervals; sleep patterns are monitored with EEG, EOG, and EMG (the protocol for each test period is standardized).[12] The mean time from lights out to the onset of sleep (mean sleep latency) for each nap opportunity provides an objective measure of daytime sleepiness, and the occurrence of REM sleep during the naps is helpful in the diagnosis of narcolepsy. The Maintenance of Wakefulness Test is a variation of the MSLT: Subjects are instructed to try to stay awake while in a darkened room and seated in a chair, rather than lying in bed (see Chap. 8).[13,14]

As with a polysomnogram, the MSLT must be interpreted in the clinical context. First, the degree of sleepiness can vary from one day to the next, and the occurrence of frequent sleep-onset REM periods does not always indicate narcolepsy (see Chap. 10). Second, the results of the test are valid only if the patient is able and willing to follow the instructions to relax and to try to sleep; thus, the findings are less useful in very young children and in acutely anxious, mentally retarded, or psychotic patients. Third, the MSLT measures the propensity to fall asleep in a setting conducive to sleep: It does not measure the likelihood of falling asleep in other settings, such as while driving or at work. Finally, although the MSLT has been validated in a number of ways and shows a strong correlation with the severity of acute sleep deprivation, it does not correlate well with subjective measures of the severity of chronic sleepiness.

Scoring of Sleep Studies

Sleep studies are scored for wakefulness and sleep stages in order to provide a way of summarizing the architecture of sleep throughout the night and to allow for comparisons between sleep studies. A standard scoring system, developed in the 1960s, has been used to score millions of nights of human sleep over the past three decades.[15] With this approach, each study is analyzed in discrete epochs of 10 to 60 seconds, usually 30 seconds, and a sleep stage is assigned to each epoch. The scoring results can then be tabulated and displayed as a *hypnogram* of sleep stages that provides a quicker overview of a night's sleep than a long list of numbers representing the sleep stages. Hypnograms can be used to correlate apneas and leg movements with sleep stages and body positions (Fig. 7–2).

The standard scoring system is reliable and accurate—in one study,[16] interscorer agreement was 88%—but it is also labor intensive because someone, usually a technologist trained to recognize sleep stages, must assign a sleep stage to each of the approximately 1000 sleep epochs that occur each night. In addition, clinical polysomnograms are usually scored for respiratory events, for periodic leg movements, and sometimes for additional events, such as oxyhemoglobin desaturations and arousals. For a patient with highly disrupted sleep, the scoring process may take several hours.

Given the amount of time required to score polysomnograms, it is not surprising that many researchers and computer scientists have tried to develop accurate computerized systems for scoring polysomnograms. Assuming that 8 bits of information were used to describe a single data point, and 100 data points were collected per second, a single 8-hour, 12-channel

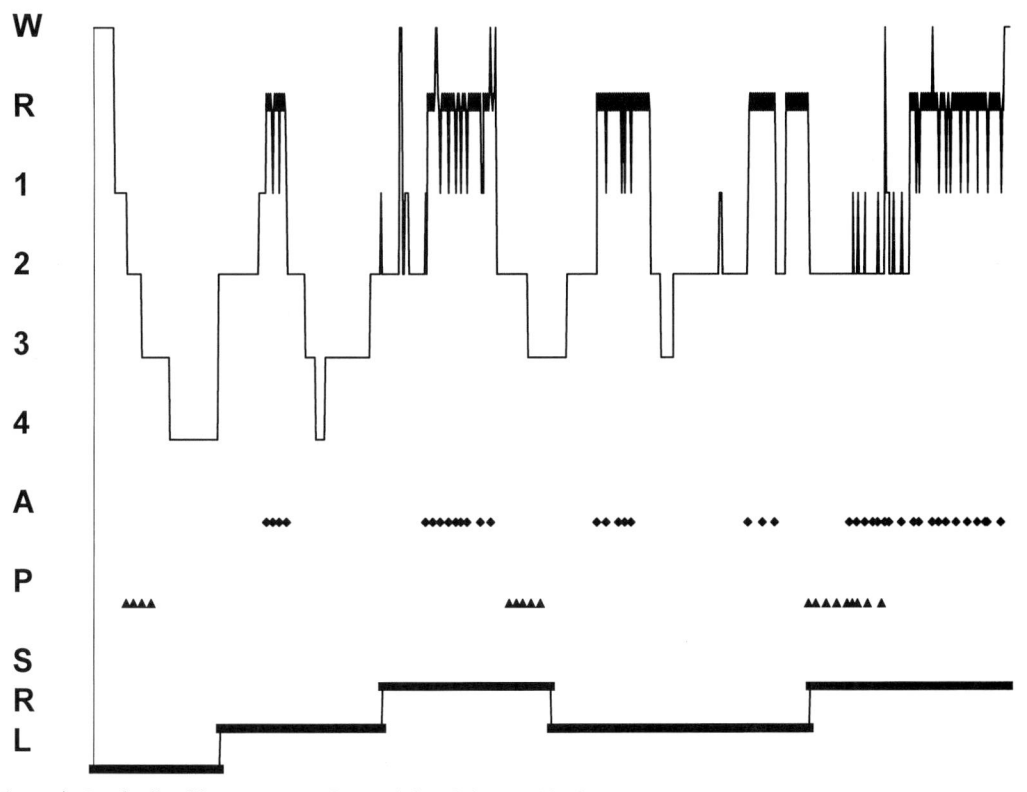

Figure 7–2. Idealized hypnogram of one night of sleep, with sleep stages (W, wake; R, REM; 1–4, NREM sleep stages); apneic episodes (A), periodic leg movements (P); and body position (S, supine; R, right lateral decubitus; L, left lateral decubitus). The hypnogram demonstrates that apneic episodes (diamonds), which occur mainly while the patient is supine or in REM sleep, are associated with transitions to stage 1 sleep, whereas periodic leg movements (triangles) are associated with little or no sleep disruption.

polysomnogram would contain more than 200 million bits of information. Currently available desktop and laptop computers can analyze this amount of information in a few minutes, and in theory, computerized signal processing with frequency and power analysis has the potential to reduce the workload of the technologist, improve scoring accuracy, and provide data for research.

In practice, the potential advantages of fully automated scoring systems have been realized only to a limited degree. Automated scoring systems tend to have difficulty identifying artifacts, brief arousals, and brief periods of wakefulness.[17,18] Thus, these systems are less valid and reliable than human scoring in cases involving abnormal baseline EEG activity, alpha-delta sleep, and age-related changes in normal background activity. In 1990, the American Sleep Disorders Association[19] concluded: ". . . to date, automatic scoring systems have not been sufficiently validated with respect to clinical sleep scoring and interpretation." Although computer programs can also prepare topographic maps of EEG activity and display power analysis and frequency analysis of EEG activity, the current clinical utility of such analyses and displays for diagnosis and management of sleep disorders is minimal.[20]

An alternative to fully automated scoring is computer-assisted analysis, in which the computer program's preliminary judgment of sleep stages is followed by human review, or the program analyzes portions of EEG or other activity, such as delta-wave activity, to assist the human scorer. Differentiation of sleep stages 2 and 3, which is based entirely on the amount of delta activity, is one of the most difficult tasks for humans, and analysis of delta-wave activity is one area in which scoring by computer is potentially superior.[21]

Computers can also be used to assist with respiratory analysis (i.e., by performing a preliminary classification of respiratory events) and to score periodic limb movements.[22] Although these systems have potential, currently available automated systems are insufficiently sensitive to be used as screening instruments for sleep-related breathing disorders. In one study, a computerized system detected only 78% of respiratory events, and the respiratory index calculated by the system was only moderately well correlated with polygraphically determined events.[23] Algorithms for scoring limb movements are generally more sensitive and specific than those used for scoring respiratory events.[24]

Non-EEG-based systems use a variety of other physiologic functions to identify sleep state. For example, motility patterns based on respiratory movements and body movements have been used in infants to differentiate quiet sleep, active sleep, and wakefulness.[25] Eye movements and body movements can be used to distinguish wakefulness from NREM sleep and from REM sleep,[26] and some systems are based on body movement alone. For example, a study of computer-analyzed wrist-actigraph patterns revealed an 88% correct classification of sleep versus wakefulness.[27]

Ancillary Tests

Radiologic and laboratory tests may clarify or refine the diagnosis, or they may indicate the condition's severity. If the diagnosis of narcolepsy is under consideration, tissue typing for specific human leukocyte antigens, implicated in the genetics of narcolepsy, is sometimes helpful. Thyroid function tests or pulmonary function tests may be indicated in some patients with suspected obstructive sleep apnea. Uremia, iron-deficiency anemia, or other metabolic abnormalities may be present in patients with suspected periodic limb movement disorder or restless legs syndrome. Pulmonary function tests, including arterial blood gases, are helpful if alveolar hypoventilation syndrome caused by obesity or neuromuscular disease is a consideration. Urine toxicology screening sometimes uncovers evidence of drug abuse that may be contributing to or causing sleep disturbance.

MANAGEMENT

To formulate a plan of management, the clinician must integrate the information obtained from laboratory testing with the clinical evaluation. If the diagnosis is clear, the clinician is in a position to recommend a particular treatment or to discuss treatment options with the patient. If the diagnosis is unknown or has not been established with reasonable certainty, the clinician must decide whether to obtain additional clinical or laboratory evaluation or to treat the patient based on the available information. Assessment of response or lack of response to treatment along with follow-up evaluation may alter the initial diagnostic impression. For each patient, the skillful clinician maintains an open mind concerning the cause of the symptoms.

SUMMARY

Although sleep problems have been recognized for centuries, they were generally viewed as consequences of other medical or psychiatric problems, rather than as primary diseases or illnesses themselves, until well into the 20th century. With the development of techniques for monitoring sleep and the subsequent discoveries of sleep apnea and periodic limb movements, it became possible to assess sleep and sleep disturbance more systematically and to develop a nosology of sleep disorders. It is now clear that sleep disorders are common, often serious, and frequently treatable.

Evaluation of the patient with sleep-related complaints begins with the chief complaint and the sleep history. As with other medical and psychiatric problems, essential information is provided by the patient's medical, psychiatric, and family history; the history of medication use; the psychosocial assessment; and the physical, neurological, and mental status examinations.

The clinical evaluation, combined with knowledge about the manifestations of sleep disturbances, provides the clinician with information that generally yields a provisional diagnosis or a set of diagnostic possibilities, which can then be confirmed or excluded based on the results of diagnostic investigations. Most patients with accurately diagnosed sleep disorders respond well to treatment.

REFERENCES

1. Gastaut, H, et al: Étude polygraphique des manifestations épisodiques (hypniques et respiratoires) du syndrome de Pickwick. Rev Neurol 112:568, 1965.
2. Jung, R, and Kuhlo, W: Neurophysiological studies of abnormal night sleep and the pickwickian syndrome. Prog Brain Res 18:140, 1965.
3. Lugaresi, E, et al: A propos de quelques manifestations nocturnes myocloniques (nocturnal myoclonus de Symonds). Rev Neurol (Paris) 115:547, 1966.
4. Association of Sleep Disorders Centers: Diagnostic Classification of Sleep and Arousal Disorders, ed 1. Prepared by the Sleep Disorders Classification Committee, HP Roffwarg, Chmn. Sleep 2:1, 1979.
5. American Sleep Disorders Association: International Classification of Sleep Disorders, Revised: Diagnostic and Coding Manual. American Sleep Disorders Association, Rochester, Mn, 1997.
6. Eyler, AE: Sexuality issues and common sexual dysfunctions: Evaluation and management in the primary care setting. In Knesper, DJ, et al (eds): Primary Care Psychiatry. WB Saunders, Philadelphia, 1997, pp 367–386.
7. Johns, MW: A new method for measuring daytime sleepiness: The Epworth sleepiness scale. Sleep 14:540, 1991.
8. Douglass, AB, et al: The Sleep Disorders Questionnaire: I. Creation and multivariate structure of SDQ. Sleep 17:160, 1994.
9. Spielman, AJ, et al: Insomnia. Neurol Clin North Am 14:513, 1996.
10. Indications for Polysomnography Task Force, American Sleep Disorders Association Standards of Practice Committee: Practice parameters for the indications for polysomnography and related procedures. Sleep 20:406, 1997.
11. Chesson, Jr, AL, et al (ASDA polysomnography task force): The indications for polysomnography and related procedures. Sleep 20:423, 1997.
12. Carskadon, MA, et al: Guidelines for the multiple sleep latency test (MSLT): A standard measure of sleepiness. Sleep 9:519, 1986.
13. Mitler, MM, et al: Maintenance of wakefulness test: A polysomnographic technique for evaluating treatment efficacy in patients with excessive somnolence. Electroencephalogr Clin Neurophysiol 53:658, 1982.
14. Doghramji, K, et al: A normative study of the Maintenance of Wakefulness Test (MWT). Electroencephalogr Clin Neurophysiol, 103:554, 1997.
15. Rechtschaffen, A, and Kales, A: A Manual of Standardized Terminology, Techniques, and Scoring System for Sleep Stages of Human Subjects. Brain Information Service/ Brain Research Institute, Los Angeles, 1968.

16. Hoelscher, TJ, et al: Two methods of scoring sleep with the Oxford Medilog 9000: Comparison to conventional paper scoring. Sleep 12:133, 1989.
17. Kubicki, S, et al: Sleep EEG evaluation: A comparison of results obtained by visual scoring and automatic analysis with the Oxford sleep stager. Sleep 12:140, 1989.
18. Palm, L, et al: Automatic versus visual EEG sleep staging in preadolescent children. Sleep 12:150, 1989.
19. American Sleep Disorders Association Position Statement: Automatic scoring. Sleep 13:284, 1990.
20. Nuwer, M: Quantitative EEG: I. Techniques and problems of frequency analysis and topographic mapping. J Clin Neurophysiol 5:1, 1988.
21. Bunnell, DE, and Horvath, SM: Evaluation of a filter/integrator system for quantifying delta activity in the sleep electroencephalogram. Sleep 9:365, 1986.
22. Lord, S, et al.: Interrater reliability of computer-assisted scoring of breathing during sleep. Sleep 12:550, 1989.
23. Gyulay, S, et al: Evaluation of a microprocessor-based portable home monitoring system to measure breathing during sleep. Sleep 10:130, 1987.
24. Kayed, K, et al: Computer detection and analysis of periodic movements in sleep. Sleep 13:253, 1990.
25. Thoman, EB, and Glazier, RC: Computer scoring of motility patterns for states of sleep and wakefulness: Human infants. Sleep 10:122, 1987.
26. Mamelak, A, and Hobson, JA: Nightcap: A home-based sleep monitoring system. Sleep 12:157, 1989.
27. Cole, RJ, et al: Automatic sleep/wake identification from wrist activity. Sleep 15:461, 1992.

Chapter 8

SLEEPINESS AND SLEEP DEPRIVATION

SLEEPINESS
Clinical Features of Sleepiness
Causes of Sleepiness
Central Nervous System Basis of Sleepiness
Measures of Sleepiness
TOTAL SLEEP DEPRIVATION
Human Studies
Animal Studies
INSUFFICIENT SLEEP SYNDROME
Clinical Features
Psychobiologic Basis
Diagnosis
Management

It is night. Nanny Varka, a girl of about thirteen, is rocking a cradle with a baby in it. . . . The baby is crying. It has long since grown hoarse and exhausted from crying, but it still shrills. . . . And Varka is sleepy. She can hardly keep her eyes open, her head is weighed down, her neck aches. She cannot move either her eyelids or her lips . . . This lulling music . . . makes you drowsy, and you must not sleep: should Varka, God forbid, fall asleep, her employers will beat her . . . The shadows are set in motion, they force their way into Varka's half-open staring eyes and form into hazy dreams . . . She sees dark clouds which chase each other across the sky . . . and a wide highway covered with watery mud . . . and forests. Varka goes into the woods and . . . suddenly someone hits her on the back of the head . . . She raises her eyes and sees her master, the cobbler . . . "What d'you think you're doing, you mangy brat? The baby is crying and you're asleep." . . . Varka feels a passionate longing for sleep . . . Outside it will soon be morning . . . but she still can hardly keep her eyes open and her head is heavy . . . "Varka, light the stove!" the mistress's voice is heard . . . Varka leaves the cradle and runs to the shed for firewood. She is glad. When you run or walk you are not so sleepy. . . . Varka scrubs the steps, tidies the rooms . . . there is a lot of work, not one free moment. But nothing is so hard as standing still in front of the kitchen table and peeling potatoes. Her head sags down to the table, the potatoes swim before her eyes, her hands can hardly hold the knife . . . The day passes . . . At last . . . the lights are put out, the master and mistress go to bed. "Varka, rock the baby!" the last order rings out . . . Varka's half-open eyes flicker, and befuddle her brain . . . through her doze she just cannot understand this force which shackles her hand and foot, weighs down on her, and interferes with her life. . . . At last, exhausted, she strains with all her strength and . . . finds the enemy that interferes with her life. That enemy is the baby. . . . The false notion takes hold of Varka. She gets up from her stool. . . . steals up to the cradle and bends over the baby. Having strangled it, she quickly lies down on the floor, laughing with joy now that she can sleep, and a minute later is already sleeping as soundly as if she were dead.

ANTON CHEKHOV, SLEEPY

The overwhelming desire for sleep, the cognitive and perceptual disturbances, and the altered judgment that accompany sleep deprivation (all shown in Chekhov's story excerpted above) have been recognized for centuries. Soldiers suffer severe sleep deprivation during military campaigns; the military science of sleep logistics balances the cost of man-hours lost to sleep against the beneficial effects of sleep on performance and morale.[1] Sleep deprivation is often a technique of torture and interrogation. Physicians, particularly those in training, generally lose a great deal of sleep.

Although nineteenth-century servant girls such as the fictional Varka often worked long hours, sleepiness caused by insufficient amounts of sleep is probably more common in the twentieth century than in earlier times. By some estimates, the average person now sleeps about 60 to 90 minutes less than in 1910, largely because of the ready availability of electrical power. Electric light bulbs and television contribute to later hours of wakefulness at home, and electric power for factories permits greater use of shift work and night work. Annual work hours increased by about 10% for the average American between 1969 and 1987, and the proportion of the population involved in shift work is probably higher now than ever before.[2,3] As a result, there is less time for sleep: Working adults sleep 7 to 8 hours per night on average, compared to 8 to 9 hours for working adults 40 to 50 years ago, and feel substantially less rested in the morning.[2] Thus, many otherwise healthy adults accumulate a sleep debt as a result of chronically insufficient amounts of sleep.[4]

SLEEPINESS

Sleepiness can be defined as a physiological state or a subjective condition. *Physiological sleepiness* is a state of the central nervous system determined by sleep need. It reflects a drive for sleep, just as hunger is a physiological manifestation of the drive for food. *Manifest sleepiness*, the subjective expression of the need for sleep, is the complex of sensations, emotions, and thoughts that accompany the need for sleep.

In the same way that a feeling of hunger does not always accompany insufficient caloric intake, a feeling of sleepiness does not always accompany reduced sleep. For example, a person who is deprived of sleep for one night and who feels sleepy while sitting quietly the next afternoon may feel less sleepy after getting up and walking around for a few minutes. The act of walking around does not alter the underlying *physiological sleepiness* or sleep need, but it does affect the subjective or manifest expression of the need. Just as hunger is normal when the organism has a need for food, sleepiness is normal when sleep need is high, such as at bedtime or after sleep deprivation.

Hunger is also normal at meal times, and the sensation of hunger can be conditioned to a variety of situations and stimuli. Similarly, sleepiness and rapid onset of sleep can be conditioned to situations and stimuli. For example, in experimental studies, animals can be conditioned to fall asleep quickly in response to specific stimuli that have been previously linked to sleep onset by being paired with sleep-inducing brain stimulation.[5] In humans, sleepiness and rapid onset of sleep can become associated with and elicited by stimuli associated with sleep onset, such as preparations for bed and the act of getting into bed. This conditioning can have positive effects, and it is part of the reason why regular sleep habits tend to promote good sleep, but it can also have harmful effects, as in a person who has developed a conditioned association of driving with drowsiness.

As with virtually all physiological functions, sleepiness has a circadian rhythm. In most persons who maintain a regular schedule of nighttime sleep, the major peak of sleepiness accompanies the nadir of the circadian temperature rhythm at about 3 to 5 am. A second, smaller peak of sleepiness occurs about 12 hours later in the middle of the afternoon (see Fig. 4–1).[6] The circadian variation in sleepiness continues even after 24 to 48 hours of sleep deprivation (see Fig. 4–2).

Clinical Features of Sleepiness

Sleepiness that occurs at inappropriate or undesirable times or that interferes with daytime activities is generally considered excessive by patients and clinicians. Symptoms associated with excessive sleepiness include too much sleep, unavoidable napping, frequent sensations of "fighting to stay awake," and impaired concentration. Sleepiness and falling asleep are often most noticeable in persons at rest while watching television, reading, riding in a car, or sitting in church, class, or a theater.

The situations associated with sleepiness help to determine its severity. Mild sleepi-

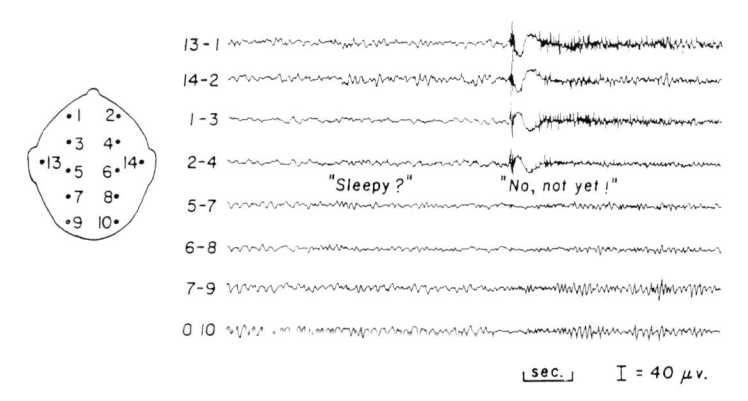

Figure 8–1. Denial of sleepiness in a patient with narcolepsy. The initial portion of the electroencephalogram shows patterns consistent with stage 1 sleep. When asked if she is sleepy, she awakens and replies, "No, no yet!" The awakening is accompanied by muscle and movement artifact in the top four channels and the appearance of alpha rhythm in the bottom two channels. (From Daly and Yoss,[8] p. 113, with permission.)

ness may be limited to sedentary situations in which falling asleep is socially acceptable, such as watching TV or reading at home. Sleepiness of moderate severity may be noticeable at work, while driving, or while sitting on the toilet. Severely sleepy patients may fall asleep while standing, during a conversation, or during sexual intercourse.

Awareness of sleepiness varies. Most persons who undergo acute sleep deprivation are fully aware that their alertness is reduced and that they may fall asleep easily. On the other hand, persons with chronic sleepiness are often less aware of it and may deny feeling sleepy (Fig. 8–1).[7,8] For a list of the various signs associated with drowsiness, see Table 8–1.

Table 8–1. Signs of Drowsiness

Eye rubbing
Decreased blinking rate
Glazed, unfocused eyes
Slow eye movements
Heavy-liddedness (ptosis)
Closed eyes
Fidgeting
Yawning
Slumped posture
Reduced activity
Slack facies
Head nodding
Sleep-seeking behavior

In surveys, 4% to 15% of the population—women more often than men—describe themselves as excessively sleepy or report that sleepiness interferes with daytime activity.[9–13] Young adults are sleepier on average than middle-aged adults, but after about age 60, daytime sleepiness begins to increase.[14] More than half of persons who report excessive sleepiness have had automobile accidents or occupational accidents caused by falling asleep.

Causes of Sleepiness

Excessive sleepiness is often a result of insufficient amounts of sleep or of disorders or conditions that cause disrupted sleep. Sleepiness can also be caused by circadian rhythm disorders, medications, and brain disturbances, which may have interactive effects as well. For example, the sedative effects of alcohol and hypnotics are more pronounced in sleepy persons; irregular sleep-wake schedules that result from jet lag, shift work, or erratic times of going to bed and arising produce sleepiness in part by disrupting normal sleep patterns and in part because of circadian rhythm disturbance.[15,16] Finally, many persons with depression describe themselves as sleepy. In a community-based study, 25% of persons with complaints of daytime sleepiness had symptoms suggestive of moderate or severe depression.[13]

INSUFFICIENT AMOUNTS OF SLEEP

Sleep deprivation, either partial or total, leads to sleepiness that increases in a predictable fashion as the duration of wakefulness increases, modified by the circadian variation in sleepiness. Even one night of total sleep deprivation, or sleep reduction by as little as 1 hour per night for several nights, can lead to levels of sleepiness that are generally considered pathologic. Conversely, most young adults become more alert if they increase their nightly amount of sleep beyond the customary 7 to 8 hours (Fig. 8–2),[17–19] and many persons who sleep late on weekends or days off do so because of insufficient sleep during the week. Falling asleep during the day, even in boring and passive situations, is not "normal" in that fully rested persons do not do so. The cause, however, may be insufficient sleep rather than another sleep disorder. Chronic partial sleep deprivation (*insufficient sleep syndrome*) is one of the most common causes of sleepiness in Western societies (see Insufficient Sleep Syndrome below).

SLEEP FRAGMENTATION

Disrupted sleep—sleep that is fragmented by repeated arousals and awakenings—is another major cause of sleepiness and is less restful than the same amount of continuous sleep. Sleepiness induced by sleep fragmentation seems to be a function of the duration of intervals of uninterrupted sleep, as periods of sleep of greater than 10 minutes are more restorative than periods of sleep of less than 10 minutes. When periods of uninterrupted sleep are reduced to less than 2 minutes, the restorative quality of sleep is markedly reduced.[20,21] In one study, subjects who were allowed to sleep continuously during stage 1 sleep were awakened after only 1 minute of stage 2 sleep or of rapid eye movement (REM) sleep. Although sleep time exceeded 4 hours per night, performance decrements and mood changes after two such nights were similar to those seen in subjects who had undergone 40 to 64 hours of sleep deprivation.[22,23]

Frequent arousals that disrupt sleep continuity are a major cause of sleepiness in patients with sleep apnea, although other factors, such as the increased metabolic rate that accompanies frequent arousals or increased sympathetic activity, may also play a role (see Chap. 13).

DIFFUSE CEREBRAL DYSFUNCTION

Depressed or aberrant cerebral function can also lead to drowsiness. When there is a diffuse disturbance of brain function, as

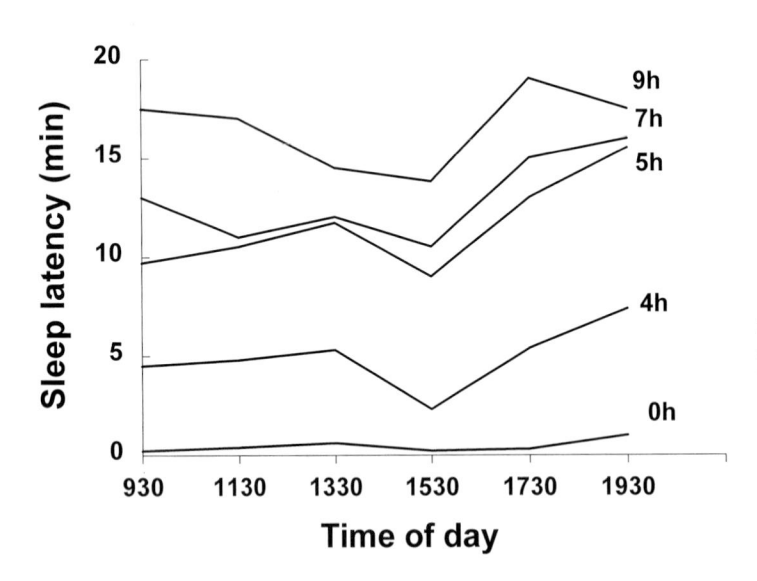

Figure 8–2. Change in sleepiness depending on hours of sleep at night. Young subjects spent two nights with sleep times as shown (h = hours of sleep at night). On MSLTs performed the following day, sleep latencies declined with fewer hours of nighttime sleep. (Redrawn from Roth et al.,[18] p 44, with permission.)

occurs for example with metabolic encephalopathies, the level of consciousness is reduced and the mental status passes through a series of stages from full alertness to coma (Fig. 8–3). Although sleep and coma are physiologically distinct, sleepiness, when operationally defined as a reduced latency to sleep, is part of the continuum from consciousness to coma just as it is part of the continuum between sleep and wakefulness. Thus, patients with progressive hepatic or uremic encephalopathy become drowsy before they develop obtundation and stupor.

Sedative drugs (e.g., hypnotics, opiates, ethanol, barbiturates) induce sleepiness through their effects on cerebral function. Although the precise mechanisms by which they produce sleepiness are not known, an overall depressant effect of cerebral function mediated by facilitation of GABAergic neurotransmission is probably partially responsible. Decreased metabolic activity of the thalamus induced by benzodiazepines may also contribute to their soporific effects.[24] Antihistamines, antidepressants, and many antihypertensives also can induce daytime somnolence.

It is difficult conceptually, and sometimes clinically, to distinguish between drug-induced sleep and drug-induced stupor or coma. Benzodiazepines, ethanol, and barbiturates promote sleep at low doses but induce coma at high doses, and there is no clear line separating the alcoholic who is "sleeping it off" from one in an alcoholic stupor. Depressed respirations, lack of arousability, and physiological changes associated with particular drugs are helpful signs in these clinical situations.

FOCAL CEREBRAL DYSFUNCTION

Focal lesions affecting areas of the brain involved in sleep-wake regulation can also cause sleepiness, either by disrupting arousal mechanisms or by enhancing sleep mechanisms. Such lesions are often bilateral, involving the brainstem or diencephalon (see Chap. 19). Narcolepsy and idiopathic hypersomnia (see Chap. 10) are special categories of focal brain disease; the lesions or abnormalities causing sleepiness in these disorders have not been identified.

PSYCHIATRIC DISORDERS

Some patients with psychiatric illnesses also complain of sleepiness. For example, patients with atypical depression may have sleepiness and increased sleep time

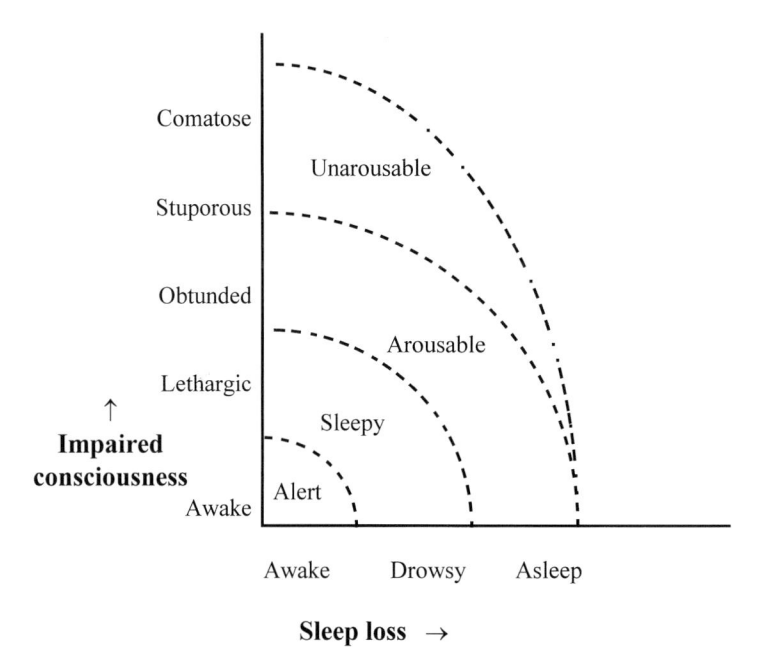

Figure 8–3. Relation between sleep loss and pathological alterations of consciousness. Sleepiness can occur with either process or with a combination of the two. Obtunded patient appear asleep but are arousable, whereas comatose patients are unarousable.

(see Chap. 16). In most such cases, sleepiness is mainly subjective, and the increased time in bed reflects a change in behavior rather than an increase in sleep need.

Central Nervous System Basis of Sleepiness

Although the neurological basis for sleepiness is unknown, conceptually it is caused by a decline in activity of systems involved in the maintenance of wakefulness, or by increased activity of systems involved in sleep initiation, or both (see Chap. 2). For example, altered neuronal traffic in monoaminergic systems or in other reticular activating system pathways could lead to reduced cortical activation and associated sleepiness. Alternatively, an increase in the activity of hypnogenic neurons of the basal forebrain could lead to sleepiness. An increase in an endogenous sleep-promoting substance or a decrease in a substance that facilitates wakefulness could also cause sleepiness.

Measures of Sleepiness

Sleepiness can be assessed with subjective measures, performance tests, and neurophysiologic procedures. The two most frequently used subjective assessments are the Epworth Sleepiness Score (Table 8–2),[25] which provides an index of sleepiness during common daily activities, and the Stanford Sleepiness Scale (Table 8–3),[26] which assesses sleepiness at a single moment. Although subjective scales usually provide an accurate assessment of acute sleepiness after sleep deprivation, they are less useful indicators of the degree of sleep need caused by chronic partial sleep deprivation or by disorders that cause chronic sleepiness, such as narcolepsy[27] or sleep apnea.[7,28] Moreover, some patients who are chronically sleepy rate themselves as alert even while in the process of falling asleep (see Fig. 8–1).[7,8] The discrepancy is probably caused by the lack of a frame of reference; persons who are chronically sleepy may not have experienced full alertness for years and may therefore be unable to assess the severity of their own sleepiness.

Table 8–2. **Epworth Sleepiness Scale**

How likely are you to doze off or fall asleep in the following situations, in contrast to just feeling tired? This refers to your usual way of life in recent times. Even if you have not done some of these things recently, try to work out how they would have affected you. Use the following scale to choose the *most appropriate number* for each situation:

  0 = would *never* doze
  1 = *slight* chance of dozing
  2 = *moderate* chance of dozing
  3 = *high* chance of dozing

Situation	Chance of dozing
Sitting and reading	_____
Watching TV	_____
Sitting, inactive in a public place (e.g., a theater or a meeting)	_____
As a passenger in a car for an hour without a break	_____
Lying down to rest in the afternoon when circumstances permit	_____
Sitting and talking to someone	_____
Sitting quietly after a lunch without alcohol	_____
In a car, while stopped for a few minutes in traffic	_____

Source: From Johns,[25] p. 541, with permission.

Table 8–3. **Stanford Sleepiness Scale**

Circle the one number that best describes your level of alertness or sleepiness right now.

SCALE	CHARACTERISTICS
1	Feeling active and vital; alert; wide awake
2	Functioning at a high level; but not at peak; able to concentrate
3	Relaxed; awake; not at full alertness; responsive
4	A little foggy, not at peak; let down
5	Fogginess, beginning to lose interest in remaining awake; slowed down
6	Sleepiness; prefer to be lying down; fighting sleep; woozy
7	Almost in reverie, sleep onset soon; lost struggle to remain awake
X	Asleep

Source: From Hoddes et al.,[26] p. 431, with permission.

PERFORMANCE TESTS

Performance tests can be used as a measure of sleepiness. The impact of sleepiness on performance is a function of the duration of the task: the longer the task, the greater the impairment induced by sleepiness. The type of task also determines the impact of sleepiness: Tasks requiring high levels of vigilance are particularly sensitive to the effects of sleep deprivation, whereas tasks that involve intense and varied visual stimulation or that require constant or near-constant movement are less susceptible to the effects of sleep loss. With one commonly used test, the Wilkinson Auditory Vigilance Test, subjects listen to a series of tones and must press a button when a tone of slightly different pitch is presented. The reaction time and the number of errors are sensitive to the effects of sleep loss. Incentives also affect performance testing: Subjects perform better on the Wilkinson Auditory Vigilance Test and similar tests if they are given a monetary or other incentive to do well.[29]

NEUROPHYSIOLOGIC TESTS

Several neurophysiologic tests can be used to assess sleepiness. With pupillometry, pupil size and reactivity are assessed under standard conditions; as sleepiness increases, the pupil becomes smaller, reacts less briskly, and may show increased spontaneous variability or hippus.[30] The amount of slow eye movement activity on the electro-oculogram, the amounts of theta and delta EEG activity, and the topography, frequency, and amplitude of the alpha rhythm are other measures that correlate with sleepiness.[30] On evoked potential studies, drowsiness is associated with a reduction in the amplitude of the P3 component and decreased amplitude in the contingent negative variation, apparently because of reduced cognitive processing of sensory stimuli.[31]

The Multiple Sleep Latency Test (MSLT) (see Chap. 7), the most widely used polygraphic measure of sleepiness, is based on the postulate that the time required to fall asleep in settings conducive to sleep is the best indicator of the internal drive to sleep. In addition to its face validity, there are many indications that the MSLT is a valid and accurate measure of sleepiness with high test-retest reliability.[32,33] Partial or total sleep deprivation and sedating drugs decrease the mean sleep latency; conversely, sleep extension increases the mean sleep latency.

On the other hand, the time required to fall asleep is affected by a number of influences. Anxiety and pain, for example, may cause difficulty in falling asleep even when sleep need is high, presumably by increasing cortical activation. At the opposite end of the spectrum, some persons are able to relax and fall asleep easily even when they have little sleep debt.[34] Thus, although the MSLT is a powerful measure of sleepiness, it is incorrect to assume that it is an exact measure of sleep need.

The mean sleep latency on the MSLT averages 15 to 20 minutes in preadolescents, 10 minutes in young adults, 11 to 12 minutes in middle-aged adults, and 9 minutes for subjects in the sixth decade.[35,36] After one night of sleep deprivation, sleep latency is less than 5 minutes; after two nights without sleep, sleep latency is usually less than 1 minute.

In the Maintenance of Wakefulness Test (MWT), a variation of the MSLT, subjects are instructed to try to stay awake rather than to try and fall asleep.[37,38] As with the MSLT, the mean of several attempts to stay awake provides an indication of the subject's ability to remain awake. As one would expect, the MWT does not correlate closely with the MSLT in a given subject,[39] probably mainly because the subject's motivation to stay awake has a marked influence on sleep latency with the MWT but not with the MSLT. Although on the surface, the MWT appears more suitable than the MSLT for assessing the ability to stay awake when trying to do so, the MWT does not accurately reflect the circumstances in which most persons find themselves when trying to stay awake, mainly because the consequences of falling asleep vary widely. For example, the potentially disastrous consequences of falling asleep while driving an automobile are much different from the consequences of falling asleep during an MWT in a sleep laboratory.

TOTAL SLEEP DEPRIVATION

Sleep deprivation is familiar to everyone, and the overwhelming sleepiness and desire for sleep that accompany sleep deprivation suggest that sleep has vital functions. Scientists have tried for more than a century to determine the function of sleep by depriving humans and animals of sleep. These efforts have produced a wealth of information about the effects of sleep deprivation, but the essential mystery of the function of sleep remains unsolved.

Human Studies

In 1896, Patrick and Gilbert[40] reported the effects of 90 hours of continuous wakefulness on one of the investigators, Allen Gilbert, and on two other subjects. Their findings include most of the major effects of sleep deprivation in humans that are currently known. Gilbert suffered no major ill effects, although he became severely sleepy during the second night of sleep loss and thereafter would fall asleep if left alone. Sleepiness was worse during nighttime hours than daytime hours, and despite the efforts of the experimenters to keep him awake, he had frequent brief naps lasting a few seconds. His performance on vigilance tasks declined over time, although it improved with encouragement. He had visual illusions and hallucinations during the latter portions of the experiment and attempted to swat away hallucinations of colored gnats. During recovery sleep, Gilbert was difficult to arouse during the first 3 hours, but he felt refreshed after 2 nights of recovery sleep even though he had made up only about one-fourth of the lost sleep.

The longest documented time without sleep by a human is 264 hours. The subject, a 17-year-old named Randy Gardner, exhibited a variety of symptoms during the 11 days without sleep. On the second day, he had trouble focusing his eyes and could no longer watch television. On the third day, he had some incoordination of body movements and speech. On the fourth day he was irritable; he had memory lapses, problems with concentration, and delusions; and his body temperature dropped by about 1°C. Gardner experienced visual illusions and hallucinations, as did Gilbert. By the sixth day, his incoordination, slurred speech, ptosis, blurred vision, and memory lapses were more pronounced, and by the ninth day he had trouble completing sentences and had continued delusions, although he did not lose contact with reality. His greatest deficits occurred during nighttime hours, but during the last night, he was able not only to play pinball but to "hold his own" in competitions with sleep researchers.[41] His strategies for remaining awake included vigorous exercise and stimulation, cold showers, and strong encouragement from his supporters. After achieving his goal, Gardner slept for almost 15 hours and awoke feeling refreshed. During the first 3 days of recovery sleep, he regained only about 25% of the sleep that had been lost, including about two-thirds of the lost stage 4 sleep, half of the lost REM sleep, and only 7% of the lost stage 1 and stage 2 sleep.[42]

Many additional studies of sleep deprivation lasting up to 9 days have generally confirmed the above findings with respect to sleepiness, performance deficits, perceptual disturbances, and mood. After a single night of sleep deprivation, subjects report feeling less energetic, less relaxed, and more sleepy, and brief sleep episodes (microsleep) may occur that are not perceived by subjects.[43] After 48 hours of sleeplessness, virtually all subjects have microsleep episodes lasting 5 to 10 seconds.[43] Sleepiness becomes severe but continues to show a circadian variation during the first 2 to 3 days of sleep loss.

Performance is affected by sleep deprivation mainly because of an impaired ability to attend to the task at hand and a decrease in motivation. Thus, performance deficits are most apparent for tasks that require close attention, such as reaction time tests and vigilance tasks, whereas visually stimulating tasks that require motor activity, such as video battle games, may be performed with little or no impairment even after 60 or more hours of sleep deprivation. During the first 36 hours of wakefulness, deficits in most performance tasks can be reduced or eliminated by increasing a subject's incentive to do well. The benefits obtained with incentives, however, decline over the next 24 hours and are virtually absent after 60 hours of wakefulness.[29]

Disturbances of vision and problems with thinking and concentration increase as the duration of sleep deprivation increases. After 48 hours of sleep deprivation, subjects may have episodes of disorientation and altered time perception. As sleep deprivation continues, they also have slowed mentation, difficulty thinking of words, difficulty maintaining a train of thought, and intrusions of dream-like mentation or fantasies.[44,45] Horne suggested that the cognitive disturbances primarily reflect frontal lobe dysfunction.[42]

On the other hand, other bodily functions are affected to only a mild degree after 2 or 3 nights without sleep. A drop in body temperature often occurs, and the amplitude of some circadian rhythms is dampened; however, other biochemical and endocrine studies are usually normal or near-normal, and the capacity for physical work is unaffected to any significant degree.[46] Although sleep deprivation for up to 64 hours appears to increase nonspecific immune responses and decrease cellular immune function, susceptibility to infection is not increased.[47-49]

Animal Studies

Sleep deprivation in animals can be carried out for longer times than in humans and therefore tends to produce more striking effects. In early animal studies, puppies kept awake for several days developed a decrease in body temperature. Other studies on puppies revealed no major pathology apart from apathy after 2 to 7 days of enforced wakefulness.[50] However, rats allowed to sleep for only 4 hours each day, one-third of normal, had slower growth and increased irritability.[51]

Early animal studies were based on behavioral observations; thus, it is almost certain that brief periods of sleep occurred occasionally. With the development of polygraphic recordings of EEG, however, it became possible to deprive animals of more than 99% of their usual amount of sleep. The most intensive modern studies of sleep deprivation in animals, carried out by Rechtschaffen and colleagues,[52,53] used a computerized sleep detection system that activated platform rotation so that rats were required to move to avoid falling off the platform into shallow water. The technique prevents sleep almost completely while yoked control rats, which have to move at the same time as the experimental rats but can sleep when the experimental rats are awake, are deprived of about 25% of their sleep.

The results of such experiments are striking. Although the sleep-deprived rats eat up to twice as much as usual, their metabolic rates increase to an even greater degree and as a result, they lose weight. Despite the increased food intake, they appear malnourished; a debilitated, disheveled appearance develops, characterized by brownish fur and ulcerative lesions of the plantar surfaces of the paws and tail. These changes are not typical of stress reactions, and the constellation of effects appears to be unique to sleep deprivation.

The increase in energy expenditure appears to be a result of increased heat loss, related mainly to REM-sleep deprivation, and to an increase in the preferred body temperature, related more to the loss of slow-wave sleep (SWS). Sleep-deprived rats given an opportunity to sleep in a thermal gradient tend to choose a high ambient temperature, suggesting that they "feel" cold. The most striking biochemical changes are an increase in plasma norepinephrine, a decrease in plasma thyroxine, and an increase in the triiodothyronine-thyroxine (T_3:T_4) ratio.

After 1 to 3 weeks, body temperature begins to fall, and after a mean of 19 days (range, 11 to 32 days) of virtually total sleep deprivation, the rats die.[52] Survival time is a function of the increase in energy expenditure: Rats that have greater increases in energy expenditure tend to die earlier. The proximal cause of death appears to be a breakdown of host defenses that leads to septicemia.[54] Postmortem brain examinations have not revealed gross or microscopic pathology.

Selective deprivation of REM sleep using this technique leads to an increased drive for REM sleep, and it soon becomes impossible to restrict deprivation to REM sleep alone because the animals tend to go directly into REM sleep when allowed to sleep. Furthermore, if allowed to recover, they have huge, immediate increases in the amounts of REM sleep. Similar results occur with selective non-rapid eye movement (NREM) sleep deprivation. Duration of survival is the main difference between total sleep deprivation and selective deprivation of REM sleep or of high-voltage SWS: Death occurs after about 35 days in the rats deprived of REM sleep or high-voltage SWS compared to about 19 days for the rats deprived of all sleep.

In summary, human and animal studies of sleep deprivation indicate that sleep loss has profound effects on mental function and, in animals, severe effects on thermoregulation and energy expenditure that lead ultimately to death. Whether an equivalent duration of total sleep deprivation in humans would show similar catastrophic effects is unknown.

Although prolonged total sleep deprivation does not occur in humans, many persons suffer from long-term chronic partial sleep deprivation, a condition referred to as insufficient sleep syndrome. Its clinical features are discussed below.

INSUFFICIENT SLEEP SYNDROME

Some persons who complain of excessive sleepiness and of repeated episodes of falling asleep during the day do not have anything wrong with their sleep; instead, they simply do not sleep enough. Although the prevalence of insufficient sleep syndrome is unknown, it seems likely that chronic inability or unwillingness to obtain sufficient amounts of sleep is the most common cause of sleepiness in the general population. The demands of modern life, the pressure to "get ahead," and the feeling of "not enough hours in the day" often lead to insufficient sleep. Indeed, afternoon drowsiness and a habit of sleeping late on weekends are so common that many people believe they are normal phenomena.

Most people who do not sleep enough are not seen at sleep centers and do not complain of sleepiness to a physician. Patients who do present to sleep centers with insufficient sleep syndrome tend to be well educated and above average socioeconomically.[55,56] However, only about 2% of all patients presenting to sleep disorders centers have a diagnosis of insufficient sleep syndrome.[57]

Clinical Features

The usual presenting complaint is daytime sleepiness, generally most apparent either in the afternoon and early evening or after meals. Confusion, disorientation, and grogginess on awakening, a complex of sensations sometimes referred to as *sleep inertia* or *sleep drunkenness*, is common, and patients may use several alarm clocks.[58] Awakening later on weekends or on days off is also common; in one series, patients with insufficient sleep syndrome slept an average of 6.4 hours during the week and 8 to 8.5 hours on weekends.[55,56] Some patients who sleep more on weekends and on vacations feel more rested at those times.

The majority of persons who do not obtain enough sleep do not complain to clinicians about sleepiness. Apparently, these persons assume that some degree of daytime sleepiness is normal, do not recognize sleepiness as a problem, do not seek help for it, or are not referred for sleep evaluations if they do. The factors that lead some persons to seek help are unknown, although the occurrence of dangerous events, such as accidents or near-accidents while driving, is probably the impetus for some patients. Patients may also find it more difficult to tolerate reduced amounts of sleep as they become older, or the development of an additional sleep disorder, such as mild sleep apnea or periodic leg movements, may add to the effects of chronically insufficient sleep and cause patients to become aware that sleepiness is a problem.

As with daytime drowsiness caused by intrinsic sleep disorders, insufficient sleep syndrome can cause decrements in daytime performance, particularly with tasks that require vigilance. Other symptoms may include irritability, difficulty with concentration, depression, fatigue, gastrointestinal disturbances, muscle aches, visual disturbance, and dry mouth.

Psychobiologic Basis

Psychosocial factors that contribute to the development of insufficient sleep syndrome include long work hours and demanding family responsibilities. Persons who work full-time and raise small children or care for elderly or infirm relatives, those who work two jobs, and those who are expected to work overtime on a regular basis are likely to curtail their sleep time. In some cases a particular life event, such as the birth of a child or a change in job responsibilities, is associated with the onset of symptoms.

Effects of chronic partial sleep deprivation on alertness, performance, and mood are similar to those seen with total sleep deprivation, but their evolution is slower. As with total sleep deprivation, attention is affected to a greater degree than motor abilities: Impaired vigilance is evident after just 2 nights with 5 hours of sleep,[17,59] whereas physical tasks may be unaffected despite up to 9 nights of no more than 3 hours of sleep.[60,61] In subjects who normally slept 7.5 to 8 hours per night, Webb and Agnew[62] observed a progressive decline in vigilance when their sleep was reduced to 5.5 hours nightly for 60 days. In addition, reduction of normal sleep hours by as little as 30 to 60 minutes can lead to increased sleepiness.[17] Subjective sleepiness and mood changes are most apparent in the first 2 weeks. With gradual voluntary sleep restriction over several months, changes in performance and mood are fairly mild until the amount of sleep is reduced to less than 5 hours per night.[63,64] Cognitive difficulties are usually greater in older persons.[65,66]

Some occupations are associated with high rates of chronic partial sleep deprivation. For example, in one study, physicians in their first year of training slept an average of slightly less than 6 hours a day for the entire year.[67] Chronic sleep loss in physicians can lead to difficulty with high-order cognitive functioning and with monitoring of critical physiologic variables.[68–70] Shift workers are particularly likely to obtain insufficient amounts of sleep, as most sleep less while working nights than they do while working days (see Chap. 12). Personality factors play a role as well: Some persons enjoy the evening hours and may stay up late despite the need to be at work early in the morning.

With partial sleep deprivation, delta sleep is relatively well preserved; most of the reduced sleep is accounted for by a reduction in stage 1–2 sleep and REM sleep.[1,63] Persons who sleep longer hours can tolerate more sleep reduction than those who sleep less, suggesting that both groups have similar "obligatory sleep" lengths. Horne proposed that sleep is composed of "core" sleep, which consists of about the first 5 hours of sleep obtained each night, and "optional" sleep, which consists of the remainder. According to Horne, optional sleep can be reduced by 1 to 2 hours in most persons with little effect on daytime function apart from an increase in sleepiness.[42]

With time, a person may adapt to some of the effects of sleep reduction.[42] Some participants in long-term studies of sleep reduction have slept for fewer hours for months after the experiments ended,[63,64] demonstrating that prolonged changes in sleep-wake patterns may follow sleep reduction in some cases.

Diagnosis

Recognition that patients with chronically insufficient sleep may present to clinicians with complaints of sleepiness and falling asleep led to the formal definition of insufficient sleep syndrome in the *International Classification of Sleep Disorders (ICSD)* as a "disorder that occurs in an individual who persistently fails to obtain sufficient nocturnal sleep required to support normally alert wakefulness. . . . The individual engages in voluntary, albeit unintentional, chronic sleep deprivation."[71]

Because persons with insufficient sleep syndrome who present for evaluation usually complain of excessive sleepiness, the differential diagnosis includes narcolepsy, sleep apnea syndromes, periodic limb movement disorder, and idiopathic hypersomnia. Short sleep times at night, the onset of sleepiness at a time of life style changes that have altered sleep patterns, such as a new job or the arrival of a baby, and increased sleep on weekends or holidays suggest a diagnosis of insufficient sleep syndrome. The diagnosis is more difficult when patients inaccurately report their usual amount of nighttime sleep.

Later hours of arising on weekends suggest either insufficient sleep syndrome or delayed sleep phase syndrome, or both; when both are present, the delayed sleep phase can contribute to problems attaining full alertness in the morning. When an inability to "get going" in the morning coexists with confusion in association with nocturnal arousing stimuli, such as telephone calls, a diagnosis of sleep drunkenness or confusional arousals is suggested. These associated problems resolve with increased sleep at night.

A habitual nightly sleep period of less than 5 hours does not necessarily mean that the person is not sleeping enough. A few people, referred to as short sleepers, apparently do not need much sleep: These persons sleep less than 5 hours per day, do not sleep longer on weekends, and show no apparent performance decrements or daytime drowsiness. Because they suffer no adverse symptoms, they are rarely seen at sleep centers or in clinical practice. In my practice, I have seen only one or two short sleepers, compared to more than 100 patients with insufficient sleep syndrome.

At the other end of the scale are long sleepers, who require 9 to 10 hours of sleep per night for full alertness. Insufficient sleep syndrome may develop in these patients, but it is difficult to diagnose because it can occur even though they sleep 8 hours each night. Diagnostic problems can also arise for patients with both insufficient sleep syndrome and another sleep disorder, such as sleep apnea, because it may be complicted to sort out which problem is the principal cause of daytime sleepiness and other symptoms.

If insufficient sleep syndrome is suspected, a trial of increased sleep at home is indicated. I usually advise patients to try and sleep for at least 8 to 8½ hours per night or at least 1 hour longer than their customary sleep time, whichever is longer; to continue this schedule for at least 1 month; and to maintain a sleep log. If daytime sleepiness resolves, then the diagnosis is clear. If sleepiness improves only partially or does not improve at all, then sleep testing is indicated to assess for possible narcolepsy, sleep apnea, or periodic limb movements. However, the clinician must consider the possibility that incomplete compliance with the recommendation for increased sleep accounts for the failure to improve.

Patients with insufficient sleep syndrome usually fall asleep quickly and sleep soundly in the sleep laboratory; as a result, sleep efficiency and total sleep time during polysomnography are generally higher than in patients with sleep apnea or narcolepsy.[57] Compared to narcoleptics, they tend to have less REM sleep, less stage 1 sleep, and more stage 3–4 sleep on polysomnography.[55] On the MSLT, they usually have short mean sleep latencies, with a dip in alertness in the afternoon corresponding to the usual circadian pattern.[55,57,72] Naps during the MSLT have few awakenings and include stage 2 sleep more than 80% of the time, a greater percentage of stage 2 sleep than occurs in narcoleptics.[57] Epochs of REM sleep occur in about 10% of naps, but rarely with a latency of less than 5 minutes.[55,57] Some persons with relatively mild sleep debts may show progressively increasing sleep latencies during the MSLT, as each nap reduces the cumulative sleep debt.

It is sometimes difficult to distinguish between insufficient sleep syndrome and idiopathic hypersomnia (see Chap. 10). Both syndromes may be associated with high sleep efficiency, difficulty awakening in the morning, and daytime drowsiness with short sleep latencies on the MSLT. Idiopathic hypersomnia, however, is associated with normal or increased amounts of sleep at night and does not respond to trials of increased sleep at night.

Management

Increased sleep time of 1 hour or more is the best treatment for insufficient sleep syndrome. The increased sleep can be added to the nighttime sleep, or it can be taken in the form of several brief naps or one long mid-afternoon "siesta" for patients whose social obligations or work do not permit increased time sleeping at night. Symptoms generally improve within a few days and resolve after a few weeks in patients who comply with these recommendations. Many patients, however, resist the idea that insufficient sleep is the problem and believe that their social responsibilities do not permit them to sleep a greater number of hours. Such patients require education about the need for sleep and the consequences of failure to obtain adequate amounts of sleep. I usually point out that they must weigh the costs and benefits of reducing sleep hours below the optimum: The costs include impaired vigilance and performance as well as an increased risk of motor vehicle and occupational accidents.

Caffeine, which improves vigilance in persons with insufficient sleep syndrome,[73,74] can be a helpful adjunct for patients who are unwilling or unable to increase their sleep time to optimal amounts. Some patients seek treatment with prescription stimulant medications, particularly if they have received such treatment in the past. In one series, 44% had been treated previously with stimulants, suggesting that they had been given a misdiagnosis of narcolepsy or idiopathic hypersomnia.[55] Although dextroamphetamine and other stimulants can reduce the performance deficits that occur with acute partial sleep deprivation in military and similar situations, they are not indicated in patients with the insufficient sleep syndrome because of their potential for inducing sleep disruption and tolerance.[75]

SUMMARY

Sleepiness can be defined as a physiologic state or a subjective condition. *Physiologic sleepiness* is determined by sleep need and reflects a drive for sleep. *Manifest sleepiness*, the subjective aspect, is the complex of sensations, emotions, and thoughts that accompany the need for sleep. Sleepiness has a circadian rhythm, with a major peak of sleepiness at about 3 to 5 AM and a second, smaller peak in the mid afternoon. The circadian variation continues even after 24 to 48 hours of sleep deprivation.

Causes of excessive sleepiness include partial or total sleep deprivation, disrupted sleep, circadian rhythm disorders, medications, and brain disturbances. In addition, many depressed persons describe themselves as sleepy, although they usually do not have objective evidence of excessive sleepiness. Sleepiness can be assessed with subjective measures, performance tests, and neurophysiologic procedures. Although subjective scales usually provide an accurate assessment of acute sleepiness after sleep deprivation, they are less useful with chronic sleepiness. The MSLT, the most widely used polygraphic measure of sleepiness, determines the time required to fall asleep in settings conducive to sleep.

The overwhelming sleepiness and desire for sleep that accompany sleep deprivation suggest that sleep has vital functions. In addition to sleepiness, sleep deprivation in humans for up to 9 days causes performance deficits, perceptual disturbances, and mood changes. Sleep deprivation for longer periods in animals has profound effects on thermoregulation and energy expenditure that ultimately lead to death. Whether an equivalent duration of total sleep deprivation in humans would show similar catastrophic effects is unknown.

Insufficient sleep syndrome is one of the most common causes of sleepiness in the general population. As with daytime drowsiness resulting from intrinsic sleep disorders, insufficient sleep syndrome can cause decrements in daytime performance, particularly with tasks that require vigilance. Short sleep times at night, the onset of sleepiness at a time of altered sleep patterns, and increased sleep on weekends or holidays suggest a diagnosis of insufficient sleep syndrome. Increased sleep time is the best treatment for this syndrome.

REFERENCES

1. Naitoh, P, and Angus, RG: Napping and human functioning during prolonged work. In Dinges, DF, and Broughton, RJ (eds): Sleep and Alertness: Chronobiological, Behavioral, and Medical Aspects of Napping. Raven Press, New York, 1989, pp 221–246.
2. Bliwise, DL: Historical change in the report of daytime fatigue. Sleep 19:462, 1996.
3. Schor, JB: The Overworked American: The Unexpected Decline of Leisure. Basic Books, New York, 1991.
4. Webb, WB, and Agnew, HW Jr: The effects on subsequent sleep of an acute restriction of sleep length. Psychophysiology 12:367, 1975.
5. Sterman, MB, et al: Forebrain inhibitory mechanisms: Conditioning of basal forebrain induced EEG synchronization and sleep. Exp Neurol 7:404, 1963.
6. Richardson, GS, et al: Circadian variation of sleep tendency in elderly and young adult subjects. Sleep 5:S82, 1982.
7. Dement, WC, et al: Excessive daytime sleepiness in the sleep apnea syndrome. In Guilleminault, C, and Dement, WC (eds): Sleep Apnea Syndromes. Alan R. Liss, New York, 1978, pp 23–46.
8. Daly, DD, and Yoss, RE: Electroencephalogram in narcolepsy. Electroencephalogr Clin Neurophysiol 9:109, 1957.
9. Janson, C, et al: Daytime sleepiness, snoring and gastro-oesophageal reflux amongst young adults in three European countries. J Intern Med 237:277, 1995.
10. Lavie, P: Sleep habits and sleep disturbances in industrial workers in Israel: Main findings and some characteristics of workers complaining of excessive daytime sleepiness. Sleep 4:147, 1981.
11. Bixler, ED, et al: Prevalence of sleep disorders in the Los Angeles metropolitan area. Am J Psychiatry 136:1257, 1979.
12. Young, T, et al: The occurrence of sleep disordered breathing among middle-aged adults. New Engl J Med 328:1230, 1993.
13. Hublin, C, et al: Daytime sleepiness in an adult, Finnish population. J Intern Med 239:417, 1996.
14. Carskadon, MA: Ontogeny of human sleepiness as measured by sleep latency. In Dinges, DF, and Broughton, RJ (eds): Sleep and Alertness: Chronobiological, Behavioral, and Medical Aspects of Napping. Raven Press, New York, 1989, pp 53–70.
15. Roehrs, T, et al: Sleepiness and ethanol effects on simulated driving. Alcohol Clin Exp Res 18:154, 1994.
16. wyghuizen-Doorenbos, A, et al: Increased daytime sleepiness enhances ethanol's sedative effects. Neuropsychopharmacology 1:279, 1988.
17. Carskadon, MA, and Dement, WC: Cumulative effects of sleep restriction on daytime sleepiness. Psychophysiology 18:107, 1981.
18. Roth, T, et al: Daytime sleepiness and alertness. In Kryger, MH, et al (eds): Principles and Practice of Sleep Medicine, ed 2. WB Saunders, Philadelphia, 1994, pp 40–49.
19. Carskadon, MA, and Dement, WC: Nocturnal determinants of daytime sleepiness. Sleep 5:S73, 1982.
20. Magee, J, et al: Effects of experimentally-induced sleep fragmentation on sleep and sleepiness. Psychophysiology 24:528, 1987.

21. Bonnet, MH: Infrequent periodic sleep disruption: Effects on sleep, performance and mood. Physiol Behav 45:1049, 1989.
22. Bonnet, MH: Effect of sleep disruption on sleep, performance, and mood. Sleep 8:11, 1985.
23. Bonnet, MH. Performance and sleepiness as a function of the frequency and placement of sleep disruption. Psychophysiology 23:263, 1986.
24. Volkow, ND, et al: Depression of thalamic metabolism by lorazepam is associated with sleepiness. Neuropsychopharmacology 12:123, 1995.
25. Johns, MW: A new method for measuring daytime sleepiness: The Epworth Sleepiness Scale. Sleep 14:540, 1991.
26. Hoddes, E, et al: Quantification of sleepiness: A new approach. Psychophysiology 10:431,1973.
27. Valencia-Flores, M, et al: Multiple sleep latency test (MSLT) and sleep apnea in aged women. Sleep 16:114, 1993.
28. Roth, T, et al: Multiple naps and the evaluation of daytime sleepiness in patients with upper airway sleep apnea. Sleep 3:425, 1980.
29. Horne, JA, and Pettitt, AN: High incentive effects on vigilance performance during 72 hours of total sleep deprivation. Acta Psychol 58:123, 1985.
30. Torsvall, L, and Akerstedt, T: Extreme sleepiness: Quantification of EOG and spectral EEG parameters. Int J Neurosci 38:435, 1988.
31. Broughton, R, et al: Excessive daytime sleepiness and the pathophysiology of narcolepsy-cataplexy: A laboratory perspective. Sleep 9:205, 1986.
32. Zwyghuizen-Doorenbos, A, et al: Test-retest reliability of the MSLT. Sleep 11:562, 1988.
33. Carskadon, MA, and Dement, WC: Effects of total sleep loss on sleep tendency. Percept Mot Skills 48:495, 1979.
34. Harrison, Y, and Horne, JA: "High sleepability without sleepiness." The ability to fall asleep rapidly without other signs of sleepiness. Neurophysiol Clin 26:15, 1996.
35. Levine, B, et al: Daytime sleepiness in young adults. Sleep 11:39, 1988.
36. Roehrs, T, and Roth, T: Multiple Sleep Latency Test: Technical aspects and normal values. J Clin Neurophysiol 9:63, 1992.
37. Mitler, MM, et al: Maintenance of wakefulness test: A polysomnographic technique for evaluating treatment efficacy in patients with excessive somnolence. Electroencephalogr Clin Neurophysiol 53:658, 1982.
38. Doghramji, K, et al: A normative study of the Maintenance of Wakefulness Test (MWT). Electroencephalogr Clin Neurophysiol, 1997, in press.
39. Sangal, RB, et al: Maintenance of wakefulness test and multiple sleep latency test: Measurement of different abilities in patients with sleep disorders. Chest 101:898, 1992.
40. Patrick, GT, and Gilbert, JA: On the effects of loss of sleep. Psychol Rev 3:469, 1896.
41. Gulevich, G, et al: Psychiatric and EEG observations on a case of prolonged (264 hours) wakefulness. Arch Gen Psychiatry 15:29, 1966.
42. Horne, J: Why we sleep. Oxford University Press, New York, 1988.
43. Lagarde, D, and Batejat, D: Evaluation of drowsiness during prolonged sleep deprivation. Neurophysiol Clin 24:35, 1994.
44. Morris, GO, et al: Misperception and disorientation during sleep deprivation. Arch Gen Psychiatry 2:247, 1960.
45. Pasnau, RO, et al: The psychological effect of 205 hours of sleep deprivation. Arch Gen Psychiatry 18:496, 1968.
46. Symons, JD, et al: Physical performance and physiological responses following 60 hours of sleep deprivation. Med Sci Sports Exerc 20:374, 1988.
47. Palmblad, J, et al: Lymphocyte and granulocyte reactions during sleep deprivation. Psychosom Med 41:273, 1979.
48. Irwin, M, et al: Partial night sleep deprivation reduces natural killer and cellular immune responses in humans. FASEB J 10:643, 1996.
49. Dinges, D, et al: Leukocytosis and natural killer cell function parallel neurobehavioral fatigue induced by 64 hours of sleep deprivation. J Clin Invest 93:1930, 1994.
50. Kleitman, N: Sleep and Wakefulness, ed 2. University of Chicago Press, Chicago, 1963.
51. Licklider, JCR, and Bunch, ME: Effects of enforced wakefulness upon the growth and the maze-learning performance of white rats. J Comp Psychol 39:339, 1946.
52. Rechtschaffen, A, et al: Sleep deprivation in the rat: X. Integration and discussion of the findings. Sleep 12:68, 1989.
53. Rechtschaffen, A, et al: Physiological correlates of prolonged sleep deprivation in rats. Science 221:180, 1983.
54. Everson, CA: Sustained sleep deprivation impairs host defense. Am J Physiol 265:R1148, 1993.
55. Roehrs, T, et al: Excessive daytime sleepiness associated with insufficient sleep. Sleep 6:319, 1983.
56. DiPhillipo, M, et al: Characterization of patients with insufficient sleep syndrome. Sleep Res 22:188, 1993.
57. Zorick, F, et al: Patterns of sleepiness in various disorders of excessive daytime somnolence. Sleep 5:165, 1982.
58. Roth, B, et al: Neurological, psychological and polygraphic findings in sleep drunkenness. Schweiz Arch Neurol Neurochir Psychiatr 129:209, 1981.
59. Wilkinson, RT, et al: Performance following a night of reduced sleep. Psychonom Sci 5:471, 1966.
60. Haslam, DR: Sleep loss, recovery sleep, and military performance. Ergonomics 25:163, 1982.
61. Haslam, DR: The military performance of soldiers in sustained operations. Aviat Space Environ Med 55:216, 1984.
62. Webb, WB, and Agnew, HW Jr: The effects of a chronic limitation of sleep length. Psychophysiology 11:265, 1974.
63. Friedman, J, et al: Performance and mood during and after gradual sleep reduction. Psychophysiology 14:245, 1977.
64. Johnson, LC, and MacLeod, WL: Sleep and awake behaviour during gradual sleep reduction. Percept Motor Skills 36:87, 1973.

65. Webb, WB, and Levy, CM: Age, sleep deprivation, and performance. Psychophysiology 19: 272, 1982.
66. Webb, WB: A further analysis of age and sleep deprivation effects. Psychophysiology 22:156, 1985.
67. Ford, CV, and Wentz, DK: The internship year: A study of sleep, mood states, and psychophysiologic parameters. South Med J 77:1435, 1984.
68. Bartle, EJ, et al: The effects of acute sleep deprivation during residency training. Surgery 104: 311, 1988.
69. Engel, W, et al: Clinical performance of interns after being on call. South Med J 80:761, 1987.
70. Denisco, RA, et al: The effect of fatigue on the performance of a simulated anesthetic monitoring task. J Clin Monit 3:22, 1987.
71. American Sleep Disorders Association: International Classification of Sleep Disorders, Revised: Diagnostic and Coding Manual. American Sleep Disorders Association, Rochester, Mn, 1997.
72. Van den Hoed, J, et al: Disorders of excessive daytime somnolence: Polygraphic and clinical data for 100 patients. Sleep 4:23, 1981.
73. Lumley, M, et al: Ethanol and caffeine effects on daytime sleepiness/alertness. Sleep 10:306, 1987.
74. Horne, JA, and Reyner, LA: Counteracting driver sleepiness: Effects of napping, caffeine, and placebo. Psychophysiology 33:306, 1996.
75. Caldwell, JA, et al: Sustaining helicopter pilot performance with dexedrine during periods of sleep deprivation. Aviat Space Environ Med 66: 930, 1995.

Chapter 9

INSOMNIA

O sleep, O gentle sleep,
Nature's soft nurse, how have I frighted thee,
That thou no more wilt weigh my eyelids down
and steep my senses in forgetfulness?

WM. SHAKESPEARE, HENRY IV, PT 2 III.I.13

CLINICAL FEATURES
Difficulty Falling Asleep
Repeated Awakenings and Difficulty
  Returning to Sleep
Early-Morning Awakening
Daytime Symptoms
Course
EPIDEMIOLOGY
PSYCHOBIOLOGIC BASIS
Predisposing Conditions
Precipitants
Perpetuating Circumstances
OVERVIEW OF THE MANAGEMENT
  OF INSOMNIA
HYPNOTIC MEDICATIONS
Benzodiazepines and Benzodiazepine-
  Receptor Agonists
Pharmacologic Alternatives to Benzodi-
  azepines and Benzodiazepine-Receptor
  Agonists
DIFFERENTIAL DIAGNOSIS
PSYCHOPHYSIOLOGIC INSOMNIA
Behavioral Treatment of Insomnia
IDIOPATHIC INSOMNIA
SLEEP-STATE MISPERCEPTION
  (PSEUDOINSOMNIA)
EXTRINSIC SLEEP DISORDERS THAT
  CAUSE INSOMNIA
Inadequate Sleep Hygiene
Environmental Sleep Disorder
Adjustment Sleep Disorder
Hypnotic-Dependent Sleep Disorder
Sleep-Onset Association Disorder
Limit-Setting Sleep Disorder
Nocturnal Eating/Drinking Syndrome
Food-Allergy Insomnia

Shakespeare, whose many references to poor sleep suggest that he may have suffered from insomnia, recognized that psychic conflict could lead to insomnia, as with Hamlet: "Sir, in my heart there was a kind of fighting that would not let me sleep."

Ethanol, opiates, and other plant derivatives were the first hypnotics. Bromides, paraldehyde, and sulfonal were used in the 19th century, but in the 20th century, after the introduction of barbital, the barbiturates became the most widely used prescription hypnotics for several decades. After the release in 1970 of flurazepam, the first benzodiazepine marketed specifically as a hypnotic, benzodiazepines largely replaced the barbiturates in the pharmacological treatment of insomnia. An estimated 62 million sedative-hypnotic prescriptions were filled in 1970,[1] but concerns about side effects, combined with the increasing popularity of nonpharmacological treatments for insomnia, probably accounted in part for the decline to 21 million prescriptions for sedative-hypnotic medications in 1989.[1]

Although they are sometimes seen by physicians as whining, neurotic, or drug-seeking persons with insoluble sleep problems, insomniacs are a diverse group unfairly characterized by the above stereotype. Most insomniacs do not whine: In fact, most never mention their sleep complaints to their physicians, and drug abuse or unauthorized dose escalation of hypnotics by patients is rare. Although psychological profiles show that many patients with insomnia also have neurotic disorders, this is not true of all such patients. Furthermore, treatment focused on the cause or causes of insomnia is often successful.

The numerous causes of insomnia, the interactions among contributing factors, and its inherently subjective nature have made it difficult to categorize insomniacs; thus, none of the many classifications of insomnia has met with universal acceptance.

CLINICAL FEATURES

Patients who complain of insomnia may report difficulty falling asleep, repeated awakenings with difficulty returning to sleep, awakening too early, or a sense that sleep is nonrestorative. Most insomniacs describe a combination of two or more of these symptoms, and up to two-thirds also complain that the sleep disturbance leads to impaired daytime functioning with fatigue, irritability, and inattention. High levels of psychological distress are common. Some patients present with acute insomnia and report being unable or virtually unable to sleep for several days, whereas others describe a gradual deterioration of sleep over weeks to months. Still others present with a complaint of years of nonrestorative sleep.

Difficulty Falling Asleep

Many insomniacs report having difficulty falling asleep, or *sleep-onset insomnia*. The time before they fall asleep may be 30 to 45 minutes or as long as several hours, and some report that they do not fall asleep at all on one or more nights each week. Some feel sleepy in the evening and may fall asleep while sitting in a recliner or while lying on the couch, yet feel wide awake and unable to sleep as soon as they climb into bed. Some avoid going to bed because of the fear of insomnia; others go to bed 1 hour or more early with the expectation that after 1 hour of sleeplessness, they will then be able to fall asleep at their usual bed time. Many patients report restlessness with "tossing and turning," and a few report that they lie quietly in bed for many hours but do not fall asleep.

Patients with sleep-onset insomnia often develop presleep rituals designed to promote relaxation. They may avoid strenuous activities in the evening or take a long bath. Some patients become convinced that sleep is impossible unless the bedroom is properly prepared with the correct illumination and temperature and an absence of noise or other distractions. In patients with obsessive-compulsive personalities, the evening rituals may last for 1 hour or more. Some patients believe that they need a type of sleep aid, such as a television or radio turned on in the bedroom, a white noise generator, an alcoholic beverage as a "nightcap," over-the-counter sleep medications, or homeopathic remedies.

Repeated Awakenings and Difficulty Returning to Sleep

Frequent nighttime awakenings with inability to return to sleep is another common complaint, particularly among older insomniacs. Such patients often describe themselves as "light" sleepers who awaken at the slightest noise. Awakenings may be scattered throughout the night or may cluster at the beginning or at the end of the night. During sleepless intervals, some patients remain in bed "tossing and turning" and worrying about their inability to sleep; others get out of bed to eat, drink, use the bathroom, take a shower, watch TV, or read. Compulsive persons who don't like to "waste time" often do housework or homework. As with sleep-onset insomniacs, some patients report that after awakenings they lie in bed for hours without returning to sleep.

Early-Morning Awakening

Early-morning awakening is a less common complaint than difficulty falling asleep or frequent awakenings. Persons with this complaint may awaken at 3 or 4 AM and then lie in bed for the rest of the night, or they may get out of bed and sit, read, walk, or watch TV. Some remain awake until 7:30 or 8 AM and then return to bed for a nap or rest period.

Daytime Symptoms

Most insomniacs do not feel rested in the morning and complain that they rarely or never obtain "good" sleep. As the day progresses, they may feel more tired and may lie down for rest periods that last a few minutes or several hours. Some patients, after feeling exhausted all day, become more energetic in the evening. Although some insomniacs report daytime sleepiness, most experience little or no daytime drowsiness and are unable to nap even if they lie down. Other daytime symptoms include irritability, mood changes, anxiety, reduced motivation, poor concentration and attention, muscle aches, and a decline in work performance. The majority of persons with insomnia believe that it affects their work performance.

Many insomniacs have treated themselves before—or instead of—seeking medical or psychiatric evaluation. Common home remedies include television viewing, reading, exercising, warm baths, warm milk, meditation, and massage. In one survey, 28% of insomniacs used alcohol to promote sleep and 29% took nonprescription sedatives.[2]

Course

The course of insomnia varies. It may become progressively worse, improve from a low point and become stationary, or improve for long intervals before recurring at times of stress. It may last for just one night or only a few nights, or it may last for years. It may vary in severity from night to night, and it is often less severe on weekend nights, when there are fewer required functions the following day. Some insomniacs also have less sleep disturbance during vacations or during certain seasons.

If insomnia does not resolve quickly but instead continues for weeks or months, patients often become preoccupied with sleep and sleep habits. They tend to ruminate at night and during the day about the ill effects of poor sleep on performance at work or at home. The anxiety and worry may lead to a vicious cycle in which anxiety about sleep leads to increased sleep disturbance and decreased ability to cope with daily activities, which in turn leads to more anxiety and even worse sleep. Occasionally, the decreased self-esteem that ensues may be accompanied by alcohol or benzodiazepine dependence or by the use of stimulants to improve daytime alertness.

Impairment of daytime function is often substantial. Insomniacs tend to have more motor vehicle accidents, worse job performance, and lower self-reported quality of life than persons without sleep complaints.[3] In a study of U.S. Navy personnel, poor sleepers were promoted less often, were less frequently recommended for re-enlistment, and were more likely than good sleepers to be discharged.[3]

Chronic insomnia is also a major risk factor for psychiatric illness. Compared to persons without sleep complaints, insomniacs who still sleep poorly 12 months after initial assessment have a fourfold greater risk for the development of a psychiatric disorder—particularly major depression, but also an anxiety disorder or an alcohol-abuse disorder.[4] Thus, insomnia may be the first indication of depression or an anxiety disorder, or it may be the cause of these disorders in some patients.

Finally, severe insomnia is a marker of increased risk of death: Those who sleep less than 4 hours per night have an increased risk of death during subsequent years compared to persons who obtain 7 to 8 hours of sleep per night.[5]

EPIDEMIOLOGY

About 10% to 20% of adults report chronic insomnia, women more than men by a ratio of about 3:2, and up to half of American workers report occasional difficulty falling asleep.[4] The prevalence of insomnia is much higher in persons over 65 years of age than in younger adults: More than 25% of all persons over the age of 65 report frequent awakenings, and 15% report sleeping less than 5 hours per night. Insomnia also occurs in a substantial proportion of children.[6-8] Insomnia is more common in unemployed people; persons with low educational levels; and persons who are divorced, separated, or widowed.[9]

Although insomnia is common, many cases are not recognized by physicians. About two-thirds of insomniacs have not discussed insomnia with their physicians, and only 5% have had physician visits focused primarily on insomnia.[2] The majority of insomniacs have never taken prescription medications to help with sleep.

PSYCHOBIOLOGIC BASIS

Insomnia occurs when the systems for sleep initiation and maintenance do not function properly, but the precise neurological basis for the failure of these systems is unknown. If sleep is viewed as a drive (see Chap. 1), then failure of sleep initiation occurs when the drive to sleep is insufficient to overcome physiologic arousal or activation, and failure of sleep maintenance occurs when exogenous or endogenous stimuli overcome the drive to sleep. From this perspective, insomnia is caused by decreased drive for sleep, increased physiological activation, increased endogenous or exogenous stimulation during sleep, increased responsiveness to stimuli, or a combination of these disturbances.

Many intrinsic and extrinsic factors contribute to insomnia (Table 9–1). In most patients, several of these factors play a role: some predispose the person to insomnia, some precipitate the condition, and others perpetuate it (Fig. 9–1).[10] The predisposing circumstances set the stage for the development of insomnia, which is then induced by one or more precipitants. The patient's response to insomnia may then perpetuate the insomnia after the precipitants are no longer active.

Predisposing Conditions

A variety of circumstances and conditions predispose patients to insomnia, including personality, age, genes, and intrinsic neurobiologic factors. Tense, anxious, nervous, and worried persons; those who tend to ruminate; those who internalize problems; and those who tend to have somatic responses to stress are at higher risk for insomnia than relaxed, phlegmatic types. Somatization may be accompanied

Table 9–1. **Causes of Insomnia**

Intrinsic Factors	Extrinsic Factors
Decreased mobility	Medications
Pain	Social isolation
Thoughts and emotions while awake in bed	Negative life events
	Poor sleep hygiene
Anxiety	Unsuitable sleeping environment
Depression	Caffeine, tobacco, and alcohol
Mental activity during sleep	High altitude
Secondary gain	Allergens and toxins
Genetic and constitutional factors	
Advanced age	
Physiologic activation	
Circadian rhythm disturbances	
Conditioned associations	
Arousals caused by sleep apnea or periodic leg movements	

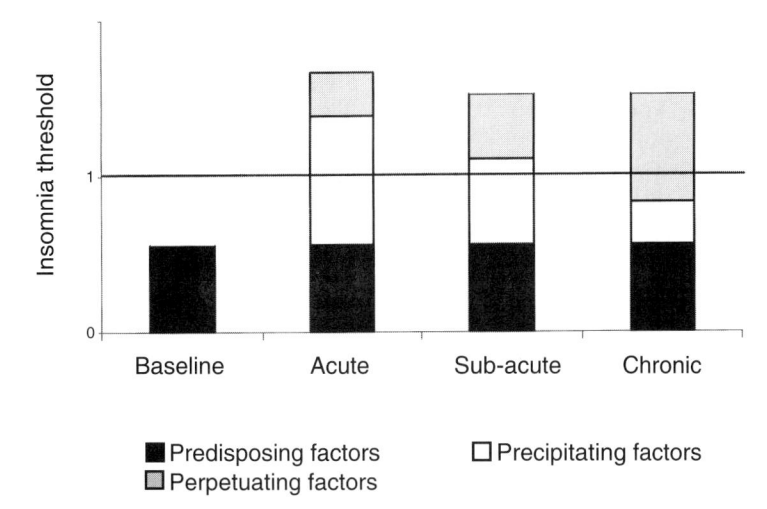

Figure 9–1. Predisposing, precipitating, and perpetuating factors contribute to insomnia in the model proposed by Spielman.[10] Predisposing factors are present at baseline, before insomnia develops. Precipitating factors, on top of predisposing ones, cause insomnia if the combination exceeds the insomnia threshold for that individual. Perpetuating factors often become increasingly important as insomnia continues. With chronic insomnia, the precipitating factors may no longer be sufficient to cause insomnia, and treatment directed at the perpetuating factors may lead to resolution of insomnia. (Adapted from Spielman,[10] pp. 13–15, with permission.)

by repression or denial of awareness of stress. Psychiatric disorders, particularly mood disorders and anxiety disorders, are probably the cause of insomnia in 40% to 50% of patients with serious insomnia.[4]

Advancing age also predisposes persons to insomnia. Central nervous system mechanisms responsible for sleep initiation and maintenance probably deteriorate with age, and the accumulation of age-related maladies often increases sleep difficulties (see Chap. 17).

Genetic factors may contribute to insomnia in some persons. One example is restless legs syndrome, which may cause severe insomnia and which appears to be inherited as an autosomal-dominant trait in some families (see Chap. 11). Studies of monozygotic and dizygotic twins suggest that both genetic and environmental factors may be responsible for the increased prevalence of insomnia in twins of insomniacs.[11]

Insomniacs tend to show increased physiologic activation—elevated galvanic skin response, increased mean sleep latency on MSLT, increased metabolic rate, increased body temperature, and increased muscle tension—compared to normal values. In contrast, normal subjects in whom insomnia is experimentally induced have decreased mean sleep latency on MSLT, decreased body temperature, decreased metabolic rate in the morning, and decreased muscle tension.[12–15] The differences between the insomniacs and the normal subjects with experimentally induced insomnia could be a result of the higher preexisting levels of physiologic activation in the insomniacs. In a model of insomnia produced by administering caffeine to normal subjects, metabolic changes and MSLT values resembled those of persons with chronic insomnia.[16]

An inherited or acquired alteration of circadian physiology can also contribute to insomnia, as suggested by the finding that some patients with sleep-onset insomnia have delayed core body temperature rhythms.[17] However, this alteration could be a result of insomnia, rather than its cause.

Precipitants

Precipitants are usually revealed by the patient's life circumstances; in one study, 74% of insomniacs reported a stressful event at the onset of insomnia,[18] and almost half of insomniacs note that worries make their sleep worse. Typical life events that precipitate insomnia include death or illness of a loved one, divorce or separation, a move to a new location, and a change in occupational status. Depression and other psychiatric disorders can precipitate insomnia, and spousal bereavement in older persons often leads to insomnia that may persist for more than 1 year (see Chap. 16). Medical illnesses may also precipitate insomnia via their effects on bodily systems (e.g., insomnia caused by cardiac failure), their symptoms (e.g., pain and stiffness of arthritic conditions), or their treatment (e.g., β-adrenergic agents for treatment of asthma). For a discussion of the medical disorders that can cause sleep disruption, see Chapter 17.

Sometimes, relatively minor life events associated with little or no obvious stress precipitate insomnia, such as a change to night work or to a rotating shift work schedule. In one study, subjects who were expected to give a brief talk on a specific topic after awakening had more difficulty falling asleep than a control group.[19] An uncomfortable bed, excessive noise, a bedroom that is too hot or too cold, or other changes in the sleeping environment may also precipitate insomnia in predisposed persons.

Perpetuating Circumstances

From a therapeutic perspective, the perpetuating factors are critical because they may be most amenable to change. Anxiety about insomnia, negative conditioning, poor sleep habits, the use of hypnotics and alcohol, and secondary gain associated with insomnia are important perpetuating factors.

Anxiety about insomnia and about its effects on daytime function often perpetuates insomnia. Concern or overconcern about the impact of insomnia on daytime function may lead to *performance anxiety*, whereby the patient feels required to perform the function or duty of falling asleep. Unfortunately, sleep cannot be willed to occur, and as the patient tries hard to fall asleep, it becomes increasingly difficult to fall asleep, which in turn leads to increased anxiety. The increased anxiety then makes falling asleep even more difficult.

Sterman and associates[20] demonstrated the role of conditioning by using classic conditioning to induce sleep in cats. The animals received paired stimuli of a tone and electrical stimulation of the preoptic basal forebrain that induced sleep. After repeated pairing of the stimuli, the experimenters found that the tone alone could induce sleep. With most people who fall asleep easily, similar conditioning probably takes place with the bed, the bedclothes, and the act of getting into bed and preparing to sleep. In patients with chronic insomnia, however, the process of preparing for sleep, getting into bed, and turning out the lights is no longer associated with falling asleep but may instead become tied to anxiety, sleeplessness, and fear of failure to fall asleep. The bed itself then becomes a source of anxiety and a stimulus for arousal. In such patients, sleep is often better away from home or on a couch, when the negative associations with the bed are absent and the expectation that sleep will be worse in an unfamiliar setting reduces the performance anxiety usually associated with attempts to sleep.

After insomnia begins, some patients develop poor sleep habits that perpetuate insomnia. They may exercise at night in order to feel more tired, spend more time in bed or go to bed at irregular times in an attempt to obtain more sleep, or increase their daytime caffeine consumption in order to feel more alert. Other patients have never had good sleep habits but were able to sleep well as young persons because of robust sleep mechanisms. With age-related impairment of sleep, their poor habits lead to or perpetuate insomnia.

Secondary gain associated with insomnia may also perpetuate the symptom. Insomnia may be used as a reason for nighttime snacks, alcohol, or TV watching, which then act as reinforcers for poor sleep. Time off work may be an additional source of secondary gain. For some, insomnia may contribute to the role of "sickly child" or dependent adult. For others, particularly those with marital or relational problems, insomnia or the associated perceived need for a quiet time before bed may provide a rationale for avoiding sexual relations.

Use of alcohol and hypnotics may perpetuate sleep disturbance. Short-acting hypnotics, such as triazolam and zolpidem, may wear off before the end of the night, leading to early-morning insomnia, and can also produce "rebound insomnia" on the following night if the hypnotic is not ingested. Insomniacs often discover that alcohol promotes sleep onset, and some insomniacs develop a conditioned association between alcohol use and falling asleep that leads them to believe they cannot sleep without a "nightcap." For heavy alcohol users, anxiety and symptoms of mild alcohol withdrawal that develop if alcohol is not consumed contribute to their impression that they cannot sleep without alcohol. Unfortunately, although alcohol can hasten the onset of sleep, it also can lead to sweaty, restless sleep during the second half of the night and frequent awakenings from dreams with difficulty returning to sleep (see Chap. 16). With chronic use at bedtime or during the night, the sleep-inducing effect of alcohol may be reduced, whereas its effects on late-night sleep continue or are increased, leading to daytime fatigue and sleepiness. Alcohol also suppresses rapid eye movement (REM) sleep and probably makes sleep less restful.

In some patients, a precipitating event is difficult to identify; it appears that poor sleep may develop gradually in these persons as the anxiety about occasional poor nights of sleep leads to progressively increasing concern about sleep.

OVERVIEW OF THE MANAGEMENT OF INSOMNIA

Treatment of insomnia is a challenging area of sleep medicine because the large number of potential causes and the high likelihood that more than one cause is operative make it difficult to define a single approach most likely to help a given patient.

Appropriate treatment may include education about better sleep habits, treatment of an underlying psychiatric disorder, adjustment of the timing or dose of a medication used to treat a medical disorder, behavioral treatment, prescription of a hypnotic, or a combination of two or more of these approaches. Education about sleep habits is discussed under the heading Inadequate Sleep Hygiene, and behavioral treatments are discussed under the heading Psychophysiologic Insomnia (see below). Treatment of insomnia caused by psychiatric disturbance is discussed in Chapter 16; insomnia caused by medical disturbance is discussed in Chapter 17. The following section focuses on the use of hypnotics and other sleep-inducing medications.

HYPNOTIC MEDICATIONS

Physicians, mainly family physicians, psychiatrists, and internists, prescribe medications to assist sleep in about one-third of visits in which the presenting complaint is insomnia, and they are more likely to prescribe hypnotics for older adults than for middle-aged adults.[21] The popularity of hypnotics reflects their initial efficacy: They usually help at first, especially with acute insomnia, regardless of the cause of sleep disturbance. The benefits include decreased sleep latency, decreased amounts of time awake after sleep onset, and increased amounts of sleep during the night.

Hypnotics can also be helpful in patients with chronic insomnia: Some patients experience symptomatic improvement with hypnotics and use them on a regular basis for years without dose escalation and with at least subjective efficacy. However, others with chronic insomnia obtain no benefit, and a few become dependent. Problems with hypnotics include side effects, development of tolerance, and the potential for dependence.

Hypnotic use should be considered in the following six kinds of patients:[22]

1. Those with acute insomnia caused by situational stress, for whom a 2- to 4-week course of a hypnotic is often helpful

2. Those with anxiety disorders, for whom hypnotics may be useful for their anxiolytic and sleep-inducing properties
3. Those with severe psychophysiologic insomnia, for whom hypnotics may be helpful as an adjunct to behavioral treatment
4. Those with periodic limb movement disorder or restless legs syndrome, in whom benzodiazepine hypnotics may reduce the frequency of arousals associated with leg movements
5. Those with insomnia associated with painful medical or neurological disorders
6. Those with insomnia refractory to nonpharmacological treatments.

In each of these settings, the possible benefits of hypnotic use must be weighed against the potential for side effects and dependence. Although many physicians are greatly concerned about the potential for drug addiction in insomniacs, unauthorized dose escalation and other forms of drug abuse are rare in insomniacs who use hypnotics.[23]

In general, the minimum effective dose of a hypnotic is the preferred dose; however, the minimum effective dose may vary among patients, and dosing must be adjusted in older people because of reduced renal and hepatic clearance. If nightly use is planned, the usual starting dose in older adults should not be more than one-half the usual initial dose for young adults. Doses above the minimum effective dose often provide little added efficacy, but they increase the risk of side effects and probably also the risk of tolerance and dependence. In most cases, if a previously effective dose of a hypnotic is no longer beneficial, the specific causes for its loss of efficacy should be explored before initiating a dose increase.

The frequency of the use of hypnotics and the choice of the hypnotic vary according to the clinical setting. Some patients benefit from the knowledge that sleeping pills are available in the medicine cabinet if needed; they may actually take the medication only once or twice per month. Others do best by taking the medication once or twice per week on nights when they feel that they are in particular need of a good night's sleep. Still others, particularly those with high levels of anxiety, seem to do best with a structured regimen of nightly use. Medication use every other night usually should be avoided because sleep is almost always worse on the nights without medication, and the difference between nights may reinforce the patient's belief that hypnotics are necessary for good sleep. On the other hand, nightly use for a limited time, such as 2 to 4 weeks, may help to promote consistently better sleep and thereby extinguish negative conditioning. Most patients should take hypnotics 15 to 30 minutes before bedtime; they should not take additional medication during the night, even if they cannot sleep, because of the greater likelihood of daytime sedation and because the anxiety evoked by waiting to determine whether another dose will be necessary tends to make sleep even more difficult and to reinforce the perceived need for hypnotics.

Short-acting hypnotics are usually preferable for most patients with insomnia because they produce less daytime sedation. However, since intermittent use of short-acting hypnotics in patients with anxiety disorders increases the likelihood of rebound insomnia and rebound anxiety, nightly use of a medium-acting agent may be preferable. For patients with early-morning awakening, a short-acting hypnotic may wear off by the middle of the night and therefore provide no benefit. In depressed patients, the benzodiazepine pills dispensed should be limited to a small number if the patient appears to be at risk for intentional overdose.

Benzodiazepines and Benzodiazepine-Receptor Agonists

The most commonly used hypnotics are benzodiazepines and benzodiazepine-receptor agonists (Table 9–2). Barbiturates are rarely if ever indicated for the treatment of insomnia because of their serious side effects, their adverse effects on sleep architecture, and their low therapeutic index. The role of tricyclic antidepressants

Table 9–2. **Pharmacologic Treatment of Insomnia**[22,30,31]

Medication	Type	Half-Life* (hours)	Absorption	Usual Dosage (mg)
Triazolam	BZ	Short (2–6)	Rapid	0.125–0.25
Midazolam	BZ	Short (2–3)	Intermediate	7.5–15
Zolpidem	BZ-receptor agonist	Short (2–3)	Rapid	10–20
Zopiclone	BZ-receptor agonist	Short (4–5)	Rapid	7.5–15
Oxazepam	BZ	Medium	Slow	15–30
Alprazolam	BZ	Medium	Intermediate	0.25–2
Estazolam	BZ	Medium (8–24)	Rapid	1–2
Lorazepam	BZ	Medium	Intermediate	0.5–2
Temazepam	BZ	Medium (8–20)	Intermediate	15–30
Chlordiazepoxide	BZ	Long (48–96)	Intermediate	10–25
Clonazepam	BZ	Long	Intermediate	0.5–2
Clorazepate	BZ	Long	Rapid	7.5–15
Diazepam	BZ	Long (48–96)	Rapid	2–10
Flurazepam	BZ	Long (48–96)	Intermediate	15–30
Halazepam	BZ	Long	Intermediate	20
Prazepam	BZ	Long	Slow	10
Quazepam	BZ	Long (20–40)	Rapid	7.5–15
Amitriptyline	TCA	Long (10–50)	Intermediate	10–100
Doxepin	TCA	Short (6–8)	Rapid	10–100
Imipramine	TCA	Medium (8–16)	Rapid	10–100
Trazodone	Triazolopyridine antidepressant	Medium	Rapid	25–100
Diphenhydramine	Antihistamine	Medium (8–12)	Rapid	25–50
Hydroxyzine	Antihistamine	Medium (16–24)	Intermediate	25–50

*Estimated for the compound and its active metabolites.
BZ = benzodiazepine; TCA = tricyclic antidepressant.

and other sedating or sleep-inducing compounds is considered later in this section.

Benzodiazepine-receptor agonists include zolpidem and zopiclone, which are imidazopyridines that appear to act at the γ-aminobutyric acid-benzodiazepine (BZ)–receptor complex. They have greater specificity for the BZ_1 receptor subtype, which is thought to mediate the sedative effects of benzodiazepines, whereas anticonvulsant and muscle relaxant effects appear to be mediated principally by BZ_2 and BZ_3 receptor subtypes. In practice, the hypnotic effects of imidazopyridines are similar to those of benzodiazepines; however, their anticonvulsant, muscle relaxant, and anxiolytic effects may be relatively less. Although receptor specificity may be important, claims that the benzodiazepine-receptor agonists have fewer side effects than benzodiazepines are similar to claims that have been made in the past with virtually every new hypnotic or class of hypnotics.

SIDE EFFECTS

The side effects of benzodiazepines and benzodiazepine-receptor agonists are dose-related and include nocturnal confusion, daytime sedation, ataxia and other motor disturbances, memory impairment, rebound insomnia, and the potential for exacerbation of sleep apnea. Nocturnal confusion and ataxia are particular concerns in patients who have to get up at night to use the bathroom or to care for small children or elderly and infirm relatives. The ataxia associated with benzodiazepines and other sedatives increases the risk of falls and associated injury. Daytime unsteadiness is also a concern because all hypnotics—even the shortest acting ones—remain in the bloodstream during waking hours. Daytime sleepiness and grogginess, especially in the morning, are also common. Moreover, performance deficits are easily detected, especially with higher doses, longer acting compounds, and in persons with impaired clearance of the drug as a result of old age, interaction with other medications, or hepatic or renal dysfunction.[22,24,25]

Memory impairment is not limited to the long half-life compounds. Some studies suggest that memory impairment is of greater clinical concern with triazolam than with some other benzodiazepines.[26,27] Rebound insomnia and rebound anxiety may occur with hypnotics, especially with short acting ones (see section titled Hypnotic-Dependent Sleep Disorder below). Although benzodiazepines can exacerbate sleep apnea and other breathing disorders during sleep in some persons, at least in part by attenuating the arousal response to apnea and hypercapnia, the degree of worsening is often minimal or clinically inconsequential unless the apnea or nocturnal hypoxemia is severe.[28] In patients with uncomplicated grief that is not overwhelming, much of the "grief work" is carried out at night, and hypnotics can interfere with this important self-treatment.

Pharmacologic Alternatives to Benzodiazepines and Benzodiazepine-Receptor Agonists

Alternatives to benzodiazepines and benzodiazepine-receptor agonists include antihistamines, sedating tricyclic antidepressants, and endogenous compounds such as melatonin and delta sleep–promoting peptide (DSIP). Over-the-counter medications, which often contain sedating antihistamines, appear to be less effective in promoting sleep and more likely to cause daytime drowsiness than benzodiazepines.[29,30]

Tricyclic antidepressants are of greatest value when insomnia is caused by or associated with depression. In such cases, the doses used are generally higher than listed in Table 9–2.[22,30,31] Tricyclic antidepressants, at doses below those used for treatment of major depression, are also prescribed frequently when insomnia has other causes. Although they are often beneficial, similar effects could probably be obtained with low doses of benzodiazepines or benzodiazepine receptor agonists. Two common reasons for prescribing tricyclics rather than benzodiazepines— their lower abuse potential and the lesser risk of respiratory depression— apply to only a small minority of insomniacs. When prescribing medications for insomnia, I tend to use benzodiazepines for more anxious patients and tricyclic antidepressants for less anxious ones.

Many patients ask about melatonin for insomnia. Although melatonin has hypnotic effects, at both physiologic doses of 0.1 to 0.3 mg and the higher doses of 1 to 75 mg that are present in melatonin sold as a health food supplement, it is far less effective than benzodiazepines.[32–34] The side effects of melatonin are not well understood, but some reports suggest that it can reduce fertility, exacerbate depression or schizophrenia, or cause constriction of pulmonary and coronary arteries. Because of concerns about the actual content of these unregulated supplements and about possible contaminants, I do not recommend melatonin to patients. DSIP, also not available as a regulated pharmaceutical in the United States, may lead to mild improvement in sleep in some patients with insomnia.[35]

DIFFERENTIAL DIAGNOSIS

Diagnostic considerations in patients with complaints of insomnia are listed in Table 9–3. Two main systems are used for diagnosis of disorders associated with a chief complaint of insomnia: the diagnostic classification used by the *International Classification of Sleep Disorders (ICSD)* and that used by the *Diagnostic and Statistical Manual, Fourth Edition (DSM-IV)* of the American

Table 9–3. **Classification of Disorders Associated with Insomnia in Adults**

ICSD Classification[36]	DSM-IV Classification[37]
Intrinsic disorders	
Psychophysiologic (conditioned) insomnia	Primary insomnia
Sleep-state misperception	Primary insomnia
Idiopathic insomnia	Primary insomnia
Restless legs syndrome	Secondary insomnia
Periodic limb movement disorder	Secondary insomnia
Sleep apnea and related disorders	Secondary insomnia
Extrinsic disorders	
Inadequate sleep hygiene	Primary insomnia
Environmental sleep disorder	Secondary insomnia
Adjustment sleep disorder	Primary unless caused by psychiatric illness
Hypnotic-dependent sleep disorder	Primary unless caused by psychiatric illness
Drug- and medication-related insomnia	Primary unless caused by psychiatric illness
Altitude insomnia	Secondary insomnia
Circadian rhythm sleep disorders	Secondary insomnia
Sleep disorders caused by medical or psychiatric illness	Secondary insomnia

Psychiatric Association.[36,37] In the *ICSD*, insomnia is viewed as a symptom that can occur with intrinsic or extrinsic sleep disorders, with circadian rhythm disturbances, and with sleep disorders secondary to medical and psychiatric illness. In the *DSM-IV*, insomnia is viewed both as a primary disorder and as a symptom secondary to a medical or psychiatric disorder.

Proponents of *DSM-IV* argue that there is insufficient evidence for differentiating the subtypes of primary insomnia on the basis of clinical presentation and pathogenesis.[38] Proponents of *ICSD* argue that the diagnosis should be based on pathogenesis to the extent that it can be determined and that there is evidence that a condition such as psychophysiologic insomnia, for example, can be differentiated from other types of insomnia.[39,40] However, even those who support subtyping of primary insomnia recognize that there is overlap among the disorders and that one patient may have the characteristics of several insomnia subtypes, such as poor sleep hygiene, maladaptive conditioning, hypnotic dependence, and a predisposition to poor sleep. The paucity of longitudinal studies of patients with a diagnosis of subtypes of insomnia makes it difficult to draw definite conclusions concerning the legitimacy of the subtypes.

In practice, a systematic history provides the most important diagnostic information (Table 9–4). Descriptions from bed partners or parents are particularly useful, as they often describe a different pattern of insomnia than does the patient. Psychiatric interviews and physical, neurological, and mental status examinations also may provide essential diagnostic information. Standard sleep questionnaires may help to identify symptoms that were not identified during the initial evaluation, and sleep charts or diaries provide a longitudinal measure of the perceived severity of the sleep disturbance (see Fig. 7–1). Psychological testing and rating scales for depression and anxiety may be helpful in some cases.

Polysomnographic studies should be considered if the diagnosis is uncertain, if the expected response to treatment does not occur, if daytime sleepiness is prominent or persists after treatment, if sleep-state misperception is suspected, or if sleep apnea or periodic limb movements are suspected.

Table 9–4. **Important Elements of the History in Patients with Insomnia**

Sleep-wake schedule: times of going to bed, turning out the lights, falling asleep, final awakening, and getting out of bed

The times and duration of awakening: differences between weekday and weekend.

Behaviors, thoughts, and emotions during sleepless intervals

Characteristics and frequencies of "good" nights and "bad" nights

Timing and frequency of social activities

Daytime light exposure

Evidence of depression or anxiety

Timing and amount of prescription and non-prescription medication use, amount of caffeine and alcohol consumption, and level of tobacco use

---

The high rate of psychiatric problems among patients with insomnia indicates that physicians must actively explore the possibility of psychiatric illness. Sadness, loneliness, loss of appetite or libido, excessive rumination, a recent major loss (e.g., death of a spouse), and a history of depressive episodes before insomnia began increase the likelihood that the insomnia is caused by depression. In contrast, racing thoughts at bedtime, nocturnal panic attacks, and anxiety in other aspects of life increase the probability that the insomnia is a result of generalized anxiety disorder. If the insomnia varies in severity with other symptoms related to anxiety or depression, a psychiatric cause is highly likely (see Chap. 16). On the other hand, although many physicians assume that virtually all cases of insomnia result from psychiatric causes, chronic insomnia that occurs in patients with stable mood and good daytime function is often caused by something other than major psychiatric illness.

If sleep-onset insomnia is the major complaint, the differential diagnosis may include an anxiety disorder, inadequate sleep hygiene, hypnotic-dependent sleep disorder, or delayed sleep phase syndrome. Impaired sleep continuity may occur with alcohol-dependent sleep disorder, periodic limb movement disorder, sleep apnea, or in association with an underlying medical disorder. Early-morning awakening is often an indication of depression, but it can also occur with the use of short-acting hypnotics and with circadian rhythm sleep disorders. Insomnia caused by a medical disorder should be considered in patients with conditions that cause pain, paresthesias, reduced mobility, nocturia, or orthopnea.

For each patient, the clinician must sort through the various elements that contribute to insomnia, try to identify the most important elements, and help the patient to overcome the treatable contributors. The following case provides an example.

CASE HISTORY: Case 9–1

A 48-year-old administrator complained of insomnia. She first had problems with insomnia in college and sometimes took secobarbital to help her sleep before examinations. The insomnia improved, but it recurred at age 35 at the time of her divorce and subsequent hospitalization for major depression. For the next 5 years, she used alcohol excessively during the day and to help her sleep. After alcohol detoxification and treatment at age 40, she joined Alcoholics Anonymous and stopped drinking alcohol; her insomnia again improved, and she slept relatively well until it recurred at age 46 at about the time of menopause. Hormonal replacement was not helpful. She usually fell asleep easily, but awakened 2 to 3 hours later and remained awake tossing and turning until about 4 AM. She then slept until she got up to go to work at 7 AM. She estimated her usual amount of sleep as 5 hours. Insomnia was worse during stressful periods at work and was better when she traveled or was on vacation, when she could sleep "24 hours straight."

She lived alone, often talked on the phone or watched TV in bed, and drank up to 8 cups of coffee per day. She described herself as a perfectionist, "type A," "go-go" kind of person. Her father and brother were alcoholics, her mother and brother had insomnia, and her daughter had episodic depression. Physical and mental status examinations were normal. In the sleep laboratory, she fell asleep in 4.5 minutes, had no periodic limb movements or sleep apnea, and was asleep for all but 28 minutes of the recording. She accurately assessed her sleep in the laboratory and described it as better than usual.

Lorazepam 0.5 to 1.0 mg at bedtime 5 to 7 nights per week was prescribed, along with recommendations to reduce coffee intake and use the bed only for sleeping. With lorazepam, her insomnia improved. It recurred and lasted several months after her daughter's suicide when the patient was 49, and her dose of lorazepam was increased to 1.5 mg at bedtime. Over the next 12 years, she took lorazepam 1.5 mg nightly. Several attempts to reduce lorazepam use to 2 nights per week led to an increase in insomnia. At age 61, she continued to drink 5 cups of coffee per day, used the bed for phone calls and for watching TV, remained abstinent from alcohol, was generally pleased with her sleep, and described only occasional nights of poor sleep related to stress at work or after watching distressing news concerning crime in her city. She was reluctant to discontinue lorazepam or coffee, although she agreed to consider dose reduction at the time of her retirement at age 62.

Numerous factors contributed to this patient's insomnia. Her "type A," perfectionist personality probably predisposed her to respond to stress with anxiety and insomnia, as she had done throughout her adult life. A family history of insomnia suggested the possible presence of constitutional or familial factors that led to high baseline levels of physiologic activation and arousal. Alcoholism contributed to her insomnia during her years of alcohol abuse and may have produced permanent impairments in sleep systems. The tendency to sleep better away from home and the improved sleep in the sleep laboratory are consistent with an element of negative conditioning. The major life stresses that precipitated episodes of insomnia included her father's and brother's alcoholism, her divorce, and her daughter's depression and suicide. These events, along with the episode of major depression and the family history of depression, suggest that a mood disorder was a major contributor to insomnia. Chronic coffee use and the use of the bed for activities other than sleep suggest that suboptimal sleep hygiene played a role in perpetuating insomnia. Finally, chronic hypnotic use and extreme reluctance to try to sleep without hypnotics suggest that hypnotic dependence contributed to insomnia. On the other hand, her use of hypnotics without dose escalation for 12 years, her ability to function at work through most of the major life stresses, and her view of her sleep as satisfactory at age 61 suggest that chronic hypnotic use had not had major deleterious effects. Furthermore, the anxiolytic effects of hypnotics may have contributed to her ability to remain abstinent from alcohol for 21 years.

I have used the *ICSD* approach in the remainder of this chapter because its emphasis on pathogenesis highlights important causes of insomnia. However, for most of the ICSD diagnoses that fall into the category of primary insomnia in DSM-IV, the approach to treatment is similar.

PSYCHOPHYSIOLOGIC INSOMNIA

Psychophysiologic insomnia is defined as "a disorder of somatized tension and learned sleep-preventing associations that results in a complaint of insomnia and associated decreased functioning during wakefulness."[36] In one series, approximately 15% of insomniacs had a diagnosis of psychophysiologic insomnia.[41] Although the diagnosis should not be used if symptoms are primarily the result of a medical or psychiatric disorder, some degree of negative conditioning is probably present in most insomniacs.

Patients with psychophysiologic insomnia tend to view good sleep as a necessity. This perceived need leads in turn to anxiety about falling asleep and about returning to sleep after an awakening. The patient fights to fall asleep and tries too hard to sleep, which results in increased arousal and increased difficulty falling asleep. Soon the bed and all its associated stimuli become associated with anxiety about falling asleep or staying asleep. Patients may feel drowsy and may fall asleep easily while lying on a couch in the living room but then become wide awake as soon as they climb into bed. A brief awakening that would ordinarily be followed by a change in position and a quick return to sleep is instead followed by worry that sleep will not quickly return, by worry about the consequences of not sleeping, and soon by full alertness with the impossibility of returning to sleep quickly. Patients may look at the clock, note that the time is, for example, 2:30 AM—the same time that an awakening the night before was followed by sleeplessness—and assume that the same pattern will recur. The assumption quickly becomes self-fulfilling, and the appearance on the clock of 2:30 AM then becomes a stimulus for poor sleep on subsequent nights (see discussion on performance anxiety above).

Patients with this form of insomnia tend to focus on sleep as the source of problems and to minimize other psychological or emotional factors. They usually prefer regular routines and respond to emotional stress with denial, repression, and somatization, the latter often in the form of tension headaches or symptoms suggesting increased sympathetic tone.[40] Dysphoria, often a symptom, may make it difficult to determine whether insomnia is caused by depression, especially a masked depression in which the patient denies sadness or hopelessness. In a comparison of patients with a diagnosis of psychophysiologic insomnia with those having a diagnosis of insomnia caused by dysthymic disorder,[40] sleep disturbance was similar, but psychological functioning differentiated the dysthymic patients from the psychophysiologic group.

Behavioral Treatment of Insomnia

Treatment of psychophysiologic insomnia is directed at extinguishing the negative conditioning that interferes with sleep and substituting positive conditioning that facilitates sleep. Behavioral treatments such as stimulus control instructions[42] and sleep restriction therapy,[43] combined with good sleep hygiene, usually provide the best approach. These treatments can also be used in other forms of chronic insomnia.

The principle of stimulus control therapy is to associate the bed and sleep environment only with sleep (Table 9–5). The

Table 9–5. **Behavioral Treatments of Insomnia**

Stimulus Control Instructions[42]	Sleep Restriction Therapy[43]
Lie down in bed only when you are sleepy.	Establish a reduced period of time to spend in bed.
Do not use the bed for anything except sleep and sexual activity.	Go to bed and get up at the same time each day, even if you have not slept or slept only a small amount.
If you are unable to sleep, get up and go into another room.	Do not abandon the sleep schedule without talking with your clinician first.
Return to bed only when you are drowsy.	Once sleep is consolidated, gradually increase the amount of time in bed.
If you still cannot sleep, get up again and go to another room.	Do not take daytime naps.
Set an alarm and get up at the same time each day.	
Do not take daytime naps	

bed is to be used only for sleep and intimacy: no reading, eating, or watching TV in bed are permitted. Patients are instructed to get out of bed and to do something relaxing if unable to sleep after 15 to 20 minutes.

With sleep restriction therapy (see Table 9–5), the goal is to maximize the ratio of time asleep to total time spent in bed. This technique is most useful for insomniacs who have increased their time in bed in an attempt to get more sleep. The result is usually more time in bed awake rather than more time asleep, and the increased time awake means more time spent worrying about sleep, more anxiety, and more negative conditioning. By eliminating naps and temporarily restricting time in bed, the patient becomes mildly sleep deprived and tends to sleep more continuously while in bed. Once sleep is improved, the time in bed can be gradually increased. The following is a case of psychophysiologic insomnia in which the patient responded well to sleep restriction therapy.

CASE HISTORY: Case 9-2

A 48-year-old woman presented with complaints of difficulty sleeping that had been present for most of her life and had worsened in the previous 5 to 6 years. She usually went to bed at 10 PM and fell asleep without difficulty. About 2 to 3 nights per week, she awoke at about 1 or 2 AM, used the bathroom, and then was unable to return to sleep. At such times she felt frustrated at her inability to sleep and became more frustrated as she watched the minutes tick by on her bedside clock. She sometimes tossed and turned in bed; at other times she lay in bed reading or got out of bed to do housework. She usually fell asleep 1 to 3 hours later; if she remained awake until 5 AM, she often had difficulty waking up when her alarm clock went off at 6 AM.

She took no medications, but she had been treated intermittently with antidepressants for depression between the ages of 34 and 40. She worked as a legal secretary and drank 3 to 4 cups of coffee in the morning. Examination was normal. I diagnosed psychophysiologic insomnia and recommended that she reduce her evening fluid intake, turn her bedside clock to the wall, limit her time in bed to 6 hours from 11:30 PM to 5:30 AM, and get out of bed and do something relaxing if she awoke and became frustrated about her inability to return to sleep.

Six weeks later, she reported that, although her insomnia was much improved while on this program, she had become more tired and sleepy toward the end of the day. Three weeks after she started the program, she slept until about 10 AM one day on a weekend and then abandoned the sleep schedule. I encouraged her to resume the program, but with an earlier bedtime of 10:30 or 11 PM. Four months later, she reported that her insomnia had resolved. She went to bed at 10:30 PM, fell asleep easily, and got up at 5:30 am. On the rare nights when she awoke during the night, she was able to return to sleep with little difficulty.

The duration of the initial time in bed must be individualized with sleep restriction therapy. Glovinsky and Spielman[43] suggested that the initial amount should be equal to the patient's estimate of the usual average amount of sleep obtained, but not less than 4.5 hours. I ask patients to estimate the average amount of sleep that they obtain and the amount that they think they need to feel rested, and then split the difference. Once I determine the amount of time in bed, I ask the patient to define the start time and end time for the sleep period.

Although sleep restriction therapy is useful for many insomniacs, patients must be advised that sleep is often worse for the first few nights. I advise them not to abandon the program without contacting me first; I can often make minor modifications to the program, usually a small increase in the amount of time in bed, that make it more tolerable. Nonetheless, some insomniacs cannot or will not comply with the restriction because of the fear of more severe insomnia during the first few days or for other reasons.

Relaxation therapies, aimed at reducing the physiologic activation that may occur at bedtime, are sometimes beneficial. Relaxation programs should be practiced thoroughly during waking hours before they are used at night. Biofeedback, hypnotic imagery, and progressive muscle relaxation are examples of such approaches, which may work primarily because by lying quietly and focusing on subjects other than the need to fall asleep, patients facilitate the onset of sleep.[44]

Patients who complain that worrying and racing thoughts interfere with sleep may find that setting aside a 20- to 30-minute block of "worry time" earlier in the day reduces nighttime anxiety. During this worry time, patients should write down several items that they worry about, along with a planned course of action.

Although behavioral treatment of insomnia is often effective, its benefits may not be apparent immediately. In a comparison of behavioral treatment and triazolam for management of insomnia, triazolam had greater initial benefit; improvements with behavioral treatment did not appear until the third week of treatment; after 4 weeks of both treatments improvement was about equal.[45] Over the 5 weeks after the end of treatment, however, benefits were maintained in the group that received behavioral treatment, but sleep returned toward baseline values in the triazolam group.

IDIOPATHIC INSOMNIA

Idiopathic insomnia is a rare disorder in which a lifelong inability to obtain adequate sleep is thought to be caused by a deficit in sleep onset and maintenance systems or overactivity of arousal systems.[36] Unlike most forms of insomnia, idiopathic insomnia requires no precipitating factor: It is caused by a constitutional predisposition to poor sleep per se. In one series of more than 1200 insomniacs, only 4 were given this diagnosis.[46]

Insomnia with difficulty getting to sleep and staying asleep may be accompanied by decreased energy, poor attention, and low motivation. Some patients have relatively mild sleep disruption and daytime symptoms; others have severe chronic insomnia associated with poor social and occupational functioning.[47] Although by definition there is no identifiable neurological disorder that accounts for the symptoms, difficult birth and prematurity may be risk factors, and some of these patients also have dyslexia or attention deficits.[47,48]

Definitive diagnosis, which requires ruling out other causes, is in my experience almost impossible in adults because it is so difficult to substantiate the onset of insomnia in childhood and to be reasonably certain that potential contributors such as child abuse, poor parenting, family discord, asthma, and allergies were not present in childhood. Furthermore, chronic difficulty with sleep may lead to poor sleep hygiene, drug or alcohol dependence, or negative conditioning that obscure the initial presentation.

Management of idiopathic insomnia is similar to the approach used with psychophysiologic insomnia, with particular emphasis on good sleep hygiene. Although hypnotics are usually of little benefit, sedating antidepressants are sometimes helpful.[49,50] Remissions are rare.

SLEEP-STATE MISPERCEPTION (PSEUDOINSOMNIA)

As typical of these accounts, we quote one from Anderson, Indiana, December 11, 1895:—"David Jones of this city, who attracted the attention of the entire medical profession two years ago by a sleepless spell of ninety-three days, and last year by another spell which extended over one hundred and thirty-one days, is beginning on another which he fears will be more serious than the preceding ones. He . . . has not slept for twenty days and nights. He eats and talks as well as usual, and is full of business and activity. He does not experience any bad effects whatever from the spell, nor did he during his one hundred and thirty-one days. During that spell he attended to all of his farm business. He says now that he feels as though he never will sleep again."

GM GOULD[51]

Although many insomniacs exaggerate the severity of insomnia, most patients who complain of poor sleep have disturbed sleep when assessed polysomnographically. On the other hand, in some patients there is a marked discrepancy between a complaint of severe insomnia and the observation of normal sleep or mild insomnia in the laboratory (*sleep-state mis-*

perception). Other terms used to describe this condition include subjective insomnia without objective findings, pseudoinsomnia, and sleep hypochondriasis. The disorder is not common: In one series, only 9% of patients with a chief complaint of insomnia had a diagnosis of sleep-state misperception.[46] Over an 11-year period at our center, only 12 of more than 6000 patients were diagnosed with this disorder.

Patients with sleep-state misperception often describe severe chronic insomnia with as little as 2 hours of sleep per night, or the frequent occurrence of two or more consecutive nights with no sleep. Some patients report long periods of "resting" in bed without sleeping. Sleep that is perceived as sleep is often viewed as nonrestorative, and the sense that sleep is not restful may be more bothersome to the patient than the nighttime difficulties. Patients often attribute disturbed social or occupational functioning to the sleep problem, and some patients visit numerous sleep specialists and physicians in an attempt to find relief of symptoms or a satisfactory explanation. The following is an illustrative case.

CASE HISTORY: Case 9-3

A 26-year-old Asian woman complained of insomnia. At age 21, while completing her undergraduate studies in Australia, she experienced difficulty falling asleep and staying asleep and was treated with a sedative with temporary benefit. After emigrating at age 23 to pursue her graduate education in the United States, she took amitriptyline 25 to 50 mg at bedtime for 2 years, but it was of little or no benefit. Psychiatric evaluation at age 25 revealed mild anxiety, focused mainly on her insomnia, without depressive symptoms. She went to bed at 11 PM and usually lay awake until 1 to 2 AM. Once asleep, she usually woke up between 4 and 6 AM, unable to return to sleep, and arose from bed at 7 to 8 AM. On weekends she went to bed between 1 and 2 AM and got up at the same time. On occasional nights she slept from 11 PM until 4 AM and felt rested during the day. On other nights she did not sleep at all, and these episodes lasted sometimes for as long as 2 weeks. During sleepless intervals she remained in bed resting.

Although she denied feeling anxious or nervous during nights that she could not sleep, she felt that sleep was likely to be worse on nights before important events. She usually felt irritable and unmotivated during the day after a night without sleep, but she did not feel sleepy or take naps.

At the time of my initial evaluation she was not taking medications. She reported good relationships with her family and did well academically. She had a steady relationship with a man; her sleep was about the same whether she slept with him or alone. She drank alcohol about once per week and did not use caffeine, tobacco, or illicit drugs. Physical and mental status examinations were normal.

My initial impression was psychophysiologic insomnia with an element of sleep misperception suggested by the long intervals "resting" in bed without sleeping. I recommended sleep restriction with no more than 6 hours in bed, and that she get out of bed when she was unable to sleep. Two months later, she had reduced her time in bed to about 6½ hours, but she had not followed the recommendation to get out of bed when unable to sleep because it made her feel irritable and uncomfortable. Her sleep log showed a regular bedtime with a perceived pattern of falling asleep within 30 minutes and sleeping 4 to 7 hours about 5 nights per week, with no sleep at all about 2 nights per week.

Over the next 9 months, this pattern continued. A trial of triazolam was not helpful. She occasionally took over-the-counter sedatives without benefit. Because of suspected sleep-state misperception, a polysomnogram was obtained; she had a sleep latency of 53 minutes and slept for just over 6 hours of the 8-hour study, with 14% REM sleep and 7% slow-wave sleep. She believed that she had slept for only 2 hours. After the results were reviewed with her, she felt frustrated and unable to understand why she felt unrefreshed on most mornings. A trial of afternoon exercise and hot baths was ineffective.

Eight months later, her sleep pattern was unchanged. She resumed seeing a psychiatrist and trials over the next 3 years of imipramine, trazodone, doxepin, and fluoxetine were not helpful. Quazepam helped for a short time, and one antidepressant "worked great" for 3 weeks during a trip home to her native country. She married, but her sleep pattern remained unchanged. Her husband slept well and could not confirm or refute the patient's reports. During pregnancy, she reported sleeping only two nights per week, a pattern that continued after the birth of her child. At age 32, she had completed her doctorate and was working as a postdoctoral fellow; she believed that her insomnia made her less productive and less "enthusiastic" about her work.

Over an 11-year period, this young woman had pronounced subjective insomnia that varied little in severity. A polysomnogram demonstrated a marked discrepancy between her perception of sleep and the polysomnographic record. Several psychiatric evaluations revealed no significant psychopathology, and she received little or no benefit from behavioral therapy or pharmacotherapy. Despite her perception of lack of sleep, she maintained good social and occupational functioning.

Sleep-state misperception should be suspected in patients with severe chronic insomnia that shows little variation over time and that does not respond to conventional treatments. Although it has been suggested that this disorder is simply an extreme example of the tendency of most insomnia patients to exaggerate their sleep problem, or that it represents an early stage of chronic insomnia,[38,52,53] in my experience these patients have a distinct disorder.

Diagnosis requires polysomnography, which usually reveals normal sleep or only a mild sleep disturbance, although the patient perceives the same night of sleep as markedly abnormal. It is essential to question the patient about sleep during the laboratory night because some insomniacs may sleep better than usual in the laboratory. If too much weight is given to polysomnographic results without considering the possibility of night-to-night variability, other forms of insomnia may be misdiagnosed as sleep-state misperception. Malingering or a psychiatric condition should also be considered when there is a marked discrepancy between the polygraphic findings and the patient's report.

The cause of sleep-state misperception is unknown, although even in persons without insomnia, unawareness of sleep is common during brief daytime naps, and almost everyone at one time or another has been nudged awake by a spouse or friend without realizing that sleep had occurred. Excessive or abnormal mentation in sleep or just before sleep may contribute to the disorder, as may ruminations that recur with each awakening during the night, giving the subject the impression of uninterrupted consciousness.[54,55] A subtle disturbance of sleep undetected by standard polysomnography could also be a factor.[56-58]

Treatment focuses on patient education about the significance of polysomnographic findings, indicating that restoration is occurring even though the patient has no subjective sense of it. This approach may reduce concerns in some patients, although many remain frustrated and dissatisfied. Behavioral treatment may help some patients; in others, hypnotics may be of value, perhaps because the amnestic effects interrupt the impression of continuous consciousness.[59-61]

EXTRINSIC SLEEP DISORDERS THAT CAUSE INSOMNIA

In extrinsic disorders causing insomnia, the ability to sleep is normal, but sleep onset or continuity is disrupted by factors originating outside the body, such as medications, noise, or stressful life events. This section includes a discussion of four extrinsic disorders that are common causes of insomnia in adults—inadequate sleep hygiene, environmental sleep disorder, adjustment sleep disorder, and hypnotic-dependent sleep disorder; as well as four disorders that are common causes of insomnia in children—sleep-onset association disorder, limit-setting sleep disorder, nocturnal eating/drinking syndrome, and food allergy insomnia.

Inadequate Sleep Hygiene

With the syndrome of inadequate sleep hygiene, poor sleep habits lead to disrupted sleep. Patients may never have had good sleep habits, perhaps because they were not required to sleep at regular times as children or because their internal sleep systems were initially so robust that they were able to sleep well under almost any circumstances. Others seem never to have recognized that sleepiness and alertness have circadian patterns; as a result, they make no attempt to maintain regular sleep hours.

With increasing age or because of other factors, some persons become less able to tolerate the effects of poor sleep hygiene. Some people who drink coffee or other caffeinated beverages in the evening do so with impunity as young adults, but they begin to have disrupted sleep after about age 40. Evening exercise and late-night parties are other activities that may have minimal effects on sleep in youths but much greater effects during middle age. Nocturnal viewing of television, sharing the bed with pets, and a habit of spending long daytime intervals in bed are other habits that contribute to poor sleep. Some patients adhere faithfully to a directive to take a medication every 6 hours and, as a result, awaken at midnight or 6 AM, or both, to take medications.

Insomnia caused initially by other factors is often complicated by poor sleep hygiene. For example, patients who increase their time in bed in an attempt to obtain more sleep, either by going to bed earlier or by going back to bed in the morning after arising, may exacerbate their insomnia by increasing time awake in bed rather than increasing time asleep. Daytime caffeine and evening alcohol use, instituted in response to insomnia, may aggravate the problem. Furthermore, midnight snacking, cigarette smoking, and drinking alcohol, often viewed by patients as relaxing activities that facilitate a return to sleep, may act as unconscious reinforcers of nocturnal awakenings and poor sleep.

Education about the nature of sleep and the reasons for maintaining good sleep habits is usually the best treatment, with emphasis on issues of particular importance for the person (Table 9–6). Secondary gains, when present, should be addressed. When poor sleep hygiene is the primary cause of insomnia, improved sleep hygiene leads to marked improvement.

Table 9–6. **Guidelines for Good Sleep Hygiene**

Maintain regular times of going to bed and arising from bed.
Avoid daytime naps if they interfere with nighttime sleep.
Avoid evening alcohol use.
Avoid consuming caffeinated drinks late in the day.
Reduce or eliminate tobacco use, especially at night or in the evening.
Exercise in moderation; avoid evening exercise.
Use the bed only for sleep and sexual activity.
Keep the bedroom dark, quiet, and cool.
Avoid stressful activities in the evening before sleep.

If prescription medications disrupt sleep (Table 9–7), it may be possible to adjust the timing of medication doses. A morning walk in the sunshine or sitting in a sunroom may help patients maintain a regular schedule by enhancing morning alertness and late-evening sleepiness. Afternoon exercise sufficient to raise body temperature and induce sweating may help to promote nighttime sleep.

Environmental Sleep Disorder

Patients with environmental sleep disorder have unsuitable sleeping environments. Noisy neighbors, a home near an airport, a barking dog, a bedroom that is too hot or

Table 9–7. **Medications That May Contribute to Insomnia**

β-Adrenergic receptor antagonists
Methylxanthines
Levodopa and dopaminergic receptor agonists
Antidepressants with alerting properties
Corticosteroids
Monoamine oxidase inhibitors
Analgesics containing caffeine
Sympathomimetics (decongestants and bronchodilators)
Psychostimulants

too cold, or behaviors of family members or neighbors may make sleep difficult. Other contributors may include bed partner movements, excessive light, and the need to be vigilant at night to care for an infant or an elderly and infirm parent. Older persons are more susceptible to the effects of an unsuitable sleeping environment than young or middle-aged persons.

Treatment should be aimed at removal of the stimulus responsible for the sleep disruption. Ear plugs, eye shades, curtains, white-noise generators, a new mattress or air conditioner, or a separate bed are beneficial in selected cases. For some patients, a move to new living quarters, if feasible, is the best solution. If the condition is time-limited, a short course of hypnotics may be useful.

Adjustment Sleep Disorder

With this disorder, which is the most common cause of acute insomnia, sleep disturbance occurs in relation to a specific stressor or in specific situations to which the patient has difficulty adjusting. The acute stress, conflict, or emotional arousal that triggers insomnia may be related to school examinations, marital strife or divorce, family illness, or work. In contrast to environmental sleep disorder, the psychological reaction is responsible for insomnia, rather than the situation itself. One of the most common causes of adjustment sleep disorder is grief following the loss of a spouse, other relative, or close friend (see Chap. 16).

For management, short-term counseling is often indicated to help the patient deal with the cause of stress, and a limited course of a hypnotic is sometimes useful. Effective early treatment of insomnia with behavioral measures, good sleep hygiene, counseling, or hypnotics may help prevent the development of chronic insomnia in patients with this disorder.

Hypnotic-Dependent Sleep Disorder

Hypnotic-dependent sleep disorder develops in a small minority of patients for whom hypnotics are prescribed. In one survey of 13,000 patients in a general medical practice, only 65 (0.5%) had been taking prescribed benzodiazepines for more than 1 year. Assuming 10% of the practice population had insomnia, 5% of the insomniacs had taken hypnotics for more than 1 year.[62] On the other hand, in a study of 1020 persons over age 65 who were living at home, more than 10% had taken hypnotics for more than 1 year.[63] Patients with hypnotic-dependent sleep disorder—most of whom take hypnotics only at bedtime and do not abuse alcohol, other sedatives, narcotics, or stimulants—should be distinguished from patients with a primary substance abuse disorder, who usually have a history of substance abuse that predates the sleep symptoms.

The syndrome often develops as follows. A patient with acute insomnia is given a prescription for a hypnotic and has a good initial response, but insomnia returns when the hypnotic is discontinued. With resumption of hypnotic use, insomnia improves, but it then returns again when the hypnotic is discontinued. The continued alternation between good sleep with the medication and poor sleep without it leads the patient to believe that good sleep is impossible without the medication. Anxiety about insomnia and rebound anxiety associated with drug withdrawal may also contribute to poor sleep, while the anxiolytic effect of the hypnotic adds to the positive conditioned association of relaxation and good sleep with hypnotic use.

Hypnotic-dependent sleep disorder was first described in patients for whom barbiturates were prescribed, who became tolerant to the hypnotic effect and then had severe rebound insomnia when the drug was discontinued.[64,65] After the introduction of benzodiazepines, similar though less severe effects were observed, usually lasting one to three nights.[66-68] High doses, short half-lives, and abrupt withdrawal increase the likelihood that rebound insomnia will occur.[67,69,70] In an experimental setting, however, the occurrence of rebound insomnia does not increase the likelihood of self-administration of hypnotics, suggesting that other factors contribute to the development of hypnotic-dependent sleep disorder, including personality traits, the severity of insomnia, and the efficacy of the hypnotic.[71] Tolerance to hypnotic effects, affective disorders, and anxiety that is exacerbated by drug withdrawal also may be factors in some patients.

To reduce the likelihood that hypnotic-dependent sleep disorder will develop, clinicians should prescribe hypnotics only when a clear indication is present, prescribe the lowest effective dose, and be particularly cautious in prescribing hypnotics for patients at risk for hypnotic dependence, such as those with a history of alcohol or substance abuse. When hypnotics are prescribed, the drug should be gradually discontinued after the precipitating event has resolved.

The mainstay of treatment for hypnotic-dependent sleep disorder is gradual dose reduction, undertaken if possible at a time of relatively low stress, combined with support and encouragement. The likelihood of success is enhanced if the planned schedule for dose reduction is one that the patient accepts as reasonable. The dose reduction may take place over several weeks, several months, or even 1 year or more. A sleep diary is often helpful to provide evidence that sleep is not deteriorating. If sleep disturbance appears during dose reduction, behavioral treatments may be helpful, along with encouragement that the sleep disruption is probably temporary and related to the dose reduction. In some cases, it is better to resume the higher dose and try again in 1 to 2 weeks.

Although many patients successfully discontinue hypnotic medications and are able to sleep well without medication, others do not. In such cases, the cause or causes of insomnia should be reassessed, along with the patient's motivation to discontinue the medication. A previously undiagnosed anxiety disorder, substance abuse disorder, or affective disorder may be present, or there may be an ongoing stressor that is contributing to insomnia.

Sleep-Onset Association Disorder

In this disorder, which is more common in children than adults, specific objects or conditions become associated with the onset of sleep, and the patient has difficulty falling asleep unless the conditions are present.[72] The most common such association is with a pacifier. As long as the pacifier is present and in the mouth, the infant or toddler is able to fall asleep quickly. If the pacifier is not in the mouth, and it usually is not if the child awakens during the night, the child has trouble returning to sleep and may not fall asleep again until the parent places the pacifier back in the mouth. Parents are often convinced that sleep for the child is impossible without the pacifier, bottle, teddy bear, or night light. In other cases, the child must be rocked or patted or requires the presence of the mother in bed in order to sleep. If sleep onset requires picking up the child for rocking, it may be difficult to place the child in the crib until sleep is sufficiently deep.

In some cases, the association is present from birth and the recognition of the problem occurs only when the infant reaches the age of 3 to 6 months and is expected to sleep without interruption. In other cases, an infant who has slept well is rocked back to sleep for several nights because of a minor illness, and then has trouble falling back to sleep without the rocking and the physical contact that accompanies it.

The key to diagnosis is the ease with which the child returns to sleep when the necessary conditions are met, which differentiates the disorder from limit-setting sleep disorder (see below). Treatment is straightforward in theory, although sometimes difficult in practice. The child must learn to fall asleep without the pacifier or without being rocked or held. To do so, the parents must decide on the appropriate sleep setting and then be willing to put up with a few nights of poor sleep by the child and the associated disruption of other household activities. Infants may learn to sleep without the associated object within one or two nights. For older children, the transition may last longer and may be complicated by issues related to limits and autonomy.

Adults who develop sleep-onset association disorder may require television, radio, special audio tapes or sound machines, pillows, or blankets to fall asleep. They often resist the idea that sleep can be achieved without the particular conditions.

Limit-Setting Sleep Disorder

With limit-setting sleep disorder, which occurs almost exclusively in children, the chief complaint is usually inability to fall asleep. Symptoms often begin when there is a change in the household or in the sleeping arrangements, as when the child is moved from a crib to a bed and the physical barriers imposed by the sides of the crib are removed.[73] The arrival of a new baby, who may share the older child's room or require the child to be quieter in the evening, may also contribute to the disorder.

The apparent inability to fall asleep is accompanied by behaviors that lead to delays in bedtime and falling asleep, such as requests for bathroom trips, drinks of water, stories, bedding adjustments, bandages, and reassurance. The inability or unwillingness of parents or caretakers to set appropriate limits for bedtime behaviors leads to inconsistent responses that tend to perpetuate the behaviors, and anxiety generated by the parents' inability to set limits may add to the child's inability to fall asleep.

Limit-setting sleep disorder should be distinguished from other causes of bedtime difficulties in children, including delayed sleep phase syndrome, too much time in bed, and separation anxiety. The key feature of the disorder is the ability of the child to fall asleep easily when limits are applied consistently. For example, a child who falls asleep for naps at school or with a baby-sitter, but not when the parents are present, often has this disorder. The parents' inability to establish appropriate and consistent limits may reflect poor parenting skills and may be part of a broader inability to set limits in all aspects of the child's life. The difficulty in setting limits may also be a result of unresolved conflicts concerning the parent's upbringing, to guilt about the amount of time spent with the child during the day, to differences between the two parents in their approach to discipline, or to unintended encouragement of the behavior resulting from an associated secondary gain for one or both parents. For the child, testing limits may reflect normal developmental issues related to autonomy, or there may be secondary gain derived from the increased attention.

The problem resolves whenever limits are handled appropriately. Parents may need counseling or psychotherapy to understand both the reason for their difficulty with setting limits and the impact of their failure on the child. Education followed by a trial of a specific plan of limit-setting helps to determine whether other issues may be involved. Without treatment, bedtime difficulties may persist for years and may lead to delayed sleep phase syndrome in adolescence (see Chap. 12).

Nocturnal Eating/Drinking Syndrome

With this disorder, which usually occurs in infants and young children, the child has difficulty returning to sleep without a bottle or other nourishment. An infant with this disorder may awaken several times per night for feedings and may drink as much as 32 ounces of formula or milk during the night. If parents believe, as they often do, that feedings are the only way to get the child back to sleep, the pattern may continue for months. The child also may become conditioned to eating at night and may develop sleep-onset association disorder with inability to fall asleep without the bottle. Increased nocturnal urination and defecation may add to the problem. The syndrome usually resolves quickly when nocturnal feedings are eliminated.

The syndrome may also occur in adults, who may develop a habit of eating once or twice during the night. Some patients may believe that a return to sleep is difficult or impossible unless a food or drink is consumed. In some cases, there appears to be an overlap with somnambulistic eating (see Chap. 15).

Management requires gradual elimination of the nighttime intake of food and drink. In infants and toddlers, a gradual increase in the interval between feedings and reduction in the amount of food will usually resolve the problem. In adults, a similar program may be helpful but may require several weeks to be effective.[74]

Food-Allergy Insomnia

Food-allergy insomnia begins in the first 2 years of life, usually whenever cow's milk is introduced. Sleep disruption may be severe, accompanied by agitation and abdominal discomfort; however, symptoms usually resolve after cow's milk is eliminated from the infant's diet. The differential diagnosis may include colic and sleep-related gastroesophageal reflux. The diagnosis is made when symptoms resolve after milk is eliminated from the diet but rapidly return when milk is reintroduced. Milk allergy usually improves or resolves in children between ages 2 and 4, but allergies to other foods sometimes cause insomnia in adults.

SUMMARY

Insomnia, a syndrome with diverse causes, may be associated with complaints of difficulty falling asleep, repeated awakenings with difficulty returning to sleep, awakening too early, or a sense that sleep is nonrestorative. Although some insomniacs report daytime sleepiness, most are unable to nap even if they lie down. Other daytime symptoms include irritability, mood changes, anxiety, and poor concentration and attention. Although 10% to 20% of adults have chronic insomnia, women more often than men, about two-thirds have not discussed insomnia with their physicians, and only 5% have had physician visits focused primarily on insomnia.

Many intrinsic and extrinsic factors contribute to insomnia. If sleep-onset insomnia is the major complaint, the differential diagnosis may include an anxiety disorder, inadequate sleep hygiene, hypnotic-dependent sleep disorder, sleep-state misperception, or delayed sleep phase syndrome. Impaired sleep continuity may occur with alcohol-dependent sleep disorder, periodic limb movement disorder, sleep apnea, or in association with an underlying medical or psychiatric disorder. Although early morning awakening is often an indication of depression, particularly in older patients, it can also occur with the use of short-acting hypnotics and with circadian rhythm sleep disorders. Insomnia caused by a medical disorder should be considered in patients with conditions that cause pain, paresthesias, reduced mobility, nocturia, or orthopnea.

In most patients, several of these factors play a role: Some predispose the person to insomnia, some precipitate the condition, and others perpetuate it. The predisposing circumstances set the stage for the development of insomnia, which is then induced by one or more precipitants. The patient's response to insomnia may then perpetuate the insomnia after the precipitants are no longer active.

Treatment is challenging because the large number of potential causes and the high likelihood that more than one cause is operative make it difficult to define a single approach most likely to help a given patient. Appropriate treatment may include education about better sleep habits, treatment of an underlying psychiatric disorder, adjustment of the timing or dose of a medication used to treat a medical disorder, behavioral treatment, prescription of a hypnotic, or a combination of these approaches. With treatment directed at the causes of insomnia, many patients improve.

REFERENCES

1. Wysowski, DK, and Baum, C: Outpatient use of prescription sedative-hypnotic drugs in the United States, 1970 through 1989. Arch Intern Med 151:1779, 1991.
2. Gallup Organization: Sleep in America. The Gallup Organization, Princeton, NJ, 1991.
3. Johnson, LC, and Spinweber, CL: Quality of sleep and performance in the navy: A longitudinal study of good and poor sleepers. In Guilleminault, C, and Lugaresi, E (eds): Sleep/Wake

Disorders. Natural History, Epidemiology, and Long-Term Evolution. Raven Press, New York, 1983, pp 13–28.
4. Ford, DE, and Kamerow, DB: Epidemiologic study of sleep disturbances and psychiatric disorders: An opportunity for prevention. JAMA 262: 1479, 1989.
5. Kripke, DF, et al: Short and long sleep and sleeping pills: Is increased mortality associated? Arch Gen Psychiatry 36:103, 1979.
6. Klackenberg, G: Sleep behavior studied longitudinally: Data from 4–16 years on duration, night-awakening and bed-sharing. Acta Pediatr Scand 71:501, 1982.
7. Lozoff, B, et al: Sleep problems seen in pediatric practice. Pediatrics 71:501, 1985.
8. Richman, N: A community survey of characteristics of one to two year olds with sleep disruptions. J Am Acad Child Psychiatry 20:281, 1981.
9. Ohayon, M: Epidemiological study on insomnia in the general population. Sleep 19:S7, 1996.
10. Spielman, AJ: Assessment of insomnia. Clin Psychol Rev 6:11, 1986.
11. Heath, AC, et al: Evidence for genetic influences on sleep disturbance and sleep pattern in twins. Sleep 13:318, 1990.
12. Freedman, R, and Sattler, H: Physiological and psychological factors in sleep-onset insomnia. J Abnorm Psychol 91:380, 1982.
13. Monroe, LJ: Psychological and physiological differences between good and poor sleepers. J Abnorm Psychol 72:255, 1967.
14. Bonnet, MH, and Arand, DL: The consequences of a week of insomnia. Sleep 19:453, 1996.
15. Stepanski, E, et al: Daytime alertness in patients with chronic insomnia compared to asymptomatic control subjects. Sleep 11:54, 1988.
16. Bonnet, MH, and Arand, DL: Caffeine use as a model of acute and chronic insomnia. Sleep 15: 526, 1992.
17. Morris, M, et al: Sleep-onset insomniacs have delayed temperature rhythms. Sleep 13:1, 1990.
18. Healey, ES, et al: Onset of insomnia: Role of life-stress events. Psychosom Med 43:439, 1981.
19. Gross, RT, and Borkovec, TD: Effects of a cognitive intrusion manipulation on the sleep-onset latency of good sleepers. Behav Ther 13:112, 1982.
20. Sterman, MB, et al: Forebrain inhibitory mechanisms: Conditioning of basal forebrain induced EEG synchronization and sleep. Exp Neurol 7: 404, 1963.
21. Walsh, JK, et al: Insomnia. In Chokroverty, S (ed): Sleep Disorders Medicine: Basic Sciences, Technical Considerations, and Clinical Aspects. Butterworth-Heinemann, Boston, 1994, pp 219–239.
22. Stepanski, EJ, et al: Pharmacotherapy of insomnia. In Hauri, PJ (ed): Case Studies in Insomnia. Plenum Medical, New York, 1991, pp 115–129.
23. Greenblatt, DJ, et al: Current status of benzodiazepines (second of two parts). N Engl J Med 309:410, 1983.
24. Greenblatt, DJ, et al: Sensitivity to triazolam in the elderly. N Engl J Med 324:1691, 1991.
25. Morgan, K: Hypnotic drugs, psychomotor performance, and aging. J Sleep Res 3:1, 1994.
26. Wysowski, DK, and Barash, D: Adverse behavioral reactions attributed to triazolam in the Food and Drug Administration's spontaneous reporting system. Arch Intern Med 151:2003, 1991.
27. Bixler, EO, et al: Next-day memory impairment with triazolam use. Lancet 337:827, 1991.
28. Hedemark, LL, and Kronenberg, RS: Flurazepam attenuates the arousal response to CO_2 during sleep in normal subjects. Am Rev Respir Dis 128:980, 1990.
29. Balter, MB, and Uhlenhuth, EH: The beneficial and adverse effects of hypnotics. J Clin Psychiatry (suppl 7)52:16, 1991.
30. Gillin, JC, and Byerley, WF: The diagnosis and management of insomnia. N Engl J Med 322: 239, 1990.
31. Shorr, RI, and Robin, DW: Rational use of benzodiazepines in the elderly. Drugs Aging 4:9, 1994.
32. MacFarlane, JG, et al: The effects of exogenous melatonin on the total sleep time and daytime alertness of chronic insomniacs: A preliminary study. Biol Psychiatry 30:371, 1991.
33. Dollins, AB, et al: Effect of inducing nocturnal serum melatonin concentrations in daytime on sleep, mood, body temperature, and performance. Proc Natl Acad Sci USA 91:1824, 1994.
34. James, SP, et al: Melatonin administration in insomnia. Neuropsychopharmacology 3:19, 1990.
35. Bes, F, et al: Effects of delta sleep-inducing peptide on sleep of chronic insomniac patients: A double-blind study. Neuropsychobiology 26:193, 1992.
36. American Sleep Disorders Association: International Classification of Sleep Disorders, Revised: Diagnostic and Coding Manual. American Sleep Disorders Association, Rochester, Mn, 1997.
37. American Psychiatric Association: Diagnostic and Statistical Manual of Mental Disorders, ed 4, Revised. American Psychiatric Association, Washington, DC, 1991.
38. Reynolds, CF, et al: Subtyping DSM-III-R primary insomnia: A literature review by the DSM-IV work group on sleep disorders. Am J Psychiatry 148:432, 1991.
39. Hauri, PJ: A cluster analysis of insomnia. Sleep 6: 326, 1983.
40. Hauri, PJ, and Fisher, J: Persistent psychophysiologic (learned) insomnia. Sleep 9:38, 1986.
41. Association of Sleep Disorders Centers, Case Series Committee, Coleman, RM (chairman): Diagnosis, treatment, and follow-up of about 8,000 sleep/wake disorder patients. In Guilleminault, C, and Lugaresi, E (eds): Sleep/Wake Disorders. Natural History, Epidemiology, and Long-term Evolution. Raven Press, New York, 1983, pp 87–97, 1983.
42. Bootzin, RR, et al: Stimulus control instructions. In Hauri, PJ (ed): Case Studies in Insomnia. Plenum Medical, New York, 1991, pp 19–28.
43. Glovinsky, PB, and Spielman, AJ: Sleep restriction therapy. In Hauri, PJ (ed): Case Studies in

Insomnia. Plenum Medical, New York, 1991, pp 49–63.
44. Hauri, PJ: Treating psychophysiological insomnia with biofeedback. Arch Gen Psychiatry 38: 752, 1981.
45. McClusky, HY, et al: Efficacy of behavioral versus triazolam treatment in persistent sleep-onset insomnia. Am J Psychiatry 148:121, 1991.
46. Coleman, RM, et al: Sleep-wake disorders based on a polysomnographic diagnosis: A national cooperative study. JAMA 247:997, 1982.
47. Hauri, P: Primary insomnia. In Kryger, M, et al (eds): Principles and Practice of Sleep Medicine. WB Saunders, Philadelphia, 1994, pp 494–499.
48. Hauri, PJ, and Olmstead, E: Childhood onset insomnia. Sleep 3:59–65, 1980.
49. Regestein, QR: Specific effects of sedative/hypnotic drugs in the treatment of incapacitating chronic insomnia. Am J Med 83:909, 1987.
50. Regestein, QR, and Reich, P: Incapacitating childhood onset insomnia. Compr Psychiatry 24: 244, 1983.
51. Gould, GM: Anomalies and Curiosities of Medicine: Being an Encyclopedic Collection of Rare and Extraordinary Cases, and of the Most Striking Instances of Abnormality in All Branches of Medicine and Surgery, Derived from an Exhaustive Research of Medical Literature from Its Origin to the Present Day, Abstracted, Classified, Annotated, and Indexed. WB Saunders, Philadelphia, 1897.
52. Carskadon, MA, et al: Self-reports versus sleep laboratory findings in 122 drug-free subjects with complaints of chronic insomnia. Am J Psychiatry 133:1382, 1976.
53. Frankel, BL, et al: Recorded and reported sleep in chronic primary insomnia. Arch Gen Psychiatry 33:615, 1976.
54. Borkovec, T: Pseudo(experiential)-insomnia and idiopathic (objective) insomnia: Theoretical and therapeutic issues. Adv Behav Res Ther 2:27, 1979.
55. Kuisk, LA, et al: Presleep cognitive hyperarousal and affect as factors in objective and subjective insomnia. Percept Mot Skills 69:1219, 1989.
56. Haynes, SN, et al: Responses of psychophysiologic and subjective insomniacs to auditory stimuli during sleep: A replication and extension. J Abnorm Psychol 94:338, 1985.
57. Mendelson, WB, et al: A psychophysiological study of insomnia. Psychiatry Res 19:267, 1986.
58. Salin-Pascual, RJ, et al: Long-term study of the sleep of insomnia patients with sleep state misperception and other insomnia patients. Am J Psychiatry 149:904, 1992.
59. Mendelson, WB: Pharmacologic alteration of the perception of being awake or asleep. Sleep 16:641, 1993.
60. Mendelson, WB: Effects of flurazepam and zolpidem on the perception of sleep in insomniacs. Sleep 18:92, 1995.
61. Mendelson, WB: Effects of flurazepam and zolpidem on the perception of sleep in normal volunteers. Sleep 18:88, 1995.
62. Wright, N, et al: Community survey of long term daytime use of benzodiazepines. BMJ 309:27, 1994.
63. Morgan, K, et al: Prevalence, frequency and duration of hypnotic drug use among the elderly living at home. BMJ 284:942, 1982.
64. Oswald, I, and Priest, R: Five weeks to escape the sleeping pill habit. Br Med J 2:1093, 1965.
65. Kales, A, et al: Chronic hypnotic-drug use: Ineffectiveness, drug withdrawal insomnia, and dependence. JAMA 227:513, 1974.
66. Schneider-Helmert, D: Why low dose benzodiazepine-dependent insomniacs can't escape their sleeping pills. Acta Psychiatr Scand 78:706, 1988.
67. Gillin, J, et al: Rebound insomnia: A critical review. J Clin Psychopharmacol 9:161, 1989.
68. Kales, A, et al: Rebound insomnia: A new clinical syndrome. Science 201:1039, 1978.
69. Greenblatt, D, et al: Effect of gradual withdrawal on the rebound sleep disorder after discontinuation of triazolam. N Engl J Med 317:722, 1987.
70. Roehrs, T, et al: Rebound insomnia: Its determinants and significance. Am J Med 88:39S, 1990.
71. Roehrs, T, et al: Rebound insomnia and hypnotic self administration. Psychopharmacology 107:480, 1992.
72. Ferber, R: Childhood insomnia. In Thorpy, MJ (ed): Handbook of Sleep Disorders. Marcel Dekker, New York, 1990, pp 435–455.
73. Ferber, R: Sleeplessness in children. In Ferber, R, and Kryger, M (eds): Principles and Practice of Sleep Medicine in the Child. WB Saunders, Philadelphia, 1995, pp 79–89.
74. Ferber, R: Childhood sleep disorders. Neurol Clin North Am 14:493, 1996.

Chapter 10

NARCOLEPSY AND RELATED DISORDERS

CLINICAL FEATURES
Sleepiness and Sleep Attacks
Cataplexy
Sleep Paralysis
Hypnagogic and Hypnopompic
  Hallucinations
Nocturnal Sleep Disruption
Automatic Behavior
Memory and Visual Disturbances
Psychiatric Symptoms and Psychosocial
  Consequences
Course
BIOLOGIC BASIS
Genetic and Familial Basis
Pathophysiology
Canine Narcolepsy
**CLINICAL VARIANTS AND RELATED
  DISORDERS**
Monosymptomatic Narcolepsy
Narcolepsy Associated with Brain Lesions
Idiopathic Hypersomnia
Post-Traumatic Hypersomnia
DIFFERENTIAL DIAGNOSIS
Sleepiness
Cataplexy
Sleep Paralysis and Hallucinations
DIAGNOSTIC EVALUATION
MANAGEMENT
Treatment of Sleepiness
Treatment of Other Symptoms

I propose to give the name of narcolepsy to a rare neurosis, or one little known up to the present time, characterized by an imperious desire to sleep, sudden and of short duration, reproducing itself at intervals more or less closely related.

JBE GELINEAU[1]

Narcolepsy is a disorder of particular interest to sleep medicine specialists because it appears to be caused by a defect in the fundamental process of sleep-wake regulation. Although the precise central nervous system (CNS) dysfunction that leads to narcolepsy is as yet unknown, it seems likely that elucidation of the dysfunction will lead to or accompany fundamental breakthroughs in the understanding of how the brain regulates sleep and wakefulness.

Narcolepsy has been recognized for more than 100 years (Table 10–1). Although Westphal[2] and Fischer[3] had recognized the syndrome, Gelineau[1] first used the term *narcolepsie* to describe it as a condition characterized by brief episodes of irresistible sleep and by falls associated with emotional stimuli. The brevity and reversibility of the attacks led Gowers[4] to distinguish narcolepsy from *trance*, a prolonged sleep from which the patient could not be roused, usually caused by hysteria (see Chap. 16); and from *sleeping sickness* (see Chap. 19), which was associated with a gradually increasing somnolence that culminated in coma and death. The term cataplexy was introduced by Lowenfeld.[5]

The condition was rarely recognized until the epidemic of encephalitis lethargica that accompanied and followed World War I led to many cases of narcolepsy, described in detail by Adie[6] and Wilson.[7] Several case series followed, the largest of which, published by Daniels[8] in 1934, included 147 patients. Interest in the disorder was further stimulated by reports in

Table 10-1. **Landmarks in Narcolepsy**

19th century	First descriptions of narcolepsy-cataplexy
1920s	Descriptions of postencephalitic narcolepsy
1930s	Introduction of amphetamine treatment
1950s	Introduction of methylphenidate treatment
1960s	Introduction of tricyclic antidepressants as treatment for cataplexy; discovery of association of narcolepsy with SOREMPs
1970s	Discovery of canine model; first consensus definition of narcolepsy
1980s	Discovery of association with HLA markers

SOREMPs = sleep-onset rapid eye movement periods;
HLA = human leukocyte antigen.

1931 that ephedrine sulfate was an effective treatment.[9,10] Treatment with amphetamine was introduced in the 1930s, followed by methylphenidate treatment in the 1950s.

In 1956, Yoss and Daly[11] defined the clinical criteria for the diagnosis of narcolepsy as a tetrad of symptoms: excessive daytime sleepiness, cataplexy, hypnagogic hallucinations, and sleep paralysis. The discovery in the 1960s that narcolepsy is associated with abnormalities of rapid eye movement (REM) sleep, principally its early onset, provided the basis for the concept that many of the symptoms of narcolepsy are caused by abnormalities of REM sleep.[12–15] With the discovery of sleep apnea and the improved characterization of other syndromes of excessive sleepiness, narcolepsy was defined as a distinct syndrome separate from other disorders of excessive sleepiness. The use of the Multiple Sleep Latency Test (MSLT) and the discovery of the association of narcolepsy with specific human leukocyte antigen (HLA) markers further clarified the clinical and biologic features of narcolepsy. The recognition of cataplexy in dogs and other species was followed by the development of a canine model of narcolepsy (see Canine Narcolepsy below) that led to major advances in the understanding of the neuropharmacology and neurochemistry of cataplexy.

Despite the advances in understanding narcolepsy, it probably is still underdiagnosed. Two factors may contribute to the failure to diagnose narcolepsy:

1. Patients may have been told by family, friends, or associates that their problem is "laziness" and therefore may not mention or may minimize their symptoms.
2. Physicians often ignore complaints of sleepiness, dismiss them as inconsequential, attribute them solely to lack of sleep, mistake sleepiness for tiredness, or diagnose depression or chronic fatigue syndrome.

CLINICAL FEATURES

The prevalence of narcolepsy in the white populations of North America and Europe is about 1 in 4000.[16] Estimates for other populations range from 1 in 600 to 1 in 10,000. Men and women are affected in approximately equal numbers; early reports that men were affected more than women were probably a result of misdiagnosis of sleep apnea as narcolepsy.

Most patients report that symptoms began in the second or third decade of life. The definite presence of symptoms before age 5 is rare, although some patients report that they were considered unusually sleepy even as infants and slept through the night at an early age. Some patients develop the syndrome in adult life, occasionally even after age 50, whereas others with early onset of symptoms remained undiagnosed into old age. I have seen 5 patients over the age of 70 whose symptoms of daytime sleepiness and cataplexy were undiagnosed for more than 50 years.

Symptoms usually develop over a few weeks, although some patients report an abrupt onset over a few days. When the onset is gradual, a decline in school performance from one year to the next may be the first indication. Sleepiness and cataplexy may develop simultaneously, or cataplexy may begin decades after the onset of sleepiness. Other symptoms related to chronic sleepiness may also occur (Table 10–2). The following is a representative case.

CASE HISTORY: Case 10–1

A 13-year-old girl was referred for evaluation. At age 11, despite obtaining 9 1/2 hours of sleep each night, she began to fall asleep easily while reading, watching television, and in school. She took 2- to 3-hour naps in the afternoon after school, and episodes developed in which, when laughing, she "went limp," her knees buckled, and she occasionally fell down; this lasted from a few seconds to 1 minute. The episodes also sometimes occurred if she became angry or frightened. At night, she had frightening nightmares, and she occasionally had sleep paralysis in the morning. Her grades in school decreased from above average to average. Physical examination was normal. Sleep latency on polysomnography was 4 minutes, with a REM-sleep latency of 4 minutes. An MSLT revealed a mean sleep latency of 1.1 minutes with four sleep-onset REM periods.

After treatment with methylphenidate 40 mg/d for sleepiness and protriptyline 10 mg/d for cataplexy, she and her mother noted marked improvement. She no longer fell asleep in school, and her grades improved to honors level; however, she still sometimes became sleepy in the late afternoon and evening and occasionally fell asleep while doing homework.

At age 16, she began to sleep only 6 to 7 hours per night, and she frequently stayed out late with friends on weekends. Daytime sleepiness increased, and she was counseled to increase her time asleep at night. During the next year, her grandmother died and she moved to a new school. Her grades deteriorated, and she complained that her classmates thought she looked "stoned" all of the time because of her drowsiness. She was turned down for a summer job at a restaurant because of narcolepsy. She applied for a driver's license but fell asleep during the driving test. At the time of her clinic visit at age 18, she appeared anxious and emotional and had several brief cataplectic episodes, with ptosis and facial weakness, during the interview. She was again advised to sleep more at night and to take a regular afternoon nap. Methylphenidate was increased to 60 mg/d and protriptyline to 15 mg/d.

After graduating from high school, she worked as a clerk, obtained a restricted driver's license, and slept 8 to 9 hours each night. Sleepiness was less of a problem, and cataplexy occurred only when she forgot to take protriptyline. At age 20, she moved in with her boyfriend and began working as a waitress in a bar from 7 PM until 2 AM 4 days per week. Her sleep habits were irregular, and she discontinued protriptyline because of its expense and side effects; without it, she had mild cataplectic episodes several times each day, mainly with laughter. During a trial of dextroamphetamine, she felt "jittery" and preferred to resume methylphenidate. At age 23, she was working as a store clerk from 8 to 4:30 PM and was maintaining more regular sleep hours. Her cataplexy had improved with regular use of protriptyline, but she continued to feel drowsy and fall asleep late in the day, and she rarely drove because of drowsiness.

Table 10–2. **Symptoms of Narcolepsy**

Excessive sleepiness and a tendency to fall asleep easily
   Sleep attacks
   Amnestic episodes with automatic behavior
   Memory disturbance, impaired concentration
   Blurred or double vision
Cataplexy
Sleep paralysis
Hypnagogic and hypnopompic hallucinations; nightmares
Disrupted nighttime sleep

Sleepiness and Sleep Attacks

Excessive sleepiness is usually the most troublesome complaint and the reason for presentation. The severity of sleepiness varies; for some it is a mild annoyance that is easily overcome, whereas others find it overwhelming and irresistible. Sleepiness in narcoleptics is not paroxysmal: Although it fluctuates in severity, it is present most of the time. As with sleepiness that follows sleep deprivation in normal persons, "circumstances normally conducive to sleep are particularly trying for narcoleptic patients."[8] Boring, sedentary situations and warm afternoons following lunch are especially difficult. Physical activity usually provides relief, but sleepiness returns as soon as the patient sits down. The striking difference from normal sleepiness induced by sleep deprivation is the inability to obtain relief with any amount of sleep.

Sleep attacks, episodes of daytime sleep that occur without warning, are common in narcolepsy. Most episodes are brief, lasting a few seconds or minutes. Although Wilson[7] thought daytime sleep episodes were paroxysmal, similar to epileptic seizures, electrophysiologic studies indicate that increasing drowsiness precedes sleep. Narcoleptics become habituated to a state of chronic drowsiness, and because the drowsiness is essentially continuous, they lack a frame of reference to help distinguish drowsiness from normal alertness; lack of awareness of the imminence of sleep probably accounts for sleep attacks. Despite the near-constant daytime sleepiness, daytime sleep episodes are usually brief and nighttime sleep is interrupted by frequent awakenings. As a result, the total amount of sleep over 24 hours is increased only slightly or not at all.

Cataplexy

Cataplexy, a symptom in about two-thirds of narcoleptics, is unique in sleep medicine because it occurs in association with only one sleep disorder—narcolepsy. It is a brief episode of bilateral weakness without altered consciousness that is usually brought on by excitement or emotion. Although it typically develops within a few months of the onset of sleepiness, cataplexy occasionally is the initial manifestation.

Perhaps the most striking feature of cataplexy is its association with specific precipitants. Laughter is the most common, but anger, embarrassment, excitement, elation, surprise, shock, amusement without actual laughter, and other forms of emotion also can induce it. Many persons with cataplexy experience it during sports activities; for example, the excitement of a fish striking the line may be enough to prevent a fisherman from reeling in his catch.

Severe cataplectic attacks produce atonic, areflexic paralysis of striated muscles, sparing only the sphincters and respiratory muscles. Patients may stagger and fall or slump into a chair; twitching around the face or eyelids often accompanies the weakness. Mild attacks that last just a second or two may cause slight weakness at the knees, drooping of the eyelids, sagging of the jaw or head, or inability to hold onto objects, walk, or rise from a chair. Some patients report a sensation of dizziness or fullness of the head accompanying the weakness; others describe momentary weakness that is imperceptible to observers. Although most episodes last just a few seconds, prolonged attacks can last several minutes. Consciousness is preserved at the onset, but prolonged episodes may be associated with auditory, visual, or tactile hallucinations and may lead directly into REM sleep. On occasion, particularly following withdrawal of medications that suppress cataplexy, attacks may last for hours or occur repeatedly over the course of a few days, a condition referred to as *status cataplecticus*.

Although cataplexy is usually benign, it can be dangerous if it occurs while the patient is driving or swimming. Narcoleptics with severe cataplexy may try to become less emotional in social situations in order to reduce the risk of cataplectic attacks.

Sleep Paralysis

Sleep paralysis, recognized for more than 100 years, is an unforgettable experience of partial or total paralysis lasting a few seconds or minutes that occurs during transitions between sleep and wakefulness. The paralysis, which prevents limb movement and eye opening, may be accompanied by a sensation of struggling to move, to speak, or to wake up. Patients often feel awake or half-awake, are usually aware of being in bed, and may believe that they are dying or unable to breathe. The sensations accompanying the first episode of sleep paralysis are often so vivid and terrifying that the person recalls them clearly for years afterwards. With repeated episodes, the experience becomes more familiar and less frightening.

Occasional twitches, slight moans, or irregular respirations may accompany the episodes, which end spontaneously, during an intense effort to move, or following brief stimulation, such as the touch of another person. Rarely, the paralysis continues despite vigorous attempts by observers to arouse the subject. After an attack, the subject usually feels entirely normal, although there may be some residual anxiety.

Hypnagogic and Hypnopompic Hallucinations

Frightening illusions or hallucinations, such as visions of animals or monsters, may accompany sleep paralysis or occur independently. These hallucinations, which occur at the interface between wakefulness and sleep, are referred to as hypnagogic when they occur at the onset of sleep and hypnopompic when they occur at the end of sleep. Visual imagery is almost always present, and auditory components are common, while tactile and olfactory elements are less frequent. The visual images may be simple forms, such as floating circles or lights, or they may be complex hallucinations of landscapes or animals. Auditory hallucinations may be simple repetitive sounds or complex musical themes. One patient reported, "I hear a scratched record which plays the same sentence over and over."[17] Multiple sensory modalities may be involved, as in a patient who felt the fur of animals and heard their chirping sounds as they ran over his body.[8] Sensations of weightlessness, falling, or body transformation may occur, or there may be a hallucination of movement associated with sleep paralysis: "I move quickly to get up and turn out the light, but then I realize that I have not moved at all and that, in fact, I am unable to move."[17] The hallucinations are sometimes accompanied by an impression that someone is nearby, standing over the bed, or threatening to enter the house. Images from the previous few hours or days may be revisualized, and sometimes the same hallucinations recur night after night.

The hallucinations resemble dreams in that their bizarre features are usually unquestioned, but they differ because some awareness of the surroundings is preserved and because of the lack of a thematic story in which the patient participates. Although some patients find the experiences pleasant and relaxing, others describe the nightly nocturnal hallucinations as the most disturbing aspect of narcolepsy. Sometimes the hallucinations are so vivid and realistic that the subject may have difficulty believing that they are not real and may attempt to escape or to block them from sight by closing windows or barricading doors. Narcoleptics with prominent complaints of hallucinations are occasionally misdiagnosed as schizophrenic, particularly if the vividness and apparent reality of the hallucinations lead to the development of a delusional system to explain the phenomena.[18]

Nocturnal Sleep Disruption

Disrupted nighttime sleep with frequent awakenings is common in narcoleptics, particularly in those with severe narcolepsy-cataplexy. In a few patients, nightmares and sleep disruption are the most disturbing symptoms, and some patients have the impression that they spend the entire night drifting in and out of sleep with nightmares, intermittent hallucinations, and episodes of paralysis. REM sleep behavior disorder (see Chap. 15) occurs with increased frequency in patients with narcolepsy and may also contribute to disrupted sleep.

Automatic Behavior

Automatic behavior refers to amnestic episodes lasting seconds to an hour or more during which patients drift in and out of sleep while engaging in aimless or semi-purposeful activity. These episodes, which occur in up to 80% of narcoleptics, usually develop in situations that are associated with sleepiness. They may be associated with brief lapses in speech, with irrelevant words or remarks, or with nonsensical activities, such as putting clothes in the dishwasher or writing words off the edge of a page. During episodes that occur while driving, patients may drive to unexpected or unknown destinations.

Memory and Visual Disturbances

Memory difficulties, noted by up to 50% of patients, are caused by drowsiness. Sleepiness also may be associated with spells of blurred vision or diplopia that are probably caused by failure of fusion induced by drowsiness in patients with exophoria.

Psychiatric Symptoms and Psychosocial Consequences

The psychosocial effects of narcolepsy are substantial. Work performance is most often affected in sedentary jobs requiring sustained vigilance. Eighty percent of narcoleptics have fallen asleep at work and two-thirds while driving, leading to an increased frequency of accidents.[19,20] Although narcolepsy was once thought to be a psychiatric disorder, it is not associated with specific psychopathology: The psychiatric symptoms identified by questionnaires and psychometric tests are usually the narcoleptic symptoms of hallucinations and automatic behavior, rather than those of an associated psychiatric illness. A prevalence of depression as high as 30% in adult narcoleptics probably reflects the psychosocial problems and the effects of chronic illness.[21] In children, the psychosocial consequences may be even greater; in one series, 75% of childhood narcoleptics had behavioral or emotional disturbances, and 25% had been initially given a misdiagnosis of a psychiatric disorder.[22]

Course

The sleepiness associated with narcolepsy usually develops over several weeks or months and then remains stable. Increasing sleepiness after a period of stability suggests the presence of an additional problem, such as sleep apnea. Although sleepiness usually remains constant, cataplexy, hypnagogic hallucinations, and sleep paralysis improve with age in about one-third of patients.

BIOLOGIC BASIS

Genetic and Familial Basis

Although a genetic basis was suspected years before research in the 1980s and 1990s clarified the genetic underpinning of narcolepsy, early descriptions of familial narcolepsy were based only on clinical symptoms, and other causes of daytime sleepiness were not well understood. For example, four obese siblings described as examples of familial narcolepsy in 1942 undoubtedly had obstructive sleep apnea rather than narcolepsy: They snored loudly, did not have cataplexy, and became cyanotic during sleep.[23] Guilleminault and colleagues,[24] who performed sleep laboratory recordings in more than 100 relatives of narcoleptics, showed that familial narcolepsy is rare. Although more than 40% of the narcoleptics in the study had first-degree relatives with complaints of sleepiness, sleep recordings and clinical evaluation revealed that only 3% of the narcoleptics had a first-degree relative with narcolepsy-cataplexy and that just 1% had more than one affected first-degree relative. Most "sleepy" relatives of narcoleptics, therefore, do not have narcolepsy. The risk for narcolepsy in children of narcoleptics and in other first degree relatives is about 1%, approximately 40 times higher than the risk for individuals without affected relatives.

The study of narcolepsy genetics was transformed in the early 1980s by the report that HLA-DR2 and -DQ1 were present in 100% of a large series of Japanese patients with narcolepsy-cataplexy (Fig. 10–1).[25] In comparison, HLA-DR2 is present in 20% to 35% of the normal population, and HLA-DQ1 is present in 60% to 80%. Although some patients with narcolepsy-cataplexy do not carry the HLA-DR2 or HLA-DQ1 antigens, the prevalence of HLA-DR2 exceeds 90% in white patients with narcolepsy-cataplexy.[26,27] The prevalence of HLA-DQ1 in narcolepsy-cataplexy is also extremely high: Its frequency equals or exceeds the frequency of HLA-DR2 in all racial groups of narcoleptics. In African-Americans with narcolepsy, the HLA-DQ1 antigen is present in more than 90%, but HLA-DR2 is found in only about 65%.[28]

The DQB1–0602 and DRB1–1501 subtypes of DQ1 and DR2 are now recognized (Table 10–3) and are highly associated with narcolepsy. Current evidence suggests that the genes that code for these antigens are responsible for narcoleptic susceptibility; no other gene in the region appears to be involved.[29,30] This haplotype is associated with susceptibility to narcolepsy, rather than with the symptom of somnolence, as its incidence is not significantly increased in sleep apnea.[31]

The low incidence of narcolepsy in family members and the frequency of the haplotype in asymptomatic relatives of narcoleptics and in control subjects make it evident that the HLA-associated genes are not sufficient to induce narcolepsy. An additional gene or genes may be required for disease expression. In some families, a gene unlinked to the HLA locus appears to confer narcoleptic susceptibility even in the absence of the HLA-associated gene.[24,32,33]

The existence of monozygotic twins discordant for narcolepsy indicates that environmental factors play an important role in the development of narcolepsy. Although the nature of the environmental factor is unknown, an immunologic pathogenesis is suggested by the associations with the HLA-D region. As immunologic abnormalities have not been detected in serum or cerebrospinal fluid (CSF) of human narcoleptics, the inciting factor may be a transient immunologic reaction that is difficult to detect in subsequent years.

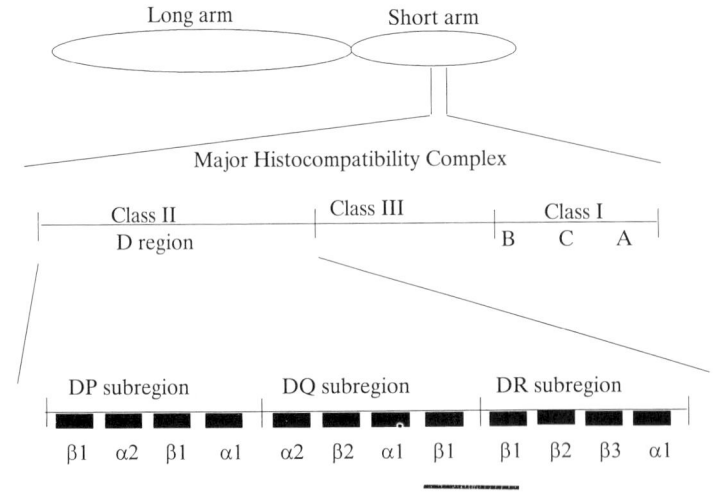

Figure 10–1. The major histocompatibility complex is located on the short arm of chromosome 6. The genes of this region control the expression of glycoprotein antigens. Class I genes encode antigens, designated HLA-A, HLA-B, and HLA-C, which are expressed on nucleated cells and platelets. Class III genes encode proteins that are involved in complement activation, as well as other proteins. Class II genes encode antigens, designated HLA-DP, HLA-DQ, and HLA-DR, that are expressed on macrophages, monocytes, and other lymphocytes. These antigens, which are also expressed on some glial cells in the CNS, are made up of combinations of alpha and beta subunits. The DNA segment closely associated with narcolepsy includes the DQ-β1 subregion and the DR-β1 subregion (underlined). (Adapted from Aldrich MS: The neurobiology of narcolepsy-cataplexy. Prog Neurobiol 41:533, 1993, p 538, with permission.)

Table 10–3. **Human Leukocyte Antigens Associated with Narcolepsy**

DR2
  DR15 (subtype of DR2)
    DRB1-1501 (subtype of DR15)
DQ1
  DQ6 (subtype of DQ1)
    DQB1-0602 (subtype of DQ6)

The onset of narcolepsy may follow sleep loss, psychological stress, head trauma, drug abuse, or pregnancy, but the temporal association of these events with symptom onset may be simply coincidental.

The DR15/DQ6 haplotype also occurs with increased frequency in narcolepsy without cataplexy (see below), suggesting that the narcoleptic susceptibility gene or genes may have variable expression. Other forms of gene expression may exist, such as idiopathic hypersomnia, which occurs in some relatives of narcoleptics.

Pathophysiology

The occurrence of REM sleep within minutes of falling asleep, referred to as a *sleep-onset REM period* (SOREMP), is one of the most striking features of narcolepsy (Fig. 10–2). The muscle atonia that occurs with sleep paralysis and cataplexy appears to be physiologically identical to the muscle atonia of REM sleep: The tendon reflexes are reduced or absent during cataplexy, as they are during REM sleep, and the H-reflex, an electrical correlate of the tendon reflexes, is inhibited during cataplectic attacks, sleep paralysis, and REM sleep.[34] Hypnagogic hallucinations, which often occur just as the narcoleptic passes from wakefulness to REM sleep or vice versa, probably represent an intrusion of REM-sleep imagery into the waking state, just as sleep paralysis and cataplexy are associated with an intrusion of REM sleep-related muscle atonia into wakefulness.

Early onset of REM sleep is part of a broader problem of impaired sleep-wake regulation. Frequent nighttime awakenings indicate that the ability to sustain sleep is also impaired, and periods of ambiguous sleep, which combine features of REM sleep with those of non-rapid eye movement (NREM) sleep, suggest that there is a blurring of boundaries between NREM and REM sleep. For example, isolated REMs may occur in stage 2 sleep, and periods of increased muscle tone may occur during REM sleep. Evidence that an element of wakefulness is often still pres-

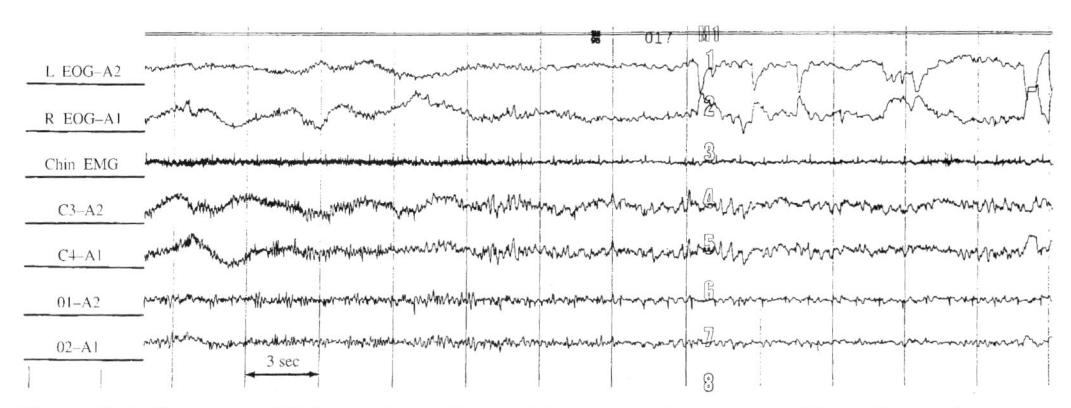

Figure 10–2. Sleep-onset REM period in a 60-year-old patient with narcolepsy. These 37 seconds were recorded during an MSLT. The patient was awake but drowsy during the first 10 seconds, with EEG alpha activity and slow eye movements. During the next 27 seconds, he entered REM sleep with the appearance of muscle atonia (channel 3), rapid eye movements (channels 1 and 2), and a low-voltage EEG dominated centrally by theta-frequency activity (channels 4 and 5). The EEG electrodes (A1, A2, C3, C4, O1, O2) were placed in accordance with the International 10-20 system. Other abbreviations: L EOG, left electro-oculogram; R EOG, right electro-oculogram; Chin EMG, submental electromyogram. (From Bassetti C, Aldrich MS: Narcolepsy. Neurol Clin North AM 14:545, 1996, p 553, with permission.)

ent during SOREMPs and the occurrence of REM sleep behavior disorder in some patients with narcolepsy (see Chap. 15) support the notion that disturbed sleep-state regulation is the essential feature of the disorder.[35,36] Despite the frequent occurrence of disturbed nighttime sleep in narcolepsy, nocturnal sleep disturbance probably does not contribute much to daytime sleepiness.[37]

Impaired sleep-wake state regulation and aberrant control of REM sleep suggest that an abnormality in the function of neural generators and modulators of sleep and REM sleep (see Chap. 2) plays an important role in the pathogenesis of narcolepsy. The cholinergic system is a candidate for the fundamental defect because of its critical role in the generation of REM sleep and the modulation of REM-sleep muscle atonia (see Chap. 2). Although cholinergic abnormalities are present in canine narcolepsy (see below), there is currently no evidence of a systemic or CNS cholinergic dysfunction in human narcolepsy.

Because monoaminergic systems, particularly those involving norepinephrine and serotonin, modulate the activity of cholinergic systems involved in REM-sleep regulation, a defect in monoaminergic activity could contribute to narcoleptic symptomatology. Amphetamines and related stimulants, used to treat sleepiness, increase the synaptic availability of norepinephrine. In addition, prazosin, an α_1-noradrenergic antagonist, increases the severity of cataplexy.[38,39] For tricyclic antidepressants, the relation between anticataplectic efficacy and inhibition of norepinephrine reuptake is consistent with an inhibitory role for norepinephrine in the control of cataplexy.[40] On the other hand, agents that increase the synaptic availability of dopamine have relatively little effect on cataplexy and other narcoleptic symptoms.

Although postmortem studies of human narcoleptic brain tissue have not shown consistent structural lesions, they have provided further indications of abnormal monoaminergic function. Dopamine D1 and D2 receptor binding is increased in the striatum, and there is evidence suggesting an increase in adrenergic α_2 receptors and increased concentrations of the norepinephrine metabolite 3-methoxy-4-hydroxyphenylglycol in a number of brain regions.[41–44] On the other hand, in vivo positron emission tomographic (PET) imaging of dopamine receptors has not demonstrated abnormalities.[45,46]

Canine Narcolepsy

Narcolepsy-cataplexy is recessively inherited in some dogs, and the development of a colony of affected animals at Stanford University has allowed investigators to perform extensive pharmacological, neurophysiologic, and genetic studies of the canine syndrome.[47] In narcoleptic dogs, a group of cells in the medial medulla, located primarily in the ventromedial and caudal portions of the nucleus magnocellularis, discharges only during cataplexy and REM sleep.[48] As these neurons are a subset of the neurons in this region that are active during REM sleep, the findings indicate that cataplexy is a behavioral state distinct from REM sleep.

Other studies suggest that cataplexy may be caused by overactivity of cholinergic systems or underactivity of catecholaminergic systems. Narcoleptic dogs have increased numbers of pontine cholinergic neurons and increased density of pontine cholinergic M2 receptors, and canine cataplexy is aggravated by cholinergic M2 receptor agonists, probably by actions at cholinoceptive sites in the pontine reticular formation and the basal forebrain.[49–52] Thus, cholinergic hypersensitivity may contribute to cataplexy and other narcoleptic symptoms. Cataplexy is also exacerbated by adrenergic α_1 receptor antagonists, by adrenergic α_2 agonists, and by dopamine D2 receptor agonists.[53–55] These findings, along with evidence of increased density of adrenergic α_1 receptors in the amygdala, adrenergic α_2 receptors in the locus ceruleus, and dopaminergic D2 receptors in the amygdala and nucleus accumbens, suggest that upregulation of these presumably inhibitory presynaptic receptors may contribute to narcoleptic symptoms by inhibition of dopaminergic and adrenergic activity.

Human narcolepsy is an autosomal-dominant disorder with low penetrance. Although canine narcolepsy is recessively inherited, identification of the canine gene and its product may provide important clues to the pathogenesis of human narcolepsy. The canine gene, designated *canarc-1*, is tightly linked to a DNA region called *the μ-switch region*, which is involved in determining the immunoglobulin class of certain activated B lymphocytes.[56] It is not, however, linked to the canine equivalent of the human HLA region. The tight linkage of *canarc-1* to the μ-switch region and the mild CSF pleocytosis that occurs in these dogs at about the time of onset of cataplexy suggest that immune-mediated mechanisms may play a role in the development of canine narcolepsy.[57] Major histocompatibility complex (MHC) class II molecules are expressed by microglia in the CNS in dogs and humans, and their increased expression during puberty may play a role in the development of narcoleptic symptoms.[58]

Although homozygosity for *canarc-1* is required for the expression of canine narcolepsy and spontaneous cataplexy, cataplexy may develop in heterozygotes when they are given a combination of physostigmine, a muscarinic agonist, and prazosin, an adrenergic α_1 antagonist.[59] Thus, one copy of *canarc-1* may have biologic effects, and a homologous gene, if present in humans, could predispose to the disease or be additive with the HLA-associated gene, or both.[59]

CLINICAL VARIANTS AND RELATED DISORDERS

Despite striking advances in understanding the genetic basis of narcolepsy, its clinical spectrum remains controversial. In 1902, Lowenfeld[5] believed that cataplexy was an essential feature of the disease, and some sleep specialists continue to reserve use of the term narcolepsy for patients who have definite cataplexy as well as sleepiness. Wilson,[7] however, referred to "the narcolepsies"— a symptom complex that could be seen in its entirety, with cataplexy, or in parts, without cataplexy. He believed that narcolepsy could occur as an idiopathic disorder or as a result of encephalitis, multiple sclerosis, head trauma, cerebral neoplasm, or psychiatric disease. When cataplexy is not a symptom, the diagnosis of *monosymptomatic narcolepsy* is sometimes given (i.e., only one of the two major symptoms is present).

The *International Classification of Sleep Disorders (ICSD)* allows the diagnosis of narcolepsy in the absence of cataplexy as well as in the presence of a structural brain lesion, as long as the lesion is determined to be the cause of the narcoleptic signs and symptoms.[60] It also differentiates narcolepsy from two other disorders that are associated with excessive sleepiness and normal or near-normal nighttime sleep: idiopathic hypersomnia and post-traumatic hypersomnia (Table 10–4).

Monosymptomatic Narcolepsy

Patients with monosymptomatic narcolepsy have excessive sleepiness and an abnormal propensity to enter REM sleep prematurely, but they do not have cataplexy. About one-third of patients with narcolepsy based on ICSD criteria fit this description. Compared to patients with cataplexy, they take fewer daytime naps and are less likely to have sleep paralysis, hypnagogic hallucinations, and disturbed nighttime sleep.[61] Although these findings suggest that the syndrome may be distinct from narcolepsy-cataplexy, the age of onset and the severity of sleepiness are similar, the HLA-DR15 antigen is more prevalent, and cataplexy eventually develops in some patients.[62–64] Thus, in many patients, monosymptomatic narcolepsy may be a less severe manifestation of the same genetic predisposition that underlies narcolepsy-cataplexy.

Patients in whom cataplexy develops years after the onset of sleepiness can be considered to have "latent cataplexy." There is currently no way to identify such patients, although the older the patient, the less likely the development of cataplexy. In canine narcolepsy, α_1-adrenergic

Table 10–4. **Diagnostic Criteria for Narcolepsy, Idiopathic Hypersomnia, and Post-Traumatic Hypersomnia**

	Symptoms	Laboratory Findings
Narcolepsy	Excessive sleepiness and daytime sleep episodes for at least 3 months, cataplexy, sleep paralysis, hypnagogic hallucinations, automatic behaviors, disrupted nocturnal sleep	Short sleep latency on polysomnography; MSLT: mean sleep latency <5 minutes and 2 or more SOREMPs
Idiopathic hypersomnia	Excessive sleepiness, excessively deep sleep, prolonged sleep episodes, or frequent daytime sleep episodes for at least 6 months	PSG: short sleep latency and normal or prolonged sleep; MSLT: mean sleep latency <10 minutes and <2 SOREMPs
Post-traumatic hypersomnia	Onset of sleepiness and frequent daytime sleep episodes temporally related to head trauma	PSG: normal sleep; MSLT: mean sleep latency <10 minutes and <2 SOREMPs

Source: Adapted from American Sleep Disorders Association,[60] pp 42, 48–51, with permission.

blockade combined with cholinergic stimulation can unmask cataplexy in heterozygotes for *canarc-1*. Similar testing might conceivably be used to unmask human cataplexy and thereby allow more definitive diagnosis.

Narcolepsy Associated with Brain Lesions

Structural brain lesions sometimes cause narcoleptic symptoms and REM-sleep abnormalities. Although postencephalitic narcolepsy was a common diagnosis after the epidemic of encephalitis lethargica in the early 20th-century, it appears to have virtually disappeared as a result of either remission of the disease or death of affected patients. The lack of polygraphic confirmation of narcolepsy and the possibility that other disorders, such as sleep apnea, may have contributed to sleepiness cast doubt on the validity of many early reports of narcolepsy caused by other types of brain lesions.

More recent reports with polygraphic confirmation and imaging studies have localized the lesions, usually in the diencephalon or upper brainstem (Table 10–5).[65–77] Various combinations of symptoms and REM-sleep findings may be found, depending on the location of the lesion. The following is a case of a patient who developed sleepiness and cataplexy in conjunction with endocrine abnormalities and a craniopharyngioma.[72]

CASE HISTORY: Case 10-2

A 36-year-old woman complained of daytime sleepiness. At age 31, amenorrhea developed; over the next 2 years, she gained 115 pounds and had bifrontal headaches and severe daytime sleepiness. At age 33, examination revealed a bitemporal hemianopia, mild right hemiparesis, and laboratory findings consistent with diabetes insipidus. Brain imaging revealed a large suprasellar mass (Fig. 10–3), and a subtotal resection of a craniopharyngioma was performed, followed by cranial irradiation.

Daytime sleepiness persisted, and 1 year after surgery she had a 1-week episode of temperature dysregulation and hypersomnolence that spontaneously improved. Two years later, daytime sleepiness became worse after an 18-hour period of continuous sleep. Brief episodes of cataplexy and sleep paralysis also developed. Polysomnography revealed a sleep latency of 1.5 minutes and a REM-sleep latency of 11 minutes. On MSLT, the mean sleep la-

Table 10–5. **Narcolepsy and Related Variants Associated with Brain Lesions**

Lesion	Sleepiness	Cataplexy	REM-sleep abnormalities	HLA
Pontine infarcts[67]	Yes	Yes	Yes	DR2
Third ventricle arteriovenous malformation[76]	Yes	Yes	Yes	DR2
Left temporal lobe lymphoma[74]	Yes	Yes	Yes	DR2
Craniopharyngioma[72]	Yes	Yes	Yes	DR2
Hypothalamic sarcoidosis[75]	Yes	Yes	Yes	DR2
Pituitary adenoma[71]	Yes	Yes	Yes	—
Third ventricle colloid cyst[71]	Yes	Yes	Yes	—
Cerebral sarcoidosis[71]	Yes	Yes	Yes	—
Midbrain glioblastoma[65]	Yes	Yes	Not tested	—
Hypothalamic syndrome[68]	Yes	Yes	Not tested	—
Third ventricle glioma[70]	Yes	Yes	Not tested	—
Craniopharyngioma[69]	Yes	Yes	No	—
Cerebral sarcoidosis[66]	Yes	No	Yes	—
Pituitary adenoma[71]	Yes	No	Yes	—
Hypothalamic syndrome[72]	Yes	No	Yes	DR2
Third ventricle sarcoidosis[72]	Yes	No	Yes	DR2
Pontomedullary multiple sclerosis[73]	No	Yes	Not tested	DR2
Pontomedullary pilocytic astrocytoma[73]	No	Yes	Not tested	—
Klüver-Bucy syndrome with normal-pressure hydrocephalus[77]	No	No	Yes	—

HLA = human leukocyte antigen.

tency was 1.2 minutes, with two SOREMPs. Tissue typing for MHC class II antigens was positive for HLA-DR2.

With methylphenidate 40 mg/d and fluoxetine 20 mg/d, sleepiness and cataplexy improved. Eight years after surgery, she continued to do well. Brain magnetic resonance imaging revealed no new tumor growth.

In many cases of symptomatic narcolepsy, including the one just described, the HLA-D antigens associated with narcolepsy are present, suggesting that certain brain injuries or lesions may trigger narcolepsy in susceptible persons, perhaps by causing a breakdown of the blood-brain barrier and allowing immune-mediated activation of the process that leads to the occurrence of narcolepsy. On the other hand, a few patients without HLA-DR2 have developed polysomnographically documented narcolepsy in close temporal association with brain lesions (see Table 10–5).

Idiopathic Hypersomnia

With idiopathic hypersomnia, initially described by Roth in the 1950s, sleepiness occurs despite adequate amounts of nighttime sleep, but REM-sleep abnormalities and cataplexy do not. Many of the cases observed by Roth[78,79] were diagnosed before the discoveries of obstructive sleep apnea, periodic limb movement disorder, and upper airway resistance syndrome; since the 1970s, with more intensive efforts at identifying other causes of sleepiness, the proportion of patients with a diagnosis of idiopathic hypersomnia has declined. Roth estimated that the ratio of narcolepsy to idiopathic hypersomnia was about 1.7:1, but more recent estimates place the ratio at about 10:1,[80,81] suggesting that the prevalence of idiopathic hypersomnia is about 2 to 6 in 100,000.

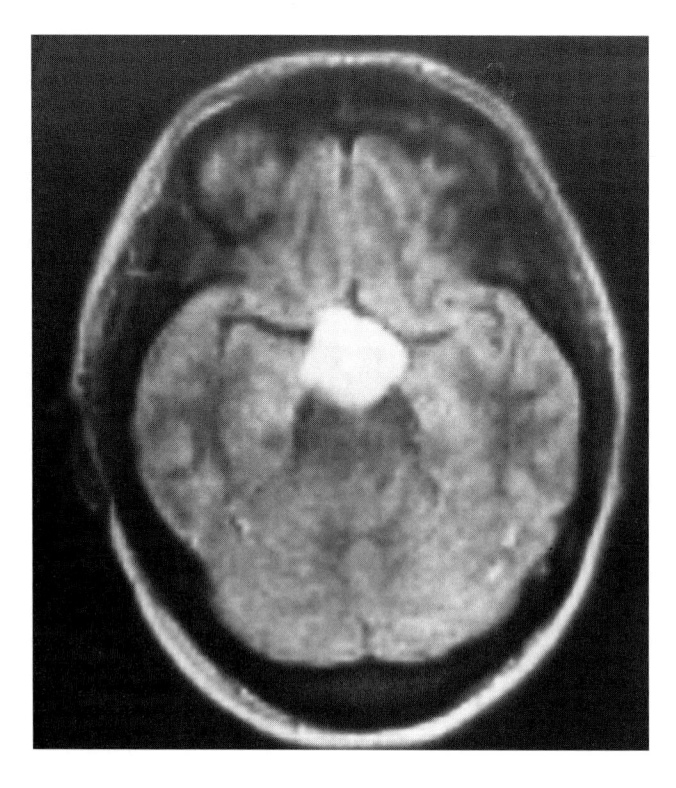

Figure 10–3. Brain magnetic resonance imaging in a patient with a 2-year history of headaches, excessive sleepiness, and endocrine abnormalities. This T_2-weighted image shows a high-signal-intensity mass lesion in the suprasellar region, which was identified at surgery as a craniopharyngioma.

Symptoms may include excessive sleepiness with daytime sleep episodes, prolonged sleep, or a failure to feel refreshed after sleep. Although Roth noted that headaches, low blood pressure, fainting episodes, and cold and clammy extremities occurred in some patients, the symptoms are probably no more frequent than in other patients with sleep complaints.[82] Some patients, including the one described below, also complain of difficulty awakening in the morning and after naps, accompanied by confusion or grogginess, slow thought and speech, or inappropriate behavior lasting from several minutes to 1 hour or more, a symptom complex called sleep drunkenness. The following case was excerpted from Bassetti and Aldrich.[81]

CASE HISTORY: Case 10–3

A 27-year old carpenter complained of a 12-year history of difficulty waking up. In high school, he sometimes fell asleep after school and did not awaken until the next morning. Although he felt sleepy during the day, he was able to resist falling asleep on important occasions. Daytime naps, which lasted up to 2 hours, were not refreshing, and he usually felt confused and sluggish for about 20 minutes after naps. He slept about 8 to 9 hours per night during the week and 12 to 18 hours per night on the weekend; in the morning, he felt sluggish for an hour or more each morning.

Examination was normal, as was a psychiatric evaluation. Several polysomnograms performed over a 9-year period revealed short sleep latencies with no other notable findings. Multiple sleep latency tests showed short sleep latencies, usually less than 5 minutes, with no sleep-onset REM periods. With methylphenidate 40 to 60 mg/day, daytime symptoms improved to only a modest degree.

Although this patient had a clinical picture typical of patients described by Roth (i.e., prolonged sleep, difficult awakening, unrefreshing naps, and sleep drunkenness), some patients with idiopathic hypersomnia experience overwhelming sleepiness—similar to that experienced by narcoleptics—but have refreshing naps and little difficulty awakening. In a series of 42 patients with idiopathic hypersomnia, sleep paralysis and hypnagogic hallucinations occurred in a number of patients, and the classic clinical picture described by Roth was present in only about one-third.[81] Another one-third had overwhelming sleepiness more typical of narcolepsy, although without a propensity for SOREMPs. The remainder had intermediate clinical characteristics. Thus, the syndrome of idiopathic hypersomnia is clinically heterogeneous, with substantial overlap between symptoms characteristic of narcolepsy and those classically associated with idiopathic hypersomnia.

The course of idiopathic hypersomnia is variable. Unlike narcolepsy, in which complete remission never or almost never occurs, idiopathic hypersomnia improves with time in a minority of cases, particularly those apparently triggered by viral illnesses.[81] Furthermore, SOREMPs may occur in some patients. The following case, in which symptoms resolved after 5 years, provides an example of clinical features typical of idiopathic hypersomnia combined with laboratory features typical of narcolepsy.

CASE HISTORY: Case 10–4

A 30-year-old electrician began to have trouble awakening at about 26 years of age. He bought louder alarms and attached a timer to a loud stereo system, but he continued to have difficulty awakening and sometimes slept until late in the afternoon or evening; one weekend he slept for 48 hours. He often fell asleep while reading, watching TV, or driving a car. Despite the fact that he drank up to 15 cups of coffee per day, his symptoms increased in severity over the next 3 to 4 years. A polysomnogram was normal, and an MSLT demonstrated a mean sleep latency of 2.8 minutes with two SOREMPs. He was negative for HLA-DR15 and positive for HLA-DQ1.

Pemoline and methylphenidate produced mild improvement, but they also caused headaches and irritability. Protriptyline was ineffective. Dextroamphetamine 20 mg/d produced modest improvement; after using it daily for about 4 months, he began to feel more alert. After discontinuing the dextroamphetamine, he continued to improve. Two years after his initial evaluation, or about 5 years after the onset of symptoms, he no longer used medications and no longer complained of daytime sleepiness or increased nocturnal sleep time.

BIOLOGIC BASIS

Although the etiology of idiopathic disorders is by definition unknown, the clinical and laboratory heterogeneity suggests that the disorder has multiple causes. Symptoms similar to those associated with idiopathic hypersomnia may develop after viral illness,[81-83] suggesting that immune-mediated changes may lead to the syndrome. However, the occurrence of HLA-D-associated antigens linked to narcolepsy-cataplexy syndrome is not increased in these patients. Increases in HLA-Cw2, -DR5, and -DR11 found by some investigators have not yet been confirmed in other series.[80,84]

Although the HLA-D antigens associated with narcolepsy are not increased in idiopathic hypersomnia patients as a whole, the gene or genes that contribute to narcolepsy may play a role in a few cases. Some patients have family members with narcolepsy, and there are rare cases of patients who initially present with clinical and laboratory findings of idiopathic hypersomnia, but who later exhibit cataplexy and REM-sleep abnormalities pathognomonic for narcolepsy.[81] As with narcolepsy, however, family members who complain of sleepiness often have neither idiopathic hypersomnia nor narcolepsy.

In some patients, the cause of hypersomnia may be an affective disorder. Although most patients with depression and complaints of excessive sleepiness do not have objective evidence of an increased propensity or need for sleep,[85] a few such patients have prolonged nighttime sleep and abnormal MSLTs,[86,87] and some young patients with depression sleep longer than controls when allowed to sleep indefinitely.[88]

The CNS changes that lead to the symptoms of idiopathic hypersomnia are unknown, although altered CNS monoaminergic activity may be a factor. In cats, destruction of mesencephalic noradrenergic neurons can produce a hypersomnolent syndrome that resembles idiopathic hypersomnia.[89]

Post-traumatic Hypersomnia

Hypersomnia that develops after head trauma may resemble idiopathic hypersomnia or narcolepsy.[90] It is often difficult to document the relation to trauma because few such patients have had objective sleep assessments before the injury, and factors such as litigation or disability issues may complicate the postinjury clinical assessment. Sleep studies in some patients with post-traumatic sleepiness are consistent with narcolepsy, suggesting that narcolepsy existed previously or that trauma can trigger the onset of narcolepsy in susceptible individuals.

DIFFERENTIAL DIAGNOSIS

Sleepiness

For patients who present with complaints of sleepiness, the differential diagnosis, in addition to narcolepsy and idiopathic hypersomnia, includes obstructive and central sleep apnea, upper airway resistance syndrome, periodic limb movement syndrome, insufficient sleep syndrome, circadian rhythm sleep disorders, depression, and sleep disorders associated with underlying medical and neurological illnesses. The severity of sleepiness and the occurrence of automatic behavior and sleep attacks do not differentiate narcolepsy from other sleep disorders: Sleepiness may be just as severe with sleep apnea as with narcolepsy, and sleep attacks and automatic behavior can occur in anyone who suffers from chronic severe sleepiness. Thus, tests that assess the severity of sleepiness, such as subjective scales, pupillometry, and evoked potentials, do not help to determine its cause. Although naps in patients with idiopathic hypersomnia are often longer and less refreshing than naps in narcoleptics, these characteristics per se are inconsistent and therefore not pathognomonic. The differentiation of sleepiness from tiredness, fatigue, and weakness is discussed in Chapter 8.

Patients who fall asleep while driving or in other dangerous situations sometimes describe the episodes as periods of loss of awareness or loss of consciousness, which can lead to extensive cardiac and neurologic evaluations for suspected seizures, syncope, or vertebrobasilar transient ischemic attacks. Premonitory drowsiness, sleeplike appearance with absence of pallor, ease of arousability, and the association with sedentary situations are helpful diagnostic features.

In patients with prominent complaints of automatic behavior, automatisms and amnesia may suggest a diagnosis of partial complex seizures, particularly if they are the primary symptoms reported by the patient. Although a single episode of automatic behavior may be clinically indistinguishable from a seizure, the association of amnestic episodes with drowsiness and with monotonous or repetitive activities is a useful distinguishing feature. In patients who complain of blurred vision or diplopia, and whose family members report observing drooping eyelids, myasthenia gravis may be suspected. The association of visual symptoms with drowsiness and with periods of inactivity should lead the physician to suspect sleepiness as the cause of the visual symptoms.

Cataplexy

Included in the differential diagnosis of cataplexy are atonic seizures, myasthenia gravis, periodic paralysis, and drop attacks associated with vertebrobasilar insufficiency. The association with emotion or laughter, the duration of episodes, and the preservation of consciousness help to differentiate cataplexy from these disorders.

Although cataplexy may occasionally occur in association with midbrain tumors ("limp man syndrome"),[65] Norrie's disease,[91] Niemann-Pick disease type C,[92] and familial isolated cataplexy,[93] a history of clear-cut cataplexy in patients with normal neurologic examinations is virtually pathognomonic for narcolepsy. However, a *feeling* of weakness is not diagnostic; nearly 30% of adults have experienced a *feeling* of weakness with emotion, and 6% have had a "sudden and abrupt feeling of weakness in both arms and legs when laughing, . . . feeling angry, or in exciting situations."[16] Furthermore, the phrase "weak with laughter" or its equivalent is common in many cultures and languages. Definite bilateral weakness, during which limbs give way, postural tone is diminished, or strength is reduced, is much more likely to represent true cataplexy.

Sleep Paralysis and Hallucinations

Sleep paralysis is not specific for narcolepsy; it can occur in otherwise normal persons, often precipitated by sleep deprivation, a change in sleep schedule, or other factors that disrupt normal sleep patterns. As many as 15% of adolescents and young adults have had sleep paralysis.

The relation of hypnagogic and hypnopompic hallucinations to the onset and end of sleep helps to distinguish them from hallucinations associated with migraine and posterior circulation ischemia. Although partial seizures arising from the occipital lobes are sometimes activated by drowsiness, the stereotyped nature of the associated hallucinations helps to distinguish them from hypnagogic hallucinations. Patients with evening and nocturnal hallucinations associated with psychosis, dementia, or toxic states often believe the hallucinations are real, whereas patients with hypnagogic hallucinations usually know that they are not. Peduncular hallucinations, which can occur in patients with midbrain lesions, are visual and often occur in the evening or at night[94]; however, such patients almost invariably have other evidence of brainstem dysfunction.

Hallucinations in narcoleptic subjects may be mistaken for evidence of psychosis.[18] They differ, however, from hallucinations associated with schizophrenia: (1) they are more likely to occur in quiet situations when the patient is drowsy; and (2) they are more likely to be visual or to involve multiple sensory modalities. Auditory hallucinations, particularly voices that converse with or instruct the patient, are more likely to be psychotic symptoms. In contrast to schizophrenics, most narcoleptics have good insight into the hallucinations; however, there are occasional cases in which narcoleptic patients construct delusional systems to account for the symptoms.[18] In such cases, the delusions tend to be relatively restricted to providing explanations of the hallucinations, whereas delusions in schizophrenics may be much more detailed and elaborate.

DIAGNOSTIC EVALUATION

Because a variety of conditions can cause excessive sleepiness, sleep studies are generally required for accurate diagnosis. Although narcolepsy can be diagnosed without polygraphic studies in patients with sleepiness and unambiguous cataplexy, sleep laboratory documentation is useful because: (1) it corroborates a diagnosis that has significant potential for inducing socioeconomic difficulties and contributing to disability; (2) it identifies malingerers who feign symptoms to obtain stimulant medications; and (3) it may disclose other causes of excessive sleepiness, such as sleep apnea.

Nocturnal polysomnography, best performed after the patient has been on a regular schedule with adequate amounts of sleep for several days, can determine the presence and severity of sleep apnea, periodic limb movements, and nocturnal sleep disturbance.

An MSLT the following day provides a measure of the severity of sleepiness and an indication of the presence or absence of early onset of REM sleep. Mean sleep latencies, an index of the propensity to fall asleep easily, are usually less than 8 minutes in narcoleptics, compared to 10 to 20 minutes in control subjects.[95] *Early onset of REM sleep*, defined as REM sleep within 15 minutes of the onset of sleep, occurs in about 40% to 50% of sleep episodes. The presence of two or more SOREMPs during the repeated nap opportunities provided by the MSLT supports a diagnosis of narcolepsy if other causes of early onset

REM sleep have been excluded (Table 10–6).[64]

Although the MSLT is a highly useful test for diagnosing narcolepsy, false-positives and false-negatives do occur. In a series of 170 narcoleptics, only 80% had two or more SOREMPs on an initial MSLT.[96] Repeat studies may increase the yield to 85% to 90%, but some patients with unequivocal narcolepsy-cataplexy do not have two or more SOREMPs, even with repeated studies.[96] MSLT findings consistent with narcolepsy—a mean sleep latency less than 5 minutes and two or more sleep-onset REM periods—may also occur with sleep apnea, periodic limb movement disorder, circadian rhythm disturbances, and REM-sleep deprivation. In a series of 229 subjects with two or more SOREMPs on MSLT, only 59% had narcolepsy: Of the remaining subjects, 28% had sleep-related breathing disorders (mainly obstructive sleep apnea) and 12% had other sleep disorders.[96]

Thus, in the absence of unambiguous cataplexy, the occurrence of pathologically short sleep latencies and two or more SOREMPs during the MSLT is not specific, and other potential causes of early onset REM sleep, particularly sleep-related breathing disturbances, should be considered. Obstructive sleep apnea syndrome may vary in severity from night to night depending on alcohol intake or the extent of upper airway edema, and the upper airway resistance syndrome (see Chap. 13) may be missed unless esophageal pressure monitoring is performed during polysomnography. In addition, patients whose obstructive sleep apnea occurs exclusively in the supine position may not exhibit apnea if they are recorded only while sleeping on their sides. If REM-sleep abnormalities occur, such persons may be given a misdiagnosis of narcolepsy. If obstructive sleep apnea is suspected strongly on clinical grounds and an initial polysomnogram is negative, a second study may show obstructive apneas in patients whose apnea is present only under certain conditions.

On the other hand, the clinician must recognize that the presence of obstructive sleep apnea or upper airway resistance syndrome does not exclude the possibility that the patient has narcolepsy as well, although it often poses a problem for definitive diagnosis. A patient with a history of sleepiness that developed in middle-age, in conjunction with an increase in snoring and body weight, is less likely to have narcolepsy than a patient who developed sleepiness in adolescence followed by the onset of snoring and weight gain in middle-age. In some cases, a trial of nasal continuous positive airway pressure treatment helps to determine the extent to which sleep-disordered breathing contributes to sleepiness.

Daytime EEGs alone are inadequate for diagnosis, since fewer than half of narcoleptics have REM sleep during such recordings.[97] Nocturnal polysomnograms alone are insufficient because SOREMPs occur in less than 50% of narcoleptics[96,98] and can occur with depression, sleep-wake schedule disturbances, drug and alcohol withdrawal, and REM-sleep deprivation from sleep apnea. HLA testing has limited diagnostic value because not all narcoleptics carry the typical haplotype and because the haplotype is common in the general population. More than 99% of persons with DRB1–1501 and DQB1–0602 do not have narcolepsy.

Diagnosis of idiopathic hypersomnia is based on history and laboratory testing and requires ruling out other causes of excessive sleepiness. In clinical practice, therefore, it is often difficult to make a definitive diagnosis. Atypical depression should always be considered in patients with complaints of increased need for sleep, especially if a mood disturbance accompanies the sleep complaint (see Chap.

Table 10–6. **Causes of Early Onset of REM Sleep**

Narcolepsy
Major depression
Circadian rhythm sleep disorders
REM sleep deprivation
Drug-induced suppression of REM sleep
REM sleep disruption associated with sleep related breathing disorders
REM sleep disruption from other causes

16). Chronic fatigue syndrome and other disorders associated primarily with fatigue rather than sleepiness should also be considered.

Standard polysomnography is usually unremarkable in idiopathic hypersomnia, showing normal proportions and timing of NREM and REM sleep. An MSLT usually demonstrates short sleep latencies without frequent sleep-onset REM periods. As with narcolepsy in the absence of cataplexy, in such cases the clinician needs to consider the possibility that the patient is not obtaining enough sleep or that the patient has obstructive sleep apnea that is present on some nights but not on others. Insufficient sleep syndrome (see Chap. 8) may develop in patients with high sleep needs (i.e., 9 hours or more) when they sleep 8 hours per night. Some patients may report that they are sleeping 8 hours per night when they are actually sleeping less.

I ask patients who have histories and sleep laboratory findings consistent with idiopathic hypersomnia to increase the usual nighttime sleep by 1 hour for at least 1 month and to keep a sleep log. If sleepiness resolves, the diagnosis is insufficient sleep syndrome. If it does not, I usually repeat the polysomnogram—with particular attention to the possibility of obstructive sleep apnea or upper airway resistance syndrome (see Chap. 13)—as well as the MSLT. If the trial of increased sleep leads to no improvement in symptoms or in the MSLT, I am more confident that the diagnosis is idiopathic hypersomnia.

Although not widely used in the United States, prolonged sleep recordings are also useful diagnostic tools. Patients are asked to remain in bed for up to 24 hours and encouraged to sleep. Those with idiopathic hypersomnia may show increased sleep duration with prolonged nighttime sleep, long naps, or both. In patients with chronic fatigue syndrome or atypical depression, prolonged recordings do not show increased daytime sleep, although patients may remain in bed for much of the recording.[86]

MANAGEMENT

As sleepiness associated with narcolepsy is a lifelong problem, the first step in management is patient and family education about the symptoms of narcolepsy and its usual clinical course. I emphasize the importance of obtaining enough sleep at night, as partial sleep deprivation aggravates symptoms. Patients should also be advised about the risks associated with sleepiness while driving and in the workplace, and about the role of medications in the treatment of narcolepsy. Management of idiopathic hypersomnia is similar to management of narcolepsy, except that patients should be advised that the disorder sometimes resolves spontaneously.

Treatment of Sleepiness

The most effective treatments for narcolepsy and idiopathic hypersomnia are psychostimulants. Naps are helpful for some patients, but they are virtually never as effective as stimulants. I usually assess response clinically and with the Epworth Sleepiness Scale[99] rather than with an MSLT or a Maintenance of Wakefulness Test (MWT).[100] The MSLT measures sleepiness in an artificial setting—the sleep laboratory—and the mean latency to sleep in this environment is not an accurate indicator of the patient's ability to stay awake while driving[19] or during work, school, and social situations. The MWT, similar to the MSLT except that patients are instructed to try to stay awake, does not fully replicate the situations in which patients wish to stay awake (see Chap. 8), and in my view, it is not indicated for routine clinical use.

Clinical symptom assessment is not without problems. Patients are not always able to assess their level of alertness accurately, and many patients who report satisfactory control of sleepiness are sleeping at intervals during the day.[101] As with other disorders causing sleepiness, the impressions of family members sometimes provide a different picture from that presented by the patient.

When sleepiness fails to respond to medications, or when increasing sleepiness develops after a period of stability, the possibility of other diagnoses such as sleep apnea, insufficient sleep, or medication effects should be considered because sleepiness rarely worsens after the first year or two after the onset of symptoms.

STIMULANTS

Stimulants, the pharmacological mainstay for treatment of sleepiness, enhance the synaptic availability of norepinephrine or dopamine, or both.[102] Unfortunately, most patients cannot obtain full alertness using currently available medications because of side effects or the development of tolerance.

Commonly used stimulants include methylphenidate (Ritalin), dextroamphetamine (Dexedrine), and pemoline (Cylert). For patients with mild sleepiness, pemoline is often helpful. An initial dose of 37.5 mg/d can be increased to 112.5 mg/d as needed. Concerns about potential hepatotoxicity, however, limit its usefulness. Methylphenidate is effective in many patients, and I usually begin with 5 to 10 mg two or three times daily and increase as needed to a total daily dose of 60 mg; in selected cases, I may increase the total daily dose of methylphenidate to 80 to 100 mg. If methylphenidate is insufficient, I switch to dextroamphetamine, again increasing as needed to up to 60 mg/d. Doses of methylphenidate above 80 mg/d, or of dextroamphetamine above 60 mg/d, rarely offer additional benefit and increase the risk of side effects (see below). Some experts prescribe methamphetamine, which has better CNS penetration than dextroamphetamine; however, methamphetamine has a longer half-life, leading to increased nocturnal sleep disruption, and it is more expensive.

Modafinil, an α1-receptor agonist that has alerting properties, is available in some countries. It has few side effects and thus may be preferable as initial treatment at 200 to 400 mg/d.[103,104] Mazindol, an imidazoline derivative, is of value for some patients and may have fewer side effects than amphetamines, although it is also less effective. Selegiline hydrochloride, a monoamine oxidase type B inhibitor that is converted to levoamphetamine, has stimulating properties and can be used when conventional stimulants are not well tolerated. Protriptyline, a tricyclic with stimulating properties, is occasionally useful.

Chronic use of high doses of stimulants can lead to irritability and insomnia, habituation and addiction, psychosis and other drug-related psychiatric conditions, and possibly hypertension and large or small vessel vasculopathy.[102] Side effects in children, reversible when the medications are withdrawn, may include anorexia, tics, and weight loss. Most of these side effects are dose-related, and there is no evidence that narcoleptics are at any higher risk for amphetamine abuse than other population groups. To reduce the likelihood of tolerance in patients taking high doses, I often recommend that patients reduce the stimulant dose or go without stimulants for 1 or 2 days per week. Long-term stimulant treatment with moderate doses does not appear to affect emotional adjustment or academic performance, and with appropriate management, most narcoleptics are able to take stimulants regularly for decades without serious side effects.

Although long-term stimulant treatment appears to be safe in children, growth needs to be monitored. For example, treatment of attention deficit-hyperactivity disorder with stimulants can cause growth retardation. These children catch up in stature, however, during summers off medication, with little or no effect on adult height.[105]

NAPS

I usually recommend naps as an adjunct to medications. One to three 10- to 60-minute naps daily can improve alertness and psychomotor performance for 1 to 2 hours without increasing nocturnal sleep disturbance; in fact, the beneficial effects of naps may be partly a result of relief of insufficient sleep.[106,107] Regular naps may also reduce the daily requirements for stimulants, thereby reducing the risk of side effects, and they reinforce the need for an adjustment in lifestyle to cope with the symptoms of this chronic disorder.

I tell patients that it is difficult for most narcoleptics to remain awake throughout a 16-hour day, that it may be more reasonable for them to try to be as alert as possible during times of the day when they most need to be alert (e.g., work, school, driving), and that medications and good sleep hygiene can help them achieve this goal. Understanding teachers and employers may allow patients to take naps during lunch or other breaks. A typical program for a teen-age patient might be a dose of methylphenidate in the morning, a nap at lunch followed by a second dose of methylphenidate, and a nap after school followed by a third dose of methylphenidate. For patients who cannot take naps during the day, the following routine can be effective: medication doses first thing in the morning, additional doses at 10:30 AM and 2 PM, a nap after work, and an additional late-afternoon dose if evening activities are planned.

SLEEPINESS AND DRIVING

The high rate of sleep-related automobile accidents raises the question of whether narcoleptics should have restricted driving privileges. I believe that the diagnosis of narcolepsy alone should not be the basis for restricting driving privileges; many narcoleptics have relatively mild sleepiness and have driven safely for years. It seems unfair to prevent such patients from driving while permitting shift workers and sleep-deprived persons, who may be just as sleepy, to drive. I advise patients with excessive sleepiness to drive with extreme caution, not to drive when they feel sleepy, and to pull off the road as soon as they begin to feel drowsy. I advise against driving at all for patients who are severely sleepy, who are unaware of their sleepiness, who fall asleep without warning, or who have severe, uncontrolled cataplexy.

Treatment of Other Symptoms

CATAPLEXY

Tricyclic antidepressants, the pharmacological treatment of choice for cataplexy and sleep paralysis, appear to act mainly through blockade of norepinephrine reuptake or serotonin reuptake, or both, rather than via anticholinergic effects. Protriptyline 5 to 30 mg/d, imipramine 50 to 250 mg/d, clomipramine 20 to 200 mg/d, or nortriptyline 50 to 200 mg/d are effective in about 80% of patients. Fluoxetine is also effective and can be used in patients who cannot tolerate the anticholinergic side effects of tricyclics.

In patients with severe cataplexy, tolerance to the effects of tricyclics may develop, requiring gradual withdrawal followed by a 2-week drug holiday to restore efficacy. Abrupt withdrawal can lead to a rebound increase in cataplexy and even to more or less continuous incapacitating cataplexy lasting for several hours or days.

SLEEP PARALYSIS AND HYPNAGOGIC AND HYPNOPOMPIC HALLUCINATIONS

Although sleep paralysis can be a frightening experience, it is not harmful and usually does not require treatment. Some narcoleptics, however, experience sleep paralysis with such regularity when they lie down to sleep that they resort to sleeping in chairs or in a semi-upright position in which sleep paralysis is less likely to occur. Tricyclic antidepressants are sometimes prescribed for these patients.

For patients with disturbing hallucinations, tricyclic antidepressants are usually effective in doses similar to those used to control cataplexy. If a delusional component is present or develops that does not respond to a reduction in stimulant dose, pimozide is often beneficial.

INSOMNIA

Occasional use of short-acting hypnotics may help some narcoleptics when disturbed nocturnal sleep is a major complaint, but whether these drugs lead to improved daytime alertness remains to be determined. The rationale for nightly use is questionable; insomnia is usually chronic, and any potential benefit is likely to be temporary. Regular sleep schedules and good sleep hygiene are helpful measures for some patients.

SUMMARY

Narcolepsy, which usually begins in the second or third decade of life, is characterized by excessive sleepiness and sleep attacks, cataplexy, and abnormalities of REM sleep. Other symptoms that are often present include sleep paralysis, sleep-related hallucinations, automatic behavior, visual and memory disturbances, and nocturnal sleep disruption.

Narcolepsy is strongly associated with HLA DQB1–0602 and DRB1–1501, indicating a genetic basis for the disorder in most cases. However, the HLA-associated genes are not sufficient to induce narcolepsy, and the existence of monozygotic twins discordant for narcolepsy indicates that environmental factors play an important role in its development.

Early onset of REM sleep, the characteristic electrophysiologic feature, is part of a broader problem of impaired sleep-wake regulation. Although the nature of the neurologic defect in narcolepsy is unknown, it likely involves cholinergic or monoaminergic systems.

Clinical variants of narcolepsy-cataplexy syndrome include monosymptomatic narcolepsy, narcolepsy associated with brain lesions, idiopathic hypersomnia, and post-traumatic hypersomnia. These disorders are more heterogeneous than the classic narcolepsy-cataplexy syndrome.

Diagnosis of narcolepsy is based on the clinical features and the results of sleep studies. Cataplexy is a particularly important symptom for diagnosis because it does not occur with other sleep disorders. Although frequent SOREMPs are characteristic of narcolepsy, they are not universal, and they may occur with other sleep disorders. Stimulants are the mainstay of treatment for patients with sleepiness; cataplexy and sleep paralysis usually respond to tricyclic antidepressants.

REFERENCES

1. Gelineau, JBE: De la narcolepsie. Gaz Hop 53:626, 1880.
2. Westphal, C: Eigenthumlich mit einschlafen verbundene Anfalle. Arch Psychiatr Nervenkr 7:631, 1877.
3. Fischer, F: Epileptoid-Schalfzundstande. Arch Psychiatr Nervenkr 8:203, 1878.
4. Gowers, WR: A Manual of Diseases of the Nervous System, ed 2. P. Blakiston, Son & Co, Philadelphia, 1893.
5. Lowenfeld, L: Über Narkolepsie. Much Med Wochenstr 49:1041, 1902.
6. Adie, WJ: Idiopathic narcolepsy: A disease sui generis, with remarks on the mechanisms of sleep. Brain 49:257, 1926.
7. Wilson, SAK: The narcolepsies. Brain 51:63, 1928.
8. Daniels, LE: Narcolepsy. Medicine 13:1, 1934.
9. Janota, O: Symptomatische Behandlung der pathologischen Schlafsucht, besonders der Narkolepsie. Med Klin 27:278, 1931.
10. Doyle, JB, and Daniels, LE: Symptomatic treatment for narcolepsy. JAMA 96:1370, 1931.
11. Yoss, RE, and Daly, D: Criteria for the diagnosis of the narcoleptic syndrome. Proc Staff Meet Mayo Clin 32:320, 1957.
12. Dement, W, et al: The nature of the narcoleptic sleep attack. Neurology 16:18, 1966.
13. Vogel, G: Studies in psychophysiology of dreams: III. The dream of narcolepsy. Arch Gen Psychiatry 3:421, 1960.
14. Takahashi, Y, and Jimbo, M: Polygraphic study of narcoleptic syndrome, with special reference to hypnagogic hallucinations and cataplexy. Folia Psychiatr Neurol Jpn (suppl)7:343, 1963.
15. Rechtschaffen, A, et al: Nocturnal sleep of narcoleptics. Electroencephalogr Clin Neurophysiol 15:599, 1963.
16. Hublin, C, et al: The prevalence of narcolepsy: An epidemiologic study of the Finnish twin cohort. Ann Neurol 35:709, 1994.
17. Ribstein, M: Hypnagogic hallucinations. In Guilleminault, C, et al (eds): Narcolepsy. Spectrum, New York, 1975, pp 145–160.
18. Douglass, AB, et al: Florid refractory schizophrenias that turn out to be treatable variants of HLA-associated narcolepsy. J Nerv Ment Dis 179:12, 1991.
19. Aldrich, MS: Automobile accidents in patients with sleep disorders. Sleep 12:487, 1989.
20. Broughton, R, et al: Life effects of narcolepsy in 180 patients from North America, Asia, and Europe compared to matched controls. Can J Neurol Sci 8:299, 1981.
21. Kales, A, et al: Narcolepsy-cataplexy: II. Psychosocial consequences and associated psychopathology. Arch Neurol 39:169, 1982.
22. Dahl, RE, et al: A clinical picture of child and adolescent narcolepsy. J Am Acad Child Adolesc Psychiatry 33:834, 1994.
23. Krabbe, E, and Magnussen, G: On narcolepsy: I. Familial narcolepsy. Acta Psychol Neurol Scand 17:149, 1942.
24. Guilleminault, C, et al: Familial patterns of narcolepsy. Lancet 2:1376, 1989.
25. Juji, T, et al: HLA antigens in Japanese patients with narcolepsy-all the patients were DR2 positive. Tissue Antigens 24:316, 1984.

26. Langdon, N, et al: Genetic markers in narcolepsy. Lancet 2:1178, 1984.
27. Billiard, M, and Seignalet, J: Extraordinary association between HLA-DR2 and narcolepsy. Lancet 2:226, 1985.
28. Neely, S, et al: HLA antigens in narcolepsy. Neurology 37:1858, 1987.
29. Holloman, JD, et al: HLA-DR restriction-fragment-length polymorphisms in narcolepsy. J Neurosci Res 18:239, 1987.
30. Olerup, O, et al: The narcolepsy-associated DRw15, DQw6, Dw2 haplotype has no unique HLA-DQA or -DQB restriction fragments and does not extend to the HLA-DP subregion. Immunogenetics 32:41, 1990.
31. Rubin, RL, et al: HLA-DR2 association with excessive somnolence in narcolepsy does not generalize to sleep apnea and is not accompanied by systemic autoimmune abnormalities. Clin Immunol Immunopathol 49:149, 1988.
32. Melberg, A, et al: Autosomal dominant cerebellar ataxia deafness and narcolepsy. J Neurol Sci 134:119, 1995.
33. Singh, SM, et al: Genetic heterogeneity in narcolepsy. Lancet 335:726, 1990.
34. Guilleminault, C, et al: A study on cataplexy. Arch Neurol 31:255, 1974.
35. Hishikawa, Y: Sleep paralysis. In Guilleminault, C, et al (eds): Narcolepsy. Spectrum, New York, 1975, pp 97–124.
36. Broughton, R, et al: Excessive daytime sleepiness and the pathophysiology of narcolepsy-cataplexy: A laboratory perspective. Sleep 9:205, 1986.
37. Broughton, R, et al: Night sleep does not predict day sleep in narcolepsy. Electroencephalogr Clin Neurophysiol 91:67, 1994.
38. Aldrich, MS, and Rogers, AE: Exacerbation of human cataplexy by prazosin. Sleep 12:254, 1989.
39. Guilleminault, C, et al: Prazosin contraindicated in patients with narcolepsy. Lancet 2:511, 1988.
40. Foutz, AS, et al: Monoaminergic mechanisms and experimental cataplexy. Ann Neurol 10:369, 1981.
41. Kish, SJ, et al: Brain neurotransmitter changes in human narcolepsy. Neurology 42:229, 1992.
42. Aldrich, MS, et al: Dopamine receptor autoradiography of human narcoleptic brain. Neurology 42:410, 1992.
43. Aldrich, MS, et al: Neurochemical studies of human narcolepsy: Alpha-adrenergic receptor autoradiography of human narcoleptic brain and brainstem. Sleep 17:598, 1994.
44. Aldrich, MS, et al: Autoradiographic studies of post-mortem human narcoleptic brain. Neurophysiol Clin 23:35, 1993.
45. Khan, N, et al: Striatal dopamine D2 receptors in patients with narcolepsy measured with PET and 11C-raclopride. Neurology 44:2102, 1994.
46. Rinne, JO, et al: Positron emission tomography study of human narcolepsy: No increase in striatal dopamine D2 receptors. Neurology 45:1735, 1995.
47. Mitler, MM, and Dement, WC: Sleep studies on canine narcolepsy: Pattern and cycle comparisons between affected and normal dogs. Electroencephalogr Clin Neurophysiol 43:691, 1977.
48. Siegel, JM, et al: Neuronal activity in narcolepsy—Identification of cataplexy-related cells in the medial medulla. Science 252:1315, 1991.
49. Reid, MS, et al: Cholinergic mechanisms in canine narcolepsy: II. Acetylcholine release in the pontine reticular formation is enhanced during cataplexy. Neuroscience 59:523, 1994.
50. Reid, MS, et al: Cholinergic mechanisms in canine narcolepsy: I. Modulation of cataplexy via local drug administration into the pontine reticular formation. Neuroscience 59:511, 1994.
51. Nishino, S, et al: Muscle atonia is triggered by cholinergic stimulation of the basal forebrain: Implication for the pathophysiology of canine narcolepsy. J Neurosci 15:4806, 1995.
52. Nitz, D, et al: Altered distribution of cholinergic cells in the narcoleptic dog. Neuroreport 6:1521, 1995.
53. Nishino, S, et al: Neuropharmacology and neurochemistry of canine narcolepsy. Sleep 17:S84, 1994.
54. Nishino, S, et al: Dopamine D2 mechanisms in canine narcolepsy. J Neurosci 11:2666, 1991.
55. Nishino, S, et al: Effects of central alpha-2 adrenergic compounds on canine narcolepsy, a disorder of rapid eye movement sleep. J Pharmacol Exp Ther 253:1145, 1990.
56. Mignot, E, et al: Genetic linkage of autosomal recessive canine narcolepsy with a mu immunoglobulin heavy-chain switch-like segment. Proc Natl Acad Sci USA 88:3475, 1991.
57. Gaiser, C, et al: Evidence for an autoimmune etiology in canine cataplexy. Sleep Res 18:230, 1989.
58. Tafti, M, et al: Major histocompatibility class II molecules in the CNS: Increased microglial expression at the onset of narcolepsy in canine model. J Neurosci 16:4588, 1996.
59. Mignot, E, et al: Heterozygosity at the canarc-1 locus can confer susceptibility for narcolepsy: Induction of cataplexy in heterozygous asymptomatic dogs after administration of a combination of drugs acting on monoaminergic and cholinergic systems. J Neurosci 13:1057, 1993.
60. American Sleep Disorders Association: International Classification of Sleep Disorders, Revised: Diagnostic and Coding Manual. American Sleep Disorders Association, Rochester, Mn, 1997.
61. Rosenthal, L, et al: Signs and symptoms associated with cataplexy in narcolepsy patients. Biol Psychiatry 27:1057, 1990.
62. Rosenthal, L, et al: HLA DR2 in narcolepsy with sleep-onset REM periods but not cataplexy. Biol Psychiatry 30:830, 1991.
63. Aldrich MS: Narcolepsy. Neurology (suppl 6)42:34, 1992.
64. Moscovitch, A, et al: The positive diagnosis of narcolepsy and narcolepsy's borderland. Neurology 43:55, 1993.
65. Stahl, SM, et al: Continuous cataplexy in a patient with a midbrain tumor: The limp man syndrome. Neurology 30:1115, 1980.

66. Rubinstein, I, et al: Neurosarcoidosis associated with hypersomnolence treated with corticosteroids and brain irradiation. Chest 94:205, 1988.
67. Rivera, VM, et al: Narcolepsy following cerebral hypoxic ischemia. Ann Neurol 19:505, 1986.
68. Gurewitz, R, et al: Recurrent hypothermia, hypersomnolence, central sleep apnea, hypodipsia, hypernatremia, hypothyroidism, hyperprolactinemia and growth hormone deficiency in a boy: Treatment with clomipramine. Acta Endocrinol 279:468, 1986.
69. Schwartz, WJ, et al: Transient cataplexy after removal of a craniopharyngioma. Neurology 34:1372, 1984.
70. Anderson, M, and Salmon, MV: Symptomatic cataplexy. J Neurol Neurosurg Psychiatry 40:186, 1977.
71. Pritchard III, PB, et al: Symptomatic narcolepsy. Neurology (suppl 2)33: 239, 1983.
72. Aldrich, MS, and Naylor, MW: Narcolepsy associated with lesions of the diencephalon. Neurology 39:1505, 1989.
73. D'Cruz, OF, et al: Symptomatic cataplexy in pontomedullary lesions. Neurology 44:2189, 1994.
74. Onofrj, M, et al: Narcolepsy associated with primary temporal lobe B-cells lymphoma in a HLA DR2 negative subject. J Neurol Neurosurg Psychiatry 55:852, 1992.
75. Servan, J, et al: Narcolepsie revelatrice d'une neurosarcoidose. Rev Neurol (Paris) 151:281, 1995.
76. Clavelou, P, et al: Narcolepsy associated with arteriovenous malformation of the diencephalon. Sleep 18:202, 1995.
77. Bromberg, MB, et al: The Kluver-Bucy syndrome preceding ventricular enlargement in normal pressure hydrocephalus. Dementia 1:169, 1990.
78. Roth, B: Functional hypersomnia. In Guilleminault, C, et al (eds): Narcolepsy. Spectrum, New York, 1976, pp 333–349
79. Roth, B: Narcolepsy and Hypersomnia. Karger, Basel, 1980
80. Billiard, M, and Besset, A: L'hypersomnie idiopathique. In Billiard, M (ed): Le Sommeil Normal et Pathologique. Masson, Paris, 1994, pp 274–280.
81. Bassetti, C, and Aldrich, MS: Idiopathic hypersomnia: A series of 42 patients. Brain 120:1423, 1997.
82. Bruck, D, and Parkes, JD: A comparison of idiopathic hypersomnia and narcolepsy-cataplexy using self report measures and sleep diary data. J Neurol Neurosurg Psychiatry 60:576, 1996.
83. Guilleminault, C, and Mondini, S: Mononucleosis and chronic daytime sleepiness: A long-term follow-up study. Arch Intern Med 146:1333, 1986.
84. Poirier, G, et al: HLA antigens in narcolepsy and idiopathic central nervous system hypersomnolence. Sleep 9:153, 1986.
85. Nofzinger, EA, et al: Hypersomnia in bipolar depression: A comparison with narcolepsy using the multiple sleep latency test. Am J Psychiatry 148:1177, 1991.
86. Billiard, M, et al: Hypersomnia associated with mood disorders: A new perspective. J Psychosom Res (suppl 1)38:41, 1994.
87. Van den Hoed, J, et al: Disorders of excessive daytime somnolence: Polygraphic and clinical data for 100 patients. Sleep 4:23, 1981.
88. Hawkins, DR, et al: Extended sleep (hypersomnia) in young depressed patients. Am J Psychiatry 142:905, 1985.
89. Petitjean, F, and Jouvet, M: Hypersomnie et augmentation de l'acide 5-hydroxy-indolacétique cérébral par lésion isthmique chez le chat. CR Soc Biol 164:2228, 1970.
90. Guilleminault, C, et al: Posttraumatic excessive daytime sleepiness: A review of 20 patients. Neurology 33:1584, 1983.
91. Vossler, DG, et al: Cataplexy and monoamine oxidase deficiency in Norrie disease. Neurology 46:1258, 1996.
92. Challamel, MJ, et al: Narcolepsy in children. Sleep (8 suppl)17:S17, 1994.
93. Hartse, KM, et al: Isolated cataplexy: A familial study. Henry Ford Hosp Med J 36:24, 1988.
94. McKee, AC, et al: Peduncular hallucinosis associated with isolated infarction of the substantia nigra pars reticulata. Ann Neurol 27:500, 1990.
95. Richardson, GS, et al: Excessive daytime sleepiness in man: Multiple sleep latency measurement in narcoleptic and control subjects. Electroencephalogr Clin Neurophysiol 45:621, 1978.
96. Aldrich, MS, et al: Value of the Multiple Sleep Latency Test (MSLT) for the diagnosis of narcolepsy. Sleep 20:620, 1997.
97. Roth, B, et al: REM sleep and NREM sleep in narcolepsy and hypersomnia. Electroencephalogr Clin Neurophysiol 26:176, 1969.
98. Zorick, F, et al: Sleep-wake abnormalities in narcolepsy. Sleep 9:189, 1986.
99. Johns, MW: A new method for measuring daytime sleepiness: The Epworth sleepiness scale. Sleep 14:540, 1991.
100. Mitler, MM, et al: Maintenance of wakefulness test: A polysomnographic technique for evaluating treatment efficacy in patients with excessive somnolence. Electroencephalogr Clin Neurophysiol 53:658, 1982.
101. Rogers, AE, et al: Patterns of sleep and wakefulness in treated narcoleptic subjects. Sleep 17:590, 1994.
102. Mitler, MM, et al: Narcolepsy and its treatment with stimulants: ASDA Standards of Practice. Sleep 17:352, 1994.
103. Broughton, RJ, et al: Randomized, double-blind, placebo-controlled crossover trial of modafinil in the treatment of excessive daytime sleepiness in narcolepsy. Neurology 49:444, 1997.
104. Laffont, F, et al: Modafinil in diurnal sleepiness: A study of 123 patients. Sleep (8 suppl)17:S113, 1994.
105. Stevenson, RD, and Wolraich, ML: Stimulant medication therapy in the treatment of children with attention deficit hyperactivity disorder. Pediatr Clin North Am 36:1183, 1989.
106. Rogers, AE, and Aldrich, MS: The effect of regularly scheduled naps on sleep attacks and excessive daytime sleepiness associated with narcolepsy. Nurs Res 42:111, 1993.
107. Roehrs, T, et al: Alerting effects of naps in patients with narcolepsy. Sleep 9:194, 1986.

Chapter 11

RESTLESS LEGS SYNDROME AND PERIODIC LIMB MOVEMENT DISORDER

RESTLESS LEGS SYNDROME
Clinical Features
Biologic Basis
Differential Diagnosis
Diagnostic Evaluation
Management
PERIODIC LIMB MOVEMENT DISORDER
Clinical Features
Diagnosis
Management

Wherefore to some, when being a Bed they betake themselves to sleep, presently in the Arms and Leggs, Leapings and Contractions of the Tendons, and so great a Restlessness and Tossings of their Members ensue that the diseased are no more able to sleep than if they were in a Place of the greatest Torture.

THOMAS WILLIS (1685)

Restless legs syndrome (RLS) was described more than 300 years ago by Thomas Willis, who conveyed its disagreeable aspects in his 17th-century textbook. The syndrome of *anxietas tibiarum* described by Wittmaack in 1861 may have been RLS, and Haskovec in 1902 described two patients with an inability to remain at rest that he termed *akathisia*. Akathisia was initially attributed to hysteria, but after the recognition of postencephalitic cases in the 1920s, an organic cause was suspected.[1,2] Akathisia is also precipitated by the use of antipsychotic medications, and since that association was reported,[3] the term akathisia has been used mainly to refer to a syndrome of inner restlessness that is not worse at night and has a less prominent sensory component than RLS.

Ekbom[4] is generally credited with the first use of the term *restless legs syndrome,* and RLS is sometimes called Ekbom's syndrome. Ekbom described the characteristic exacerbation at rest or at night of internal "creeping or crawling sensations" and distinguished the sensations from more superficial paresthesias associated with peripheral neuropathies.

The periodic leg movements that occur at night in most patients with RLS were first reported in detail by Symonds,[5] who used the term *nocturnal myoclonus* to describe them. However, polygraphic recordings, first performed by Lugaresi and colleagues[6,7] in the 1960s, demonstrated that the movements are usually slow contractions, rather than quick myoclonic jerks, and that they occur periodically about every 20 to 30 seconds.[8,9] Currently used terms include periodic leg movements (PLMs), periodic limb movements, and periodic movements of sleep.

Although PLMs are present during sleep in most patients with RLS, they also occur in persons without other symptoms of RLS. When the leg movements lead to or contribute to disrupted sleep but are not accompanied by sensory disturbances and waking dyskinesias, the diagnosis of *periodic limb movement disorder* is used.

RESTLESS LEGS SYNDROME

Clinical Features

RLS is characterized by uncomfortable sensations, usually affecting the legs and sometimes also the arms, associated with an urge to move. The sensations are worse at rest, in the evening, and at night. The mean age of onset is in middle age, although many patients do not present for evaluation until 10 to 20 years after the onset of symptoms.[10] About 2% to 5% of the population is affected, men and women equally.

Many patients find it difficult to describe the disagreeable sensations, apparently because they are unlike other types of sensory disturbance. The sensations may be described as creeping, crawling, pulling, drawing, tingling, itching, aching, or pins and needles. They usually seem to be located underneath the skin, affecting the muscle or bone in the calves and thighs; less commonly, they may be felt in the knees, feet, arms, shoulders, or head. As the sensations build, patients feel an accompanying urge to move that eventually becomes irresistible; with movement of the affected limbs or body parts, the sensations temporarily abate.

The sensations often begin while the patient is sitting still or trying to sleep, and the consequent insomnia is often severe, with difficulty getting to sleep, frequent awakenings, and reduced sleep efficiency. Patients may flex and extend their legs, rock back and forth in bed, or turn over repeatedly. Some get up at night to pace the floor or rub their legs to relieve the sensation, only to have it begin again as soon as they return to bed. The sensations and accompanying urge to move are often made worse by confined situations, such as sitting in an automobile or airplane.

Many patients also have leg movements during wakefulness, which increase in frequency during attempts to remain still.[11,12] The movements during wakefulness, referred to as *waking dyskinesias*, may occur in association with or in response to the unpleasant sensations, or they may occur independently.[13] Although some of these patients fidget during the interview, neurologic examinations are usually normal.

The syndrome is usually chronic, with unpredictable fluctuations. Some patients have prolonged remissions; others slowly become worse over years or decades. An illustrative case is presented below.

CASE HISTORY: Case 11-1

A 70-year-old woman complained of unpleasant sensations in the legs. She first had them during pregnancy at age 23, after which they subsided for several years. At about age 45, they recurred, at first only occasionally but eventually every night, and they became progressively more bothersome over the next two decades. She described the feeling as a deep, gnawing, aching, "arthritis-like" sensation in the thighs and calves of both legs and around the knees, more pronounced on the right. The sensations typically began when she got into bed and prevented her from falling asleep for several hours. When she was in bed and awake, the sensations were often accompanied by proximal flexor "jumps" of her legs, and her husband noted kicking movements during sleep that were sometimes so severe that he had to sleep on the couch. Although she was able to relieve the sensations by moving her legs or walking, they recurred a few minutes after she returned to bed.

She had hypertension complicated by glomerulosclerosis and renal failure and had been on hemodialysis since age 64. After beginning dialysis, she required intermittent transfusions, and the leg sensations became more frequent and bothersome. At age 68, she first noted similar sensations in her arms and also had occasional daytime symptoms in her legs. Treatment with acetaminophen with codeine, chlorpheniramine, and amitriptyline provided only occasional relief. Her mother and her daughter had similar sensations at night. Neurologic examination revealed findings consistent with a mild sensorimotor polyneuropathy.

Levodopa 100 mg combined with carbidopa 25 mg provided more effective relief than other medications she had tried, although symptoms remained present and interfered with sleep onset. She was unable to tolerate higher doses of levodopa.

About 80% to 90% of patients with RLS have PLMs during sleep (Fig. 11–1). The

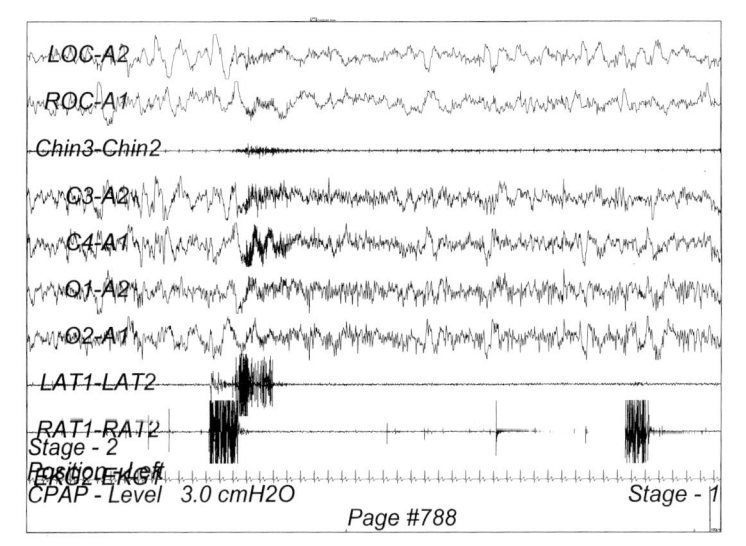

Figure 11–1. Periodic leg movements. During this 1-minute epoch, two periodic movements of the right leg occur: The first is accompanied by a movement of the left leg and a brief awakening, whereas the second is associated with no change in the EEG pattern. The EEG electrodes (A1, A2, C3, C4, O1, O2) were placed in accordance with the International 10-20 system. Other abbreviations: LOC, left outer canthus; ROC, right outer canthus; Chin3–Chin2, submental EMG; LAT1–LAT2, surface EMG over the left anterior tibialis; RAT1–RAT2, surface EMG over the right anterior tibialis.

PLMs, which may occur several hundred times in one night in episodes of several minutes to 1 hour or more, usually have an intermovement interval of about 20 to 30 seconds (range, 5 seconds to 2 minutes). Extension of the great toe and variable degrees of ankle extension, knee extension, and hip extension or flexion are characteristic movements. The arms may be involved as well, and flexion at the elbow may either accompany the leg movements or occur independently. The most vigorous movements resemble the triple flexion withdrawal response that occurs with pyramidal tract dysfunction. The movements are most frequent during stage 1 and stage 2 sleep, less frequent during delta sleep, and generally absent during rapid eye movement (REM) sleep. They may be associated with a K-complex, a burst of alpha activity on the electroencephalogram (EEG), or a transient increase in heart rate or respiratory tidal volume; or with no discernible effect on other polygraphic measures.

Biologic Basis

The cause of RLS is unknown. Although a genetic factor is suspected because up to 50% of patients have relatives with symptoms suggestive of RLS, systematic studies of family members have not been performed, and some relatives reported by patients to have the disorder probably have other conditions. Because 2% to 5% of the general population is affected, a positive family history may sometimes occur on the basis of chance alone. On the other hand, some families have several affected persons in a pattern consistent with an autosomal-dominant inheritance. Furthermore, there is a wide spectrum in the severity of the disorder, and many mildly affected family members may be undiagnosed.

In susceptible persons, RLS may be induced or aggravated by a variety of conditions (Table 11–1). Iron deficiency may account for the associations of RLS with anemia and chronic renal failure. Symptoms occur in 10% to 20% of pregnant women and usually resolve postpartum.[14]

Although the site of nervous system dysfunction associated with RLS and PLMs is unknown, sensory leg discomfort in patients with RLS is associated with bilateral activation of the cerebellum and contralateral activation of the thalamus, whereas the combination of sensory discomfort with a PLM is associated with additional activation of the red nuclei and areas of the pons and mesencephalon.[15] These findings, in conjunction with the absence of cortical potentials associated with PLMs,[16,17] suggest that the generators for

Table 11–1. Agents or Conditions That May Contribute to Periodic Limb Movements and Restless Legs Syndrome

Probable Associations
Iron deficiency
Anemia, especially iron-deficiency anemia
Chronic renal failure
Pregnancy

Possible Associations
Systemic disorders
    Gastric diseases
    Rheumatoid arthritis
    Congestive heart failure
    Chronic obstructive pulmonary disease
    Peripheral vascular occlusive disease
    Varicose veins
Neurological disorders
    Parkinson's disease
    Peripheral neuropathy
    Myelopathy
Deficiency disorders
    Folate deficiency
    Vitamin B deficiency
Drug-related disorders
    Barbiturate withdrawal
    Tricyclic antidepressants
    Fluoxetine
    Lithium carbonate
    Amphetamines
    Caffeine
    Verapamil

the sensory discomfort and the PLMs are located in the brainstem and diencephalon. In addition, the similarity of PLMs to the triple flexion response suggests that the pathways involved in generating the movement include the pyramidal tracts or the dorsal reticulospinal tracts, or both.[18,19] Furthermore, a number of other biological functions—blood pressure, respiration, pupil diameter, intraventricular fluid pressure—show 20- to 30-second periodicity that may be modulated by the reticular formation, suggesting that disinhibition of a central nervous system (CNS) pacemaker regulating reticular excitability may also facilitate the appearance of PLMs.

Pharmacologic studies suggest that the CNS dysfunction associated with PLMs probably involves dopamine systems or opioid systems, or both. Levodopa, dopamine agonists, and opioids suppress PLMs, and in patients with RLS treated with opiates, naloxone causes the reappearance of PLM-like movements.[20,21] In addition, symptoms of RLS become worse with pimozide, a dopamine-receptor blocker, and with γ-hydroxybutyrate, which blocks dopamine release.[22] As iron deficiency makes RLS symptoms worse, and the dopamine D2 receptor and mu-opiate receptor are iron dependent, abnormal iron metabolism affecting opioid and dopaminergic systems may contribute to RLS and PLMs.[23]

Although peripheral nerve function is clinically normal in most patients with RLS, those with coexisting peripheral neuropathy are less likely to have a positive family history for RLS, suggesting that peripheral nerve dysfunction may contribute to pathogenesis, perhaps through abnormal sensory input to CNS structures.[24] In a study of eight patients with RLS and no clinical evidence of neuropathy, six had abnormal sural nerve biopsy results.[25]

Differential Diagnosis

Differential diagnosis of unpleasant sensations of the limbs and associated movements includes akathisia, small-fiber peripheral neuropathies, polymyalgia rheumatica, claudication, nocturnal leg cramps, the syndrome of painful legs and moving toes, and Vesper's curse.

Although the term *akathisia* was used to describe RLS in the past, it is now generally reserved for a syndrome of restlessness associated with a compulsion to move that occurs in association with phenothiazine use or Parkinson's disease. The sensory component of akathisia is less than in RLS; the restlessness is more likely to affect the body as well as the extremities, and episodes in the evening or at night are less common. Marching in place is more common and sleep disturbance less pronounced with akathisia than with RLS.[26]

Small-fiber neuropathies associated with diabetes mellitus, amyloidosis, or other conditions may cause burning paresthesias that are usually more distal than proximal and are felt more on the surface than in deeper structures. The discomfort is usually not accompanied by an urge to move. Rubbing the legs may relieve RLS discomfort, whereas tactile stimulation often exacerbates the pain of sensory neuropathy.

Polymyalgia rheumatica is characterized by aching and stiffness of the neck and shoulders, sometimes involving the thighs and arms, associated with an elevated erythrocyte sedimentation rate. Pain is usually worse in the morning, and aching may increase with activity.

Claudication refers to deep muscular pain caused by arterial insufficiency. It is brought on by exercise and relieved by rest; conversely, RLS is worse at rest and relieved by movement.

Nocturnal leg cramps (see Chap. 15) are sudden, painful muscle contractions, usually involving the calf or foot, that may occur during awakenings at night. They are associated with visible and palpable muscle cramps, which are not present with RLS.

Painful legs and moving toes syndrome is associated with aching pain of the feet associated with involuntary irregular wriggling movements of the toes and feet. Some patients have lumbar radiculopathies or peripheral nerve lesions, but many have no identifiable pathology. The movements are not increased during the evening and night.[27]

Vesper's curse refers to lumbosacral and leg pain with calf cramps and fasciculations, often accompanied by an urge to move the legs, that awakens the patient from sleep. The cause is thought to be increased paraspinal venous volumes caused by lumbar spinal stenosis in patients with congestive heart failure. The symptoms are relieved by treatment of the congestive heart failure.[28]

Diagnostic Evaluation

In most cases, the diagnosis of RLS is made clinically, based on four major clini-

Table 11–2. **Diagnostic Criteria for Restless Legs Syndrome**[29]

Paresthesias or dysesthesias of the limbs associated with a desire to move the limbs
Motor restlessness
Worsening of symptoms at rest
Worsening of symptoms in the evening and at night

Source: From Walters et al.,[29] p 638, with permission.

cal aspects (Table 11–2).[29] Polysomnography has limited value because PLMs are common in persons without RLS, especially older persons, and about 10% to 20% of persons with RLS do not have PLMs. Complete blood count, serum iron, ferritin, blood urea nitrogen, and creatinine should be obtained if there is a suspicion of anemia, iron deficiency, or renal disease. Electromyography and nerve conduction studies should be obtained if neuropathy or radiculopathy is suspected, but they are not necessary if the neurologic examination is normal.

Management

It is important to provide education and counseling to patients with RLS, especially because they often have had symptoms for many years before being diagnosed and may be relieved to know that others suffer from the same disorder. Support groups are beneficial for many patients as well. Of the array of medications sometimes used for RLS, only three types provide substantial benefit: levodopa and dopaminergic agonists, benzodiazepines, and opiates. Unfortunately, the efficacy of these medications is not always sustained, and many patients require repeated dosage adjustments or medication changes.

Most authorities recommend levodopa and dopaminergic agonists, introduced as treatment for RLS in the 1980s, as the initial drugs of choice because these drugs are usually more effective and have fewer side effects than benzodiazepines or opiates. More than half of patients obtain substantial, lasting relief with dopaminergic agonists, and as many as 85% may obtain at least short-term benefit,[30] with fewer arousals and awakenings, fewer PLMs, and better sleep continuity.[31-33]

Levodopa 100 mg combined with carbidopa 25 mg—which prevents peripheral decarboxylation of levodopa—at bedtime is a typical initial regimen; however, some patients obtain relief for only one-third to one-half of the night and require either additional doses during the night or the use of a controlled-release formulation. Dopamine agonists such as bromocriptine 7.5 mg or pergolide 0.05 to 0.25 mg at bedtime are also helpful. The medications are usually well tolerated by older people; nausea or other gastrointestinal symptoms are rarely severe enough to prevent treatment. Tolerance to levodopa-carbidopa develops in about one-third of patients, who eventually require higher doses.

Unfortunately, new or more prominent daytime symptoms of restless legs develop in the majority of patients with RLS who use levodopa-carbidopa for several months or more.[34,35] The increase in daytime symptoms may include the emergence of morning symptoms, the earlier onset of evening symptoms, or an increase in the severity of evening as well as nighttime symptoms. The frequency of this complication is high enough that patients should be advised of this possibility before beginning treatment with levodopa. The cause of the increase in daytime symptoms is uncertain. Although it tends to develop in patients who take more than 200 mg of levodopa daily,[35] this association could be a result of greater disease severity in patients on higher doses rather than an indication of levodopa toxicity. Even if augmentation is caused by levodopa, it is unknown whether it is a temporary side effect or a permanent change in the condition; the tendency of RLS to fluctuate in severity makes it difficult to answer this important question with certainty.

For most patients, when an increase in daytime symptoms develops, levodopa should be discontinued and replaced by a dopamine agonist. Currently, the emergence of daytime symptoms appears to be less frequent with dopaminergic agonists than with levodopa. Therefore, I usually recommend an initial trial of pergolide or bromocriptine for patients who have not taken dopaminergic agents. If these are ineffective, I then suggest a trial of levodopa, provided that the patient is willing to take the risk that daytime symptoms will emerge. For patients who obtain effective relief only with levodopa, increases in daytime symptoms can be treated with additional levodopa, and some patients take this medication several times per day with total daily intake of 1500 mg or more. If levodopa or dopamine agonists produce unacceptable side effects, temporary withdrawal and reinstitution of therapy a few weeks later may improve the efficacy of treatment. Increased symptoms during the withdrawal phase may be treated with benzodiazepines.

High doses of levodopa and dopaminergic agonists may cause dyskinesias, and in patients with severe RLS who take large amounts of levodopa, it is sometimes difficult to determine whether daytime restlessness and dyskinesias are caused by the medication or the disorder. In such cases, reducing the dose usually provides an answer.

Of the benzodiazepines, clonazepam is used most frequently, although temazepam, triazolam, diazepam, or other benzodiazepines may be effective in selected patients. At a dose of 0.5 to 2 mg per night, clonazepam reduces leg discomfort and improves subjective sleep quality in many patients.[36,37] Although benzodiazepines can provide significant relief, older patients with RLS may have problems with daytime sedation and nocturnal confusion, and tolerance and loss of efficacy occur with benzodiazepine therapy in a significant number of patients.

Thomas Willis noted the therapeutic effects of opium in RLS more than 300 years ago. The benefits of opiates are derived from opiate-receptor actions; their effects are prevented by simultaneous administration of the opiate-receptor blocker naloxone.[21] Propoxyphene 65 mg, codeine 30 mg, methadone 10 mg, and hydrocodone 5 mg or oxycodone 5 mg are often helpful.[38] Although opiates may reduce arousals and improve subjective sleep quality, they appear to have little effect on the number of PLMs.[31,39,40] The benefits associated with opiates must be balanced against their potential side effects, particularly in older people.

Other medications sometimes used for RLS include gabapentin, clonidine, baclofen, carbamazepine, quinine, vitamins B_{12} and E, and folic acid.[37] Of these, gabapentin 300 to 1000 mg at bedtime appears to be the most useful. Gabapentin is particularly suitable for use in older persons because it has virtually no interactions with other medications. Sedation, however, can be a significant side effect. I sometimes use gabapentin in combination with levodopa or a dopaminergic agonist. Baclofen, a γ-aminobutyric acid (GABA)-ergic agonist, is probably helpful only because of its sedating effects, since the number of PLMs is not reduced.[41] Carbamazepine 200 to 600 mg reduces leg sensations in some patients, but it has little effect on PLMs or associated arousals.[42–44] Although beneficial effects from clonidine have been reported,[45] I have not found it helpful.

PERIODIC LIMB MOVEMENT DISORDER

Clinical Features

In this disorder, complaints of difficulty falling asleep, frequent awakenings, daytime sleepiness, or a combination of these symptoms are associated with PLMs and sleep disruption; however, the sensory symptoms and waking dyskinesias characteristic of RLS do not occur. Although patients with the disorder usually have sleep-related complaints, in some the chief complaint comes from the spouses, who describe kicking movements that can be vigorous enough to force them into another bed.

Severely affected patients have movements every night throughout sleep. Others may have substantial night-to-night variability, with frequent PLMs on some nights and few or no movements on others.[46,47] Body position during sleep appears to affect PLM frequency and may contribute to night-to-night variation.[48]

PLMs become more frequent with age. Uncommon in childhood and in young adults, they occur in about 5% of 30- to 50-year-olds, 30% of 50- to 65-year-olds, and in 30% to 45% of persons over age 65.[49–55]

By definition, periodic limb movement disorder is associated with symptoms, reported either by the patient or the bed partner. However, some patients with PLMs are asymptomatic: PLMs are a common finding in older persons without sleep complaints and in persons undergoing evaluation for suspected obstructive sleep apnea. PLMs sometimes increase in frequency after treatment with nasal continuous positive airway pressure (Fig. 11–2), but it is unknown whether the movements in such persons have any clinical significance.[56]

Diagnosis

When the chief complaint is kicking or twitching of the limbs that is observed by the bed partner, the differential diagnosis includes sleep starts, phasic jerks of REM sleep, partial seizures, REM sleep behavior disorder, and other causes of myoclonus. A history of periodic recurrence of leg movements is the most useful feature differentiating PLMs from these disorders. Sleep starts (see Chap. 15) may cause sudden arousal, but they are not periodic and are not associated with discomfort or inner restlessness. The association of PLMs with sleep and their predilection for the legs help to differentiate them from other types of daytime myoclonus. Polysomnography is indicated when PLMs are the suspected cause of sleep-related complaints because polysomnography is the only reliable method of determining their presence.[57]

Management

Determining the presence or absence of PLMs is easier than determining their significance. Although they are common in older insomniacs, they are also common in asymptomatic older persons.[49–55] Furthermore, older persons, who are more likely to have PLMs, are also more likely to have psychiatric, medical, or neurological prob-

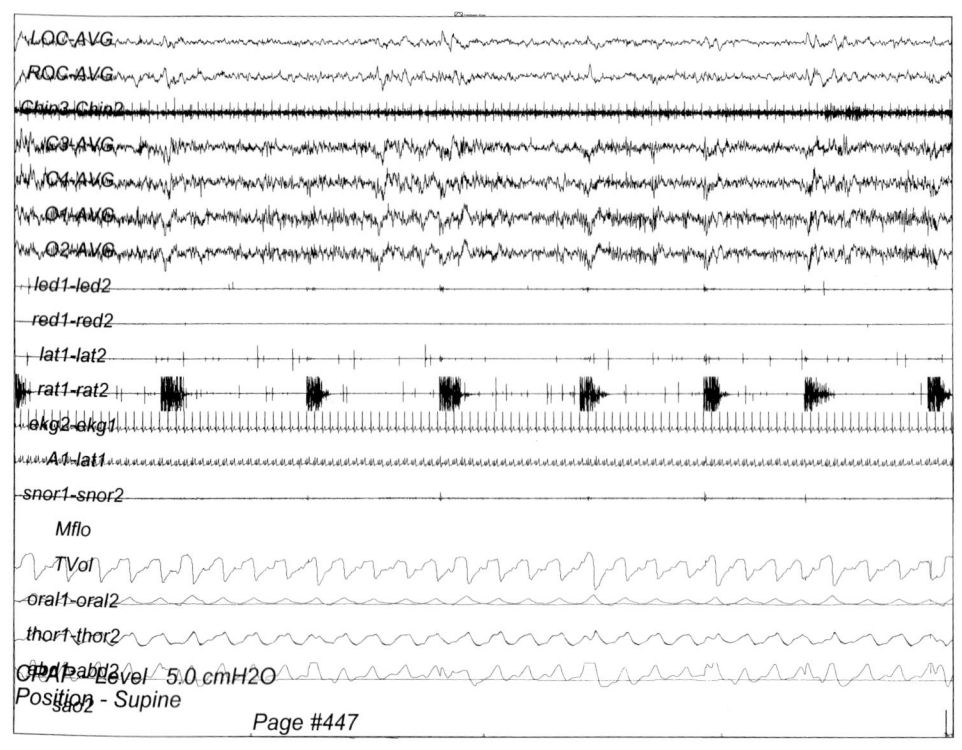

Figure 11–2. Periodic leg movements during a trial of nasal continuous positive airway pressure in a patient with obstructive sleep apnea. Although sleep apnea was well controlled during this 2-minute segment of the study, eight periodic leg movements were recorded from the right leg with little or no change in the appearance of the EEG. The EEG electrodes (C3, C4, O1, O2) were placed in accordance with the International 10-20 system. Other abbreviations: AVG, average reference electrode; LOC, left outer canthus; ROC, right outer canthus; Chin3–Chin2, submental EMG; led1–led2, surface EMG electrodes over the left arm extensor compartment; red1–red2, surface EMG over the right arm extensor compartment; Lat1–Lat2, surface EMG electrodes over the left anterior tibialis; Rat1–Rat2, surface EMG electrodes over the right anterior tibialis.

lems that can affect sleep. As with RLS, the PLMs of periodic limb movement disorder may occur in association with an awakening, an arousal, or a K-complex, or may not be associated with any visually apparent change in the EEG. Although many clinicians assume that PLMs are contributing to insomnia or daytime sleepiness if arousals accompany the PLMs, in a series of patients diagnosed with PLMs, the frequency of arousals associated with PLMs did not correlate with severity of sleepiness, complaints of insomnia, or reports of feeling unrefreshed in the morning.[58]

Because of the uncertain significance of PLMs, other possible causes of insomnia or daytime sleepiness should be considered in patients with PLMs before concluding that the PLMs are the basis for the symptoms (see Chaps. 8 and 9). Treatment is warranted only if the PLMs and associated sleep disturbance appear to be responsible. In patients with obstructive sleep apnea and PLMs, first the apnea should be treated, and then the patient should be reassessed before the PLMs are treated. If the patient is asymptomatic after the apnea is treated, then the PLMs do not need to be treated. Similarly, in patients with symptoms suggestive of inadequate sleep hygiene, psychophysiologic insomnia, or an affective disorder, usually the suspected disorder should be treated first, and then the patient should be reassessed before treatment of PLMs is undertaken.

The most helpful treatments for PLMs are those used for RLS: levodopa-carbidopa and dopamine agonists, opiates, and benzodiazepines.[59–65] Levodopa 100 mg combined with carbidopa 25 mg, taken at bedtime, is often effective and can be repeated once during the night if needed. Higher doses may be necessary in some patients. Bromocriptine 2.5 or 5 mg at bedtime is an alternative. The benefits of clonazepam (0.5 to 2 mg at bedtime) and other benzodiazepines may be related more to their effects on arousals than to the effect on the leg movements themselves.[59–62] Opiates such as codeine 30 to 60 mg or propoxyphene 65 to 100 mg at bedtime can also be used.

SUMMARY

RLS, which affects 2% to 5% of the population, is characterized by uncomfortable sensations, usually affecting the legs, associated with an urge to move. Because the sensations are worse at rest, in the evening, and at night, insomnia is often a prominent complaint. Many patients also have leg movements during wakefulness, called waking dyskinesias, which increase in frequency during attempts to remain still.

About 80% to 90% of patients with RLS have PLMs during sleep, characterized by extension of the great toe and variable degrees of ankle extension, knee extension, and hip extension or flexion. Arousals accompanying the movements lead to sleep disturbance. Although the cause of RLS is unknown, in some families it appears to have an autosomal-dominant pattern of inheritance. Iron deficiency exacerbates symptoms, and dysfunction of dopamine or opioid systems caused by abnormal iron metabolism may be the basis for the disorder.

The diagnosis of RLS is based on its characteristic clinical features. Treatments include levodopa and dopaminergic agonists, benzodiazepines, and opiates. Unfortunately, the efficacy of these medications often is not sustained.

In periodic limb movement disorder, complaints of difficulty falling asleep, frequent awakenings, daytime sleepiness, or a combination of these symptoms are associated with PLMs and sleep disruption; however, the sensory symptoms and waking dyskinesias that are characteristic of RLS do not occur. Diagnosis of this disorder is often difficult because PLMs are not always symptomatic and, if present, may not be the cause of symptoms. The most helpful treatments are those used for RLS.

REFERENCES

1. Bing, R: Uber einige bemerkenswerte begleiterscheinungen der 'extrapyramidalen rigidat.' Schweiz Med Wochenschr 4:167, 1923.
2. Sicard, JA: Akathisie and tasikinesie. Presse Med 31:265, 1923.
3. Sigwald, J, et al: Le traitement de la maladie de Parkinson et des manifestations extrapyramidales par le diethylaminoethyl n-thiophenylalanine (2987 RP): Resultats d'une annee d'application. Rev Neurol 79:683, 1947.
4. Ekbom, K-A: Restless legs. Acta Med Scand Suppl 158:5, 1945.
5. Symonds, CP: Nocturnal myoclonus. J Neurol Neurosurg Psychiatry 16:166, 1953.
6. Lugaresi, E, et al: A propos de quelques manifestations nocturnes myocloniques (nocturnal myoclonus de Symonds). Rev Neurol (Paris) 115:547, 1966.
7. Lugaresi, E, et al: Restless legs syndrome and nocturnal myoclonus. In Gastaut, H, et al (eds): The Abnormalities of Sleep in Man. Aulo Gaggi Editore, Bologna, 1968, pp 285–294.
8. Coleman, RM, et al: Periodic movements in sleep (nocturnal myoclonus): Relation to sleep disorders. Ann Neurol 8:416, 1980.
9. Coleman, RM: Periodic movements in sleep (nocturnal myoclonus) and restless legs syndrome. In Guilleminault, C (ed): Sleeping and Waking Disorders: Indications and Techniques. Addison-Wesley, Menlo Park, Ca, 1982, pp 265–295.
10. Ondo, W, and Jankovic, J: Restless legs syndrome: Clinicoetiologic correlates. Neurology 47:1435, 1996.
11. Pollmächer, T, and Schulz, H: Periodic leg movements (PLM): Their relationship to sleep stages. Sleep 16:572, 1993.
12. Walters, AS, et al: Review and videotape recognition of idiopathic restless legs syndrome. Mov Disord 6:105, 1991.
13. Pelletier, G, et al: Sensory and motor components of the restless legs syndrome. Neurology 42:1663, 1992.
14. Goodman, JD, et al: Restless leg syndrome in pregnancy. Br Med J 297:1101, 1988.
15. Bucher, SF, et al: Cerebral generators involved in the pathogenesis of the restless legs syndrome. Ann Neurol 41:639, 1997.
16. Trenkwalder, C, et al: Bereitschafts potential in idiopathic and symptomatic restless legs syndrome. Electroencephalogr Clin Neurophysiol 89:95, 1993.
17. Lugaresi, E, et al: Nocturnal myoclonus and restless legs syndrome. In Fahn, S, et al (eds): Myoc-

lonus. Advances in Neurology, Vol 43. Raven Press, New York, 1986, pp 295–207.
18. Smith, RC: Relationship of periodic movements in sleep (nocturnal myoclonus) and the Babinski sign. Sleep 8:239, 1985.
19. Yokota, T, et al: Sleep-related periodic leg movements (nocturnal myoclonus) due to spinal cord lesion. J Neurol Sci 104:13, 1991.
20. Walters, A, et al: Dominantly inherited restless legs with myoclonus and periodic movements of sleep: A syndrome related to the endogenous opiates? Adv Neurol 43:309, 1986.
21. Hening, WA, et al: Dyskinesias while awake and periodic movements in sleep in restless legs syndrome: Treatment with opioids. Neurology 36:1363, 1986.
22. Bedard, M-A, et al: Nocturnal gamma-hydroxy butyrate effect on periodic leg movements and sleep organization of narcoleptic patients. Clin Neuropharmacol 12:29, 1989.
23. Stefano, G: Comparative aspects of opioid-dopamine interaction. Cell Mol Neurobiol 2:167, 1982.
24. Salvi, F, et al: Restless legs syndrome and nocturnal myoclonus: Initial clinical manifestation of familial amyloid polyneuropathy. J Neurol Neurosurg Psychiatry 53:522, 1990.
25. Iannaccone, S, et al: Evidence of peripheral axonal neuropathy in primary restless legs syndrome. Mov Disord 10:2, 1995.
26. Walters, AS, et al: A clinical and polysomnographic comparison of neuroleptic-induced akathisia and the idiopathic restless legs syndrome. Sleep 14:339, 1991.
27. Spillane, JD, et al: Painful legs and moving toes. Brain 94:541, 1971.
28. LaBan, MM, et al: Restless legs syndrome associated with diminished cardiopulmonary compliance and lumbar spinal stenosis—a motor concomitant of "Vesper's curse." Arch Phys Med Rehabil 71:384, 1990.
29. Walters, AS, et al (The International Restless Legs Syndrome Study Group): Towards a better definition of the restless legs syndrome. Mov Disord 10:634, 1995.
30. Becker, PM, et al: Dopaminergic agents in restless legs syndrome and periodic limb movements of sleep: Response and complications of extended treatment in 49 cases. Sleep 16:713, 1993.
31. Kaplan, PW, et al: A double-blind, placebo-controlled study of the treatment of periodic limb movements in sleep using carbidopa/levodopa and propoxyphene. Sleep 16:717, 1993.
32. Brodeur, C, et al: Treatment of restless legs syndrome and periodic movements during sleep with L-Dopa: A double-blind, controlled study. Neurology 38:1845, 1988.
33. Akpinar, S: Restless legs syndrome treatment with dopaminergic drugs. Clin Neuropharmacol 10:69, 1987.
34. Guilleminault, C, et al: Dopaminergic treatment of restless legs and rebound phenomenon. Neurology 43:445, 1993.
35. Allen, RP, and Earley, CJ: Augmentation of the restless legs syndrome with carbidopa/levodopa. Sleep 19:205, 1996.
36. Boghen, D, et al: The treatment of the restless legs syndrome with clonazepam: a prospective controlled study. Can J Neurol Sci 13:245, 1986.
37. Montplaisir, J, et al: The treatment of the restless leg syndrome with or without periodic leg movements in sleep. Sleep 15:391, 1992.
38. Walters, AS, et al: Successful treatment of the idiopathic restless legs syndrome in a randomized double-blind trial of oxycodone versus placebo. Sleep 16:327, 1993.
39. Allen, RP, et al: Double-blinded placebo controlled comparison of propoxyphene and carbidopa/levodopa for treatment of periodic limb movements in sleep. Sleep Res 20:199, 1991.
40. Allen, RP, et al: Doubleblinded, placebo controlled comparison of high dose propoxyphene and moderate dose carbidopa/levodopa for treatment of periodic limb movements in sleep. Sleep Res 21:166, 1992.
41. Guilleminault, C, and Flagg, W: Effect of baclofen on sleep related periodic leg movements. Ann Neurol 15:234, 1984.
42. Lundvall, O, et al: Carbamazepine in restless legs: A controlled pilot study. Eur J Clin Pharmacol 25:323, 1983.
43. Telstad, W, et al: Treatment of the restless legs syndrome with carbamazepine: A double-blind study. Br Med J 288:444, 1984.
44. Zucconi, M, et al: Nocturnal myoclonus in restless legs syndrome: Effect of carbamazepine treatment. Funct Neurol 4:263, 1989.
45. Handwerker, JV, and Palmer, RF: Clonidine in the treatment of restless leg syndrome. N Engl J Med 313:1228, 1985.
46. Bliwise, DL, et al: Nightly variation of periodic leg movements in sleep in middle aged and elderly individuals. Arch Gerontol Geriatr 7:273, 1988.
47. Mosko, SS, et al: Night to night variability in sleep apnea and sleep related periodic leg movements in the elderly. Sleep 11:340, 1988.
48. Dzvonik, ML, et al: Body position changes and periodic movements in sleep. Sleep 9:484, 1986.
49. Ancoli-Israel, S, et al: Sleep apnea and nocturnal myoclonus in a senior population. Sleep 4:349, 1981.
50. Bixler, EO, et al: Nocturnal myoclonus and nocturnal myoclonic activity in a normal population. Res Commun Chem Pathol Pharmacol 36:129, 1982.
51. Ancoli-Israel, S, et al: Sleep apnea and periodic movements in an aging sample. J Gerontol 40:419, 1985.
52. Ancoli-Israel, S, et al: Periodic limb movements in sleep in community-dwelling elderly. Sleep 14:496, 1991.
53. Coleman, RM, et al: Epidemiology of periodic movements during sleep. In Guilleminault, C, and Lugaresi, E (eds): Sleep/Wake Disorders; Natural History, Epidemiology, and Long-Term Evolution. Raven Press, New York, 1983, pp 217–229.
54. Mosko, SS, et al: Sleep apnea and sleep-related periodic leg movements in community resident seniors. J Am Geriatr Soc 36:502, 1988.
55. Roehrs, T, et al: Age-related sleep-wake disorders at a sleep disorders center. J Am Geriatr Soc 31:364, 1983.

56. Fry, JM, et al: Periodic leg movements in sleep following treatment of obstructive sleep apnea with nasal continuous positive airway pressure. Chest 96:89, 1989.
57. Indications for Polysomnography Task Force, American Sleep Disorders Association Standards of Practice Committee: Practice parameters for the indications for polysomnography and related procedures. Sleep 20:406, 1997.
58. Mendelson, WB: Are periodic leg movements associated with clinical sleep disturbance? Sleep 19:219, 1996.
59. Peled, R, and Lavie, P: Double-blind evaluation of clonazepam on periodic leg movements in sleep. J Neurol Neurosurg Psychiatry 50:1679, 1987.
60. Ohanna, N, et al: Periodic leg movements in sleep: Effect of clonazepam treatment. Neurology 35:408, 1985.
61. Montagna, P, et al: Clonazepam and vibration in restless legs syndrome. Acta Neurol Scand 69:428, 1984.
62. Mitler, MM, et al: Nocturnal myoclonus: Treatment efficacy of clonazepam and temazepam. Sleep 9:385, 1986.
63. Doghramji, K, et al: Triazolam diminishes daytime sleepiness and sleep fragmentation in patients with periodic leg movements in sleep. J Clin Psychopharmacol 11:284, 1991.
64. Bonnet, MH, and Arand, DL: The use of triazolam in older patients with periodic leg movements, fragmented sleep, and daytime sleepiness. J Gerontol 45:M139, 1990.
65. Bonnet, MH, and Arand, DL: Chronic use of triazolam in patients with periodic leg movements, fragmented sleep and daytime sleepiness. Aging 3:313, 1991.

Chapter 12

CHRONOBIOLOGIC DISORDERS

JET LAG SYNDROME
Clinical Features
Biologic Basis
Management
SHIFT-WORK SLEEP DISORDER
Clinical Features
Biologic Basis
Diagnosis
Management
DELAYED SLEEP PHASE SYNDROME
Clinical Features
Psychobiologic Basis
Diagnosis
Management
ADVANCED SLEEP PHASE SYNDROME
IRREGULAR SLEEP-WAKE PATTERN
  DISORDER
Clinical Features
Biologic Basis
Diagnosis
Management
NON–24-HOUR SLEEP-WAKE SYNDROME
Clinical Features
Biologic Basis
Diagnosis
Management

Thus are poor servitors, when others sleep upon their quiet beds, constrain'd to watch in darkness, rain, and cold.

WM. SHAKESPEARE, HENRY VI PT 1 II.i.15

The circadian pacemaker exerts a powerful influence on sleep and wakefulness (see Chap. 4), and when it is not in phase with the desired sleep-wake schedule, sleep-wake disturbances can occur. Symptoms include sleepiness when alertness is desired, insomnia when sleep is desired, or both. The six chronobiologic disorders listed in the *International Classification of Sleep Disorders*[1] are discussed in this chapter. Because knowledge of chronobiology is essential for understanding these disorders, the reader may benefit from reading Chapter 4 before continuing with this chapter.

JET LAG SYNDROME

Except for the few persons living in extreme northern or southern latitudes, rapid travel across time zones—the major cause of jet lag—was impossible before the invention of the airplane, and it was rare before the use of jet aircraft for commercial travel after World War II. Travel across several time zones is now commonplace, with millions of trips taking place each year, and most persons who travel east or west by jet across three or more time zones have disturbances of sleep or wakefulness.[2]

Sleep problems after jet travel, however, are not caused entirely by time zone changes: Anxiety, excitement, and the unsuitability of aircraft seats for sleep contribute to sleep loss and poor sleep quality before or during north-south as well as east-west jet travel.

Clinical Features

After travel through several time zones, sleep-related symptoms are often more severe and longer lasting in older persons,[3] may include difficulty falling asleep, difficulty remaining asleep, or excessive sleepiness during the new daytime. The severity and duration of symptoms are affected by the number of time zones traversed, the direction of travel, and the amount of sleep disruption and sleep loss during travel. Symptoms, which are minimal with trips across one or two time zones, may last for 3 to 5 days in persons who cross three or four time zones and as long as 7 to 10 days after trips across six or more time zones. Generally, symptoms are more severe and resynchronization of circadian rhythms after arrival takes longer with eastbound flight—from, for example, North America to Europe—than with westbound flight.[2,4] Although sleepiness after eastbound night flights is often exacerbated by sleep loss during the flight, jet lag symptoms tend to be worse after daytime eastbound flights as well.

Sleep loss during travel may lead to relatively good sleep on the first night at the destination, with worse sleep on subsequent nights. Some persons alternate for several nights between good and poor nights of sleep.[5] With westbound flight, evening sleepiness may occur, even after an initial night of good sleep, until the traveler's biologic clock has adjusted to the new time.

Chronic jet lag may occur in airline personnel and others who make frequent transcontinental or transoceanic flights. Malaise and irritability may accompany chronic sleep disruption and irregular sleep patterns, and the associated performance decrements may be potentially dangerous in pilots and other air crew who have made several transoceanic trips over a period of a few days.[6,7]

Biologic Basis

Jet lag syndrome is caused mainly by desynchronization between the internal clock and the external clock. Sleep loss and sleep disruption associated with stress, anxiety, and excitement also contribute to sleep complaints, and alcohol consumption during the flight may aggravate symptoms by increasing sleep disturbance during the flight.

The range of external schedules to which the internal clock can adjust in one cycle—*the range of entrainment*—is about 3 to 4 hours in theory (see Chap. 4), but in practice, it is closer to 1 to 2 hours in either direction.[8-10] Thus, jet lag is likely to occur, at least to a mild degree, after trips through as few as three time zones. The duration of desynchronization depends on how quickly the internal pacemaker shifts its phase, which in turn is mainly a function of the timing and duration of light exposure during and after travel. In persons who make no particular attempt to regulate their exposure to light, the internal clock shifts by about 60 minutes per day after eastbound flights and about 90 minutes per day after westbound flights.[11]

The severity of insomnia is a function of the timing of attempts to sleep as well as the direction of travel and the number of time zones crossed, because sleep latency and sleep duration is related to the phase of the temperature rhythm at which sleep is attempted (see Chap. 4).[12-14] The relation between sleep and circadian phase is best understood by considering the hypothetical constructs, process C and process S (see Chap. 4). *Process S* is a homeostatic process that accumulates during wakefulness and discharges during sleep. *Process C* is a circadian drive for wakefulness that, under usual conditions, increases during the day, thereby helping to maintain wakefulness, and decreases during sleep, thereby helping to maintain sleep.

After an eastbound flight across six time zones and an attempt to sleep at 11 PM in the new time zone, sleep onset may be difficult and sleep likely to be disrupted during the first part of the night because the attempt to sleep occurs at the peak of the temperature cycle, when process C is high. Delaying sleep onset until 1 or 2 AM may not make sleep onset any easier because, even though process S increases with the increased duration of wakefulness, process C tends to counteract this effect. After about 4 AM, however, sleep is likely to improve because body temperature and process C begin to drop.

After a westbound flight across six time zones and an attempt to sleep at 10 PM (4 AM of the original time zone), sleep onset is likely to be rapid because process S is higher than usual and process C, at or just after the nadir of the temperature cycle, is low. Sleep duration, however, is likely to be short as process S discharges and process C begins to rise.[11]

Although symptoms after jet travel are caused principally by the phase shift, dissociation of other internally timed functions from the sleep-wake rhythm may also contribute to insomnia and daytime sleepiness and may account in part for the greater difficulty in adjusting after eastbound flights than after westbound flights.[4,15,16] After westbound flights across seven to nine time zones, virtually all physiologic rhythms shift in the same "westbound" direction, with lengthened periods, until they are synchronized with the new time zone. With eastbound flights of similar length, however, some rhythms shift in the "eastbound" direction, with shortened periods, while others shift in the "westbound" direction, with lengthened periods that produce a phase delay of 15 to 17 hours, before becoming synchronized with the new time zone.[11,17]

Management

The approach to management depends on the number of time zones crossed and the length of time at the destination. In all cases, symptoms can be reduced by altering the home sleep-wake schedule in the direction of the anticipated destination schedule during the days preceding flight. For trips of a few days across three or four time zones, eating and sleeping on the home time schedule after arrival is a simple although sometimes impractical approach that minimizes jet lag. For longer stays, or for travel across five or more time zones, immediate adoption of the new time zone schedule is usually the best approach.

Light exposure can reduce the duration of symptoms, but the timing and intensity of light exposure in relation to the body temperature rhythm is critical (see Fig. 4–3), and improperly timed light exposure can increase symptoms. The periods of maximal phase-shifting effects of bright light occur 2 to 3 hours before or after the minimum of the endogenous temperature rhythm, which for clinical purposes can be approximated as the midpoint of the usual sleep period. Thus, for someone traveling from North America to Europe who usually sleeps from 11 PM to 7 AM Eastern Standard Time (EST), the temperature minimum is at about 3 AM and the maximal phase advance occurs with bright light exposure between 4 AM and 6 AM EST. With a 6-hour time zone change, bright light exposure should be timed to occur at about 5 AM EST (11 AM European time) the first day and 1 to 2 hours earlier each day for the next 2 to 3 days. Exposure to 10,000 Lux for 30 minutes is sufficient and can be obtained by sitting with eyes open two feet in front of four 40-watt fluorescent light bulbs. Several commerical light boxes are available, and home-made light boxes also can be used. Spending time outdoors on sunny or bright cloudy days has similar effects, whereas indoor lighting is usually of insufficient brightness. In addition, light exposure should be minimized from midnight to 2 AM EST (6 to 8 AM European time) the day of arrival, and 1 to 2 hours earlier on subsequent days.

For someone traveling from eastern North America to Japan with a 10-hour time zone change, bright light exposure should occur at about midnight to 2 AM EST (2 to 4 PM Japanese time) the first day, with subsequent exposure 1 to 2 hours later each day for about 5 days. Light exposure should be minimized between 4 AM and 6 AM EST (6 to 8 PM) initially and 1 to 2 hours earlier on subsequent days.

Although hypnotics may be helpful for insomnia, high doses of short-acting agents may produce next-day amnestic effects in air travelers and may increase the possibility of insomnia on the subsequent night.[18] Melatonin, which produces phase shifts in the opposite direction from bright light, can reduce jet lag symptoms and shorten their duration.[19,20] Thus, after eastbound flights, 10 mg of melatonin in the evening may be helpful; however, because the manufacture and sale of melatonin is not regulated in the United States, and because of concerns about possible side effects, I do not recommend melatonin use to patients.

SHIFT-WORK SLEEP DISORDER

Unlike jet lag, sleep disturbance associated with shift work is not a new problem: Sailors, herdsmen, and soldiers have worked at night for thousands of years. Night-shift work became more common with the industrial revolution and became even more prevalent in the early 20th century with the availability of electric light and power. By the 1980s, more than 20 million persons in the United States—one-fourth of employed men and one-sixth of employed women—performed some shift work.[21,22] Unfortunately, night-shift and rotating-shift work often lead to sleep problems: 40% to 80% of night-shift workers have sleep complaints, and 5% to 20% report moderate to severe sleep problems.[23,24]

Sleep disturbance with shift work can be more than just a nuisance: The most common cause of nighttime single-vehicle accidents is falling asleep at the wheel, and the human errors that contributed to the Chernobyl and Three Mile Island nuclear accidents occurred during the night.

Clinical Features

Sleep-wake symptoms in night workers and those who work rotating shifts may include daytime insomnia with short or disrupted daytime sleep, sleepiness at work, and sleepiness while commuting to and from work. Difficulty falling asleep and staying asleep at night during nights off work is a problem for some shift workers, although many obtain their best sleep at such times. An additional problem for some night-shift workers is sleep paralysis, which may occur in the early morning when the propensity for rapid eye movement (REM) sleep is high.[25]

Those who consider themselves "morning persons" and those who have difficulty functioning during drowsy periods are more likely to complain of sleep problems.[26] Other risk factors for sleep disturbance associated with shift work include psychiatric illness, unstable family situations, and excessive caffeine and alcohol use. Shift-work–related sleep disturbance is more likely to develop in those who are over age 40, and many persons who cope fairly well with shift work as young adults find it more difficult to obtain good sleep as they become older, as in the following case.

CASE HISTORY: Case 12–1

A 39-year-old dispatcher complained of sleepiness and difficulty concentrating at work. He reported that the problems had increased over the previous 2 to 3 years and had led to poor work evaluations. Difficulty sleeping had begun 11 years earlier, when he first began work as a dispatcher, but it had not affected his work initially. His work schedule was as follows: Sunday 3 to 11 PM, Monday 2 to 10 PM, Tuesday 7 AM to 3 PM, Wednesday 6 AM to 2 PM, and Wednesday 11 PM to Thursday 7 AM.

He took triazolam 0.25 mg on Wednesday afternoons to help him sleep in preparation for the Wednesday night shift. On Thursday, Friday, Saturday, and Sunday nights, he went to sleep at 11:30 PM and slept until 8:30 AM. On Monday and Tuesday nights, he went to bed at 11 PM and arose at 5 AM, although he sometimes slept poorly, awakening at 2:30 AM and sleeping fitfully the rest of the night. On Thursday mornings, he usually slept for 2 to 3 hours after work. He drank 6 to 7 cups of coffee and smoked 40 to 60 cigarettes each day. He complained of being irritable and having difficulty interacting with others at work. He also reported that he obtained less enjoyment from recreational activities, but he denied feeling sad or tearful. Although he and his wife got along well, his relationship with his teen-age stepdaughter was strained. Examination was normal. A polysomnogram and Multiple Sleep Latency Test (MSLT), obtained after a 2-week vacation, were normal apart from a slightly above average number of nocturnal awakenings.

My impression was that the patient had a triad of problems: (1) shift-work sleep disorder; (2) mild depression, probably caused partly by family difficulties and rapidly changing work schedules; and (3) excessive caffeine use. The patient then negotiated with his union, employer, and coworkers for a more regular schedule; reduced his caffeine intake; discontinued triazolam; and began counseling. Five months later, he worked mainly afternoon shifts and slept from 12:30 AM to 9 AM. His alertness at work and his sleep had improved, but he continued to see a psychotherapist for mild depression.

Biologic Basis

Sleep disturbance in shift workers has several causes (Table 12–1), the most important of which is the sudden change in the sleep schedule at the time of shift rotation; this change leads to reduced sleep time, particularly in the last third of the sleep period, to reduced amounts of slow-wave sleep, and to increased amounts of stage 1 sleep.[27–30]

Mainly for social reasons, most night-shift workers attempt to sleep at night (rather than days) on their nights off; this alternation prevents them from synchronizing their internal clock to the night-shift schedule.[31] In addition, the frequent changes between day sleep (after a work night) and night sleep (on nights off work) probably increase the degree of internal desynchronization and thereby contribute to reduced sleep quality and continuity during night-shift work and during the first few nights after rotation to the day or evening shift.[32]

Table 12–1. **Sleep Disturbance in Shift Workers**

Causes
- Sudden changes in the sleep schedule
- Different sleep times on work days and days off
- Reduced total amount of sleep
- Associated medical and psychiatric conditions

Feature of shift work that increase the likelihood of sleep disturbance[30]
- Several consecutive night shifts
- Shift rotation once or more each week
- "Backward" shift rotation (days to nights to evenings)
- Regular double shifts or overtime

Reduced amounts of sleep also contribute to symptoms. In each 24-hour period, night workers sleep about 1 to 1½ hours less than day workers, even after a year or more of night work.[33,34] Gastrointestinal disorders (e.g., dyspepsia, reflux esophagitis, gastritis, peptic ulcer disease) are increased in shift workers, probably because of their increased tobacco and alcohol use and irregular eating habits; these disorders, in turn, may also contribute to sleep disturbance.[23,35–38]

Diagnosis

In shift workers who complain of sleep disturbance, the differential diagnosis includes inadequate sleep hygiene, delayed sleep phase syndrome (DSPS), insufficient sleep syndrome, and intrinsic sleep disorders. Temporal association with night or rotating-shift work and remission during vacations point to a diagnosis of shift-work sleep disorder. On the other hand, sleep problems that develop in shift workers after years of stable functioning may be caused by depression, marital or family problems, job dissatisfaction, or other psychosocial factors. In some patients, issues related to retirement or potential disability complicate the picture.

The history, supplemented by a diary of work hours and sleep hours, provides the most useful information for diagnosis. If sleep apnea is suspected, polysomnography may be helpful. An MSLT can document the presence of excessive sleepiness; however, even when sleep-onset REM periods occur, they are nonspecific because they may reflect the schedule disturbance rather than narcolepsy. If narcolepsy is strongly suspected, the MSLT is best performed after the patient has been on a regular schedule of night sleep for at least 2 weeks. Psychiatric assessment is indicated if depression appears to be a contributing factor.

Management

The ideal treatment for shift-work sleep disorder is a sleep schedule that allows the patient to obtain restful sleep 7 days or nights per week. Appropriately timed exposure to bright light may help to maintain a stable circadian phase during permanent night or evening work, and it can assist with phase shifts during changes from one shift to another. In practice, the ideal schedule for sleep and timing of bright light exposure is often unobtainable because of social obligations or the unavailability of an appropriate sleep environment during the day. However, the development of an idealized schedule allows the clinician and patient to focus on ways to accommodate social obligations in order to maintain a schedule close to the ideal.

In theory, persons who work nights for several weeks or longer sleep best if they sleep during the day 7 days per week, whether they are working or not. Persons who maintain the same schedule 7 days per week, with sleep times during the day, should be advised to keep the bedroom as dark as possible during the day and to avoid light exposure during the times of peak phase-shift capability. For someone who sleeps from 10 AM to 6 PM and has a temperature minimum at 2 PM, light should be avoided between about 11 AM and 5 PM. Increased amounts of light at night in the workplace may enhance alertness during work hours.[39–42]

In practice, most night workers do not comply with such a recommendation because for social reasons they prefer to be awake during the day and asleep at night during nights off. Nonetheless, it may be possible to design a schedule that allows them to obtain a major block of sleep of 4 to 5 hours' duration at approximately the same time of day 7 days per week. Typically, the major sleep period should occur just before going to work or upon return home. Shift workers who have school-age children sometimes prefer to sleep from about 9 AM to 2 PM, when the children are at school, and then take a 2- to 3-hour nap in the evening before leaving for work. Others do best with a 5-hour block of sleep in the evening before going to work, supplemented by a 2- to 3-hour nap in the morning when they return home. Sleeping in the evening before going to work, rather than in the morning just after work, may improve daytime sleep, but it also may increase the phase shift between day sleep on work nights and night sleep on nights off.

For workers who rotate shifts, although appropriately timed bright light exposure can shift the period of peak alertness, the degree of phase shift is rarely more than 1 to 1½ hours in each 24-hour cycle. Thus, if shift rotation occurs more often than once per week, the phase shift will be incomplete, even with properly timed bright light.

Shift rotations from days to evenings to nights are usually better tolerated than rotations in the opposite direction.[43] Most persons who move from day shift to evening shift go to bed later and get up later. If they have been sleeping from 10 PM to 6 AM and the temperature minimum is at 2 AM, bright light exposure between about 11 PM and 1 AM should assist with the phase shift. In preparation for the change from evening shift to night shift, they should go to bed later and obtain bright light exposure at progressively later hours. Once the night shift begins, the major sleep period should occur as soon as the person arrives home, and bright light exposure must be avoided during sleep hours. With the change from night shift to day shift, the sleep period shifts backward by 8 hours. To promote this phase shift, bright light exposure should occur 1 to 3 hours after the temperature minimum. If the person has fully accommodated to the night schedule and is sleeping from 9 AM to 4 PM with a temperature minimum at noon, bright light exposure will be most effective at 1 to 3 PM, with earlier exposure on subsequent days.

The phase-shifting properties of melatonin suggest that it may be helpful for shift workers. Daytime sedative use is usually not helpful, as it does not appear to promote adaptation to a daytime sleep schedule when administered during the first few days of night-shift work.[44] For those who do not obtain relief of symptoms with the above measures, the most effective approach is a change to a permanent day shift. In some cases, it may be possible for the clinician to assist in the development of more rational schedules, such as longer periods spent on individual shifts.

DELAYED SLEEP PHASE SYNDROME

In DSPS, first described by Weitzman and colleagues,[45] the sleep phase occurs later than desired, leading to difficulty falling asleep and difficulty waking up at the desired times. The disorder is most common in adolescents and young adults, and some degree of DSPS affects a substantial proportion of high school and college students.[46,47]

Clinical Features

In patients with DSPS, difficulty awakening and getting up in the morning is the usual presenting complaint. Sleepiness and sluggishness during morning hours is another common complaint: Some children and adolescents fall asleep during morning classes, whereas others seek help because their parents find it difficult or impossible to awaken them in time for school. Common reasons for seeking evaluation are chronic absenteeism and tardiness, which require some patients to repeat grades. Leaving home for college often exacerbates DSPS, as parents are no longer available to enforce awakenings. Difficulty falling asleep at night is almost always present, but it is usually not the main complaint.

Some patients with DSPS get up at conventional hours to attend school or work during the week, but they sleep until late morning or into the afternoon on weekends. Such persons usually have normal uninterrupted sleep on weekends, and some may sleep for 12 to 18 hours to catch up on sleep lost during the week. Others adopt a full-time schedule of going to bed between 2 and 6 AM and getting up between 11 AM and 4 PM. College students with DSPS may choose to sign up only for classes that meet in the afternoon or evening.

During highly scheduled experiences away from home, such as summer camps, sleep problems are often much better. On the other hand, adults with DSPS often wind up in occupations, such as computer programming, that do not always require regular hours. Chronic dysthymia or depression are common in adults with DSPS, as is illustrated in the following case.

CASE HISTORY: Case 12–2

A 42-year-old part-time librarian complained of daytime sleepiness, difficulty getting out of bed in the morning, and difficulty falling asleep at night. She had always been a "night owl" and had difficulty getting up for school as early as seventh grade. In high school, if she got up too late to take the bus, her parents drove her to school. In college, she rarely attended her morning classes and studied best between 8 PM and 2 AM. At the time of initial evaluation, she went to bed at about 1:30 AM and fell asleep between 2 and 3 AM. She awakened at 6 AM and sometimes remained awake for 1 to 2 hours before falling back asleep until noon. After work, she sometimes took a nap from 7 PM to 9 PM. She described her naps as "luxuries" that helped her to relax.

She had been hospitalized twice for depression, once at age 24 and once at age 41. She took fluoxetine 20 mg/d and saw a psychiatrist regularly. In the past, she had taken diphenhydramine 100 mg and diazepam, up to 40 mg, to help with sleep onset. Neurological examination was normal. A polysomnogram, beginning at 11:50 PM, revealed a sleep latency of 72 minutes and was otherwise normal. During an MSLT, she did not fall asleep.

My diagnosis was DSPS and chronic affective disorder. She was advised to maintain a sleep log, to sleep only between 1 and 9 AM, to put her alarm clock on the other side of the room so that she would have to get out of bed to turn it off, and to obtain bright light exposure on awakening. At follow-up several weeks later, she had complied with none of these recommendations. During a vacation in Europe, she felt that her sleep habits had ruined her trip because she was unable to get out of bed to shop and visit museums as she had planned.

Over the next 4 years, she continued to have sleep difficulties despite trials of triazolam and several antidepressants. During periods of little or no depression, she went to bed at about 1 AM and usually remained awake until 4 AM. Her alarm went off at 10 AM, and although she sometimes got up at that time, she often remained in bed until noon. During periods of more significant depression, her sleep was more irregular and she often took late-afternoon or evening naps. Despite repeated attempts, she was unable to comply with programs of regular sleep hours and bright light exposure.

Psychobiologic Basis

The cause or causes of DSPS are not well understood. Biologic bases are favored by some: Weitzman and colleagues[45] proposed that impaired phase-resetting capability of the internal pacemaker in response to external stimuli is the basis for the syndrome. According to this theory, a few days of delayed sleep onset and delayed awakening, such as during a weekend or a short vacation, cause a phase delay, and the impaired phase-shifting capacity prevents the person from resuming a normal schedule. Subsequent weekends exacerbate the problem as the subject sleeps later and later.

The model proposed by Weitzman and colleagues was based on studies performed in the 1960s and 1970s; these suggested that the endogenous pacemaker had a period of about 25 hours and that it was reset each day to maintain entrainment. Current evidence suggests that the pacemaker's endogenous rhythm is closer to 24.2 hours and that the clock continuously speeds up or slows down depending on the timing and intensity of light exposure.[48] To provide a biologic explanation for DSPS using the current model, the phase-response curve to light would have to be asymmetrical (Fig. 12–1). A second biologic theory suggests that lengthening of the intrinsic circadian period during puberty leads to less sleepiness in the evening and promotes later bedtimes and later wake-up times.

Although both theories are plausible, little solid evidence currently exists to support either, and stable entrainment during camp or other structured living situations suggests that an asymmetrical phase-shift response or an increased intrinsic period, if present, is not sufficient to induce DSPS.

On the other hand, substantial evidence suggests that depression, other psychological factors, and social influences play a major role in DSPS. Up to 75% of patients with DSPS are depressed at the time of evaluation or have a history of depression.[49] In some patients with poor social adaptation, DSPS probably provides sec-

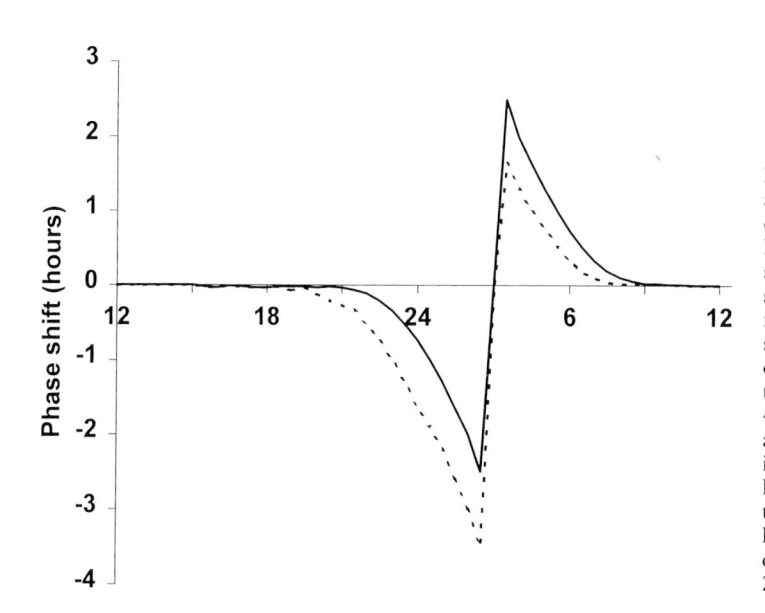

Figure 12–1. Hypothetical alteration of the phase-shifting effects of light that could account for delayed sleep phase syndrome. The solid line represents the normal shift of phase of the circadian pacemaker induced by light, which is strongly dependent on the timing of light exposure as well as its intensity and duration (see Figure 4–3). The dotted line represents an asymmetrical response to light in which phase delays induced by light in the evening are greater than phase advances induced by light in the morning. (Normal curve adapted from Takahashi and Zatz,[48] p 1106, with permission.)

ondary gain as a means to reduce or avoid social interaction. In addition, bedtimes and wake-up times can become focal points for family discord related to adolescent rebellion. Unemployed or disabled persons are also at risk for DSPS because of the lack of a regular work schedule.

The increased disparity between sleep need and actual sleep time that occurs with puberty may account, in part, for the association of DSPS with adolescence.[50] Despite the increased sleepiness that accompanies the disparity, peer pressure and greater independence concerning bedtime lead to later bedtimes, particularly on weekends. As a result, many adolescents phase-delay the internal clock by up to 1 to 2 hours each weekend day.[50,51]

The importance of psychosocial factors does not mean that biologic ones do not play a role. DSPS does not occur in all adolescents with depression, perhaps because of varying biologic vulnerability to phase shifts.

Diagnosis

Delayed sleep onset may occur during the manic phase of bipolar (manic-depressive) disorder and during an acute psychotic episode in schizophrenics. However, manic patients usually have short sleep periods and arise at the same time or earlier than usual. Delayed sleep onset in adult patients with major psychiatric disorders is usually transient and varies in severity with the underlying disorder.

Diagnosis can usually be made from the history, with confirmation by a sleep log. The key features of the history are difficult awakening in the morning and late bedtimes or difficulty with sleep onset. Although excessive daytime sleepiness may suggest obstructive sleep apnea or narcolepsy, delayed sleep onset is not typical of these disorders. If sleep apnea or periodic leg movements are suspected as a complicating factor, polysomnography is indicated, preferably on or close to a schedule that matches the subject's usual one. Long sleep latency with otherwise normal sleep are typical findings on polysomnography.

Management

The treatment of DSPS is theoretically simple because it usually involves a phase shift of less than 8 hours. With an appropriately timed sleep schedule and bright light exposure, the phase shift can be accomplished in a few days.[52] The greater challenge for patient, family, and physician is to develop a plan that allows the patient to maintain the new schedule. No matter what program is used initially to achieve the phase shift, a regular sleep-wake schedule must be maintained 7 days per week thereafter. In most cases, naps should be prohibited during the phase shift interval and thereafter because they may make sleep initiation at night more difficult and thereby trigger a new episode of delayed sleep phase. If the syndrome is a manifestation of adolescent depression or severe family conflict, as it often is, scheduling should be combined with psychiatric treatment.

To achieve the phase shift, I usually suggest progressive phase advances of bed and arising times by 15 to 30 minutes every night or two, combined with bright light exposure. For a patient with a sleep period from 5 AM to 1 PM, the temperature minimum is likely to be at about 9 AM, and bright light will be most effective initially between 10 AM and 2 PM. As the sleep period advances, the bright light exposure should be shifted to earlier hours. With this program, a phase shift of 4 to 6 hours can be accomplished in 2 to 3 weeks. Melatonin can be used to assist with phase shifts in countries in which it is available as a regulated pharmaceutical; in one study of patients with DSPS, 5 mg daily for 4 weeks significantly advanced the sleep period.[53]

If a rapid phase shift is desired, two more drastic and socially disruptive protocols can be used. With the first, patients are prescribed a schedule of 5 to 7 days of progressive, daily, 3-hour delays of bedtime and arising time until the desired bedtimes and wake-up times are reached.[54] With the second, patients are asked not to sleep at all on a Friday night, followed by a 90-minute advance of the sleep period the following week. The sleepless Friday night and 90-minute advances are then repeated each week until the desired schedule is obtained.[55]

Outcome is a function of the cause of the problem and its severity, the family's support and understanding of the treatment, and most importantly, the patient's motivation to overcome the problem. The outlook is best for a patient who understands the rationale for maintaining a regular schedule, is motivated to succeed, and is supported by other family members. On the other hand, no program will succeed if the patient is not motivated to achieve and maintain the new schedule. Unfortunately, as demonstrated in Case 12-2 (see above), many patients continue with symptoms for years despite a variety of approaches to treatment.

In designing a treatment program, the clinician must determine the degree to which family members or others will assist the patient to achieve the desired schedule. Need for support from family or friends must be balanced against the value of putting control of the program in the hands of the patient. Well-meaning parents may attempt to help by managing aspects of the schedule, but such efforts may lead the patient to sabotage the program.

When an initial treatment program fails, the reasons for the failure sometimes provide new insight into family dynamics, which may, in turn, lead to more effective intervention. Teen-age patients who resist even an attempt to maintain a regular sleep schedule may have underlying depression or severe family problems, or both.[56] If motivational problems appear to be the primary cause, it is often useful to try an initial approach of maintaining a regular schedule with only a minimal phase shift in the desired direction. If patients are unwilling or unable to comply with such recommendations, further exploration of motivational issues is indicated. In some cases, the failure of the patient to make a serious attempt at maintaining the proposed schedule helps physician, family, and patient realize that inadequate motivation is the major reason for treatment failure and that it must be addressed before additional scheduling attempts are undertaken.

ADVANCED SLEEP PHASE SYNDROME

Advanced sleep phase syndrome, much less common than DSPS, is associated with complaints of evening sleepiness and early morning awakening. Bedtime is usually earlier than desired, and morning awakening occurs between 1 and 3 AM.[57-59] Most cases occur in older persons.

The cause is unknown. Although endogenous changes in pacemaker function (e.g., defect in the capacity for phase delay, overly active phase advance capacity, reduction in the period of the internal clock) could cause the syndrome, social factors such as isolation, loss of scheduled activities, and reduced exposure to sunlight may play a more significant role.

The occurrence of marked mood disturbance and vegetative symptoms helps to distinguish advanced sleep phase syndrome from major depression, the most important other diagnosis to consider. Although the diagnosis is usually apparent from the history, sleep logs can be used to confirm the clinical impression. A sleep study is rarely required; if performed, it is likely to show a short sleep latency and premature end of sleep.

The syndrome can be treated with daily 15- to 30-minute phase advances of the sleep schedule combined with evening bright light exposure, analogous to the procedure used for DSPS, followed by maintenance of the new schedule 7 days per week.

IRREGULAR SLEEP-WAKE PATTERN DISORDER

Clinical Features

In this disorder, which occurs most commonly in patients with severe retardation or dementia, sleep episodes occur at irregular intervals and for varying lengths of time. In mild cases, sleep timing maintains some regularity, with sleep more likely to occur during the 2 to 6 AM block than at other times.[60] In severe cases, however, sleep patterns are completely arrhythmic, and sleep periods, usually of 2 to 4 hours each, are scattered over the 24-hour period, with marked variability in the timing and duration of sleep from one day to the next.[61,62]

Patients or their caretakers may report difficulty falling asleep, difficulty remaining asleep, daytime somnolence, or combinations of these symptoms. Nocturnal wandering and agitation and the use of sedatives or physical restraints to control such behavior may complicate the clinical picture, particularly in institutionalized patients.

Biologic Basis

The usual cause of the syndrome is degeneration or destruction of the suprachiasmatic nucleus, its outflow pathways, or the brain systems mediating sleep and wakefulness. Reduced responsiveness to social cues, resulting from brain injury or degeneration, and reduced retinal input to the suprachiasmatic nucleus, resulting from cataracts or other ocular pathology, may also contribute to irregular sleep-wake patterns. As a result of pacemaker dysfunction, core temperature rhythms may be reduced in amplitude or completely absent.[63]

In patients with irregular sleep-wake pattern disorder and normal or near-normal cognitive functioning, social isolation or avoidance, irregular activity patterns and eating schedules, and increased time in bed, usually because of medical illness, trauma, or psychiatric illness, may be contributing causes (Fig. 12–2).[64]

Diagnosis

The differential diagnosis may include inadequate sleep hygiene, non–24-hour sleep-wake syndrome (see below), DSPS, and medication toxicity. Sleep-wake logs, the most useful diagnostic tool, usually show polyphasic sleep patterns with no consistent times for the onset or termination of sleep. If the patient is unable to complete a sleep log, actigraphy and visual assessment of state may also be used.[65] Sleep-wake patterns should be assessed for at least 1 month, as shorter logs may not reveal the long cycles characteristic of the non–24-hour sleep-wake syndrome (see below). Polysomnography may be helpful if sleep apnea is a consideration. An MSLT, if performed, may show marked variation in sleep latency from one nap test to the next.

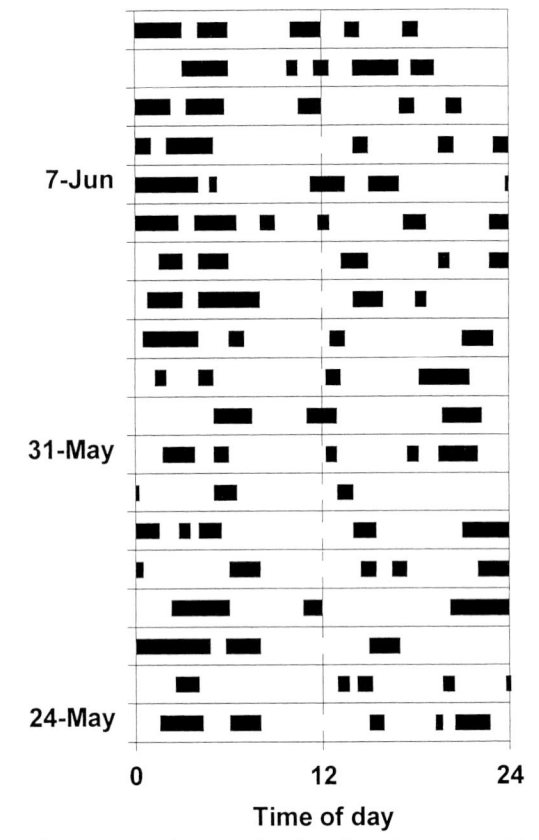

Figure 12–2. Nineteen-day sleep log of a 63-year-old woman with multiple sclerosis, schizoaffective disorder, and an irregular sleep-wake schedule. The patient had few daytime activities, ate meals irregularly, and made no attempt to maintain a regular sleep schedule. Periods of sleep are indicated in black. Although sleep was fragmented, some circadian rhythmicity was present, as the patient was usually awake between 8 AM and noon, and the longest periods of sleep occurred between 6 PM and 8 AM.

Management

For institutionalized patients, education of caretakers, increased daytime activity and social interaction, and increased daytime light exposure may be helpful. Scheduled afternoon naps and bright light therapy may be useful in some cases; however, hypnotics, neuroleptics, and stimulants are usually of little value and may cause added problems of nocturnal confusion and daytime somnolence.

Although cognitively intact patients may benefit from a regular sleep-wake schedule, morning bright light, and consistent meal times, those who obtain secondary gain from the irregular schedule are often resistant to change.

NON–24-HOUR SLEEP-WAKE SYNDROME

Clinical Features

With the non–24-hour sleep-wake syndrome, also called the hypernycthemeral syndrome (from the Greek word *nycthemeron*—a night and a day), the internal pacemaker is not entrained to the 24-hour day-night cycle and runs based on a shorter or longer period.

Patients complain of daytime sleepiness and nighttime insomnia that fluctuate in severity depending on the phase angle between the pacemaker and the day-night cycle. When the phase difference is relatively small, patients may be able to maintain sleep hours that are close to the societal norm. As the phase difference increases, the sleep period becomes progressively later for several nights. When the internal clock becomes maximally out of phase with the environment, sleep is irregular for several days and nights, and symptoms are most prominent with difficulty falling asleep in the evening, problems waking up in the morning, and drowsiness during the day. As the phase declines, some patients have a prolonged period of wakefulness lasting up to 40 to 50 hours, followed by sleep period of 12 to 20 hours that terminates with a return to a relatively normal schedule and a small phase difference. The following is an illustrative case.

CASE HISTORY: Case 12–3

A 42-year-old blind woman complained of a long history of sleep problems. Her vision began to deteriorate at age 4 as a result of chronic uveitis, and by age 17, she was unable to perceive light. Sleep problems, characterized by periods of insomnia mixed with periods of relatively normal sleep, began at the time of visual loss. During periods of normal sleep, which typically lasted about 2 weeks, she went to bed at 9 PM and awakened at 7 AM. During periods of insomnia, which usually lasted 1 to 3 weeks, she went to bed about 9 PM, fell into a "deep" sleep until about midnight or 1 AM, and then was awake for several hours. She then fell back asleep at about 5 or 6 AM and slept for 1 to 5 hours. During periods of insomnia, she sometimes took a 30- to 60-minute nap in the morning, from which she awakened feeling exhausted.

Examination was normal except for complete blindness. She was advised to begin light therapy with 2500 lux exposure from 5:30 to 7:30 AM, to avoid daytime naps, and to go to bed at 9:30 PM and get up at 5:30 AM each day. With this program, her sleep improved, and the long periods of wakefulness during the middle of the night no longer occurred. She still sometimes awakened at 4:30 AM and was unable to return to sleep.

Biologic Basis

Congenital or acquired pregeniculate blindness with loss of retinal input to the suprachiasmatic nucleus is the most common cause of non-24-hour sleep-wake syndrome.[66–68] The cause may be optic or retinal pathology, interruption of the retinohypothalamic tract, lack of responsiveness of the suprachiasmatic nucleus to transmitted impulses, or failure of the suprachiasmatic nucleus to entrain sleep-wake rhythms.

In the absence of photic input, the internal clock runs with a rhythm that is almost always longer than 24 hours (see Chap. 4). The cycle length is usually about 24.2 to 25 hours, although it may be as long as 27 hours. As a result, the internal clock moves in and out of phase with the environmental clock. For patients with cycle lengths of 25 hours, the cyclic fluctuation has a period of about 24 days.

In free-running conditions, the length of time a person is asleep is a function of the duration of prior wakefulness and the phase of the temperature rhythm at the time of sleep onset. As a result, the timing and duration of sleep vary as the temperature rhythm moves in and out of phase with the desired sleep-wake schedule.[69,70]

Most patients with non–24-hour sleep-wake syndrome are blind. Although the prevalence of the syndrome in blind persons is unknown, most people with total blindness have chronic sleep-wake complaints, and the often cyclical nature of their symptoms is consistent with a non–24-hour sleep-wake rhythm.[72,73] Furthermore, circadian rhythms of melatonin and cortisol secretion are commonly free-running in blind persons.[74–76]

Patients with lesions affecting the optic chiasm (e.g., pituitary adenoma) may present with this syndrome even when visual loss is incomplete, presumably because the lesions interrupt the retinohypothalamic tract but spare some portion of the optic pathways.[71] Postchiasmal blindness does not lead to this syndrome because input to the suprachiasmatic nucleus via the retinohypothalamic tract remains intact.

Non–24-hour sleep-wake syndrome also occurs in some persons without ocular pathology. For example, it may contribute to *midwinter insomnia syndrome* in persons who live at extreme latitudes. The insomnia and associated daytime sleepiness typically last through the dark period of winter, when the sun remains below the horizon. In Tromsö, Norway, at latitude 69 degrees, the dark period lasts about 2 months, and approximately 18% of women and 9% of men have insomnia only during the dark period.[77] Presumably, limited or erratic light exposure during these months leads to disentrainment. In the occasional patient with non–24-hour sleep-wake syndrome who neither has ocular pathology nor lives in an extreme latitude, social isolation and lack of sufficient illumination at appropriate times may account for the disorder. Such persons often have schizoid personalities or other personality traits that lead to reduced social interaction.

Diagnosis

The diagnosis should be considered in any blind person with sleep complaints, especially if the sleep problems fluctuate in severity. In patients who are not blind, the differential diagnosis often includes inadequate sleep hygiene and DSPS. In institutionalized patients who are mentally retarded and blind, an irregular sleep-wake pattern may also be a consideration. A sleep-wake log maintained for several weeks is the key to diagnosis because it reveals the cyclical variation of disturbed sleep. Magnetic resonance imaging of the suprasellar region is indicated when the syndrome develops in persons with normal vision; this procedure is not necessary for persons with long-standing blindness.

Management

Management of this disorder is difficult. Although strict 24-hour scheduling of sleep and waking activities is probably the best approach, many patients who try to maintain the schedule continue to have cyclical variations in sleep disturbance.[67,78,79] High-intensity bright light is of value for those blind persons who have an intact circadian variation of melatonin secretion, presumably because photic information in such patients reaches the suprachiasmatic nucleus despite the absence of light perception.[80] Pharmaceutical-quality melatonin (0.5 to 7.5 mg/d, if available) may be effective in some patients.[79,81,82] Sedatives and stimulants appear to have little value, probably in large part because these medications tend to be most effective when taken during the appropriate phase of the rhythm of endogenous pacemaker output.

SUMMARY

The circadian pacemaker exerts a powerful influence on sleep and wakefulness, and when it is not in phase with the desired sleep-wake schedule, sleep-wake disturbances can occur. Jet lag occurs in most persons who travel east or west by jet across three or more time zones; symptoms can be reduced by altering the home sleep-wake schedule in the direction of the anticipated destination schedule during the days preceding travel.

Sleep-wake disturbances in shift workers are caused by sudden changes in the sleep schedule, by different sleep times on workdays and days off, and by a reduction in the total amount of sleep. Although shift workers sleep best if they sleep at the same time of day 7 days per week, whether they are working or not, most night workers prefer to be awake during the day and asleep at night during nights off.

In DSPS, the sleep phase occurs later than desired, leading to difficulty falling asleep and difficulty waking up at the desired times. The disorder is most common in adolescents and is often complicated by depression or family problems. In advanced sleep phase syndrome, much less common than DSPS, evening sleepiness and early morning awakening are the usual complaints.

In irregular sleep-wake pattern disorder, which occurs most commonly in patients with severe retardation or dementia, sleep episodes occur at irregular intervals and for varying lengths of time. The usual cause is degeneration or destruction of the suprachiasmatic nucleus.

In the non–24-hour sleep-wake syndrome, the internal pacemaker is not entrained to the 24-hour day but has either a shorter or longer period. Patients complain of daytime sleepiness and nighttime insomnia that fluctuate in severity depending on the phase angle between the pacemaker and the day-night cycle. Congenital or acquired pregeniculate blindness with loss of retinal input to the suprachiasmatic nucleus is the most common cause.

Most circadian sleep-wake disorders are diagnosed on the basis of the patient's history and sleep-wake log. Many patients reduce or even eliminate their symptoms by maintaining a regimen of properly timed bright light exposure as well as a regular sleep schedule.

REFERENCES

1. American Sleep Disorders Association: International Classification of Sleep Disorders, Revised: Diagnostic and Coding Manual. American Sleep Disorders Association, Rochester, MN, 1997.
2. Leger, D, et al: The prevalence of jet-lag among 507 traveling businessman. Sleep Res 22:409, 1993.
3. Monk, TH, et al: Inducing jet lag in the laboratory: Patterns of adjustment to an acute shift in routine. Aviat Space Environ Med 59:703, 1988.
4. Aschoff, J, et al: Re-entrainment of circadian rhythms after phase shifts of the zeitgeber. Chronobiologia 2:23, 1975.
5. Seidel, WF, et al: Jet lag after eastbound and westound flights. Sleep Res 16:639, 1987.
6. Lauber, JK, and Kayten, PJ: Keynote Address: Circadian dysrhythmia, and fatigue in transportation system accidents. Sleep 11:503, 1988.
7. Graeber, RC: Sleep and wakefulness in international aircrews. Aviat Space Environ Med 57:B63, 1986.
8. Gander, PH, et al: Simulating the action of zeitgebers on a coupled two-oscillator model of the human circadian system. Am J Physiol 16:R418, 1984.
9. Gander, PH, et al: Modeling the action of zeitgebers on the human circadian system: Comparisons of simulations and data. Am J Physiol 16:R427, 1984.
10. Aschoff, J: Circadian rhythms within and outside their ranges of entrainment. In Assenmacher, J, and Farner, DS (eds): Environmental Endocrinology. Springer-Verlag, Berlin, 1978, pp 172–181.
11. Graeber, RC: Jet lag and sleep disruption. In Kryger, MH, et al (eds): Principles and Practice of Sleep Medicine, ed 2. WB Saunders, Philadelphia, 1994, pp 463–470.
12. Borbély, AA: Sleep homeostasis and models of sleep regulation. In Kryger, MH, et al (eds): Principles and Practice of Sleep Medicine, ed 2. WB Saunders, Philadelphia, 1994, pp 309–320.
13. Dijk, DJ, and Czeisler, CA: Paradoxical timing of the circadian rhythm of sleep propensity serves to consolidate sleep and wakefulness in humans. Neurosci Lett 166:63, 1994.
14. Dijk, DJ, and Czeisler, CA: Contribution of the circadian pacemaker and the sleep homeostat to sleep propensity, sleep structure, electroencephalographic slow waves, and sleep spindle activity in humans. J Neurosci 15:3526, 1995.
15. Mills, JN, et al: Adaptation to abrupt time shifts of the oscillator(s) controlling human circadian rhythms. J Physiol 285:455, 1978.
16. Winget, CM, et al: A review of human physiological and performance changes associated with desynchronosis of biological rhythms. Aviat Space Environ Med 55:1085, 1984.
17. Klein, KE, and Wegmann, HM: Significance of circadian rhythms in aerospace operations. NATO AGARDograph number 247. Neuilly sur Seine, France NATO AGARD, 1980.
18. Morris, HH, and Estes, ML: Traveler's amnesia, transient global amnesia secondary to triazolam. JAMA 258:945, 1987.
19. Arendt, J, et al: Alleviation of "jet-lag" by melatonin: Preliminary results of a controlled double-blind trial. Br Med J 292:1170, 1986.
20. Petrie, K, et al: Effect of melatonin on jet lag after long haul flights. Br Med J 298:705, 1989.
21. Gordon, NP, et al: The prevalence and health impact of shiftwork. Am J Public Health 76:1225, 1986.
22. Kogi, K: Introduction to the problems of shiftwork. In Folkard, S, and Monk, TH (eds): Hours

of Work: Temporal Factors in Work Scheduling. John Wiley & Sons, Chichester, 1985, pp 165–197.
23. Moore-Ede, MC, and Richardson, GS: Medical implications of shift-work. Ann Rev Med 36:607, 1985.
24. Rutenfranz, J, et al: Shiftwork research issues. In Johnson, LC, et al (eds): The Twenty-Four Hour Workday: Proceedings of a Symposium on Variations in Work-Sleep Schedules. DHHS (NIOSH) publication no. 81-127, U.S. Department of Health and Human Services, Washington, D.C., 1981, pp 221–259.
25. Folkard, S, et al: Night shift paralysis. Experientia 40:510, 1984.
26. Wagner, DR: Disorders of the circadian sleep-wake cycle. Neurol Clin North Am 14:651, 1996.
27. Weitzman, ED, et al: Acute reversal of the sleep-waking cycle in man: Effect on sleep stage patterns. Arch Neurol 22:483, 1970.
28. Webb, WB, et al: Effect on sleep of a sleep period time displacement. Aerospace Med 42:152, 1971.
29. Kripke, DF, et al: Sleep of night workers: EEG recordings. Psychophysiology 7:377, 1971.
30. Monk, TH: Shift work. In Kryger, MH, et al (eds): Principles and Practice of Sleep Medicine, ed 2. WB Saunders, Philadelphia, 1994, pp 471–476.
31. Knauth, P, and Rutenfranz, J: Experimental shift work studies of permanent night and rapidly rotating shift systems: 1. Circadian rhythm of body temperature and reentrainment at shift change. Int Arch Occup Environ Health 37:125, 1976.
32. Strogatz, SH, et al: Circadian pacemaker interferes with sleep onset at specific time each day. Am J Physiol 253:R1, 1987.
33. Knauth, P, et al: Duration of sleep depending on the type of shift work. Int Arch Occup Environ Health 46:167, 1980.
34. Walsh, JK, et al: The EEG sleep of night and rotating shift workers. In Johnson, LC, et al (eds): The Twenty-Four Hour Workday: Proceedings of a Symposium on Variations in Work-Sleep Schedules. DHHS (NIOSH) publication no. 81-27. U.S. Department of Health and Human Services, Washington, D.C., 1981, pp 451–465.
35. Segawa, K, et al: Peptic ulcer is prevalent among shift workers. Digest Dis Sci 32:449, 1987.
36. Angerspach, D, et al: A retrospective cohort study comparing complaints and diseases in day and shift workers. Int Arch Occup Environ Health 45:127, 1980.
37. Koller, M: Health risks related to shift work: An example of time-contingent effects of long-term stress. Int Arch Occup Environ Health 53:59, 1983.
38. Scott, AJ, and Ladou, J: Shiftwork: Effects on sleep and health with recommendations for medical surveillance and screening. Occup Med 5:273, 1990.
39. Dawson, D, and Campbell, SS: Timed exposure to bright light improves sleep and alertness during simulated night shifts. Sleep 14:511, 1991.
40. Eastman, CL: High-intensity light for circadian adaptation to a 12-h shift of the sleep schedule. Am J Physiol 262:R428, 1992.
41. Eastman, CL: Dark goggles during daylight and high intensity light during night shifts for circadian adaptation to simulated night shift work. Sleep Res 22:402, 1993.
42. Czeisler, CA, et al: Exposure to bright light and darkness to treat physiologic maladaptation to night work. N Engl J Med 322:1253, 1990.
43. Czeisler, CA, et al: Rotating shift work schedules which disrupt sleep are improved by applying circadian principles. Science 217:460, 1982.
44. Walsh, JK, et al: Acute administration of triazolam for the daytime sleep of shift workers. Sleep 7: 223, 1984.
45. Weitzman, ED, et al: Delayed sleep phase syndrome, a chronobiological disorder with sleep-onset insomnia. Arch Gen Psychiatry 38:737, 1981.
46. Pelayo, RP, et al: Prevalence of delayed sleep phase syndrome among adolescents. Sleep Res 17:392, 1988.
47. Carskadon, MA, et al: Pubertal changes in daytime sleepiness. Sleep 2:453, 1980.
48. Takahashi, JS, and Zatz, M: Regulation of circadian rhythmicity. Science 217:1104, 1982.
49. Regestein, QR, and Monk, TH: Delayed sleep phase syndrome: A review of its clinical aspects. Am J Psychiatry 152:602, 1995.
50. Carskadon, MA, and Dement, WC: Sleepiness in the normal adolescent. In Guilleminault, C (ed): Sleep and Its Disorders in Children. Raven Press, New York, 1987, pp 53–66.
51. Carskadon, MA, et al: Sleep, sleepiness and mood in college-bound high school seniors. Sleep Res 20:175, 1991.
52. Rosenthal, NE, et al: Phase-shifting effects of bright morning light as treatment for delayed sleep phase syndrome. Sleep 13:354, 1990.
53. Dahlitz, M, et al: Delayed sleep phase syndrome response to melatonin. Lancet 337:1121, 1991.
54. Czeisler, CA, et al: Chronotherapy: Resetting the circadian clock of patients with delayed sleep phase insomnia. Sleep 4:1, 1981.
55. Thorpy, MJ, et al: Delayed sleep phase syndrome in adolescents. J Adolesc Health Care 9:22, 1988.
56. Ferber, R, and Boyle, P: Phase shift dyssomnia in early childhood. Sleep Res 12:242, 1983.
57. Kamei, R, et al: Advanced-sleep phase syndrome studied in a time isolation facility. Chronobiologia 6:115, 1979.
58. Moldofsky, H, et al: Treatment of a case of advanced sleep phase syndrome by phase advance chronotherapy. Sleep 9:61, 1986.
59. Billiard, M, et al: A case of advanced-sleep phase syndrome. Sleep Res 22:109, 1993.
60. Okawa, M, et al: Disorders of circadian body-temperature rhythm in severely brain-damaged patients. Chronobiol Int 1:67, 1984.
61. Allen, SR, et al: Seventy-two hour polygraphic and behavioral recordings of wakefulness and sleep in a hospital geriatric unit: Comparison between demented and nondemented patients. Sleep 10:143, 1987.
62. Okawa, M, et al: Disturbance of circadian rhythms in severely brain-damaged patients correlated with CT findings. J Neurol 233:274, 1986.

63. Okawa, M, et al: Sleep-wake rhythm disorder and phototherapy in elderly patients with dementia. Biol Psychiatry (suppl)29:161S, 1991.
64. Hauri, PJ: Sleep Disorders. Scope Publication (Upjohn), Kalamazoo, 1992, pp 33–34.
65. Bliwise, DL, et al: Systematic 24-hr behavioral observations of sleep and wakefulness in a skilled-care nursing facility. Psychol Aging 5:16, 1990.
66. Miles, LE, et al: Blind man living in normal society has circadian rhythms of 24.9 hours. Science 198:421, 1973.
67. Okawa, M, et al: Four congenitally blind children with circadian sleep-wake rhythm disorder. Sleep 10:101, 1987.
68. Weber, AL, et al: Human non–24-hour sleep-wake cycles in an everyday environment. Sleep 2:347, 1980.
69. Czeisler, CA, et al: Human sleep: Its duration and organization depend on its circadian phase. Science 210:1264, 1980.
70. Czeisler, CA, et al: Timing of REM sleep is coupled to the circadian rhythm of body temperature in man. Sleep 2:329, 1980.
71. Kokkoris, CP, et al: Long term ambulatory monitoring in a subject with a hyper-nychthemeral sleep-wake cycle disturbance. Sleep 1:177, 1978.
72. Miles, LE, and Wilson, MA: High incidence of cyclic sleep-wake disorders in the blind. Sleep Res 6:192, 1977.
73. Martens, H, et al: Sleep/wake distribution in blind subjects with and without sleep complaints. Sleep Res 19:398, 1990.
74. Lewy, AJ, and Newsome, DA: Different types of melatonin circadian secretory rhythms in some blind subjects. J Clin Endocrinol Metab 56:1103, 1983.
75. Orth, DN, et al: Free-running plasma cortisol rhythm in a blind human subject. Clin Endocrinol 10:603, 1979.
76. Sack, RL, et al: Circadian rhythm abnormalities in totally blind people; incidence and clinical significance. J Clin Endocrinol Metab 75:127, 1992.
77. Husby, R, and Lingjaerde, O: Prevalence of reported sleeplessness in northern Norway in relation to sex, age and season. Acta Psychiatr Scand 81:542, 1990.
78. Klein, T, et al: Circadian sleep regulation in the absence of light perception: Chronic non-24-hour circadian rhythm sleep disorder in a blind man with a regular 24-hour sleep-wake schedule. Sleep 16:333, 1993.
79. Palm, L, et al: Correction of a non–24-hour sleep/wake cycle by melatonin in a blind retarded boy. Ann Neurol 29:336, 1991.
80. Czeisler, CA, et al: Suppression of melatonin secretion in some blind patients by exposure to bright light. N Engl J Med 332:6, 1995.
81. Arendt, J, et al: Synchronisation of a disturbed sleep-wake cycle in a blind man by melatonin treatment. Lancet 1:772, 1988.
82. Sack, RL, et al: Melatonin administration to blind people: Phase advances and entrainment. J Biol Rhythms 6:249, 1991.

Chapter 13

OBSTRUCTIVE SLEEP APNEA SYNDROME

CLINICAL FEATURES
Snoring, Snorting, Gasping, and Choking Sounds During Sleep
Restless Sleep
Sleepiness and Tiredness
Other Symptoms
Symptoms in Infants and Children
Examination
EPIDEMIOLOGY
Comorbid Conditions
BIOLOGIC BASIS
Narrowing of the Upper Airway
Causes of Airway Narrowing
Snoring
Causes of Arousals from Apneas and Hypopneas
Effects of Alcohol and Sedative Medications
Causes of Sleepiness
SYSTEMIC EFFECTS OF OBSTRUCTIVE SLEEP APNEA
Cognitive and Psychosocial Function
Hypoxemia
Sympathetic Activity
Systemic and Pulmonary Arterial Pressure
Cardiac Function
Intracranial Hemodynamics and Cerebral Perfusion
Renal Function
Endocrine Effects
COMPLICATIONS
Accidents
Vascular Morbidity and Mortality
Anesthetic Complications
Complications in Infants and Children
DIFFERENTIAL DIAGNOSIS
DIAGNOSTIC EVALUATION
MANAGEMENT
Continuous Positive Airway Pressure

Surgery
Dental Appliances
Weight Loss
Medications
Oxygen

Peto: Falstaff!—Fast asleep behind the arras, and snorting like a horse.
Prince Henry: Hark, how hard he fetches breath.
WM. SHAKESPEARE, HENRY IV PART 1

The recognition that obstructive sleep apnea (OSA) is a common disorder with disabling symptoms and substantial associated morbidity and mortality has had a profound impact on the field of sleep medicine. For most sleep specialists, sleep apnea is the principal diagnosis in at least two-thirds of their patients, and without sleep apnea, sleep medicine as a field probably would not exist. Few comparable situations exist in other fields of medicine. The development of mechanical ventilators, which led to the development of critical care medicine as a discipline, was a technological development, rather than a clinical one. The effect of tobacco smoking on pulmonary medicine is perhaps similar but the increase in lung cancer and chronic obstructive pulmonary disease has occurred over nearly a century, whereas OSA was almost unrecognized as recently as 1964.

Although it was not understood until the 1960s that the symptoms of OSA are caused by abnormal breathing during sleep, the clinical picture was identified much earlier. Charles Dickens'[1] description in *The Posthumous Papers of the Pickwick Club* of Joe—a fat, sleepy boy who snored loudly—is the most famous literary example, although Shakespeare's description of Falstaff, cited above, suggests that he, too, recognized the essentials of the syndrome. In the 19th century, Caton[2] and Lamacq[3] described excessive sleepiness accompanied by intermittent upper airway obstruction during sleep. In 1936, Kerr and Lagen[4] observed the cardiovascular and pulmonary consequences, and the term "pickwickian," introduced by Osler[5] to refer to obese and hypersomnolent patients, was widely used after the report of Burwell and colleagues[6] in 1956. The term is now more commonly used to describe patients with OSA combined with obesity-hypoventilation syndrome (see Chap. 14).

In 1965, Gastaut and associates[7] and Jung and Kuhlo[8] reported electrophysiologic studies of patients with "pickwickian syndrome" and demonstrated that these patients had repetitive apneas during sleep. Subsequently, studies by Lugaresi and associates[9] and Guilleminault and co-workers[10,11] outlined the essential clinical framework of the disorder, including its occurrence in children as well as adults. In the 1960s and 1970s, the only effective treatment was tracheostomy, which was often unacceptable to patients. Then, in the 1980s, the application of devices that permitted administration of continuous positive airway pressure by a mask worn over the nose (nasal CPAP) revolutionized the treatment of OSA.[12] Epidemiological studies in the 1980s and 1990s showed that OSA causes daytime symptoms in millions of men, women, and children throughout the world and suggested that OSA contributes substantially to morbidity and mortality associated with vascular diseases and occupational and motor vehicle accidents.

CLINICAL FEATURES

The usual presenting symptoms of OSA are snoring and excessive daytime sleepiness (Table 13–1). Some patients seek help because of episodes of falling asleep while

Table 13–1. **Symptoms of Obstructive Sleep Apnea in Adults**

Proportion Reporting the Symptom Often or Almost Always	Controls (Gender Matched)	OBSTRUCTIVE SLEEP APNEA		
		Mild (AHI 10–20)	Moderate (AHI 20–50)	Severe (AHI > 50)
Number of subjects	143	313	412	339
Major problem with sleepiness	10%	54%	51%	61%
Major problem with fatigue even when not sleepy	7%	57%	49%	54%
Major problem at work or school from fatigue or sleepiness	1.4%	35%	26%	30%
Major problem with driving because of sleepiness	0%	23%	18%	21%
Snoring	13%	77%	83%	88%
Loud snoring	4%	70%	76%	85%
Apnea observed by others	3%	44%	54%	66%
Restless sleep	16%	60%	58%	65%

AHI = apnea-hypopnea index.
Data from patients seen at the University of Michigan Sleep Disorders Center.

driving or during other potentially hazardous activities, as in the following case.

CASE HISTORY: Case 13-1

A 40-year-old obese man fell asleep while driving his car and hit a tree. Injuries included bilateral sixth-nerve palsies, multiple fractures, and spinal cord trauma at the midthoracic level with complete and permanent paraplegia. The accident occurred during the afternoon on his way home from work; he had driven for 1 hour and was 3 miles from his home. He reported a long history of snoring and daytime naps, and he frequently pulled off the road because of drowsiness. A polysomnogram showed severe OSA.

Some patients complain principally of interrupted sleep or a sense that sleep is not restful. Other patients have no complaints and present because the bed partner or parent has noted restless sleep, gasping respirations, or periods of apnea during sleep. Still others are referred because of a suspicion that OSA is contributing to cardiac arrhythmias, systemic hypertension, pulmonary hypertension, polycythemia, or impotence.

Snoring, Snorting, Gasping, and Choking Sounds During Sleep

Alice . . . checked herself in some alarm, at hearing something that sounded to her like the puffing of a large steam-engine in the wood near them, though she feared it was more likely to be a wild beast. "Are there any lions or tigers about here?" she asked timidly. "It's only the Red King snoring," said Tweedledee.

LEWIS CARROLL, THROUGH THE LOOKING GLASS

Loud snoring, which occurs in more than 80% of persons with OSA, usually begins years before the onset of sleep apnea. The volume of snoring can exceed 80 dB and may disturb the bed partner, family members, and even neighbors. On the other hand, some bed partners sleep so soundly that they are not bothered by snoring. Some patients report that they cannot go camping with others and must have separate hotel rooms on business trips because of the obnoxious quality of their snoring.

Most patients are unaware of their snoring and many minimize its significance; it is only by questioning the bed partner that an accurate description is obtained. Although snorts and choking sounds are usually noted only by the parent or bed partner, some patients report that they can hear their own snoring or that snoring awakens them, and others awaken feeling short of breath.

Although snoring is the norm, patients with OSA may snore softly or not at all. Patients with neuromuscular disease affecting muscles of the chest wall and diaphragm may not be able to generate sufficient inspiratory force to cause vibration of soft tissues. Other patients, such as those who have undergone uvulopalatopharyngoplasty, may not have sufficient floppy airway tissue to vibrate, despite airway narrowing.

Although a complaint of loud snoring that cannot be verified with objective assessment is occasionally an indication of marital discord, some patients who snore loudly at home do not snore during laboratory sleep studies. In such cases, snoring may be intermittent and related to depth of sleep, sleeping position, alcohol use, allergies, or other factors that affect nasal and upper airway patency and diameter.

Restless Sleep

Restless sleep is characteristic of sleep apnea. The restlessness is caused by arousals at the end of apneas, which may be accompanied by jerks, twitches, limb movements, and repeated changes in body position in apparent attempts to find a sleeping position that permits airway patency. Restlessness is generally the spouse's complaint, since the patient usually has little or no recollection of the arousals.

With severe OSA, apneas occur continually throughout the night and in all body positions, interrupted only by brief arousals accompanied by snorting and choking sounds. Some patients may sleep in a semirecumbent position in a chair or on a couch; some sleep sitting up on the side of the bed or even standing and leaning against a wall.

Despite the loud snoring and pronounced restlessness, patients typically awaken with the impression that sleep has been continuous or interrupted only by trips to the bathroom, yet has not produced the anticipated sense of restoration and refreshment. Patients who describe repeated awakenings and complain mainly of insomnia usually have mild or moderate, rather than severe, OSA.

Sleepiness and Tiredness

Sleepiness and tiredness are characteristic symptoms of OSA. Some patients describe themselves as "slow starters" who feel mentally dull, groggy, confused, or disoriented for minutes to an hour or more each morning. Others report difficulty remaining awake during the day. When daytime sleepiness is mild, it is usually most apparent during boring, sedentary situations in the afternoon or evening; difficulty remaining awake after lunch or while driving or reading are common complaints. With more severe sleepiness, patients may fall asleep during conversation, while talking on the telephone, or during sexual intercourse. As in narcolepsy (see Chap. 10), sleepiness caused by sleep apnea may be accompanied by episodes of automatic behavior together with amnesia, during which complex activities are performed, such as driving for miles with no recollection, writing illegibly, or performing nonsense operations at a computer terminal. Complaints of memory difficulties and inability to concentrate may accompany the concerns about sleepiness.

Sleepiness often develops gradually over a number of years as patients gain weight and apnea worsens. Perhaps because of the gradual onset of OSA, patients often minimize or deny having difficulties with alertness. For example, a patient may report that sleepiness rarely occurs or that it is no worse than in colleagues or peers, while the spouse reveals that the patient has unplanned sleep episodes on a daily basis. Daytime sleepiness may worsen with seasonal allergies, alcohol use, or other factors that exacerbate OSA. Although many patients with OSA have daytime sleepiness, this is not a universal symptom and its absence does not rule out the disorder.

Some patients with OSA complain of fatigue, tiredness, and lack of energy in addition to or instead of daytime sleepiness. Such complaints are less specific, as they may occur in association with a variety of medical, neurological, and psychiatric disorders.

Other Symptoms

Other symptoms that occur in some persons with OSA include mood disturbances, impotence or reduced libido, irritability or personality change, morning headaches, dry mouth or sore throat, frequent nighttime urination or nocturnal enuresis, heavy nighttime sweating, and inability to sleep in a supine position. Symptoms of gastroesophageal reflux disease (GERD) may be present if the repeated episodes of negative intrathoracic pressure associated with apneas lead to passage of gastric contents through the lower esophageal sphincter.

Symptoms in Infants and Children

In infants, symptoms of OSA may include snoring, inability to sleep supine because of choking or gasping, and failure to thrive.[13] Failure to thrive in infancy and slow growth in childhood may be caused partly by increased energy expenditure during sleep as a result of the increased work of breathing.[14]

In children, snoring and restless sleep are common symptoms; enuresis and nocturnal sweating may also occur, and some parents report that their child assumes unusual sleeping postures. Many children with OSA have labored, noisy breathing at night with few or no complete apneas, probably accounting for the less frequent reports of witnessed apneas. Daytime symptoms in children may include poor attention, decreased performance in school, and hyperactivity. Although complaints of difficulty staying awake during the day are less common in children than in adults, this difference does not necessarily mean that children with OSA are less sleepy. Sleepiness in children is probably more often manifested as impaired attention or hyperactivity than as daytime sleep episodes. Cognitive deficits are attributable largely to problems with attention and concentration, which are most likely caused by nocturnal sleep disturbance.

Examination

Increased weight and high blood pressure are common in persons with OSA. Increased neck circumference with fatty infiltration of the neck suggests that excess retropharyngeal adipose tissue may be contributing to upper airway obstruction during sleep. Examination of the head and neck occasionally reveals major craniofacial anomalies that predispose persons to OSA, including mandibular hypoplasia and craniosynostosis. More commonly, despite the absence of major craniofacial abnormalities, there may be one or more of the following craniofacial features: small jaw, overbite or overjet, long face, and triangular chin. Allergic rhinitis and associated edema of the nasal mucosa or deviation of the nasal septum after a trauma may compromise nasal patency and contribute to nasal obstruction and chronic mouth breathing, which are common in persons with OSA.

Oropharyngeal findings may include an elongated soft palate and uvula; a high, arched palate; edema and erythema of oropharyngeal soft tissues; redundant pharyngeal mucosa; an enlarged tongue; or enlarged tonsillar tissue. The position of the teeth during occlusion is also informative. In normal (class I) occlusion, the first lower molar is approximately 2 mm anterior to the upper molar. In class II occlusion, which is associated with retrognathia and increased risk of OSA, the first lower molar is immediately under or even posterior to the upper molar during occlusion.

Patients with severe OSA may have findings consistent with right ventricular enlargement and right-sided heart failure, including a cardiac thrust at the left sternal border, a prominent second heart sound, and ankle edema.

EPIDEMIOLOGY

Most persons with OSA snore, and thus the prevalence of habitual snoring, which can be assessed quickly in large numbers of persons, can be used as a starting point for studies of the prevalence of OSA. Habitual snoring appears to be more common among North Americans and Australians than among Europeans; for example, in Wisconsin, habitual snoring was reported in 28% of middle-aged women and 44% of middle-aged men, whereas in a survey from three European countries (Iceland, Sweden, and Belgium), habitual snoring was reported in 5% of men and 2% to 3% of women.[15,16] In an Australian study, the prevalence of snoring, including habitual and occasional snorers, in a middle-aged to elderly population was 66%.[17] Differences among these populations probably reflect, in part, differences in the prevalence of obesity.

The prevalence of OSA among middle-aged men ranges from 1% to 4% in various populations (Table 13–2); estimates for middle-aged women range from 1.2% to 2.5%.[16-25] The prevalence is higher in African Americans than in white Americans, independent of the effects related to obesity.[26] In adults, the prevalence of habitual snoring and OSA increases with advancing age, at least in part because of the tendency to gain weight during middle age.[27,28] On the other hand, snoring may remit in as many as 35%, especially in those over age 65.[29]

Habitual snoring also occurs in 6% to 12% of children.[30-32] The prevalence of sleep apnea in Icelandic children was estimated to be at least 2.9%, and in a study of British 4- to 5-year-olds, approximately 1% had symptoms attributable to snoring or sleep apnea.[33,34] Although a high proportion of adults with OSA are men, the gender ratio in children is close to 1:1.

Comorbid Conditions

OSA is associated with four common conditions: obesity, hypertension, GERD, and cigarette smoking. In obese patients, 40% of men and a somewhat smaller propor-

Table 13–2. **Selected Epidemiological Studies of Sleep Apnea in Adults**

Population	Criteria*	Proportion (%)
U.S. middle-aged men[16]	AHI > 5 and EDS	4%
U.S. middle-aged women[16]	AHI > 5 and EDS	2%
Australian middle-aged men[17]	AHI > 15	≥ 5.7%
Australian middle-aged women[17]	AHI > 15	≥ 1.2%
Australian men[18]	AHI > 5 and EDS	≥ 3%
Italian men[19]	AHI > 10	2.7%
Finnish middle-aged men[20]	ODI > 10	1%
Israeli men[22]	AI > 5	≥ 1%
Icelandic middle-aged women[24]	Snoring + EDS + AHI ≥ 30	≥ 2.5%
U.S. older adults (age > 65 years)[25]	AI > 5	24%

*AI = apnea index; AHI = apnea-hypopnea index; ODI = oxygen desaturation index; EDS = excessive daytime sleepiness.

tion of women have OSA.[35–37] The prevalence of OSA is greater than 25% in patients with hypertension, and in untreated severe OSA, the prevalence of hypertension may be as high as 50%.[38–40] Although several studies suggest that OSA is an independent risk factor for hypertension, several others suggest that the association is related to confounding variables, particularly obesity.[41,42] Although the association with GERD is probably mainly a result of the association of both OSA and GERD with obesity, the negative intrathoracic pressure generated during attempts to breathe against a closed airway may contribute to reflux symptoms in some patients. Sleep-disordered breathing is also associated with smoking, probably in part because of upper airway irritation and edema associated with smoking and because nocturnal hypoxemia leads to an increased likelihood of hypopneas.[43]

BIOLOGIC BASIS

Narrowing of the Upper Airway

Narrowing and closure of the upper airway during sleep is the basis for OSA. The pharynx, which lacks supporting cartilage and bone because of its additional functions related to swallowing and phonation, is the site of occlusion, which may occur at the level of the velopharynx, the oropharynx, or the hypopharynx. In some patients, the occlusion is at more than one level, or at different sites during different stages of sleep, owing to the differential activity of muscles involved in maintaining airway patency. Although other portions of the upper airway are less compliant and therefore less collapsible than the pharynx, obstruction can occur at the epiglottis, the larynx, or even the trachea in patients with tracheomalacia.

Once airway closure occurs, the patient attempts to breathe without success. Breathing resumes only when the airway reopens, usually as a result of an arousal (Fig. 13–1). With repeated episodes of apnea, the frequent arousals disrupt sleep and contribute to sleepiness and other daytime symptoms.

In patients with OSA, the pharyngeal airway is usually small, even during wakefulness, and on average is more compliant than in normal subjects.[44–47] The narrowing occurs predominantly in the lateral dimension; the cross-sectional area of the oropharynx is often less than 50 mm^2 in adults with OSA compared to about 110 mm^2 or more in control subjects.[48,49] Because of the narrower airway, OSA patients tend to have increased inspiratory resistance while awake and greater waking activity of the genioglossus and tensor palatini muscles, particularly in the supine

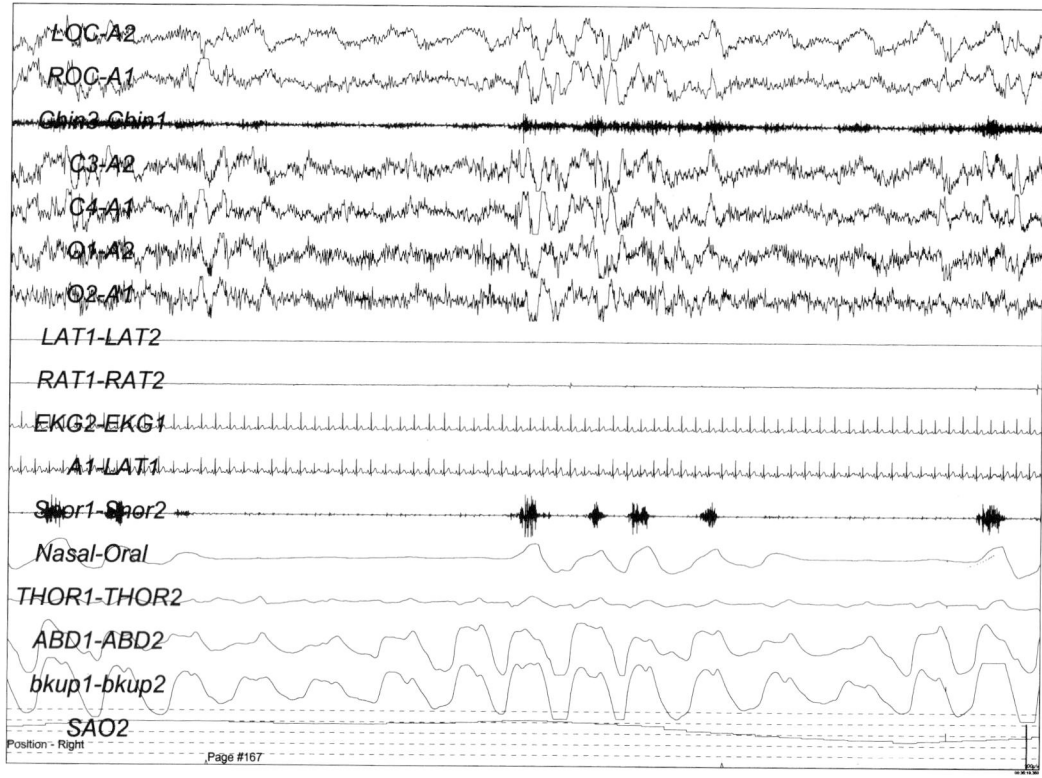

Figure 13–1. One-minute segment of a polysomnogram with two episodes of OSA during light NREM sleep. Snoring accompanies the four breaths during the arousal that separates the two episodes. The EEG electrodes (A1, A2, C3, C4, O1, O2) were placed in accordance with the International 10-20 system. Other abbreviations: LOC, left outer canthus; ROC, right outer canthus; Chin3–Chin1, submental EMG; LAT1–LAT2, surface EMG electrodes over the left anterior tibialis; RAT1–RAT2, surface EMG electrodes over the right anterior tibialis; EKG2–EKG1, electrocardiogram; A1–LAT1, electrocardiogram; Snor1–Snor2, snoring sound; Nasal-Oral, nasal-oral airflow; THOR1–THOR2, thoracic wall motion; ABD1–ABD2, abdominal wall motion; bkup1–bkup2, backup respiratory effort (thoracoabdominal wall motion); SAO2, pulse oximetry by finger probe.

position, in order to maintain airway patency.[50–52]

With sleep, pharyngeal muscle relaxation leads to further narrowing of the pharyngeal airway. In control subjects, the decrement is fairly small; in OSA patients, who have augmented activity of pharyngeal muscles during wakefulness, the decrement is substantial.[51] The airway narrowing that occurs with sleep as a result of loss of pharyngeal dilator activity may lead to several types of respiratory events: (1) obstructive apnea with airway occlusion that begins during expiration; (2) obstructive apnea with airway occlusion that begins during inspiration; (3) central apnea; (4) mixed apnea with a combination of central and obstructive components; (5) obstructive hypopnea with partial airway collapse during inspiration, leading to reduced airflow; (6) increased upper airway resistance with airway narrowing, but with compensatory increased inspiratory effort that minimizes changes in airflow and ventilation. Each of these events is discussed in the following sections.

OBSTRUCTIVE APNEAS BEGINNING DURING EXPIRATION

With episodes of obstructive apnea that begin during expiration, pharyngeal muscle relaxation associated with sleep is enough to allow airway closure. In such cases, pharyngeal obstruction begins during the late phase of expiration as the positive pressure developed during early expiration decreases and is no longer sufficient to maintain airway patency.

Once airway occlusion occurs, surface tension between the opposed surfaces of the pharyngeal mucosa adds to the negative pressure generated within the thorax to prevent reopening of the upper airway during inspiratory efforts. With succeeding breaths, inspiratory effort increases, but in the absence of effective activation of upper airway dilator muscles, the increased negative airway pressure simply distorts the lower pharynx and does not produce airflow. Downward movement of the diaphragm produces paradoxical inward movement of the chest wall during inspiratory attempts because of the absence of airflow (Fig. 13-2). Ultimately, the airway opens only after an arousal or brief awakening leads to increased activity of pharyngeal dilators. The patient then takes a few deep breaths and returns to sleep, whereupon the cycle repeats itself.

OBSTRUCTIVE APNEAS BEGINNING DURING INSPIRATION

With episodes of obstructive apnea that begin during inspiration, passive relaxation of the pharynx produces narrowing, but not complete closure, during expiration. The narrowing leads to increased airflow resistance and increased airflow velocity during subsequent inspiration, which, by Bernoulli's principle, leads to increased inward (negative) pressure on the upper airway. As inspiratory muscle activity increases to compensate for the increased airway re-

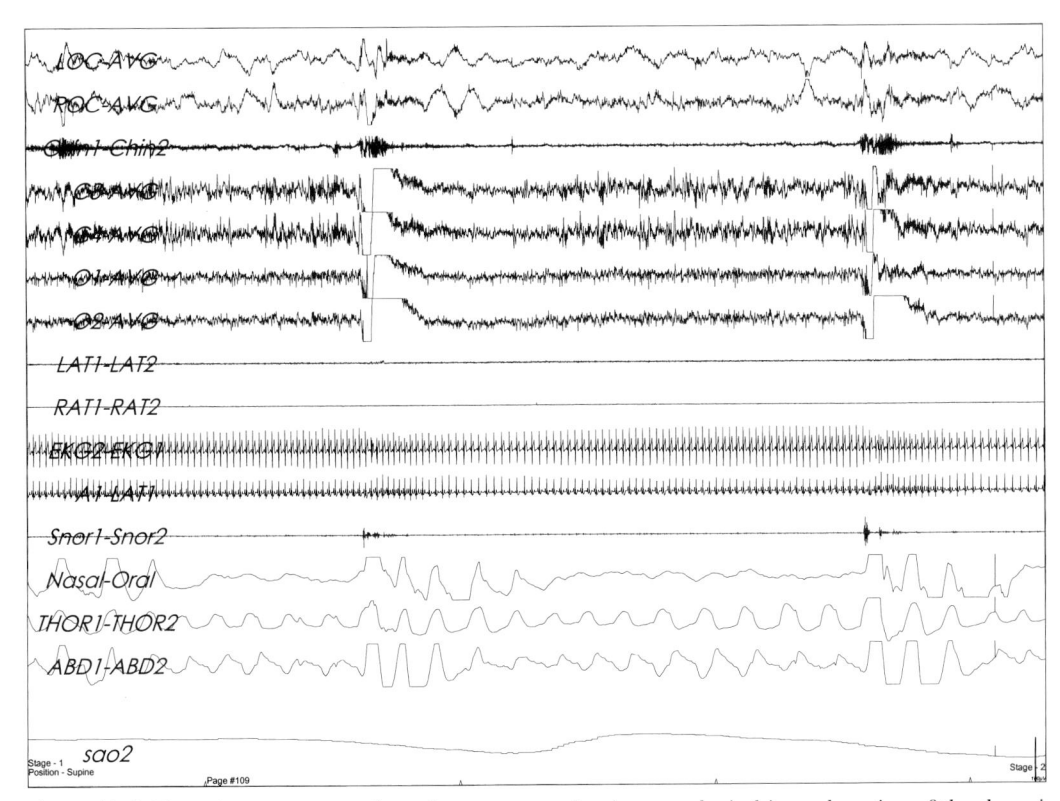

Figure 13-2. Two-minute segment of a polysomnogram showing paradoxical inward motion of the thoracic wall during obstructed breathing. The Nasal-Oral airflow channel demonstrates two episodes of nearly complete airflow cessation accompanied by breathing effort. The expansion of the thorax (THOR1-THOR2) and abdomen (ABD1-ABD2) are out of phase during obstructed breathing and in phase during arousals. Such paradoxical movements are not always recorded during obstructive apneas, depending on the placement of the sensors of wall motion. (AVG, average reference.) For other abbreviations, see Figure 13-1.

sistance, the negative pressure may exceed the closing pressure of the pharynx, resulting in occlusion (see Chap. 3).

One common pattern consists of an inward collapse of the lateral walls of the pharynx, leading to occlusion of the oropharynx, hypopharynx, or both. Another pattern consists of occlusion at the oropharyngeal level as the negative force generated by inspiratory muscles sucks the uvula downward. Continued negative pressure may then lead to caudal extension of the occluded segment into the hypopharynx.[53]

CENTRAL APNEAS CAUSED BY UPPER AIRWAY OBSTRUCTION

With central apneas caused by upper airway obstruction, complete or partial closure of the airway leads to a reflex inhibition of inspiratory efforts, probably mediated by pharyngeal afferents. The apnea terminates, usually with an arousal, at the time that inspiratory efforts resume (Fig. 13–3). Central apneas can also result from other causes (see Chap. 14).

MIXED APNEAS

With mixed apneas, reflex inhibition of inspiratory effort occurs, just as with central apneas. The period of inhibition lasts for several seconds and is then followed by gradually increasing efforts to breathe, just as with obstructive apneas (Fig. 13–4).

OBSTRUCTIVE HYPOPNEAS

With obstructive hypopneas, the negative pressure generated during inspiration,

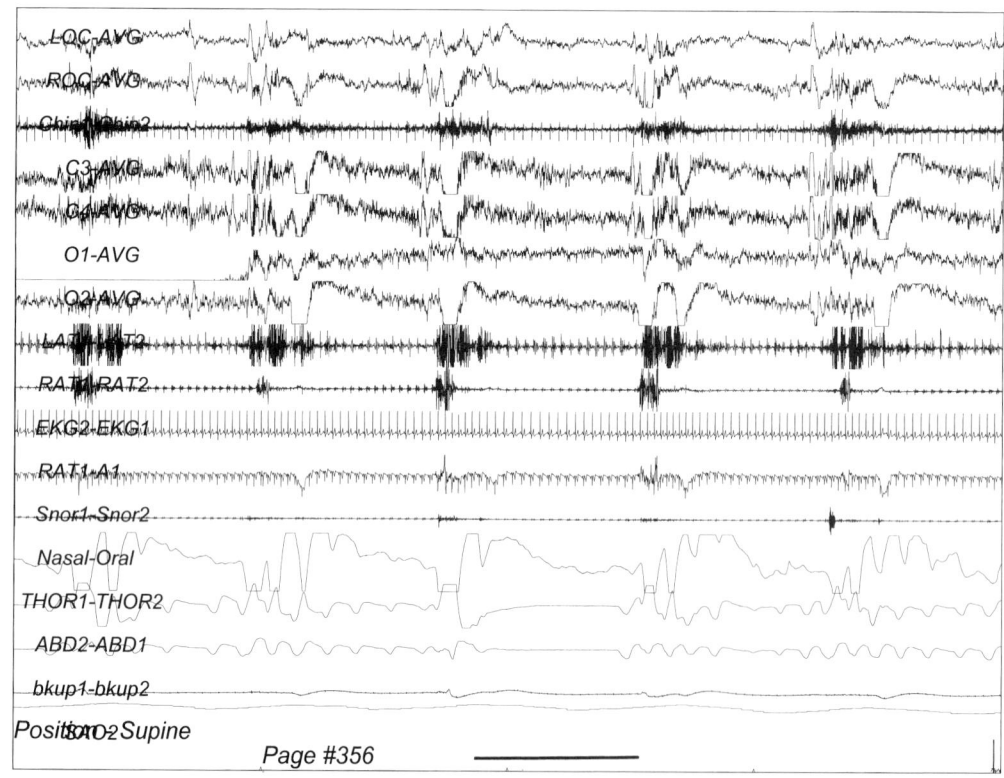

Figure 13–3. Central apnea in a patient with repetitive OSA. This 2-minute segment of a polysomnogram shows repetitive obstructive apneas and periodic leg movements in a 54-year-old man with a history of snoring, restless sleep, and daytime sleepiness. During one of the apneas (marked with an underline), no respiratory effort was apparent for more than 25 seconds. This central apnea was not accompanied by a Cheyne-Stokes pattern of breathing before or after the apnea and was most likely a result of upper airway obstruction. The apnea was treated effectively with nasal CPAP, and the patient's symptoms improved despite the persistence of periodic leg movements. For abbreviations, see Figures 13–1 and 13–2.

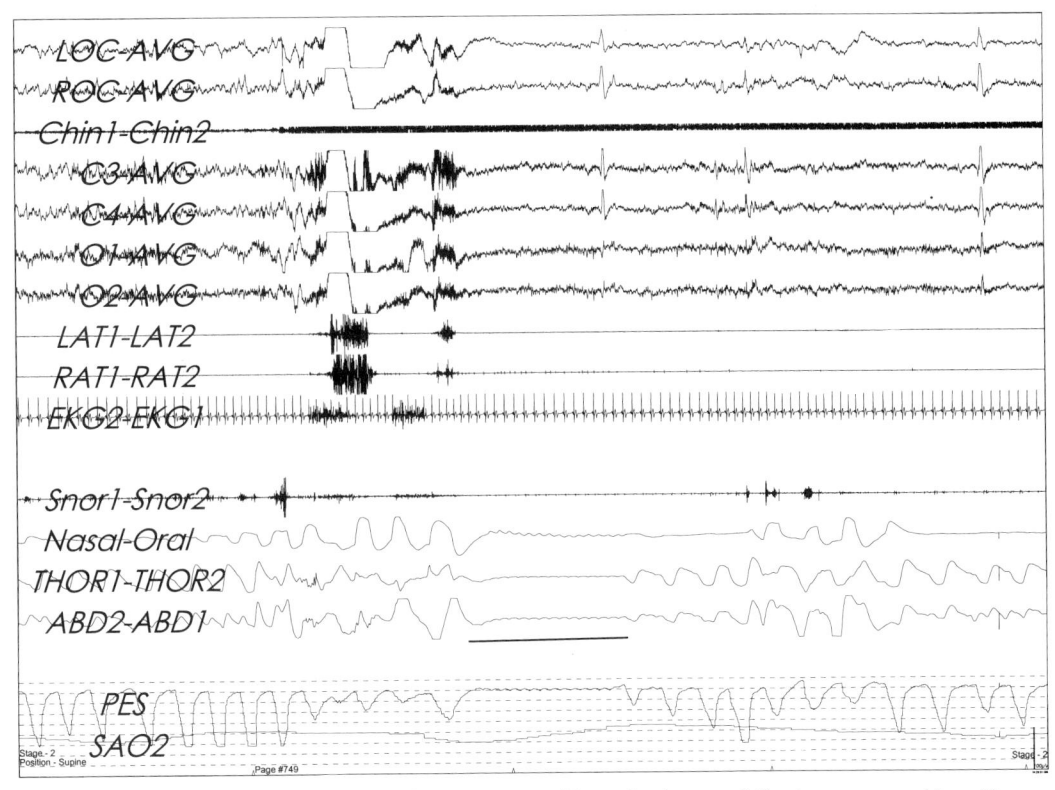

Figure 13–4. Two-minute segment of a polysomnogram with a mixed apnea following an arousal in a 43-year-old man with a history of snoring and apnea observed by his wife. During the underlined initial 15- to 20-second portion of the apnea, no respiratory effort occurs, as shown by the respiratory effort channels (THOR1–THOR2 and ABD1–ABD2) and confirmed by the intraesophageal pressure monitor (PES). During the second half of the apnea, increasing respiratory effort occurs with each attempted breath. For abbreviations, see Figures 13–1 and 13–2.

combined with high resistance to airflow, leads to partial rather than complete collapse of the airway. Airflow is impaired, and the increasing efforts to breathe through the excessively narrow airway are similar to those that occur with complete occlusion (Fig. 13–5). The hypopneas appear to have the same or nearly the same functional effect on sleep continuity as apneas. Most patients with OSA have a combination of obstructive apneas and obstructive hypopneas.

INCREASED UPPER AIRWAY RESISTANCE WITH LITTLE OR NO CHANGE IN VENTILATION

In some patients, particularly those with relatively noncompliant upper airways, high resistance to airflow is associated with compensatory increased respiratory effort that minimizes changes in tidal volume, ventilation, and oxygenation. However, the increased respiratory effort associated with airway narrowing still leads to arousals and sleep fragmentation (Fig. 13–6). The effect on sleep continuity is similar to that occurring with apneas or hypopneas, but because ventilation is preserved, hypoxemia does not occur or is minimal. When such episodes are the predominant type of respiratory event, the disorder is referred to as *upper airway resistance syndrome*.[54] The body weight of patients with upper airway resistance syndrome is often normal or near-normal. Compared to OSA, this syndrome appears to be more prevalent among women, probably because the female upper airway is less compliant and therefore more resistant to collapse. Children also tend to have less compliant upper airways and therefore appear to be susceptible to this form of sleep-disordered breathing.[55]

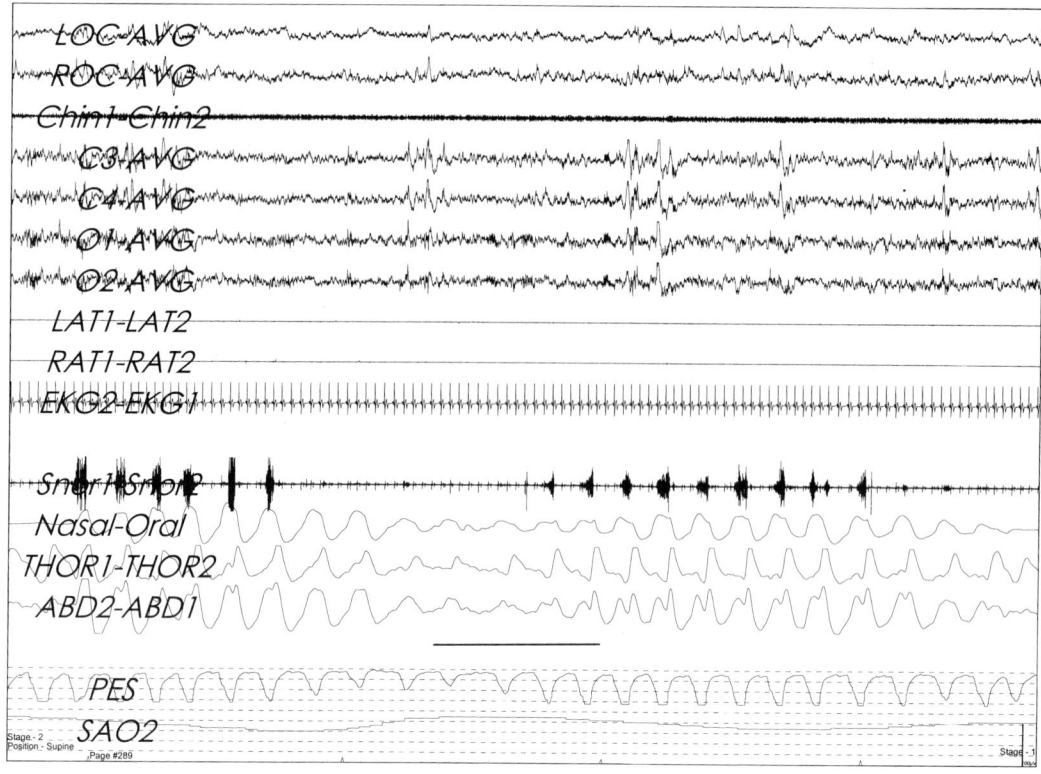

Figure 13–5. Hypopnea during NREM sleep in a 43-year-old man with snoring and daytime sleepiness. During the underlined portion of the hypopnea, there is little change in airflow (Nasal-Oral channel) despite the increasing effort to breathe, as shown in the respiratory effort channels (THOR1–THOR2, ABD1–ABD2) and the intraesophageal pressure monitor (PES). The decline in oxyhemoglobin saturation (SaO_2 channel) is delayed by several seconds because of the time required for the deoxygenated blood to reach the pulse oximetry finger probe. For abbreviations, see Figures 13–1 and 13–2.

EFFECT OF SLEEP STATE, BODY POSITION, AND SLEEP LOSS ON AIRWAY NARROWING AND APNEA

OSA is usually more severe during rapid eye movement (REM) sleep than non–rapid eye movement (NREM) sleep because the increased muscle atonia associated with REM sleep affects upper airway muscles and increases the likelihood of airway collapse. In addition, apneas are usually longer in REM sleep because the arousal threshold is higher during REM sleep compared to NREM sleep. The supine sleeping position tends to precipitate OSA or make it worse because the effects of gravity on the tongue and soft palate lead to airway narrowing (Fig. 13–7). Sleep deprivation can also exacerbate OSA, probably because the arousal threshold is elevated and the degree of muscle relaxation during sleep is greater.[56]

Causes of Airway Narrowing

Airway narrowing that predisposes persons to OSA may occur at any age. The Pierre-Robin syndrome, characterized by micrognathia, glossoptosis, and usually cleft palate, is one of the most common causes of OSA in infancy. In this syndrome, the tongue is displaced posteriorly by the small jaw and obstructs the airway during sleep. Apnea is often most apparent in the supine position, and therefore the presenting complaint is usually the infant's inability to sleep in this position. A variety of other craniofacial syndromes

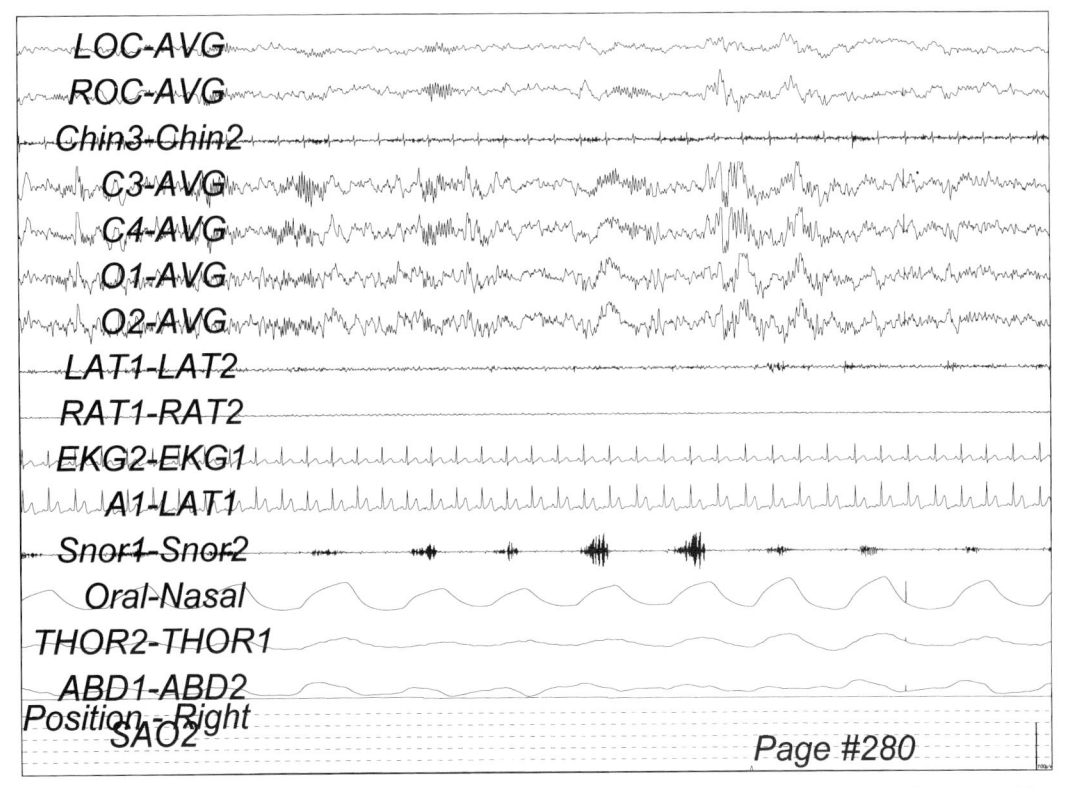

Figure 13–6. Crescendo snoring followed by a K-complex and a brief arousal in a 28-year-old woman with a history of snoring and daytime fatigue. The progressive increase in snoring, accompanied by slight diminution in the amplitude of the signals from the thorax and abdomen, is consistent with increased upper airway resistance. For abbreviations, see Figures 13–1 and 13–2.

and structural lesions can also cause a narrow upper airway and OSA in infancy (Table 13–3).

In children, enlargement of the tonsils and obesity are common causes of airway narrowing. Conditions associated with lymphoid hyperplasia, such as sickle cell anemia, are associated with an increased risk of OSA because they increase the likelihood of tonsillar enlargement. Neuromuscular syndromes causing pharyngeal hypotonia increase the risk of OSA, as do genetic and congenital syndromes associated with midface hypoplasia, a small nasopharynx, or micrognathia.[57] In children without major malformations, certain craniofacial features predispose them to airway narrowing and occlusion during sleep, including a small, triangular chin; retroposition of the mandible; a steep mandibular plane; a high hard palate; and a long soft palate.[55] Allergic rhinitis proba-

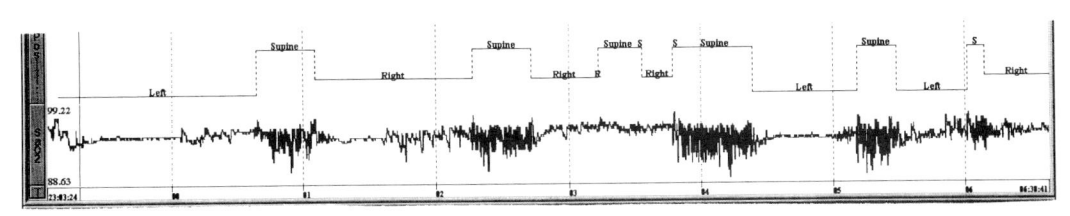

Figure 13–7. Sleep-disordered breathing occurring almost exclusively during sleep in the supine position in a 47-year-old man with complaints of snoring and daytime sleepiness. The figure shows the relation between body position and oximetry for the 7.5-hour polysomnogram. Respiratory events were mainly hypopneas.

Table 13–3. Causes of Upper Airway Narrowing and Obstructive Sleep Apnea

Mucosal edema and inflammation: Allergies, sinusitis, gastroesophageal reflux

Anatomic malformations of the jaw and pharynx: Pierre-Robin syndrome, Crouzon's syndrome, Apert's syndrome, Treacher Collins syndrome, Möbius' syndrome

Enlarged tongue, soft palate, or uvula: Down's syndrome, hypothyroidism, acromegaly

Infiltration of pharyngeal tissue: Obesity, Cushing's syndrome, Prader-Willi syndrome, mucopolysaccharidoses

Structural lesions: Enlarged tonsils and adenoids, webbed pharynx, tumors of the pharynx

Surgical correction of cleft palate or prognathism

Cranial base abnormalities: Arnold-Chiari malformation, achondroplasia, rheumatoid arthritis, Klippel-Feil syndrome

Weakness of pharyngeal and laryngeal dilator muscles: Congenital myopathies, muscular dystrophies, medullary lesions, cranial neuropathies, myasthenia gravis, motor neuron diseases, vocal cord paralysis

Dyscoordinated breathing due to altered feedback and abnormal central control: Arnold-Chiari malformation, Shy-Drager syndrome, dysautonomias, brainstem lesions

Increased airway compliance: Marfan's syndrome, laryngomalacia, tracheomalacia

---

bly exacerbates OSA in many children because of associated nasal obstruction and pharyngeal edema.

In adults, obesity with fatty infiltration of the neck and associated airway narrowing is the most common cause of OSA. Obesity also may contribute to OSA and nocturnal hypoxemia through fat deposition in and around the abdomen and rib cage that decreases thoracic compliance, increases the work of breathing, and reduces functional residual capacity. The low functional residual capacity combined with atelectasis may lead to ventilation-perfusion mismatch that increases hypoxemia. The abdominal and thoracic mass of adipose tissue increases the work of breathing, especially in the supine position. This work is increased even further during REM sleep, when the diaphragm must perform virtually all of the work of breathing. This mechanical load further reduces total body oxygen stores and leads to more severe hypoxemia during periods of apnea.

As in children, genetic and developmental factors that influence craniofacial anatomy and airway diameter are major contributors to OSA in adults.[58] A genetic effect is suggested by studies that indicate an increased risk of OSA in family members of persons with OSA.[59-61] Acquired conditions that produce a large tongue, a large soft palate, retrognathia or micrognathia, nasal obstruction, or a combination of these factors contribute to reduced airway diameter (see Table 13–3).

Enlarged palatine tonsils, a common cause of OSA in childhood, may contribute to airway narrowing. Some patients' airway narrowing has more than one cause. For example, patients with Down's syndrome and OSA may have multiple sites of occlusion caused by one or more of mandibular or midface hypoplasia, large tongue, lymphoid hyperplasia, a narrow nasopharynx, and laryngomalacia.[62]

In addition to the craniofacial malformations outlined in Table 13–3, some patients with OSA have *long face syndrome*, characterized by a narrow, high, arched hard palate; retrognathia; and a long, narrow facial appearance.[63] If nasal obstruction is present, mouth breathing during sleep leads to posterior displacement of the mandible, further narrowing of the pharynx with increased airflow resistance, and a mechanically less efficient action of the genioglossus and geniohyoid muscles. The result is a greater likelihood of airway collapse. Narrowing of the nasal passages with increased resistance to airflow may increase the amount of negative inspiratory pressure developed by the diaphragm and other inspiratory muscles, leading in turn to pharyngeal collapse.

Chronic nasal obstruction early in life, caused by allergies, enlarged adenoids, or other factors, may contribute to the development of OSA in later years.[64] In experimental models, nasal obstruction leads to mandibular deficiency, presumably as a result of the effects of altered facial muscle activity.[65] The mandibular deficiency is partially reversible if nasal obstruction is relieved at an early enough age. Edema of the pharynx caused by snoring, allergies, or upper respiratory infections also may narrow the airway and cause or exacerbate OSA.

Neurological disorders associated with pharyngeal weakness may lead to functional narrowing of the upper airway with subsequent development of OSA. Respiratory muscle incoordination may also contribute to OSA. In neurological diseases such as syringobulbia, olivopontocerebellar degeneration, multiple system atrophy, and extrinsic and intrinsic brainstem diseases, incoordination of respiratory muscles affecting the preinspiratory activation of upper airway dilators may destabilize the airway during initial inspiration (see Chaps. 18 and 19).

Although the reasons for the more frequent occurrence of OSA in men than in women are not known with certainty, differing patterns of fat deposition and responsiveness to inspiratory loading probably contribute. Partly because of the effects of testosterone, men tend to have more upper body obesity for a given body-mass index, which leads to greater fatty infiltration of upper airway structures, increased pharyngeal resistance, and increased pharyngeal compliance.[66,67] In addition, the augmentation of genioglossus activity in response to increased airway resistance is less pronounced in men than in women, which tends to make the airway less stable and more collapsible.[68] As a result of these differences, men tend to have narrower, more compliant upper airways that close at lower pressures for a given airway diameter.

Snoring

Snoring can occur during inspiration or expiration, or both.[69] The sound, which is produced by vibration of the soft tissues of the upper airway, may be heard with nasal or oral breathing, or both. The uvula is often involved, but some patients may snore as a result of vibration of the epiglottis, soft palate, or pharyngeal walls.

Snoring varies depending on the time of the night, the stage of sleep, the position of the body, the rate of airflow, and the anatomic structure of the person's nose and throat. The intensity of snoring is a function of the amount of floppy tissue that can vibrate and the velocity of airflow, which is in turn a function of the negative intrathoracic pressure and the diameter of the airway. Increased volume of snoring in a patient who has snored for many years usually indicates that the airway has become narrower.

Causes of Arousals from Apneas and Hypopneas

The precise mechanisms that lead to arousal from apnea are not defined, but the degree of inspiratory effort appears to be a major factor.[70] Although both hypercapnia and hypoxia lead to increased ventilatory drive, arousal from a given stage of sleep in a given subject tends to occur at the same level of inspiratory effort, independent of the cause of increased ventilatory drive.[70,71] Upper airway receptors probably detect the changes in airway pressure; if these receptors are anesthetized, the duration of apneas increases and the maximum negative intrathoracic pressure before arousal also increases.[72]

Apneas tend to be longer in REM sleep than in NREM sleep, partly because the increase in ventilatory drive that occurs with hypoxia and hypercapnia is less pronounced in REM sleep than in NREM sleep (see Chap. 3), which means that inspiratory effort increases more slowly and more time passes before the arousal threshold is reached. Sleepiness associated with sleep apnea itself may increase both the duration and the maximal effort during apneas, probably because the increased drive for sleep raises the arousal threshold.[73]

Effects of Alcohol and Sedative Medications

Alcohol has three major effects that tend to make OSA and its consequences worse. First, it causes airway narrowing. Second, it increases the duration of apneas. Third, it can make daytime sleepiness worse. Although benzodiazepines can cause narrowing of the upper airway, their effect on apnea duration and hypoxemia, except in patients with severe OSA, is modest with the usual hypnotic doses.[74,75]

Alcohol causes narrowing of the upper airway by reducing the tone of upper airway dilator muscles, which leads to increased resistance to breathing as well as increased airflow velocity and turbulence. Nasal and pharyngeal resistance may increase in men by more than 50% after alcohol consumption,[76] and when alcohol is consumed in the evening, airway resistance increases during sleep, particularly during the first part of the night, when blood alcohol levels are highest.[77,78] The increased airway resistance leads to louder snoring and increased inspiratory effort, which may in turn cause airway collapse and obstructive apneas in persons who do not usually have them. Obstructive apneas are more likely to develop after alcohol consumption in snorers because their pharyngeal airways tend to be smaller than those of nonsnorers. Similarly, men appear to be more susceptible to the effects of alcohol on breathing because they tend to have smaller upper airways than women.[79]

As little as 0.5 mL/kg of alcohol (about two mixed drinks) can increase sleep-disordered breathing in persons with OSA, but has little effect on breathing during sleep in normal subjects.[80–82] Alcohol also blunts chemoreceptor responses to changes in blood gases, thus prolonging the time required to arouse or awaken after an apnea occurs and, in turn, leading to increased duration of apneas and increased hypoxemia.[83] After consuming six or more alcoholic drinks, OSA patients may experience much more severe hypoxemia, especially during the first hour of sleep.[80]

Daytime alcohol use tends to make sleepiness worse in patients with OSA because of the sedative effects of alcohol. The combined effects of alcohol and sleep apnea on alertness may impair performance during activities requiring vigilance, such as driving a motor vehicle. Prior alcohol use may also interact with sleep apnea to affect driving performance. In one study, the proportion of subjects with one or more sleep-related motor vehicle accidents was 6.7% in a control group who drank fewer than two drinks per day, 15% in a group of subjects with OSA who drank fewer than two drinks per day, and 26% in those with OSA who consumed two or more alcoholic drinks per day.[84]

Causes of Sleepiness

Sleep disruption produced by repeated arousals to resume breathing appears to be the major cause of daytime sleepiness in OSA. Although experimental studies confirm that frequent nocturnal arousals lead to daytime sleepiness (see Chap. 8), and numerous studies of sleepiness in OSA have demonstrated positive correlations between arousal frequency and severity of sleepiness, the correlation is not strong, and much of the variation in sleepiness among patients with OSA is not explained by arousals. Respiratory effort, increased metabolic rate, increased sympathetic activity, or hypoxemia during apneas may also contribute to daytime sleepiness.[85,86] In addition, subcortical arousals associated with transient increases in heart rate or other physiologic changes may affect the restorative quality of sleep in patients with OSA.

SYSTEMIC EFFECTS OF OBSTRUCTIVE SLEEP APNEA

Breathing abnormalities during sleep and the associated arousals from sleep lead to a number of systemic disturbances.

Cognitive and Psychosocial Function

Complaints of poor memory or impaired attention are common in patients with OSA, and neuropsychological studies have shown impaired planning abilities, decreased ability to initiate new mental processes, and deficits of learning, memory, and attention.[87] These cognitive disturbances, which are largely a result of sleepiness, improve with treatment. Impaired work performance, increased use of sick time, and marital problems and divorce are more frequent among persons with OSA.[88]

Hypoxemia

Nocturnal hypoxemia is common with OSA, and arterial oxyhemoglobin saturation often falls below 70% in patients with severe OSA. The rate of oxygen desaturation during apneas varies from 0.1% to 1.6% per second depending on baseline oxygenation and lung oxygen stores (functional residual capacity), and the severity of hypoxemia during apnea is a function of this rate and apnea duration.[89,90] Because of the lower functional residual capacity and the longer duration of apneas during REM sleep, hypoxemia is usually more severe in REM sleep than in NREM sleep. Hypoxemia leads to increased sympathetic activity, vasoconstriction, and cardiac arrhythmias, and it may also contribute to increased erythropoietin levels. Severe, prolonged hypoxemia during sleep may contribute to the development of right heart failure in some patients with coexisting chronic obstructive pulmonary disease.[91]

Sympathetic Activity

Sympathetic activity is increased during apneas, mainly but not entirely as a result of hypoxemia, with a further increase associated with the arousal at the end of the apnea.[92–94] The increased sympathetic activity leads to vasoconstriction and contributes to increased nocturnal blood pressure and possibly to daytime hypertension.[95] Some studies suggest that daytime sympathetic activity increases with OSA and diminishes with effective treatment.[96,97]

Systemic and Pulmonary Arterial Pressure

Unlike normal persons, OSA patients generally do not have a decline in blood pressure during the night.[98] Instead, systemic blood pressure increases during repetitive apneas, peaking (sometimes to greater than 200/110 mm Hg) upon resumption of ventilation (Fig. 13–8).[99,100] Although there is considerable inter-individual variability, pulmonary artery pressure is also increased during apneas to as high as 80/50 mm Hg, mainly because of arteriolar constriction caused by hypoxemia and acidosis. Tachycardia and increased sympathetic activity, which accompany arousals, contribute to increases in systemic arterial pressure, which are most pronounced during REM sleep when hypoxemia is usually greatest.

Cardiac Function

Cardiac rate, rhythm, and output are all affected by OSA. Bradycardia, common during apneas, is a result of increased vagal tone caused by (1) fluctuations in intrathoracic pressure, and (2) stimulation of the carotid body receptors by hypoxemia. Hypoxemia in the absence of apnea induces an increase in respiration that

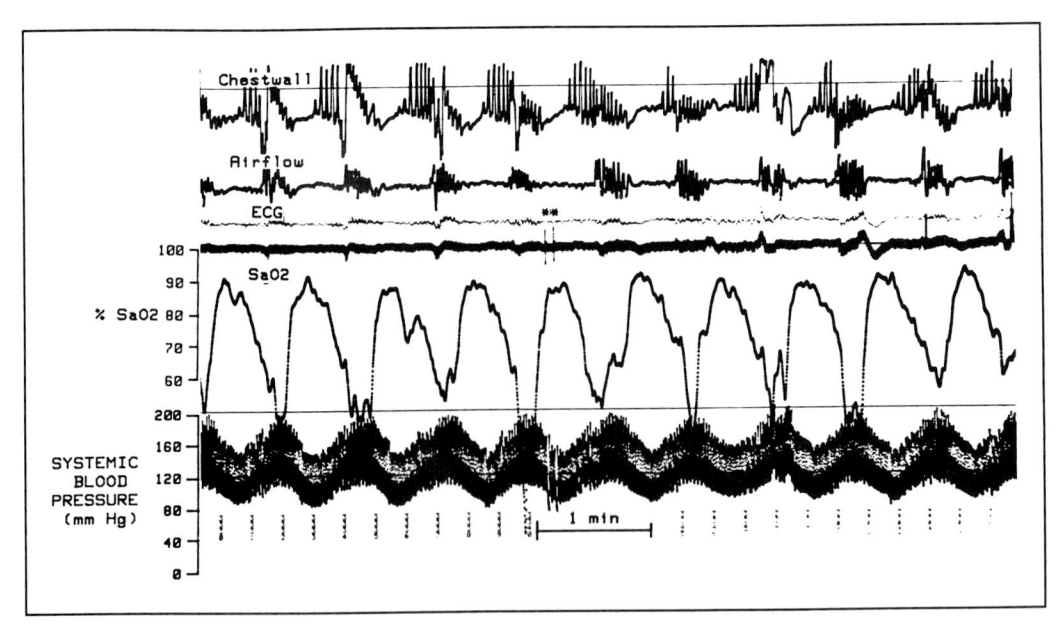

Figure 13–8. Cyclic changes in blood pressure and oxyhemoglobin saturation (SaO$_2$) associated with repetitive mixed apneas. Blood pressure rises as the SaO$_2$ falls and reaches a peak during the arousal that accompanies resumption of breathing; it then declines during the initial portion of the subsequent apnea. The fluctuation in blood pressure exceeds 40 mmHg with each apnea (From Shepard,[100] p. 1255, with permission.)

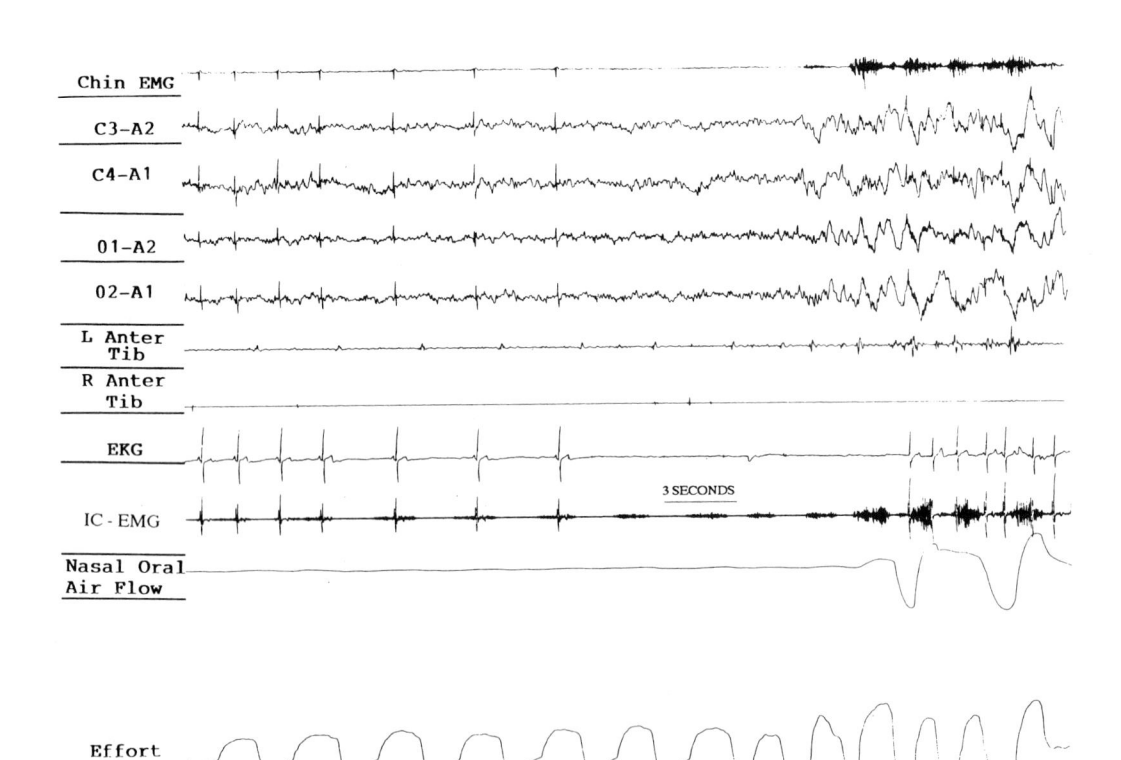

Figure 13–9. Thirteen-second asystole in a 35-year-old man with OSA and no history of heart disease. The figure shows the last 30 seconds of a 55-second episode of OSA during REM sleep. The asystolic period is preceded by progressive sinus bradycardia. A single p-wave can be seen in the electrocardiographic (EKG) tracing midway through the asystolic period. Several other 5 to 10-second asystolic periods occurred with other apneas; bradycardia and asystolic intervals were abolished by treatment with nasal CPAP. Other abbreviations: IC-EMG, intercostal muscle surface EMG; Anter Tib, surface EMG electrodes over the left (L) and right (R) anterior tibialis. For other abbreviations, see Figures 13–1 and 13–2.

causes lung distention, which then inhibits vagal activity (the Hering-Breuer reflex) and permits cardiac acceleration to occur. During apnea, however, lung distention does not occur, and consequently the increased vagal tone induced by hypoxia leads to bradycardia. The bradycardia during the apnea is followed by tachycardia during the arousal; repetitive cycles of bradycardia and tachycardia are characteristic of OSA.

Other arrhythmias that may occur include sinus arrest lasting for several seconds (Fig. 13–9), premature ventricular contractions, second- or third-degree atrioventricular block (Fig. 13–10), and ventricular tachycardia. Arrhythmias, which are particularly common during REM sleep, may be caused by hypoxemia or altered autonomic activity; however, structural lesions of the cardiac conduction system are generally absent.[101]

Cardiac stroke volume is reduced during apneas because the increased negative intrathoracic pressure associated with attempts to breathe against a closed airway leads to increased venous return to the right side of the heart.[102] The increased right ventricular volume causes a right-to-left shift of the cardiac interventricular septum that leads to reduced filling of the left ventricle. The reduced stroke volume, combined with bradycardia, leads to a 30% to 50% reduction in cardiac output.[103] In addition, the negative intrathoracic pressure leads to increased cardiac afterload, which increases myocardial oxygen requirements. The increased myocardial work during apneas may contribute, along with systemic hypertension, to the high prevalence of ventricular hypertrophy in patients with OSA. In one study, 50% of patients with apnea indices greater than 20 had left ventricular hypertrophy and 20% had right ventricular hypertrophy.[104] Right ventricular hypertrophy also may occur in children with OSA.[105]

Intracranial Hemodynamics and Cerebral Perfusion

Intracranial pressure may exceed 50 mm Hg during obstructive apneas in patients

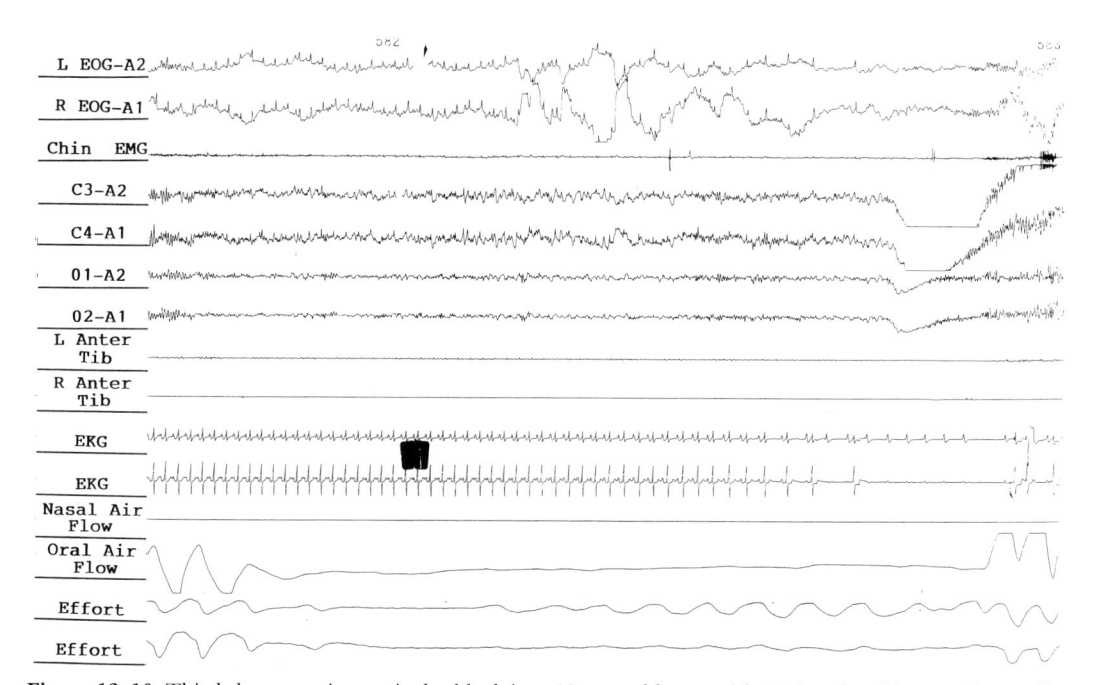

Figure 13–10. Third-degree atrioventricular block in a 46-year-old man with OSA and no history of heart disease. The heart block occurred during the last several seconds of a 32-second episode of OSA. Note the persistence of p-waves in the upper EKG channel. (Effort = thoracic [upper channel] and abdominal [lower channel] wall motion.) For other abbreviations, see Figures 13–1, 13–2 and 13–9.

with severe OSA, with associated reduction in cerebral perfusion pressure.[106] Cerebral blood flow velocity may vary substantially during obstructive apneas and subsequent arousals, potentially contributing to chronic vascular stress and stroke.[107] The altered intracranial hemodynamics may also contribute to morning headaches and cognitive impairment.

Renal Function

OSA is associated with increased release of atrial natriuretic peptide during sleep, probably because of atrial distention caused by increased negative intrathoracic pressure during inspiratory efforts. The resulting increase in urine output is probably the major cause of complaints of nocturia and enuresis.[108,109]

Endocrine Effects

Reduced nocturnal growth hormone secretion occurs in some patients with OSA, probably because of disruption of slow-wave sleep, and may contribute to the growth retardation exhibited in some children with OSA.[110,111] Dysmenorrhea and amenorrhea are complaints in some women with OSA and may improve after treatment.[112] Hyperinsulinemia occurs in some patients with OSA, probably because of coexisting obesity,[113] but this condition is perhaps related also to increased sympathetic activity. Treatment of OSA improves insulin responsiveness in some patients with OSA and non-insulin-dependent diabetes mellitus.[114]

COMPLICATIONS

Accidents

Life-threatening vehicular and occupational accidents are serious complications of OSA. Risk rates for automobile accidents are two to three times greater in OSA patients than in matched control populations, and the relative risk for single car accidents is probably even higher.[115,116] Accidents are caused by sleepiness, impaired vigilance, falling asleep while driving, and impaired judgment about the risks of driving when drowsy. On a computer-simulated driving task, some OSA patients are more impaired than normal subjects who have had several drinks of alcohol.[117,118]

Vascular Morbidity and Mortality

OSA syndrome is associated with increased risk of stroke, myocardial infarction, and cardiovascular death during sleep.[119–123] Stroke patients are three to four times more likely to have habitual snoring and sleep apnea than matched control subjects, and sleep-disordered breathing in patients who have suffered a stroke is associated with higher mortality and worse functional outcome.[119,124–126] The association of increased risk of stroke with self-reports of sleeping more than 8 hours per night also may be largely a result of OSA.[127]

Compared to normal subjects, patients with angina or myocardial infarction have at least a twofold higher prevalence of sleep-disordered breathing.[123,128] In one study, 9 of 10 patients with nocturnal angina had OSA; angina diminished after the OSA was treated.[129]

In males cardiovascular deaths and death while sleeping are more common among male snorers than nonsnorers; in one series, patients with moderate to severe untreated OSA had an estimated 37% chance of dying within 8 years compared to 4% for treated patients.[120,130] In a second series of patients with OSA, 11% of patients conservatively treated with weight-reduction diets had died after 5 years, compared to 0% of patients treated with tracheostomy.[131]

These studies clearly demonstrate an association of OSA with vascular morbidity and mortality. A question of critical importance is whether OSA is an *independent* risk factor for vascular disease. To date, this question has no definitive answer, and proponents on either side of the issue have evidence to support their position.[132,133] For example, in a cohort of snoring and nonsnoring men followed for 3 years, habitual snorers had a twofold increased risk of new ischemic heart disease or new stroke after adjusting for confounding variables.[134] In another case-control study, snoring was associated with an increased adjusted risk of stroke of 2.1, whereas snoring combined with witnessed apnea, obesity, and daytime sleepiness was associated with an adjusted stroke risk of 8.0.[135] On the other hand, a different case-control study found no increased stroke risk associated with habitual snoring.[136] In a community-based study, the adjusted odds ratio for subjects with sleep-disordered breathing versus nonsnorers was 1.5 for stroke and peripheral vascular disease and 1.4 for coronary artery disease, but the increases were not statistically significant.[42] Finally, a study from Denmark found little or no association between snoring and ischemic heart disease in men after adjusting for confounding variables.[137]

Although studies to date do not allow us to identify OSA as an unequivocal independent risk factor for stroke, I find compelling evidence that young and middle-aged persons with moderate or severe OSA are at increased risk for stroke. Most studies have found an association, even when potentially confounding variables have been assessed. In my view, the evidence that OSA is an independent risk factor for myocardial infarction and premature death is also suggestive, although less compelling, and in older persons with mild OSA, the risk may be small or absent.[138] Epidemiological studies currently in progress should provide clearer answers to questions about the risks associated with OSA in various age groups.

If OSA is an independent risk factor for vascular disease, vascular stress related to changes in blood pressure and blood flow velocity, nocturnal increases in platelet activation and aggregability, and autonomic disturbances may all contribute to the increased risk.[139] The increased risk of hypoxemia and the alterations in sympathetic activity associated with alcohol use may increase the risk that vascular morbidity will occur in association with sleep-disordered breathing.[140]

Although some patients with OSA have mildly elevated daytime pulmonary artery pressures,[141] probably from the effects of nocturnal hypoxemia, sleep apnea alone rarely, if ever, leads to fixed pulmonary hypertension. However, pulmonary hypertension can develop in patients with OSA who also have daytime hypoxemia and hypercapnia from coexisting obesity-hypoventilation syndrome or chronic obstructive pulmonary disease, or both.[91,142] In such patients, pulmonary hypertension and associated right heart failure may improve after the OSA is treated.

Anesthetic Complications

The narrow airway characteristic of patients with OSA increases the risk of anesthetic complications. Intubation may be difficult, and airway obstruction may occur during premedication for anesthesia, at the time of induction of anesthesia, or immediately after extubation. In addition, sedative medications administered during anesthesia may cause prolonged apneas with severe hypoxemia several hours later. Complications are less likely to develop if the anesthesiologist is aware that the patient has OSA, is familiar with the condition, and has equipment available to provide positive pressure, if needed, to support the airway.

Complications in Infants and Children

Failure to gain weight and to grow may occur in infants and children with severe OSA.[143] The causes may include (1) impaired growth hormone secretion during slow-wave sleep and (2) increased work of breathing associated with high airway resistance. Rarely, cardiomegaly, cor pulmonale, and cardiac failure may develop in untreated severe OSA. Stroke may develop in patients with OSA who have other predisposing conditions, as illustrated in the following case.

CASE HISTORY: Case 13–2

A 6-year-old girl with sickle cell anemia, large tonsils, and a history of snoring had several small strokes during sleep over a period of several weeks. Polysomnography showed severe OSA with oxygen desaturations as low as 47% during REM sleep. After tonsillectomy and adenoidectomy, breathing and oxygenation were normal. Nocturnal hypoxia may have produced intravascular sickling and thus may have contributed to the infarctions.[144]

DIFFERENTIAL DIAGNOSIS

Snoring that disturbs others, witnessed apneas, reports of gasping or choking sounds, finding the bedclothes in disarray in the morning, hypertension, increased body mass index, and a neck circumference greater than 16.5 inches are useful predictors of OSA in the general population.[145] In patients referred to sleep centers, predictors of OSA are similar but not identical and include increased neck circumference, increased body-mass index, witnessed apneas, hypertension, and habitual snoring.[146] Although more than 80% of persons with OSA snore loudly, the absence of snoring does not rule out this diagnosis: If the airway does not contain tissue that can vibrate, no snoring will occur even with a very narrow airway.

Certain aspects of snoring increase the likelihood of OSA: (1) increasing volume of snoring, or snoring that has changed in character; (2) snoring that is heard by the patient, which is usually an indication of an arousal at the end of an apnea; (3) loud, rhythmic snoring that is punctuated by snorts, gasps, choking noises, and body jerks; (4) snoring accompanied by restless sleep and repeated changes of body position; and (5) snoring that occurs in all body positions. Snoring patients who are said to "hold their breath" during sleep usually have apnea, and among children who snore, daytime mouth breathing, witnessed apnea, and struggling to breathe during sleep are suggestive of OSA.

In patients with complaints of excessive daytime sleepiness, OSA must be differentiated from other sleep disorders (e.g., narcolepsy, idiopathic hypersomnia, periodic leg movements of sleep, central sleep apnea, insufficient sleep syndrome) as well as atypical depression. Sleepiness that began at a time of weight gain and increased volume of snoring is more likely to be a result of OSA than is sleepiness that began years before the onset of snoring. In patients who complain of nocturnal choking, gasping, coughing, or shortness of breath, the differential diagnosis includes GERD, nocturnal asthma, congestive heart failure, central sleep apnea, nocturnal panic attacks, and sleep-related laryngospasm.

DIAGNOSTIC EVALUATION

Of the approaches to diagnosis, the simplest is to make a clinical diagnosis based on a history of snoring, apneas observed by others, and excessive sleepiness. However, a clinical diagnosis is not always correct, particularly in mild or moderate cases. In one study of children with either primary snoring or OSA, a clinical symptom assessment misclassified about one-fourth of the children.[147] Scores on the Epworth Sleepiness Scale (see Chap. 8), based on answers given by the patient or by the bed partner, provide an assessment of subjective sleepiness, but they are not good predictors of the severity of OSA.[148]

Although ambulatory home studies can be used, they also do not have a strong predictive value for the diagnosis of OSA. Actigraphic studies and sound-activated tape recorders are not reliable, although the latter may be useful to assess snoring at home, especially when snoring reported to occur at home does not occur during laboratory studies.[149] Although a nocturnal oximetry study with a combination of 15 or more oxygen desaturations of 4% or more per hour and a baseline SaO_2 greater than 90% is strongly suggestive of OSA, patients who do not have these findings may still have OSA.[150] In patients with mild OSA, clinical features and oximetry are not good predictors of the presence or absence of OSA.[151] Ambulatory studies that include oximetry along with one or more airflow, respiratory effort, or electrocardiographic (ECG) monitors provide a more detailed assessment and are more accurate than oximetry alone, but they do not identify all patients with clinically significant OSA. I use ambulatory recorders when my suspicion for OSA is high; in other cases, a laboratory polysomnogram is indicated.[152]

Polysomnography helps to determine the presence and type of apnea and its relation to sleep stage and body position, defines the severity of hypoxemia and sleep disturbance, and identifies complications such as arrhythmias. Assessment of the severity of OSA is based on several factors: (1) the frequency and duration of apneas, (2) the body positions and stages of sleep during which they occur, (3) the type and frequency of associated cardiac arrhythmias, (4) the severity of associated hypoxemia, and (5) the degree of sleep fragmentation. With mild OSA syndrome, apneas may occur only when the patient is supine or only after alcohol ingestion. Thus, the absence of apnea on a polysomnogram does not exclude the diagnosis, and one or more additional recording nights may be required in some patients.

Standard polysomnography may be supplemented by monitoring of intraesophageal pressure or end-tidal PCO_2. Intraesophageal pressure recordings, which provide a sensitive and quantitative measure of respiratory effort, are particularly useful in cases of suspected upper airway resistance syndrome because they may show increased respiratory effort and increased work of breathing, findings that are not always apparent with standard

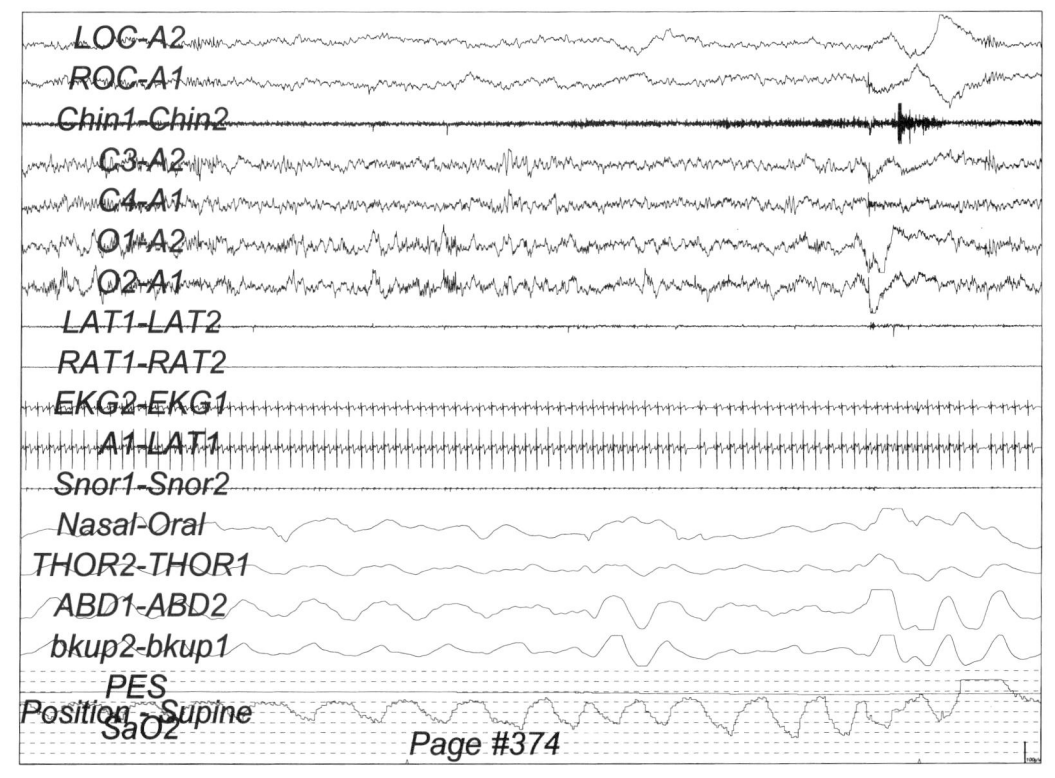

Figure 13–11. One-minute segment of a polysomnogram with intraesophageal pressure (PES) monitoring in an 8-year-old girl with snoring and enuresis. During the latter two-thirds of the segment, the amplitude of inspiratory-expiratory fluctuations in intraesophageal pressure (lowest channel) increased from 8 to 24 cm H_2O. The increase in respiratory effort, which was not accompanied by snoring, is more readily apparent from the PES tracing than from the other respiratory measures. For abbreviations, see Figures 13–1 and 13–2.

measures of respiratory effort (Fig. 13–11). Their increased sensitivity to respiratory effort may also help to identify hypopneas as obstructive or nonobstructive and may reveal subtle indications of increased respiratory effort preceding an arousal that suggest that the arousal was caused by airway narrowing. The mild discomfort associated with their placement, however, makes them unacceptable to some patients. In addition, normative values have not been established for esophageal pressures during sleep. The lowest negative esophageal pressure during the night may reach -20 cm H_2O in healthy children.[153] Monitoring of end-tidal $P{CO}_2$ is helpful when hypoventilation is suspected; however, end-tidal $P{CO}_2$ may not accurately reflect arterial $P{CO}_2$, especially during rapid, shallow breathing. In one study, esophageal pressure monitoring was more useful than transcutaneous or end-tidal $P{CO}_2$ monitoring in children.[55]

Additional studies may be indicated in some patients. A Multiple Sleep Latency Test (MSLT) helps to assess the presence and severity of excessive sleepiness and assists with the differential diagnosis; however, the MSLT is usually unnecessary for patients with sleepiness that began in conjunction with increased snoring, apneas observed by others, and weight gain. Lateral cephalometric radiographs, obtained and analyzed in accordance with a standard protocol, and fiberoptic endoscopy of the upper airway can be used to assess airway anatomy (Fig. 13–12).[154,155] Arterial blood gases and pulmonary function tests obtained during the waking state are important if obesity-hypoventilation syndrome or other causes of hypoventilation are suspected. For patients with clinical evidence of right or left heart failure, a complete blood count, ECG, and chest radiograph are indicated. Endocrine studies and esophageal pH monitoring are occasionally helpful. If craniofacial or brainstem malformations or airway neoplasms are contributing to OSA, magnetic resonance imaging or computerized tomography of the head or neck may be indicated.

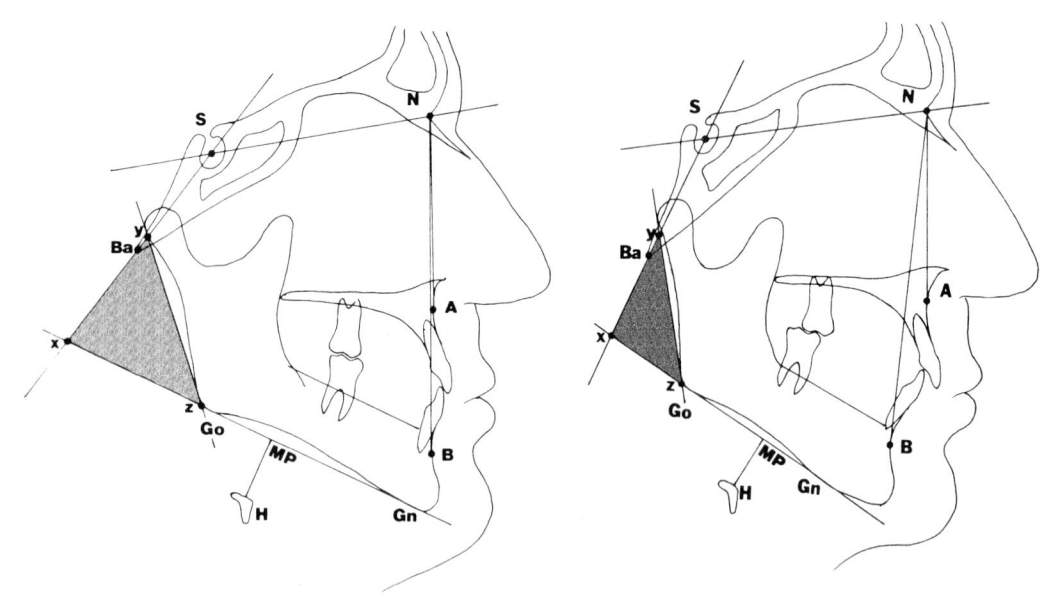

Figure 13–12. Measurements derived from lateral cephalometric radiographs taken from a normal subject (*left*) and one with OSA (*right*). The SNA angle indicates the degree of retroposition of the maxilla; the SNB angle, more acute in the patient with OSA than in the control subject, indicates retroposition of the mandible. Abbreviations: S, sella turcica; N, nasion; A, deepest point on the premaxillary outer contour; B, deepest point on the outer mandibular contour; Go, gonion; Gn, gnathion; MP, mandibular plane; H, hyoid; Ba, basion. (From Jamieson et al.,[63] p. 472, with permission.)

MANAGEMENT

There are three principal reasons to treat OSA:
1. To relieve the patient of symptoms
2. To provide relief from the effects of OSA on others
3. To reduce the risk of future illness or death from complications of OSA.

The immediate goal of treatment is to keep the airway open during sleep, leading to improved sleep, better oxygenation, and enhanced daytime alertness. Factors that precipitate or exacerbate OSA, such as alcohol, allergies, the supine sleeping position, and inadequate amounts of sleep,[56] should be avoided. For many patients, the best available treatment is nasal CPAP worn at night (Fig. 13–13). Other available treatments include upper airway surgery, oral appliances, weight loss, and supplemental oxygen.

Continuous Positive Airway Pressure

The introduction of nasal CPAP in the early 1980s revolutionized the management of OSA.[12] The mask covers the nose only, and although patients must learn to keep their mouths closed during sleep for optimal effectiveness, the mouth need not be sealed. Nasal CPAP, which functions as an air splint to maintain positive intraluminal pressure, is effective in approximately 80% to 90% of patients and can be used in children as well as adults. Infants can also use CPAP, but successful treatment in an infant requires close coordination among the family, pediatrician, and sleep specialist.[156] Most patients with severe OSA and many with OSA of mild to moderate severity experience dramatic improvement, as shown in the following case.

CASE HISTORY: Case 13–3

A 49-year-old man complained of excessive drowsiness, difficulty with concentration, and nonrestorative sleep. During the previous 10 years, he had gained about 35 pounds and presented with a weight of 210 pounds. His wife complained about his snoring and reported episodes of gasping during sleep. Although he slept for 9 to 9½ hours each night and often took a nap in the late morning, he did not feel rested.

Physical examination revealed mild obesity and a long soft palate. A polysomnogram demonstrated 44 respiratory events per hour of

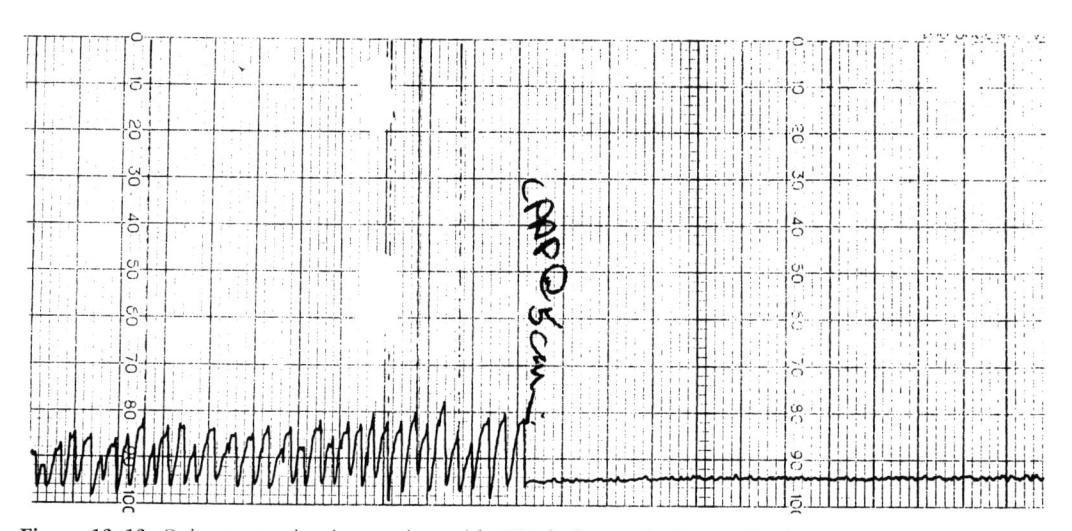

Figure 13–13. Oximetry tracing in a patient with OSA before and after application of nasal CPAP. This 45-minute segment of an overnight study demonstrated repetitive obstructive apneas before CPAP was initiated, with associated decreases in oxyhemoglobin saturation from 96% to 80–85%. With CPAP at 5 cm H_2O, apneas and hypoxemia were completely eliminated. (From Aldrich and Chervin,[133] p. 204, with permission.)

sleep, mainly obstructive hypopneas and obstructive apneas with a minimum oxyhemoglobin saturation of 81%. With nasal CPAP at 11 cm of water pressure, he had almost complete resolution of OSA. The minimum oxyhemoglobin saturation was 92%. He began using CPAP at home, and 3 months later, he reported that he was "doing great" while using nasal CPAP nightly without difficulty. His daytime sleepiness had resolved, and he no longer snored. He slept from midnight to 8 AM and no longer took naps during the day.

It is best to determine the optimal CPAP level via a sleep study, since the treatment may be ineffective—even if snoring is reduced—if the level of positive pressure is too low. Titration of CPAP should be performed by experienced technologists because the pressure requirement may differ depending on the stage of sleep and the body position of the patient. The pressure requirement is usually, but not always, highest when the patient is supine and in REM sleep. On the other hand, it is important not to prescribe too high a pressure, because such pressures may induce new arousals and periods of central sleep apnea, particularly in patients with a predisposition to central sleep apnea (Fig. 13–14). In addition, patients with cardiopulmonary disease may experience arrhythmias or hypoventilation with CPAP if it is not titrated to the optimum level.

Although a full night of recording without CPAP is usually required to determine the severity of apnea, in patients with severe OSA, a few hours of recording with-

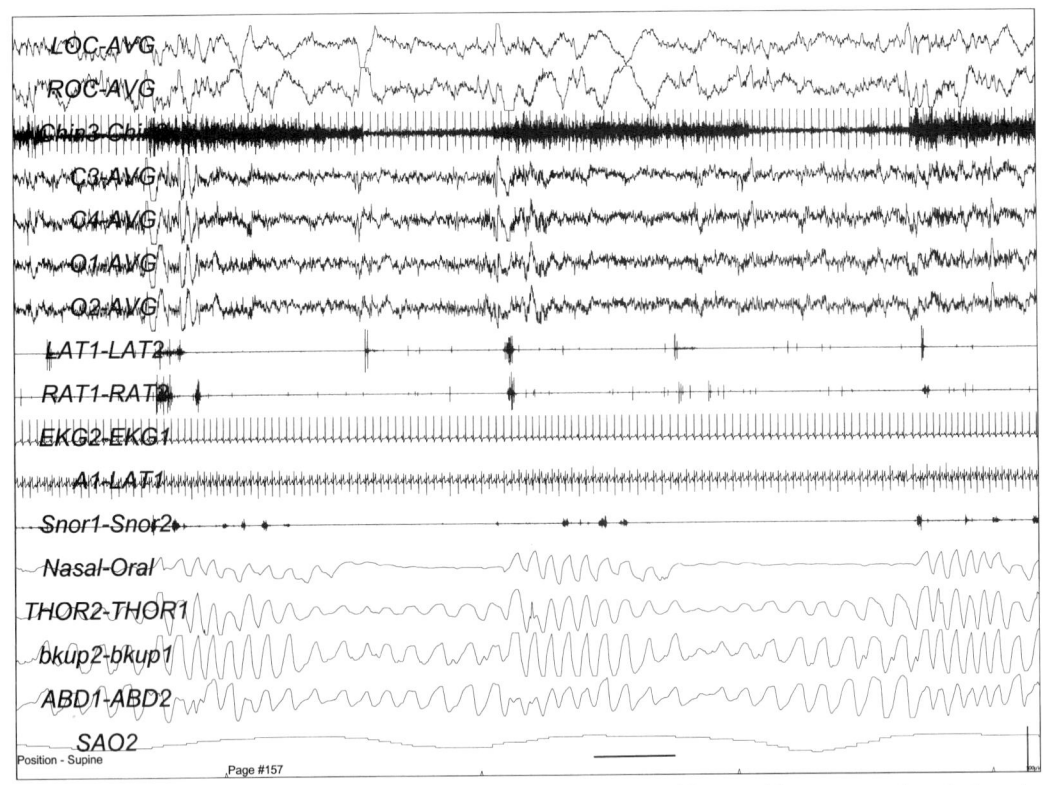

Figure 13–14. Two-minute segment of a polysomnogram in a 67-year-old man with an acute subcortical stroke and a history of snoring and apneas observed by his wife. Two obstructive apneas are shown; however, the decline in the amplitude of respiratory effort just preceding the second apnea (underlined) and the marked increase in respiratory rate with the arousals suggests an underlying Cheyne-Stokes breathing pattern with a propensity for central apneas, and an increased likelihood that central apneas will be observed during treatment with nasal CPAP. For abbreviations, see Figures 13–1 and 13–2.

out CPAP can be combined with several hours of CPAP titration. Although split-night studies are less expensive than the practice of recording a full night without CPAP followed by a second night with CPAP titration, they have two disadvantages. First, in patients with apnea that occurs only during REM sleep or only in certain body positions, it may not be possible to assess the severity of OSA during the first half of the night, which may include only one or two periods of REM sleep. Second, in patients with severe OSA, or OSA combined with central sleep apnea, it may not be possible to determine the optimal CPAP level in half a night. With bilevel positive airway pressure (BPAP) equipment (see below), the complexity of the pressure titration increases, which makes it more difficult to determine optimal pressure settings.

The subjective benefits of CPAP include improved alertness, fewer traffic accidents, less time taken off work, and better work efficiency.[157] In patients with severe OSA who use CPAP regularly, alertness is reduced after just one night without CPAP.[158] With these benefits, which are often apparent within the first few days, it is not surprising that some patients and their partners view CPAP as practically miraculous. The objective benefits include improved cognitive function, fewer nocturnal cardiac arrhythmias, lower blood pressure, fewer depressive symptoms, and better sexual function.[159-162]

For CPAP users, the median nightly use is about 5½ hours. Most patients who discontinue CPAP do so during the first year of use, and patients who experience symptomatic improvement obviously are much more likely to use CPAP regularly than those who do not. Noncompliance with CPAP mainly occurs not because of loss of efficacy, but because of side effects (e.g., dry nose and mouth, sneezing, nasal congestion, skin irritation, abrasions of the bridge of the nose, eye irritation from mask leaks). Nasal congestion in some cases is caused by mouth or mask leaks: These leaks are sensed by the CPAP machine and induce marked increases in airflow, which may cause nasal irritation and congestion.[163] A chin strap to ensure mouth closure, a better fitting mask, or humidification may help.

Treatment failures occur in patients who cannot breathe through the nose, who cannot tolerate the apparatus because of discomfort or claustrophobia, who cannot sleep with their mouths closed, or who remove the mask during the night while in a semiconscious state. Repeated sinus or other upper respiratory infections may prevent some patients from using the apparatus. Other complications of nasal CPAP include sleep disruption caused by noise from the machine, facial discomfort from the mask, air swallowing, and allergic rhinitis. Occasional patients complain of discomfort from misdirection of airflow into the eyes. Barotrauma to the lungs appears to be an extremely rare complication. Other rare complications are increased left ventricular dysfunction in patients with heart failure, and hypoventilation in patients with expiratory muscle weakness or severe lower airway obstructive disease. For patients who cannot use nasal CPAP because of nasal obstruction, options include nasal septoplasty or other nasal surgical procedures, or an oronasal mask.

BPAP devices provide different pressures during inspiration and expiration. Although they are more expensive than CPAP devices, they are sometimes better tolerated, particularly by patients who have a sensation of being unable to breathe with CPAP. However, in a study that randomly assigned patients to either CPAP or BPAP, there was no difference in mean hours of use per night or in complaints related to mask discomfort, machine noise, or nasal stuffiness.[164]

Devices that automatically adjust the level of CPAP during the night are recent innovations. With these systems, information from pressure, flow, and vibration sensors is used to vary positive pressure during the night in order to match the amount required to overcome obstruction. The reduction in breathing disturbance and the improvement in sleep architecture may be comparable to manual CPAP titration,[165] and the mean hours of use per night appear to be similar.[166] The ability to match pressure requirements to changing needs during the night and the potential opportunity to eliminate costly laboratory studies are the major advantages of these devices. The mean positive pressure for the night is usually less than with manual systems, but whether a reduced pressure requirement will lead to better compliance is unknown. Some patients find the frequent changes in pressure to be annoying and prefer the manual systems. As with other unattended monitoring studies, the inability to deal with technical problems, such as mask leakage, is a major disadvantage. Because arrhythmias can develop during manual or automated CPAP titration in patients with cardiopulmonary disease, such patients should not undergo unattended monitoring.[167]

Surgery

Patients who prefer surgery to CPAP include those who cannot tolerate CPAP and those who are unwilling to use it regularly because of their lifestyles. Young patients with OSA may be interested in surgery as an alternative to decades of CPAP use.

Surgical procedures for OSA include tonsillectomy with adenoidectomy, uvulopalatopharyngoplasty (UPP), laser-assisted uvulopalatoplasty (LAUP), genioglossus advancement with hyoid suspension (GAHS), maxillomandibular osteotomy (MMO), and tracheostomy. Although surgery should be directed at the anatomic site of obstruction, determining the site of obstruction is not always easy. Patients with large uvulas may have obstruction at the hypopharyngeal level instead of or in addition to obstruction at the velopharynx. Although evaluation of the airway with nasopharyngolaryngoscopy, with endoscopy combined with the Müller maneuver, or with cephalometric radiographs may help to determine the site or sites of obstruction, the ability to predict surgical success is limited because the dynamics of airway collapse differ between sleep and wakefulness, and even between REM and NREM sleep.[168]

Tonsillectomy with adenoidectomy can be effective treatment for children with OSA who have large tonsils or adenoids, because they are often the only (or the major) site of obstruction. Adults with large tonsils may also benefit, but the likelihood of success is less than with children. Although tonsillectomy is often performed as an outpatient procedure, clinically significant airway edema can occur in up to one-fourth of OSA patients[169]; therefore, inpatient observation overnight in the hospital is recommended for patients who are less than 2 years old and for those who have major craniofacial anomalies, failure to thrive, hypotonia, cor pulmonale, severe obesity, or severe OSA.[169]

UPP with removal of the uvula, portions of the soft palate, and redundant pharyngeal tissue is commonly performed for OSA. Because of postoperative airway edema, patients are usually hospitalized for one night after the procedure. Unfortunately, it is beneficial in only 40% to 60% of cases, more often in younger patients than older patients, and complete cure occurs in no more than one-fourth.

The ability to predict UPP outcome successfully would make the procedure much more useful. Preoperative dynamic pharyngoscopy, during which patients are asked to perform a reverse Valsalva maneuver (Müller maneuver), is sometimes used to try to predict the likelihood of UPP success; if collapse occurs at the level of the base of the tongue and epiglottis, then UPP probably will fail. On the other hand, some patients who have collapse at the velopharynx during dynamic pharyngoscopy also do not do well after surgery because the site of obstruction is below the velopharynx, or because they still have obstruction at the retropalatal level.[168,170] Thus, the predictive value of the Müller maneuver is uncertain, and there is evidence for and against its usefulness.[155,171,172] Cephalometric evaluation appears to provide better predictors: A low hyoid position, an increased craniocervical angle, and a short mandible length all predict a poor outcome.[155] The operation is more effective for snoring: Based on reports from bed partners, the cure rate is 80% to 90%, and more than 80% have improvement from preoperative baseline measurements at the 5-year follow-up.[173] Objective studies of snoring, however, suggest that the improvement in snoring is less than subjective reports indicate.[174]

Some patients with initially good results later relapse.[175] Subjective assessment is unreliable, as nonresponders may report complete resolution of daytime drowsiness. In some patients, UPP may make it more difficult to use CPAP because of increased risk of mouth air leaks.[176]

LAUP, which involves sequential treatments with laser pulses to remove portions of the soft palate and uvula, can be performed under local anesthesia in an office setting. Although snoring is often improved, current evidence, which is limited, suggests that the success rate of LAUP for OSA treatment is not better, and may even be worse, than that for conventional UPP.[177] In the absence of controlled studies demonstrating efficacy, I do not recommend it for patients with OSA.

GAHM combines a limited mandibular osteotomy with hyoid surgery. The genioid tubercle of the mandible, which serves as the anterior attachment of the tongue, is freed from the remainder of the mandible, advanced and rotated, and then reattached to the mandible. The result is a tongue advancement of several millimeters, which tends to widen the hypopharyngeal airway. The hyoid bone is usually stabilized anteriorly and inferiorly by attachment to the thyroid cartilage or the mandible. This procedure is designed for patients with obstruction at the hypopharyngeal level and can substantially improve OSA in the majority of patients with OSA associated with obstruction confined to the retrolingual area.[178]

MMO, with advancement of the mandible or maxilla, or both, is helpful for selected patients. Mandibular advancement increases the size of the airway at the level of the hypopharynx and may also increase airway diameter more rostrally, at the velopharyngeal level. Although the surgery is more extensive than UPP, most patients are able to leave the hospital after one night. In one series, MMO was as effective as nasal CPAP in patients with severe OSA.[179]

Some surgeons adopt a staged approach to surgical treatment, beginning with nasal surgery, UPP, GAHM, or all three, followed by MMO if needed. Because many patients experience more subjective improvement of symptoms with treatment than objective changes in apnea severity, the outcome of treatment in all but the mildest cases should be assessed with a nocturnal polysomnogram.

Tracheostomy, effective in almost all cases, is used less frequently than in the past because of the effectiveness of nasal CPAP. It still has a role for patients with severe OSA who cannot tolerate CPAP and are not good candidates for other surgical approaches. Tracheostomy may be the only treatment likely to succeed in mentally impaired patients with OSA, but the risks associated with tracheostomy in such patients are significant and may outweigh any potential therapeutic benefit. Infants with severe OSA may also require tracheostomy.

Dental Appliances

Dental appliances are often useful for patients who are unwilling or unable to use nasal CPAP, particularly in those with OSA of mild to moderate severity. Snoring improves or resolves in many cases, and OSA improves in the majority, but it is often not completely eliminated.[180,181] When these appliances are effective, most patients prefer them over CPAP; the major side effect, discomfort of the jaw or mouth, as in the following case, is uncommon.[182]

CASE HISTORY: Case 13-4

A 40-year-old man had no complaints apart from an occasional sore throat in the morning, but his wife was bothered by his snoring and had observed periods of apnea. A polysomnogram demonstrated 46 apneas per hour of sleep with a minimum oxyhemoglobin saturation of 76%. Sleep apnea and hypoxemia resolved with nasal CPAP at 5 cm H_2O. The patient used nasal CPAP intermittently, but he found it cumbersome and did not use it at all when he was on business trips away from home. A dental appliance was prescribed for nighttime use, but he had morning jaw pain when he used it, although his wife noted that his snoring and apneas did not occur. A second appliance was prepared that caused less mandibular advancement; with nightly use of this device, he no longer had jaw pain and his wife noted only soft snoring without apneas. A polysomnogram performed while he wore the appliance demonstrated nine hypopneas per hour of sleep with a minimum oxyhemoglobin saturation of 85%.

Comment: The patient had minimal symptoms and sought treatment mainly because his wife was bothered by his snoring. His sleep apnea was severe enough to suggest that it might increase the risk of stroke or myocardial infarction. As with many patients with mild or no symptoms, he did not use CPAP regularly. He found a dental appliance acceptable, and it led to marked improvement in his sleep apnea.

Unfortunately, the success rate of appliances for patients with severe OSA is low.[183] In addition, although dozens of different types of appliances are available, the likelihood of success probably varies among them, and few have been assessed systematically.

Weight Loss

Weight loss has variable effectiveness and is often difficult to achieve. Some patients treated with CPAP find it easier to lose weight, perhaps because the increased alertness leads to more activity during the day or because there is less desire to eat snacks to increase alertness. OSA usually improves in patients who lose weight, but it is difficult to predict how much improvement will be achieved. Some patients are substantially better after a loss of just 20 lb, whereas others gain minimal benefit from a loss of 50 to 75 lb.

Medications

Protriptyline 5 mg at bedtime may help in mild cases by (1) reducing pharyngeal secretions, which may produce a slightly larger airway lumen; (2) increasing muscle tone, which may stiffen the airway and raise the closing pressure; and/or (3) reducing the amount of REM sleep. It is of little benefit in moderate to severe OSA. Other medications have even less value.

Oxygen

Although supplemental oxygen can improve oxygenation during the night, particularly in patients with respiratory events that are primarily hypopneas rather than apneas, and may produce modest reductions in the number of apneas and hypopneas, the clinical benefits are minimal. Sleepiness and other daytime symptoms are usually not affected. On the other hand, early concerns that the duration of apneas would increase with supplemental oxygen because of loss of hypoxic drive to arousal have not been realized, although occasional patients may develop mild nocturnal CO_2 retention.[184] Oxygen therapy should be considered for patients who are unwilling or unable to tolerate more effective alternatives, or for patients for whom hypoxemia is the major concern. For patients with chronic obstructive pulmonary disease and OSA who do not tolerate CPAP, transtracheal oxygen may be an option.[185]

SUMMARY

OSA is a common disorder with disabling symptoms and substantial associated morbidity and mortality. Infants and children, as well as adults, may be affected. Although in adulthood it is more common in men, in childhood the gender distribution is approximately equal. The usual presenting symptoms of OSA are snoring and excessive daytime sleepiness. Obesity, elevated blood pressure, and craniofacial features suggestive of a narrow airway are common findings on examination.

Sleep-related relaxation of pharyngeal dilator muscles, leading to narrowing or collapse of an already small upper airway, is the basis for the disorder; thus, any anatomic or neuromuscular disorder that leads to a small airway increases the risk of OSA. Arousals, required to terminate the apneas, lead to disrupted sleep and contribute to daytime sleepiness. Hypoxemia during apneas, hemodynamic changes, autonomic disturbances, and sleep disruption lead to a number of systemic disturbances, including impaired cognition, cardiac arrhythmias, and blood pressure alterations.

OSA is associated with an increased risk of automobile accidents, stroke, myocardial infarction, and cardiovascular death. Although not conclusive, evidence suggests that the disorder is an independent risk factor for these consequences.

Diagnosis is based on characteristic clinical symptoms along with laboratory findings, mainly polysomnography. The immediate goal of treatment is to keep the airway open during sleep, which leads to improved sleep, better oxygenation, and enhanced daytime alertness. For many patients, the best available treatment is nasal CPAP worn at night. Other available treatments include upper airway surgery, oral appliances, and weight loss.

REFERENCES

1. Dickens, C: The Posthumous Papers of the Pickwick Club. Chapman & Hall, London, 1837.
2. Caton, R: A case of narcolepsy. Br Med J 1:358, 1889.
3. Lamacq, L: Quelque cas de narcolepsie. Rev Med 17:699, 1897.
4. Kerr, WJ, and Lagen, JB: The postural syndrome of obesity leading to postural emphysema and cardiorespiratory failure. Ann Intern Med 10:569, 1936.
5. Osler, W: The Principles and Practice of Medicine. Appleton, New York, 1918.
6. Burwell, CS, et al: Extreme obesity associated with alveolar hypoventilation. A pickwickian syndrome. Am J Med 21:811, 1956.
7. Gastaut, H, et al: Etude polygraphique des manifestations épisodiques (hypniques et respiratoires) du syndrome de Pickwick. Rev Neurol 112:568, 1965.
8. Jung, R, and Kuhlo, W: Neurophysiological studies of abnormal night sleep and the pickwickian syndrome. Prog Brain Res 18:140, 1965.
9. Lugaresi, E, et al: Hypersomnia with periodic apneas. In Weitzman, E (ed): Advances in Sleep Research, Vol 4. Spectrum, New York, 1978, pp 1–151.
10. Guilleminault, C, et al: Sleep apnea in eight children. Pediatrics 58:23, 1976.
11. Guilleminault, C, et al: The sleep apnea syndromes. Annu Rev Med 27:465, 1976.
12. Sullivan, CE, et al: Reversal of obstructive sleep apnoea by continuous positive airway pressure applied through the nares. Lancet 1:862, 1981.
13. Freezer, NJ, et al: Obstructive sleep apnoea presenting as failure to thrive in infancy. J Paediatr Child Health 31:172, 1995.
14. Marcus, CL, et al: Determinants of growth in children with the obstructive sleep apnea syndrome. J Pediatr 125:556, 1994.
15. Janson, C, et al: Daytime sleepiness, snoring and gastro-oesophageal reflux amongst young adults in three European countries. J Intern Med 237:277, 1995.
16. Young, T, et al: The occurrence of sleep disordered breathing among middle-aged adults. N Engl J Med, 328:1230, 1993.
17. Olson, LG, et al: A community study of snoring and sleep-disordered breathing: Prevalence. Am J Respir Crit Care Med 152:711, 1995.
18. Bearpark, H, et al: Snoring and sleep apnea: A population study in Australian men. Am J Respir Crit Care Med 151:1459, 1995.
19. Cirignotta, F, et al: Prevalence of every night snoring and obstructive sleep apnoeas among 30–69-year-old men in Bologna, Italy. Acta Neurol Scand 79:366, 1989.
20. Telakivi, T, et al: Periodic breathing and hypoxia in snorers and controls: Validation of snoring history and association with blood pressure and obesity. Acta Neurol Scand 76:69, 1987.
21. Gislason, T, et al: Prevalence of sleep apnea syndrome among Swedish men: An epidemiological study. J Clin Epidemiol 41:571, 1988.
22. Lavie, P: Sleep apnea in industrial workers. In Guilleminault, C, and Lugaresi, E (eds): Sleep-Wake Disorders: Natural History, Epidemiology and Long-Term Evolution. Raven Press, New York, 1983, pp 127–135.
23. Stradling, JR, and Crosby, JH: Predictors and prevalence of obstructive sleep apnoea and snoring in 1001 middle aged men. Thorax 46:85, 1991.
24. Gislason, T, et al: Snoring, hypertension, and the sleep apnea syndrome: An epidemiologic survey of middle-aged women. Chest 103:1147, 1993.
25. Ancoli-Israel, S, et al: Sleep-disordered breathing in community-dwelling elderly. Sleep 14:486, 1991.
26. Ancoli-Israel, S, et al: Sleep-disordered breathing in African-American elderly. Am J Respir Crit Care Med 152:1946, 1995.
27. Martikainen, K, et al: Natural evolution of snoring: A 5-year follow-up study. Acta Neurol Scand 90:437, 1994.
28. Lugaresi, E, et al: Some epidemiological data on snoring and cardiocirculatory disturbances. Sleep 3:221, 1980.
29. Honsberg, AE, et al: Incidence and remission of habitual snoring over a 5- to 6-year period. Chest 108:604, 1995.
30. Hultcrantz, E, et al: The epidemiology of sleep related breathing disorder in children. Int J Pediatr Otorhinolaryngol (suppl)32:S63, 1995.

31. Owen, GO, et al: Snoring, apnoea and ENT symptoms in the paediatric community. Clin Otolaryngol 21:130, 1996.
32. Ali, NJ, et al: Natural history of snoring and related behaviour problems between the ages of 4 and 7 years. Arch Dis Child 71:74, 1994.
33. Ali, NJ, et al: Snoring, sleep disturbance and behaviour in 4–5 year olds. Arch Dis Child 68:360, 1993.
34. Gislason, T, and Benediktsdottir, B: Snoring, apneic episodes, and nocturnal hypoxemia among children 6 months to 6 years old: An epidemiologic study of lower limit of prevalence. Chest 107:963, 1995.
35. Vgontzas, AN, et al: Sleep apnea and sleep disruption in obese patients. Arch Intern Med 154:1705, 1994.
36. Sloan, EP, and Shapiro, CM: Obstructive sleep apnea in a consecutive series of obese women. Int J Eat Disord 17:167, 1995.
37. Richman, RM, et al: The prevalence of obstructive sleep apnoea in an obese female population. Int J Obesity 18:173, 1994.
38. Kales, A, et al: Sleep apnoea in a hypertensive population. Lancet 2:1005, 1984.
39. Fletcher, EC, et al: Undiagnosed sleep apnea in patients with essential hypertension. Ann Intern Med 103:190, 1985.
40. Williams, AJ, et al: Sleep apnea syndrome and essential hypertension. Am J Cardiol 55:1019, 1985.
41. Coy, TV, et al: The role of sleep-disordered breathing in essential hypertension. Chest 109:890, 1996.
42. Olson, LG, et al: A community study of snoring and sleep-disordered breathing: Health outcomes. Am J Respir Crit Care Med 152:717, 1995.
43. Wetter, DW, et al: Smoking as a risk factor for sleep-disordered breathing. Arch Intern Med 154:2219, 1994.
44. Suratt, PM, et al: Collapsibility of the nasopharyngeal airway in obstructive sleep apnea. Am Rev Respir Dis 132:967, 1985.
45. Haponik, E, et al: Computerized tomography in obstructive sleep apnea: Correlation of airway size with physiology during sleep and wakefulness. Am Rev Respir Dis 127:221, 1983.
46. Brown, IG, et al: Pharyngeal compliance in snoring subjects with and without obstructive sleep apnea. Am Rev Respir Dis 132:211, 1985.
47. Bradley, TD, et al: Pharyngeal size in snorers, non-snorers, and patients with obstructive sleep apnea. New Engl J Med 315:1327, 1986.
48. Avrahami, E, and Englender, M: Relation between CT axial cross-sectional area of the oropharynx and obstructive sleep apnea syndrome in adults. Am J Neuroradiol 16:135, 1995.
49. Schwab, RJ, et al: Upper airway and soft tissue anatomy in normal subjects and patients with sleep-disordered breathing: Significance of the lateral pharyngeal walls. Am J Respir Crit Care Med 152:1673, 1995.
50. Stauffer, JL, et al: Pharyngeal size and resistance in obstructive sleep apnea. Am Rev Respir Dis 136:623, 1987.
51. Mezzanotte, WS, et al: Influence of sleep onset on upper-airway muscle activity in apnea patients versus normal controls. Am J Respir Crit Care Med 153:1880, 1996.
52. Martin, SE, et al: The effect of posture on airway caliber with the sleep-apnea/hypopnea syndrome. Am J Respir Crit Care Med 152:721, 1995.
53. Levy, P, et al: Dynamique des structures pharyngees au cours des apnees obstructives (en ventilation spontanee, pression positive continue et BIPAP). Neurophysiol Clin 24:227, 1994.
54. Guilleminault, C, et al: A cause of excessive daytime sleepiness: The upper airway resistance syndrome. Chest 104:781, 1993.
55. Guilleminault, C, et al: Recognition of sleep-disordered breathing in children. Pediatrics 98:871, 1996.
56. Persson, HE, and Svanborg, E: Sleep deprivation worsens obstructive sleep apnea: Comparison between diurnal and nocturnal polysomnography. Chest 109:645, 1996.
57. American Thoracic Society: Standards and indications for cardiopulmonary sleep studies in children. Am J Respir Crit Care Med 153:866, 1996.
58. Guilleminault, C, et al: Familial aggregates in obstructive sleep apnea syndrome. Chest 107:1545, 1995.
59. Redline, S, et al: The familial aggregation of obstructive sleep apnea. Am J Respir Crit Care Med 151:682, 1995.
60. Pillar, G, and Lavie, P: Assessment of the role of inheritance in sleep apnea syndrome. Am J Respir Crit Care Med 151:688, 1995.
61. Mathur, R, and Douglas, NJ: Family studies in patients with the sleep apnea-hypopnea syndrome. Ann Intern Med 122:174, 1995.
62. Jacobs, IN, et al: Upper airway obstruction in children with Down syndrome. Arch Otolaryngol Head Neck Surg 122:945, 1996.
63. Jamieson, A, et al: Obstructive sleep apneic patients have craniomandibular abnormalities. Sleep 9:469, 1986.
64. Shapiro, PA: Effects of nasal obstruction on facial development. J Allergy Clin Immunol 81:967, 1988.
65. Vargervik, K, and Harvold, EP: Experiments on the interaction between orofacial function and morphology. Ear Nose Throat J 66:201, 1987.
66. Ryan, CF, and Love, LL: Mechanical properties of the velopharynx in obese patients with obstructive sleep apnea. Am J Respir Crit Care Med 154:806, 1996.
67. White, DP, et al: Pharyngeal resistance in normal humans: Influence of gender, age, and obesity. J Appl Physiol 58:365, 1985.
68. Popovic, RM, and White, DP: Influence of gender on waking genioglossal electromyogram and upper airway resistance. Am J Respir Crit Care Med 152:725, 1995.
69. Hoffstein, V: Snoring. Chest 109:201, 1996.
70. Berry, RB, et al: Effect of hypercapnia on the arousal response to airway occlusion during sleep in normal subjects. J Appl Physiol 74:2269, 1993.

71. Gleeson, K, et al: Arousal from sleep in response to ventilatory stimuli occurs at a similar degree of ventilatory effort irrespective of the stimulus. Am Rev Respir Dis 142:295, 1989.
72. Berry, RB, et al: Effect of upper airway anesthesia on obstructive sleep apnea. Am J Respir Crit Care Med 151:1857, 1995.
73. Berry, RB, et al: Sleep apnea impairs the arousal response to airway occlusion. Chest 109:1490, 1996.
74. Berry, RB, et al: Triazolam in patients with obstructive sleep apnea. Am J Respir Crit Care Med 151:450, 1995.
75. Camacho, ME, and Morin, CM: The effect of temazepam on respiration in elderly insomniacs with mild sleep apnea. Sleep 18:644, 1995.
76. Robinson, RW, et al: Moderate alcohol ingestion increases upper airway resistance in normal subjects. Am Rev Respir Dis 132:1238, 1985.
77. Mitler, MM, et al: Bedtime ethanol increases resistance of upper airways and produces sleep apneas in asymptomatic snorers. Alcohol Clin Exp Res 12:801, 1988.
78. Dawson, A, et al: Effect of bedtime ethanol on total inspiratory resistance and respiratory drive in normal nonsnoring men. Alcohol Clin Exp Res 17:256, 1993.
79. Block, AJ, et al: Effect of alcohol ingestion on breathing and oxygenation during sleep: Analysis of the influence of age and sex. Am J Med 80:595, 1986.
80. Issa, FG, and Sullivan, CE: Alcohol, snoring and sleep apnoea. J Neurol Neurosurg Psychiatry 45:353, 1982.
81. Scrima, L, et al: Increased severity of obstructive sleep apnea after bedtime alcohol ingestion: Diagnostic potential and proposed mechanism of action. Sleep 5:318, 1982.
82. Scrima, L, et al: Effect of three alcohol doses on breathing during sleep in 30–49 year old nonobese snorers and nonsnorers. Alcohol Clin Exp Res 13:420, 1989.
83. Taasen, VC, et al: Alcohol increases sleep apnea and oxygen desaturation in asymptomatic men. Am J Med 71:240, 1981.
84. Aldrich, MS, and Chervin, RD: Alcohol use, obstructive sleep apnea, and sleep-related motor vehicle accidents. Sleep Res 26:308, 1997.
85. Bedard, MA, et al: Nocturnal hypoxemia as a determinant of vigilance impairment in sleep apnea syndrome. Chest 100:367, 1991.
86. Zamagni, M, et al: Respiratory effort: A factor contributing to sleep propensity in patients with obstructive sleep apnea. Chest 109:651, 1996.
87. Naegele, B, et al: Deficits of cognitive executive functions in patients with sleep apnea syndrome. Sleep 18:43, 1995.
88. Grunstein, RR, et al: Impact of self-reported sleep-breathing disturbances on psychosocial performance in the Swedish Obese Subjects (SOS) Study. Sleep 18:635, 1995.
89. Shepard, JW, Jr: Cardiorespiratory changes in obstructive sleep apnea. In Kryger, MH, et al (eds): Principles and Practice of Sleep Medicine, ed 2. WB Saunders, Philadelphia, 1994, pp 657–666.
90. Strohl, KP, and Altose, MD: Oxygen saturation during breath-holding and during apneas in sleep. Chest 85:181, 1984.
91. Bradley, TD, et al: Role of diffuse airway obstruction in the hypercapnia of obstructive sleep apnea. Am Rev Respir Dis 134:920, 1986.
92. Leuenberger, U, et al: Surges of muscle sympathetic nerve activity during obstructive apnea are linked to hypoxemia. J Appl Physiol 79:581, 1995.
93. Shimizu, T, et al: Muscle sympathetic nerve activity during apneic episodes in patients with obstructive sleep apnea syndrome. Electroencephalogr Clin Neurophysiol 93:345, 1994.
94. Smith, ML, et al: Role of hypoxemia in sleep apnea-induced sympathoexcitation. J Auton Nerv Syst 56:184, 1996.
95. Lofaso, F, et al: Sleep fragmentation as a risk factor for hypertension in middle-aged nonapneic snorers. Chest 109:896, 1996.
96. Somers, VK, et al: Sympathetic neural mechanisms in obstructive sleep apnea. J Clin Invest 96:1897, 1995.
97. Waravdekar, NV, et al: Influence of treatment on muscle sympathetic nerve activity in sleep apnea. Am J Respir Crit Care Med 153:1333, 1996.
98. Weiss, JW, et al: Hemodynamic consequences of obstructive sleep apnea. Sleep 19:388, 1996.
99. Schroeder, JS, et al: Hemodynamic studies in sleep apnea. In Guilleminault, C, and Dement, W (eds): Sleep Apnea Syndromes. AR Liss, New York, 1978, pp 177–196.
100. Shepard, JW, Jr: Cardiopulmonary consequences of obstructive sleep apnea. Mayo Clin Proc 65:1250, 1990.
101. Grimm, W, et al: Electrophysiologic evaluation of sinus node function and atrioventricular conduction in patients with prolonged ventricular asystole during obstructive sleep apnea. Am J Cardiol 77:1310, 1996.
102. Tolle, FA, et al: Reduced stroke volume related to pleural pressure in obstructive sleep apnea. J Appl Physiol 55:1718, 1983.
103. Guilleminault, C, et al: Obstructive sleep apnea and cardiac index. Chest 89:331, 1986.
104. Noda, A, et al: Cardiac hypertrophy in obstructive sleep apnea syndrome. Chest 107:1538, 1995.
105. Hunt, CE, and Brouillette, RT: Abnormalities of breathing control and airway maintenance in infants and children as a cause of cor pulmonale. Pediatr Cardiol 3:249, 1982.
106. Jennum, P, and Borgesen, SE: Intracranial pressure and obstructive sleep apnea. Chest 95:279, 1989.
107. Siebler, M, et al: Cerebral blood flow velocity alterations during obstructive sleep apnea syndrome. Neurology 40:1461, 1990.
108. Follenius, M, et al: Obstructive sleep apnea treatment: Peripheral and central effects on plasma renin activity and aldosterone. Sleep 14:211, 1991.
109. Krieger, J, et al: Atrial natriuretic peptide release during sleep in patients with obstructive

109. sleep apnea before and during treatment with nasal continuous positive airway pressure. Clin Sci 77:407, 1989.
110. Goldstein, SJ, et al: Reversibility of deficient sleep entrained growth hormone secretion in a boy with achondroplasia and obstructive sleep apnea. Acta Endocrinol 116:95, 1987.
111. Cooper, BG, et al: Hormonal and metabolic profiles in subjects with obstructive sleep apnea syndrome and the acute effects of nasal continuous positive airway pressure (CPAP) treatment. Sleep 18:172, 1995.
112. Guilleminault, C, et al: Upper airway sleep-disordered breathing in women. Ann Intern Med 122:493, 1995.
113. Stoohs, RA, et al: Insulin resistance and sleep-disordered breathing in healthy humans. Am J Respir Crit Care Med 154:170, 1996.
114. Brooks, B, et al: Obstructive sleep apnea in obese noninsulin-dependent diabetic patients: Effect of continuous positive airway pressure treatment on insulin responsiveness. J Clin Endocrinol Metab 79:1681, 1994.
115. Findley, L, et al: Automobile accidents involving patients with obstructive sleep apnea. Am Rev Respir Dis 138:337, 1988.
116. Wu, H, and Yan-Go, F: Self-reported automobile accidents involving patients with obstructive sleep apnea. Neurology 46:1254, 1996.
117. George, CF, et al: Simulated driving performance in patients with obstructive sleep apnea. Am J Respir Crit Care Med 154:175, 1996.
118. Findley, L, et al: Vigilance and automobile accidents in patients with sleep apnea or narcolepsy. Chest 108:619, 1995.
119. Spriggs, DA, et al: Historical risk factors for stroke: A case control study. Age Aging 19:280, 1990.
120. Seppala, T, et al: Sudden death and sleeping history among Finnish men. J Intern Med 229:23, 1991.
121. Partinen, M: Epidemiology of sleep disorders. In Kryger, MH, et al (eds): Principles and Practice of Sleep Medicine, ed 2. WB Saunders, Philadelphia, 1994, pp 437–452.
122. Partinen, M, and Guilleminault, C: Daytime sleepiness and vascular morbidity at seven-year follow-up in obstructive sleep apnea patients. Chest 97:27, 1990.
123. Hung, J, et al: Association of sleep apnoea with myocardial infarction in men. Lancet 336:261, 1990.
124. Dyken, ME, et al: Investigating the relationship between stroke and obstructive sleep apnea. Stroke 27:401, 1996.
125. Neau, JP, et al: Habitual snoring as a risk factor for brain infarction. Acta Neurol Scand 92:63, 1995.
126. Good, DC, et al: Sleep-disordered breathing and poor functional outcome after stroke. Stroke 27:252, 1996.
127. Qureshi, AI, et al: Habitual sleep patterns and risk for stroke and coronary heart disease: A 10-year follow-up from NHANES I. Neurology 48:904, 1997.
128. Mooe, T, et al: Sleep-disordered breathing in men with coronary artery disease. Chest 109:659, 1996.
129. Franklin, KA, et al: Sleep apnoea and nocturnal angina. Lancet 345:1085, 1995.
130. He, J, et al: Mortality and apnea index in obstructive sleep apnea. Experience in 385 male patients. Chest 94:9, 1988.
131. Partinen, M, et al: Long-term outcome for obstructive sleep apnea syndrome patients: Mortality. Chest 94:1200, 1988.
132. Wright, J, et al: Health effects of obstructive sleep apnoea and the effectiveness of continuous positive airways pressure: A systematic review of the research evidence. Br Med J 314:851, 1997.
133. Aldrich, MS, and Chervin, RD: Obstructive sleep apnea: Threat to health or innocent bystander? Neurol Network Commentary 1:201, 1997.
134. Koskenvuo, M, et al: Snoring as a risk factor for ischemic heart disease and stroke in men. BMJ 294:16, 1987.
135. Palomaki, H: Snoring and the risk of ischemic brain infarction. Stroke 22:1021, 1991.
136. Qizilbash, N, et al: Fibrinogen and lipid concentrations as risk factors for transient ischaemic attacks and minor ischaemic strokes. BMJ 303:605, 1991.
137. Jennum, P, et al: Risk of ischemic heart disease in self-reported snorers: A prospective study of 2937 men aged 54 to 74 years: The Copenhagen male study. Chest 108:138, 1995.
138. Mant, A, et al: Four-year follow-up of mortality and sleep-related respiratory disturbance in non-demented seniors. Sleep 18:433, 1995.
139. Bokinsky, G, et al: Spontaneous platelet activation and aggregation during obstructive sleep apnea and its response to therapy with nasal continuous positive airway pressure: A preliminary investigation. Chest 108:625, 1995.
140. Dolly, FR, and Block, AJ: Increased ventricular ectopy and sleep apnea following ethanol ingestion in COPD patients. Chest 83:469, 1983.
141. Laks, L, et al: Pulmonary hypertension in obstructive sleep apnoea. Eur Respir J 8:537, 1995.
142. Chaouat, A, et al: Pulmonary hemodynamics in the obstructive sleep apnea syndrome: Results in 220 consecutive patients. Chest 109:380, 1996.
143. Brouillette, RT, et al: Obstructive sleep apnea in infants and children. J Pediatrics 100:31, 1982.
144. Robertson, PL, et al: Stroke associated with obstructive sleep apnea in a child with sickle cell anemia. Ann Neurol 23:614, 1988.
145. Olson, LG, et al: A community study of snoring and sleep-disordered breathing: Symptoms. Am J Respir Crit Care Med 152:707, 1995.
146. Flemons, WW, et al: Likelihood ratios for a sleep apnea clinical prediction rule. Am J Respir Crit Care Med 150:1279, 1994.
147. Carroll, JL, et al: Inability of clinical history to distinguish primary snoring from obstructive sleep apnea syndrome in children. Chest 108:610, 1995.

148. Kingshott, RN, et al: Self assessment of daytime sleepiness: Patient versus partner. Thorax 50:994, 1995.
149. Middlekoop, HAM, et al: Wrist actigraphic assessment of sleep in 116 community based subjects suspected of obstructive sleep apnoea syndrome. Thorax 50:284, 1995.
150. Ryan, PJ, et al: Validation of British Thoracic Society guidelines for the diagnosis of the sleep apnoea/hypopnoea syndrome: Can polysomnography be avoided? Thorax 50:972, 1995.
151. Deegan, PC, and McNicholas, WT: Predictive value of clinical features for the obstructive sleep apnoea syndrome. Eur Respir J 9:117, 1996.
152. American Sleep Disorders Association: Practice parameters for the treatment of obstructive sleep apnea in adults: The efficacy of surgical modifications of the upper airway. Sleep 19:152, 1996.
153. Gaultier, C: Sleep-related breathing disorders: 6. Obstructive sleep apnoea syndrome in infants and children: Established facts and unsettled issues. Thorax 50:1204, 1995.
154. Guilleminault, C, et al: Obstructive sleep apnea and abnormal cephalometric measurements: Implications for treatment. Chest 86:793, 1984.
155. Petri, N, et al: Predictive value of Müller maneuver, cephalometry and clinical features for the outcome of uvulopalatopharyngoplasty: Evaluation of predictive factors using discriminant analysis in 30 sleep apnea patients. Acta Otolaryngol (Stockh) 114:565, 1994.
156. Guilleminault, C, et al: Home nasal continuous positive airway pressure in infants with sleep-disordered breathing. J Pediatr 127:905, 1995.
157. Engleman, HM, et al: Self-reported use of CPAP and benefits of CPAP therapy. Chest 109:1470, 1996.
158. Sforza, E, and Lugaresi, E: Daytime sleepiness and nasal continuous positive airway pressure therapy in obstructive sleep apnea syndrome patients: Effects of chronic treatment and 1-night therapy withdrawal. Sleep 18:195, 1995.
159. Borak, J, et al: Effects of CPAP treatment on psychological status in patients with severe obstructive sleep apnoea. J Sleep Res 5:123, 1996.
160. Becker, H, et al: Reversal of sinus arrest and atrioventricular conduction block in patients with sleep apnea during nasal continuous positive airway pressure. Am J Respir Crit Care Med 151:215, 1995.
161. Karacan, I, and Karatas, M: Erectile dysfunction in sleep apnea and response to CPAP. J Sex Marital Ther 21:239, 1995.
162. Millman, RP, et al: Depression as a manifestation of obstructive sleep apnea: Reversal with nasal continuous positive airway pressure. J Clin Psychiatry 50:348, 1989.
163. Richards, GN, et al: Mouth leak with nasal continuous positive airway pressure increases nasal airway resistance. Am J Respir Crit Care Med 154:182, 1996.
164. Reeves-Hoche, MK, et al: Continuous versus bilevel positive airway pressure for obstructive sleep apnea. Am J Respir Crit Care Med 151:443, 1995.
165. Teschler, H, et al: Automated continuous positive airway pressure titration for obstructive sleep apnea syndrome. Am J Respir Crit Care Med 154:734, 1996.
166. Meurice, JC, et al: Efficacy of auto-CPAP in the treatment of obstructive sleep apnea/hypopnea syndrome. Am J Respir Crit Care Med 153:794, 1996.
167. Juhasz, J, et al: Unattended continuous positive airway pressure titration: Clinical relevance and cardiorespiratory hazards of the method. Am J Respir Crit Care Med 154:359, 1996.
168. Shepard, JW, and Thawley, SE: Localization of upper airway collapse during sleep in patients with obstructive sleep apnea. Am Rev Respir Dis 141:1350, 1990.
169. Rosen, GM, et al: Postoperative respiratory compromise in children with obstructive sleep apnea syndrome: Can it be anticipated? Pediatrics 93:784, 1994.
170. Hudgel, DW, et al: Uvulopalatopharyngoplasty in obstructive apnea: Value of preoperative localization of the site of upper airway narrowing during sleep. Am Rev Respir Dis 143:942, 1991.
171. Sher, AE, et al: Predictive value of Müller maneuver in selection of patients for uvulopalatopharyngoplasty. Laryngoscope 95:1483, 1985.
172. Katsantonis, GP, et al: The predictive efficacy of the Müller maneuver in uvulopalatopharyngoplasty. Laryngoscope 99:677, 1989.
173. Friberg, D, et al: UPPP for habitual snoring: a 5-year follow-up with respiratory sleep recordings. Laryngoscope 105:519, 1995.
174. Miljeteig, H, et al: Subjective and objective assessment of uvulopalatopharyngoplasty for treatment of snoring and obstructive sleep apnea. Am J Respir Crit Care Med 150:1286, 1994.
175. Larsson, H, et al: Long-time follow-up after UPPP for obstructive sleep apnea syndrome: Results of sleep apnea recordings and subjective evaluation 6 months and 2 years after surgery. Acta Otolaryngol (Stockh) 111:582, 1991.
176. Mortimore, IL, et al: Uvulopalatopharyngoplasty may compromise nasal CPAP therapy in sleep apnea syndrome. Am J Respir Crit Care Med 154:1759, 1996.
177. Walker, RP, et al: Laser-assisted uvulopalatoplasty for snoring and obstructive sleep apnea: Results in 170 patients. Laryngoscope 105:938, 1995.
178. Sher, AE, et al: The efficacy of surgical modifications of the upper airway in adults with obstructive sleep apnea syndrome. Sleep 19:156, 1996.
179. Riley, RW, et al: Maxillofacial surgery and nasal CPAP: A comparison of treatment for obstructive sleep apnea syndrome. Chest 98:1421, 1990.
180. Schmidt-Nowara, WW, et al: Treatment of snoring and obstructive sleep apnea with a dental orthosis. Chest 99:1378, 1991.
181. Schmidt-Nowara, W, et al: Oral appliances for the treatment of snoring and obstructive sleep apnea: A review. Sleep 18:501, 1995.
182. Clark, GT, et al: A crossover study comparing the efficacy of continuous positive airway pres-

sure with anterior mandibular positioning devices on patients with obstructive sleep apnea. Chest 109:1477, 1996.
183. American Sleep Disorders Association: Practice parameters for the treatment of snoring and obstructive sleep apnea with oral appliances. Sleep 18:511, 1995.
184. Marcus, CL, et al: Supplemental oxygen during sleep in children with sleep-disordered breathing. Am J Respir Crit Care Med 152:1297, 1995.
185. Chauncey, JB, and Aldrich, MS: Preliminary findings in the treatment of obstructive sleep apnea with transtracheal oxygen. Sleep 13:167, 1990.

Chapter 14

CENTRAL SLEEP APNEA AND HYPOVENTILATION DURING SLEEP

CENTRAL SLEEP APNEA
Clinical Features
Biologic Basis
Differential Diagnosis
Diagnostic Evaluation
Management
SUDDEN INFANT DEATH SYNDROME
Clinical Features
Biologic Basis
Diagnosis
Prevention
CENTRAL ALVEOLAR HYPOVENTILATION SYNDROME
Clinical Features
Biologic Basis
Diagnosis
Management
OBESITY-HYPOVENTILATION SYNDROME
Clinical Features
Biologic Basis
Diagnosis
Management
HYPOVENTILATION ASSOCIATED WITH NEUROMUSCULAR DISORDERS
Clinical Features
Biologic Basis
Diagnosis
Management

For several days his breathing was irregular; it would entirely cease for a quarter of a minute, then it would become perceptible, though very low, then by degrees it became heaving and quick, and then it would gradually cease again.

JA CHEYNE[3]

The decline in the length and force of the respirations [is] as regular and remarkable as their progressive increase. The inspiration becomes each one less deep than the preceding, until they are all but imperceptible, and then the state of apparent apnea occurs. This is at last broken by the faintest possible inspiration; the next effort is a little stronger, until so to speak, the paroxysm of breathing is at its height, again to subside by a descending scale.

W STOKES[4]

Central sleep apnea (CSA) is a condition in which absence of respiratory effort leads to cessation of breathing. Gastaut and coworkers[1] first used the term *central sleep apnea* to distinguish central apneas from obstructive apneas, in which respiratory effort persists and the apnea is caused by airway occlusion. The term *periodic breathing* is also used to describe regularly recurring periods of central apnea and other breathing patterns in which variations in rate or amplitude occur in a regular, repetitive fashion.

Socrates described periodic breathing more than 2000 years ago,[2] and since the early 19th century, it has been known that lesions of the medulla and cervical spinal cord can lead to abnormal breathing patterns or to cessation of breathing. Although the use of the word "central" implies a central nervous origin, non-neurologic causes of CSA have also been recognized for centuries. In 1818, Cheyne described a crescendo-decrescendo pattern of breathing in patients with congestive heart failure, and Stokes expanded on

237

the topic in 1854.[3,4] This periodic pattern, often accompanied by periods of apnea, is now called *Cheyne-Stokes breathing* or *Cheyne-Stokes respirations*.

Despite the many possible causes of CSA, the condition is much less common than obstructive sleep apnea (OSA); CSA unaccompanied by OSA probably accounts for less than 5% of all cases of sleep apnea.

CENTRAL SLEEP APNEA

Clinical Features

Although the principal cause of OSA is narrowing of the upper airway, CSA has a variety of causes, and the clinical presentation of the syndrome may differ depending on the cause and the age of the patient. Insomnia with difficulty falling asleep and frequent awakenings, sometimes associated with gasping for air, may be the chief complaint in patients with CSA caused by metabolic disturbances. On the other hand, patients with CSA related to paralysis of the diaphragm may complain that they are unable to sleep in a horizontal position and must sleep in recliner chairs. When CSA is caused by upper airway narrowing, snoring and excessive daytime sleepiness are prominent symptoms, and morning headaches are sometimes a complaint in persons who hypoventilate during sleep. Despite these general characteristics, however, the presenting symptoms usually do not provide a strong indication of the cause of the disorder. In the following case, sleepiness was the principal manifestation of CSA, presumably caused by a static neurological disorder.

CASE HISTORY: Case 14–1

A 34-year-old woman with mild mental retardation and a seizure disorder was referred for evaluation of sleepiness that had been present for 10 to 15 years. She lived in a group home and worked at a workshop for the mentally handicapped, where she usually fell asleep three to five times during an 8-hour shift. She also fell asleep regularly during meals and while watching television. She slept from 9 PM to 6:30 AM and did not snore. She had been on phenytoin 300 mg/d for many years and for the past several years had not experienced any seizures. During the examination, she was talkative, but while sitting quietly, she fell asleep. Except for the mild mental retardation, the examination was normal. Brain magnetic resonance imaging (MRI) was also normal.

A polysomnogram demonstrated a sleep latency of 1 minute and 645 apneas, of which more than 90% were central apneas. Sleep was markedly fragmented by arousals associated with apneas. Treatment with acetazolamide, medroxyprogesterone, nasal continuous positive airway pressure (CPAP), and supplemental oxygen, separately or in combination, produced no improvement. With nasal bilevel positive airway pressure (BPAP), however, central apneas and associated arousals were entirely eliminated. Daytime sleepiness was reduced with home use of BPAP, although the patient did not like the equipment and usually took it off during the night. Over the next several years, with the application of behavioral techniques to reward compliance, her use of BPAP gradually increased. At age 42, a polysomnogram showed almost continuous central apneas without BPAP and complete elimination of central apneas with BPAP.

Like OSA patients, some patients with CSA are asymptomatic and are referred because the spouse or bed partner has noted apneas or rhythmic variations in breathing during sleep.

Some infants with CSA present because of an apparent life-threatening event (ALTE). Parents or other caregivers may observe a frightening or apparently potentially fatal episode of apnea, sometimes associated with pallor or cyanosis, that terminates after stimulation, shaking, or resuscitation of the infant.

Biologic Basis

With CSA, although the breathing disturbance is caused by aberrant function of systems involved in respiratory control (see Chap. 3), respiratory function during wakefulness is usually normal or near-normal. In the 1950s, Plum and Swanson,[5] who observed that medullary lesions often disrupt breathing during sleep without affecting breathing during wakefulness, emphasized the importance of two systems for breathing control: (1) metabolic control of breathing, mediated by the medulla; and (2) a cortically mediated "behavioral" system. In their model, the latter permitted normal breathing to occur during wakefulness, but it became inactive during sleep, leading to the occurrence of central apneas in patients with medullary lesions.

Although studies and clinical observations during the subsequent 40 years led to a more complex model of breathing regulation (see Chap. 3), the etiology of CSA still can be viewed as a disturbance of metabolic control of breathing that is usually caused by a metabolic, cardiac, neurological, or pulmonary disorder (Table 14–1). In addition to these recognized causes, some cases are idiopathic, and although an impairment of respiratory control is presumed to exist in such cases, it may not be apparent despite extensive evaluation. In premature infants, apneas may occur because the systems involved in respiratory control are not fully developed, or for other reasons (Table 14–2).

METABOLIC CONTRIBUTORS TO CENTRAL SLEEP APNEA

Four major features of the metabolic system for respiratory control determine whether central apneas are present: (1) sleep-wake state, (2) arterial CO_2 tension ($PaCO_2$), (3) arterial O_2 tension (PaO_2), and (4) responsiveness to arterial blood gas (ABG) changes.

Because most patients with CSA do not have apnea while awake, the changes in breathing that occur during sleep account for the development of CSA. During non–rapid eye movement (NREM) sleep, these include an increase in airway resistance, withdrawal of the wakefulness stim-

Table 14–1. **Causes of Central Sleep Apnea in Children and Adults**

Causes	Examples
Upper airway narrowing	Obstructive sleep apnea
Hypoxia with respiratory alkalosis	High altitude; pulmonary congestion
Metabolic acidosis	Renal failure
Increased loop gain	Idiopathic central sleep apnea (?)
Delayed chemoreceptor feedback	Congestive heart failure
Abnormal chemoreceptor function	Congenital and acquired central alveolar hypoventilation
Abnormal afferent pathways for chemoreceptor information	Shy-Drager syndrome, Riley-Day syndrome, diabetic autonomic neuropathy
Tracheal or nasal stimulation with reflex inhibition of inspiration	Obstructive sleep apnea; treatment of apnea with excessively high levels of CPAP
Abnormal function of medullary controller	Medullary tumors, infarcts, and hemorrhages; encephalitis, syringobulbia, Arnold-Chiari malformation, Leigh's syndrome
Abnormal integration of behavioral and metabolic control of breathing; or dysfunction of respiratory upper motor neurons in the cervical spinal cord	Trauma, tumors, multiple sclerosis
Motor neuron diseases and motor neuropathies	Amyotrophic lateral sclerosis, phrenic nerve palsies
Neuromuscular junction disorders	Myasthenia gravis
Muscle diseases	Duchenne's muscular dystrophy

CPAP = continuous positive airway pressure.

Table 14–2. **Causes of Central Sleep Apnea in Infants**

Causes	Examples
Incomplete development of respiratory control systems	Apnea of prematurity, idiopathic apnea of infancy
Cardiac	Patent ductus arteriosus; congestive heart failure
Hematologic	Anemia
Infectious	Pertussis; respiratory syncytial virus; botulism; meningitis
Metabolic	Genetic metabolic disorders; hypoglycemia; electrolyte disorders; placental transfer of sedating drugs
Neurologic	Intracranial hemorrhage; seizures; congenital central alveolar hypoventilation
Pulmonary	Hypoxemia
Gastrointestinal	Vagal stimulation with feeding; gastroesophageal reflux
Neuromuscular	Congenital myopathies; neonatal myasthenia gravis; diaphragmatic paralysis

ulus, a change in breathing mechanics, and a reduction in load compensation (see Chap. 3). The wakefulness stimulus, which includes the effects of the behavioral system for breathing control described above, prevents apnea during wakefulness unless there is a severe disturbance of respiratory control.

With sleep onset, the wakefulness stimulus ceases to act on the neuronal systems involved in the generation of rhythmic breathing. As a result, breathing during sleep is more dependent on chemical stimuli and on the chemoreceptors, respiratory muscles, and central respiratory control systems that make up a negative feedback loop to maintain consistent ventilation (see Chap. 3). The increased dependence on metabolic control leads to greater vulnerability to changes in $PaCO_2$. For example, a decrease in $PaCO_2$ of 3 to 6 mm Hg produced by passive positive-pressure hyperventilation during sleep can lead to CSA in normal subjects, whereas a similar reduction of $PaCO_2$ during wakefulness has little effect on breathing.[6] During sleep, the level of $PaCO_2$ that leads to apnea—referred to as the *apneic threshold*—may be only 1 to 2 mm Hg lower than the usual $PaCO_2$ during wakefulness.[6]

Variations in the responsiveness to ABG changes can also lead to CSA. Some patients with CSA have high ventilatory responses to CO_2 and chronically hyperventilate during sleep and while awake, apparently because of increased respiratory drive.[7] As a result of the lower concentrations of CO_2, they are closer to the threshold for apnea during sleep.

Responsiveness to ABG changes is also a function of the gain of the feedback loop involved in respiratory control. Persons with *high loop gain* have large increases in ventilation in response to hypoxia and hypercapnia that drive $PaCO_2$ down to low levels. High loop gain can lead to respiratory instability and oscillations of respiration with periods of central apnea because of the time required for blood to pass from the pulmonary capillary bed, where gas exchange takes place, to the carotid and aortic chemoreceptors. When the hypocapnic blood reaches the chemoreceptors, apnea occurs if the $PaCO_2$ is lower than the apneic threshold.

The tendency for respiration to oscillate in response to ABG changes, which is increased when loop gain is high, is reduced by stabilizing influences. The wakefulness stimulus described above is one stabilizing influence. An additional stabilizer is residual gas in the lungs, which reduces breath-to-breath fluctuations in arterial gas tensions.

On the other hand, hypoxia is a destabilizing influence. During NREM sleep, and to a lesser degree during REM sleep or wakefulness, experimentally induced hypoxia leads to hyperventilation and consequent hypocapnia, followed by apnea and then by hyperventilation again. Thus, periodic breathing that is induced by hypoxia is still dependent on hypocapnia. In such circumstances, administration of supplemental CO_2 to eliminate hypocapnia also eliminates periodic breathing despite continued hypoxia.

A decrease in $Paco_2$ during sleep and the destabilizing influence of hypoxia are the causes of CSA and associated insomnia in persons sleeping at high altitude. In normal subjects, CSA appears at altitudes as low as 2000 m, and it is almost always present above 4000 m. Central apneas and sleep disruption, which are worst during the first few nights, usually improve after the first week in persons who remain at high altitude because of compensatory changes.[8,9] On the other hand, conditions associated with metabolic acidosis, such as acute or chronic renal failure, may lead to CSA in the absence of hypoxia because of the hypocapnia associated with compensatory respiratory alkalosis (see Chap. 17).

Some central apneas are normal events that occur in response to sudden application and withdrawal of the wakefulness stimulus. For example, an awakening or arousal from NREM sleep may be associated with a sigh and two or three large breaths as the wakefulness stimulus leads to increased ventilation. With the return to sleep, the wakefulness stimulus is withdrawn, and a brief *sleep-onset central apnea or postarousal apnea* occurs. In persons with arousals from any cause, including periodic leg movements and obstructive apneas, sleep-onset central apneas may occur as the patient returns to sleep after the arousal.

CARDIAC CONTRIBUTORS TO CENTRAL SLEEP APNEA

With congestive heart failure, poor left ventricular function leads to decreased cardiac output and to increased circulation time. As a result, chemoreceptors receive delayed information regarding gas exchange and acid-base balance, so the reduction in ventilation that occurs in response to hypocapnia is delayed, leading to further hypocapnia and eventually to apnea if the apneic threshold is reached. Similarly, the rise in alveolar Pco_2 induced by apnea is not transmitted to the controller in a timely fashion, leading to hypercapnia followed by hyperventilation and a repetition of the cycle. The length of the cycle, but not the length of the apneas themselves, is proportional to the circulation time.[10] With advanced cardiac failure, central apneas and the Cheyne-Stokes respiratory pattern may occur during wakefulness because $Paco_2$ falls below the waking apneic threshold.

Although delayed feedback leads to periodic breathing, circulation times usually are not long enough to account for the full extent of breathing changes, suggesting the presence of additional contributing factors. Pulmonary congestion associated with left heart failure is an added influence; congestion stimulates lung parenchymal stretch receptors and thereby induces tachypnea and increased ventilation, which produces hypocapnia. Congestion also reduces lung volume and thus reduces body stores of O_2 and CO_2, further impairing the damping effect. Hypoxemia, if present, increases ventilatory drive and contributes to further instability of respiration; it may also reduce cardiac output and increase circulation time.

NEUROLOGIC CONTRIBUTORS TO CENTRAL SLEEP APNEA

The behavioral system for respiratory control, which originates in the cortex, and the metabolic system for respiratory control, which originates in the medulla, are integrated in the cervical spinal cord.[11] Neurons of the metabolic (i.e., automatic) system traverse the ventrolateral quadrant of the cord, whereas neurons of the behavioral system pass via the dorsolateral quadrant. Neurons of these pathways control the lower motoneurons of the cervical and thoracic spinal cord that pass via the phrenic and intercostal nerves to the respiratory muscles (see Chap. 3). Dysfunction of the neural elements of the metabolic system for breathing control may lead to CSA, whereas dysfunction of the behavioral system for breathing control is more likely to lead to irregular breathing patterns during wakefulness.

Abnormalities in any of the neural elements of the metabolic system for breathing control—chemoreceptors, autonomic pathways, medulla, descending motor pathways, motoneurons, neuromuscular junction, or respiratory muscles—can lead to CSA and other abnormal breathing patterns during sleep (see Table 14–1). Breathing patterns during wakefulness are less affected because the wakefulness stimulus and the behavioral system blunt the impact of metabolic system dysfunction.

Chemoreceptor dysfunction is the cause of CSA in patients with congenital or acquired central hypoventilation syndromes (see section on this topic below). In patients with autonomic dysfunction, information from chemoreceptors may not be transmitted correctly to the central nervous system (CNS) respiratory controller or may not be integrated properly, or both, leading to irregular breathing patterns during sleep, with intermittent central apneas and hypopneas. Lesions of the medulla and cervical spinal cord can cause CSA or hypoventilation, or both, sometimes without other significant neurological abnormalities.[12] Lower motoneuron dysfunction, neuromuscular junction disorders, and muscle diseases can also cause CSA or hypoventilation (see below).

Incomplete development of the metabolic control system for breathing is the cause of central apneas in premature infants. Brief central apneas may occur several times per hour, triggered by neck flexion or other body movement, hiccups, and regurgitation. They become less frequent with increasing age and are usually absent by term. Obstructive and mixed apneas also occur in premature infants, in part because preinspiratory activation of upper airway dilators, which stabilizes the airway, may not occur or may not be properly timed. CSA in otherwise normal infants born at term, sometimes called *idiopathic apnea of infancy*, is presumed to be caused by immaturity of CNS control of breathing (i.e., incomplete connections of brainstem neuronal groups involved in inspiration, expiration, and respiratory rhythm generation). With maturation of the CNS, the apneas become less frequent and eventually disappear.

Brief central apneas often occur in normal persons during REM sleep, particularly during phasic REM sleep, probably because the increased tonic inhibition associated with phasic activity affects brainstem respiratory neurons as well as spinal α-motoneurons. These apneas usually last 4 to 8 seconds and are almost never of clinical significance.

UPPER AIRWAY OBSTRUCTION AS A CAUSE OF CENTRAL SLEEP APNEA

Upper airway obstruction can lead to central apneas, particularly in patients who also have obstructive apneas. In such cases, airway muscle relaxation during expiration leads to airway occlusion, and upper airway afferents activated by the occlusion trigger reflex inhibition of inspiration. The inhibition of inspiratory effort may continue until an arousal occurs and breathing resumes. The arousal is often accompanied by a snore or a snort, similar to the termination of an obstructive apnea. If breathing efforts resume before the arousal occurs, the result is a mixed apnea (see Fig. 13–4)

Differential Diagnosis

ADULTS AND CHILDREN

In many cases of CSA in children and adults, the main differential diagnosis is OSA. Loud snoring and daytime sleepiness are usually symptoms of OSA, whereas soft breathing with reports of rhythmic variation in breathing effort are more suggestive of CSA with Cheyne-Stokes respiration. A prominent complaint of insomnia with frequent awakenings is also more suggestive of CSA than of OSA. Clinical differentiation, however, is difficult, particularly because many patients with central apneas also have obstructive apneas, and monitoring during sleep is almost always required.

INFANTS

Apnea in an infant is a diagnostic challenge because it may be a normal event associated with sleep, a feature of a benign disorder that will resolve with time, a sign of a serious disorder, or an indication of increased risk for sudden infant death syndrome (SIDS) (see section on this topic, below). For parents, apnea may be a source of great anxiety because of concerns about the possibility of sudden death during sleep. For apneas occurring during sleep, diagnostic considerations include OSA; idiopathic apnea of infancy; apneas associated with feeding or reflux; apneas associated with CNS disorders, particularly brainstem disorders; and apneas induced by seizures. Choking or gasping that accompanies or follows the apnea suggests the possibility of OSA or gastroesophageal reflux. Breath-holding spells, a consideration for apneas that occur during wakefulness, do not occur during sleep.

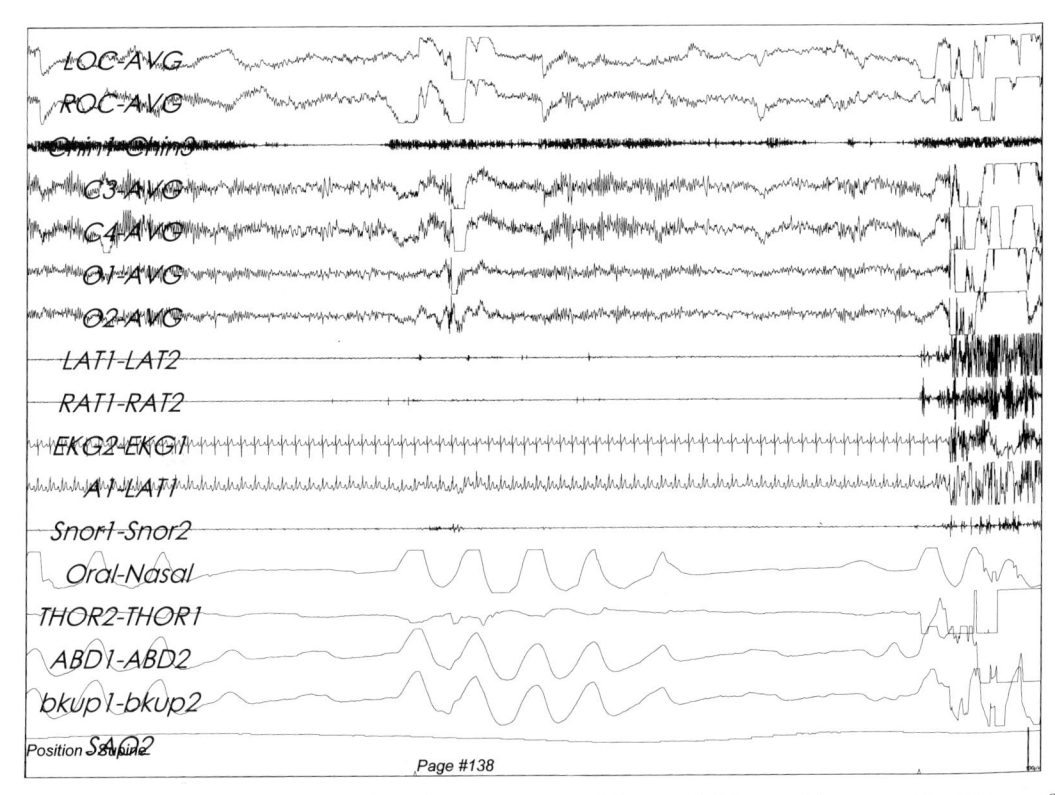

Figure 14–1. One-minute segment of a polysomnogram in a 340-pound, 38-year-old woman with a history of snoring and daytime sleepiness. Two obstructive apneas are shown. Because of the low-amplitude signal in the effort channels (THOR–THOR2, ABD1–ABD2, bkup1–bkup2), the respiratory events could be mistaken for central apneas, especially if the effort channels were recorded at a lower sensitivity. The EEG electrodes (A1, A2, C3, C4, O1, O2) were placed in accordance with the International 10-20 system. Other abbreviations: LOC, left outer canthus; ROC, right outer canthus; AVG, average reference; Chin1–Chin3, submental EMG; LAT1–LAT2, surface EMG electrodes over the left anterior tibialis; RAT1–RAT2, surface EMG electrodes over the right anterior tibialis; EKG2–EKG1, electrocardiogram; A1–LAT1, electrocardiogram; Snor1–Snor2, snoring sound; Nasal-Oral, nasal-oral airflow; THOR1–THOR2, thoracic wall motion; ABD1–ABD2, abdominal wall motion; bkup1–bkup2, backup respiratory effort (thoracoabdominal wall motion); SAO2, pulse oximetry by finger probe.

Diagnostic Evaluation

ADULTS AND CHILDREN

On polysomnography, the main tool for diagnosis, the pattern of respiratory effort differentiates obstructive apneas from central apneas. In markedly obese individuals, respiratory effort is sometimes difficult to detect because the increase in circumference of the thoracoabdominal cavity associated with inspiration is proportionately less than in persons of normal weight, as is the change in impedance across the thoracoabdominal contents. The result is a reduced signal in transducers used to assess respiratory effort (Fig. 14–1). In addition, obese individuals may have reduced tidal volume as a result of the increased load on the respiratory muscles. For these reasons, the respiratory effort during apneas, particularly during rapid eye movement (REM) sleep, may be almost undetectable with standard transducers. Esophageal pressure monitoring may be helpful in such cases because these transducers are sensitive to small changes in intrathoracic pressure; however, extremely obese individuals may not generate much intrathoracic pressure because of their thoracoabdominal mass. Some patients have mixed apneas, which, if thoracoabdominal movements are reduced in amplitude, may simulate central apneas (Fig. 14–2).

Patients with CSA caused by metabolic or cardiac disorders have Cheyne-Stokes breathing with regular waxing and waning of breathing effort that is most prominent during light NREM sleep. The pe-

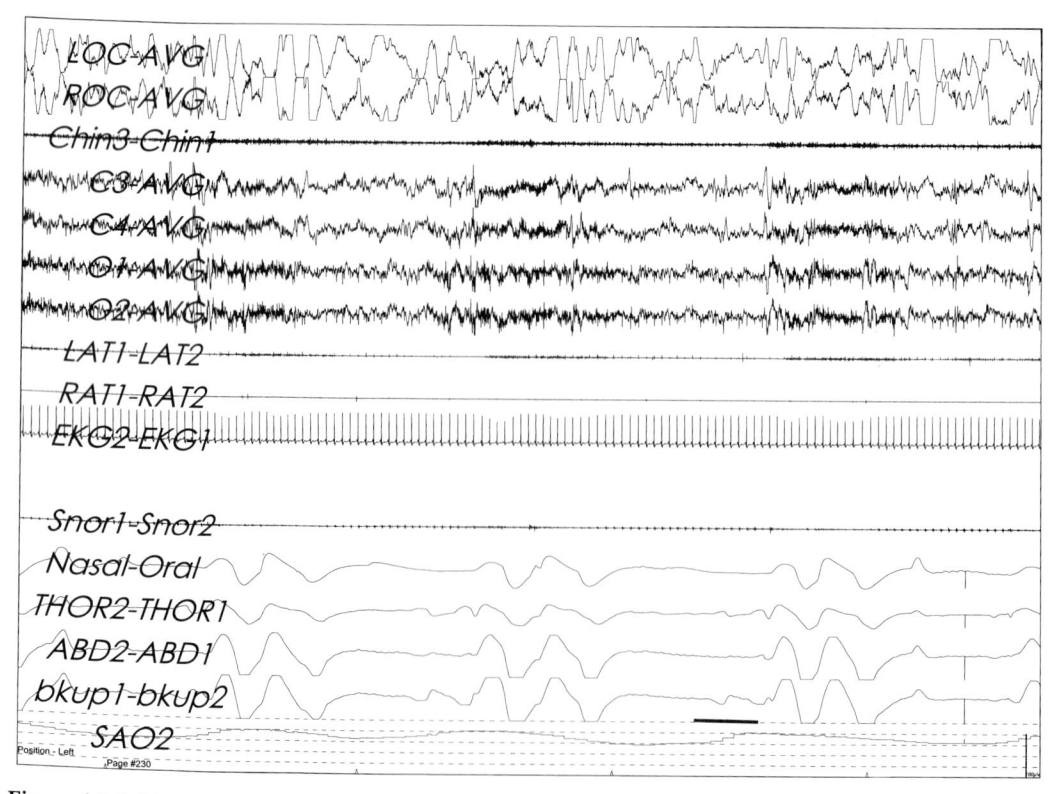

Figure 14–2. Two-minute segment of a polysomnogram in a 260-pound, 45-year-old man with a history of snoring and daytime sleepiness. A series of mixed apneas is apparent; however, the amplitude of respiratory wall motion during obstructed breaths (underlined) is low in the thoracic effort channel (THOR2–THOR1) and almost imperceptible in the abdominal effort channel (ABD2–ABD1), indicating that the event could be mistaken for a central apnea. A backup respiratory effort channel (bkup1–bkup2), with sensors placed between the thoracic and abdominal sensors, shows the respiratory effort more clearly. For abbreviations, see Figure 14–1.

riod of apnea is followed by a slow increase in respiration, with an arousal that usually occurs at the apex of the cycle, rather than concurrent with resumption of breathing (Fig. 14–3).[13] In CSA caused by upper airway occlusion, however, the arousal usually occurs with the initial resumption of breathing (Fig. 14–4). Thus, correlation of the electroencephalogram (EEG) with indices of respiratory effort and flow is required for accurate diagnosis.

Additional laboratory studies are often needed if a metabolic, cardiac, neurological, or pulmonary disturbance is the suspected cause of CSA. Daytime ABGs can be used to determine whether waking respiratory acidosis is present; if so, a disorder affecting lung function, respiratory muscles, or ventilatory drive is probably contributing to CSA.[14] Daytime hypercapnia without acidosis may be a response to metabolic alkalosis brought about by compensation for nocturnal respiratory acidosis. Spirometry in the supine and sitting positions helps to determine whether diaphragm dysfunction is present; a reduction of vital capacity by more than 25% in the supine position strongly suggests bilateral diaphragm weakness or paralysis. Electromyography (EMG) is indicated if a neuromuscular disease is suspected, and ventilatory response to hypercapnia and hypoxia should be assessed if central alveolar hypoventilation is suspected (see section on this topic, below). If clinically occult congestive heart failure is suspected, two-dimensional echocardiography may be indicated. Although computerized tomography or MRI studies of the brain may be useful in selected cases, they are not required for a patient with a normal

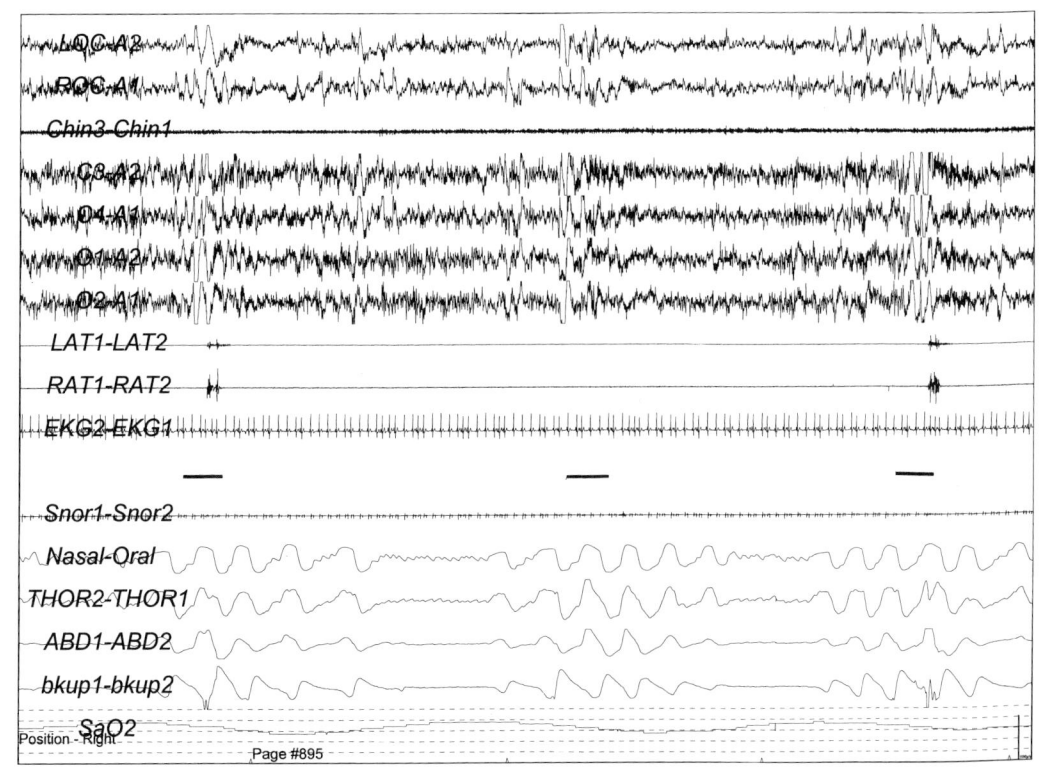

Figure 14–3. Two-minute segment of a polysomnogram in a 54-year-old man with congestive heart failure, atrial fibrillation, and a history of snoring and daytime sleepiness. A series of central apneas with Cheyne-Stokes breathing is shown. The arousals between apneas (underlined) tend to occur at the apex of breathing. In contrast, arousals occur more often at the time of resumption of breathing in central apneas caused by upper airway obstruction (see Figures 13–3 and 14–4. For abbreviations, see Figure 14–1).

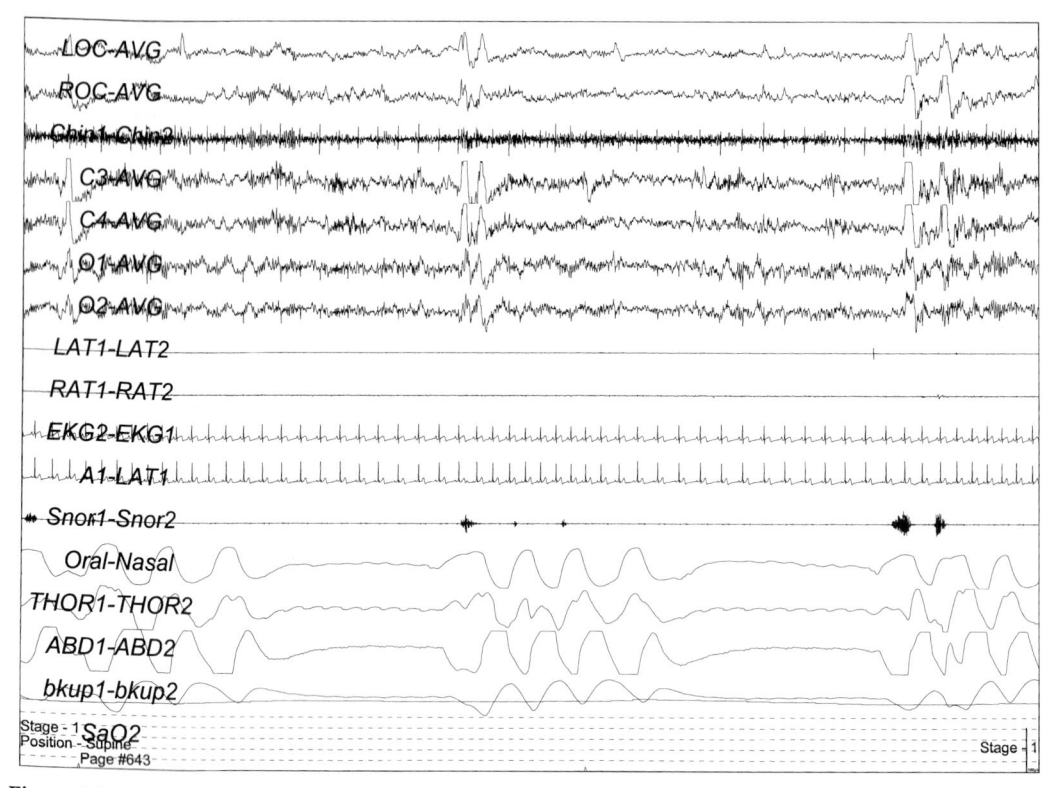

Figure 14–4. One-minute segment of a polysomnogram in a 49-year-old man with a history of snoring and daytime sleepiness. The segment shows two central apneas. In contrast to Figure 14–3, arousals with snoring occur at the resumption of breathing, a pattern that is often due to upper airway obstruction. For abbreviations, see Figure 14–1.

neurological examination and uncomplicated CSA.

INFANTS

For infants with apnea, the history and examination help to determine whether extensive evaluation is required.[15] The time of day or night during which the apnea occurred, its relationship to sleep or wakefulness, its duration, the types of interventions used, and the infant's response to the interventions should be determined. The coincident occurrence of snoring, choking, gasping, limpness, or stiffening should be elicited from the parents or caretakers, who should also be questioned about the relationship of the event to body position, feeding, or infection. For infants with recurrent events, home audio or video recordings may be useful.

Events of greater concern include prolonged apneas, frequent brief apneas, periodic breathing, apnea with color change, and apnea with tonic postures or jerking movements. For infants with such events, a complete blood count, chest radiograph, electrocardiogram (ECG), and EEG are often indicated, and additional tests for infection may also be required. Urine tests for toxicology may be indicated, and if a metabolic disorder is suspected, biochemical testing is indicated. If abuse is a consideration, funduscopic examination as well as long-bone examination and radiographs may be required, and appropriate social agencies should be notified.

Sleep recordings are often used to assess apnea in infants. Two-channel pneumocardiograms, introduced in the 1970s,[16] are useful for detecting central apneas and bradycardias, but they do not detect obstructive apneas reliably. Four-channel monitors, which typically include air flow and oxygen saturation, are better able to detect obstructive apneas, but they do not provide information about snoring, body position, or sleep stage. Twelve-channel and 18-channel recorders may include all of the above, along with esophageal pH monitoring, which helps to determine whether apneas are associated with episodes of reflux. Laboratory polysomnographic studies permit even more detailed assessments, such as monitoring of P_{CO_2}, intrathoracic pressure, behavior, and multichannel EEG.

No consensus currently exists on the indications for the various types of studies, and the clinical significance of findings on these studies is often uncertain. Infants with ALTEs who later die of SIDS are more likely to have obstructive or mixed apneas on a nocturnal recording than infants with ALTEs who survive,[17] but the group differences are not large enough to make accurate predictions. A single prolonged apnea is probably of greater significance in infants than several short ones, partly because infants have low total body oxygen stores and therefore become hypoxemic during apneas much more easily than adults, particularly with apneas that begin at the end of expiration. Most authorities agree that apneas greater than 20 seconds are abnormal. Although some believe that apneas of greater than 15 seconds are abnormal,[18] normal neonates may have apneas lasting up to 18 seconds.[19] Short apneas are more likely to be significant if accompanied by bradycardia, cyanosis, pallor, or oxygen desaturation.

Unfortunately, abnormal results of pneumograms, sleep polygraphs, and EEGs are not strongly predictive of the severity of subsequent episodes,[20] and therefore the potential value of a given study for diagnosis and management should be carefully considered before the study is obtained.[21]

Management

Treatment of CSA is most successful if it is directed at the underlying cause. Thus, for CSA associated with congestive heart failure, optimal management of ventricular dysfunction should be the initial therapy. Treatment for CSA that is not accompanied by daytime hypoventilation is discussed in this section. Treatment of CSA that is associated with hypoventilation is discussed in the following sections on obesity-hypoventilation syndrome and neuromuscular disorders.

POSITIVE AIRWAY PRESSURE

Nasal CPAP is helpful when CSA is caused by upper airway obstruction,[22] and it also may help in some patients with CSA associated with congestive heart failure (see Chap. 17).[23,24] Some patients with CSA (see Case 14–1, above) respond better to BPAP than to nasal CPAP. BPAP produces a higher pressure during inspiration than during expiration[25]; differential stimulation of upper airway pressure-sensitive receptors during inspiration and expiration or the mild inflation of the lungs induced by the pressure difference may stimulate inspiratory efforts and contribute to the increased likelihood that CSA will respond.

Although nasal CPAP and BPAP are sometimes helpful for CSA, they can also induce central apneas, presumably related to an increased expiratory load or to stimulation of the upper airway (Fig. 14–5). The occurrence of central apnea with CPAP or BPAP can make it difficult to determine the best pressure settings for the equipment, as there is sometimes only a small range of effective pressures.

OXYGEN AND CARBON DIOXIDE

Supplemental oxygen is helpful in CSA if hypoxia during apneas contributes to respiratory instability or impairs cardiac output. In patients with Cheyne-Stokes breathing related to low cardiac output, oxygen can reduce the amount of time spent in Cheyne-Stokes breathing during the night, reduce the number of arousals, and reduce the amount of stage 1 sleep.[26] For CSA that is not associated with congestive heart failure or significant hypoxemia, supplemental oxygen has little value.

Although 3% inhaled CO_2 appears to be effective in eliminating CSA and periodic breathing associated with congestive heart failure, the equipment required to deliver controlled amounts of CO_2 is cumber-

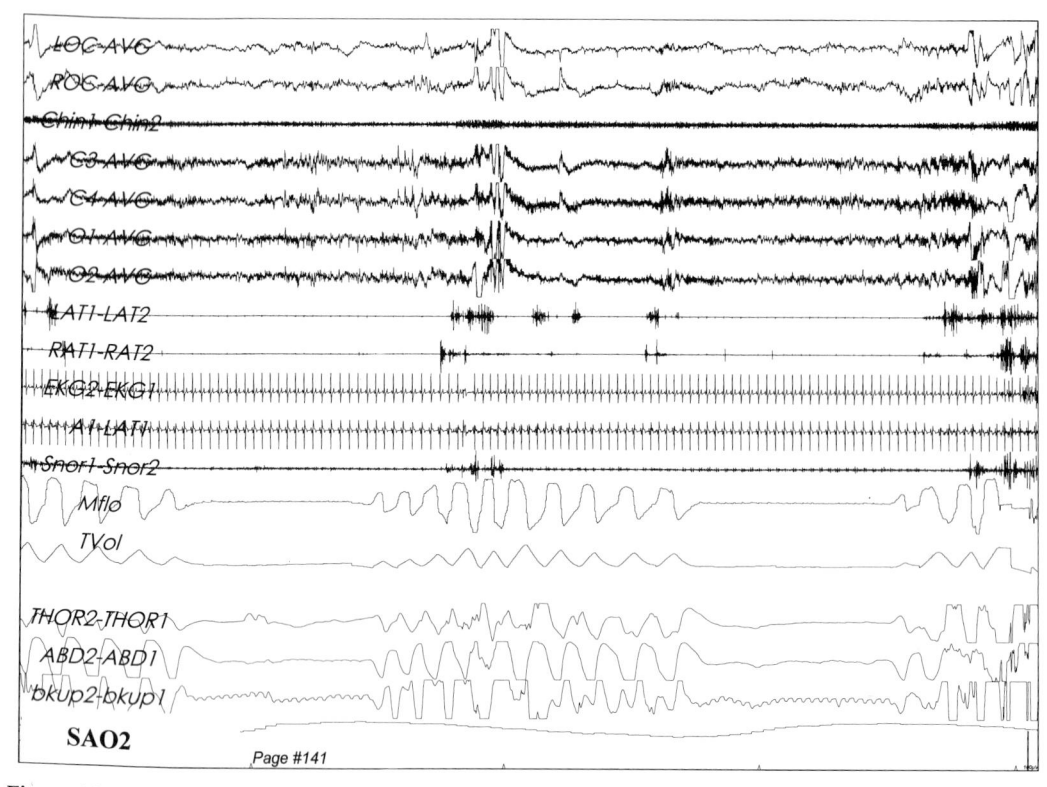

Figure 14–5. Two-minute segment of a polysomnogram with a trial of nasal CPAP in a 61-year-old man with hypertension, coronary artery disease, diabetes mellitus, and a history of snoring and apneas observed by others. An initial polysomnogram showed frequent apneas and hypopneas (67 respiratory events per hour of sleep) which included central apneas, mixed apneas, and obstructive apneas. With nasal CPAP, the obstructive apneas were eliminated; however, central apneas persisted. The segment, recorded with nasal CPAP at 11 cm H_2O, shows two central apneas. In addition, a pattern of Cheyne-Stokes breathing is evident, with an arousal at the apex of breathing. Central apneas were reduced in frequency, although not entirely eliminated, with nasal BPAP. (Mflo=metered air flow rate; Tvol=tidal volume.) For other abbreviations, see Figure 14–1.

some, uncomfortable, and for the most part unavailable.[27,28]

MEDICATIONS

Acetazolamide inhibits carbonic anhydrase and facilitates bicarbonate diuresis, thus reducing the respiratory alkalosis and hypocapnia that can trigger central apneas. Some patients with CSA obtain lasting benefit from a dose of 250 mg at bedtime.[29,30] In addition, for CSA associated with high altitude, acetazolamide 250 mg three times daily beginning before ascent reduces the amount of CSA and associated sleep disruption.[8,31,32]

For patients with CSA caused by frequent arousals and associated ventilatory instability, benzodiazepine hypnotics are occasionally useful, provided that the apneas are associated with little or no hypoxemia.[33–35] However, hypnotics must be used cautiously in such patients because of the potential to exacerbate coexisting OSA.

Other medications are of limited value. Although medroxyprogesterone reduces periodic breathing, it has significant side effects and does not improve sleep continuity. Almitrine, a respiratory stimulant that is not available in the United States, enhances the vasoconstriction response to hypoxia and improves ventilation-perfusion matching and oxygenation, but it also augments the hypoxic response and can increase periodic breathing.[8,31] Theophylline and naloxone are generally not helpful.

MANAGEMENT OF INFANT APNEA

In most cases, the course of idiopathic apnea of infancy is benign, with gradual resolution during the first year of life. Methylxanthines, such as theophylline or caffeine, may be of value during the symptomatic period.[36] If gastroesophageal reflux contributes to apnea, small feedings and a prokinetic agent such as cisapride may be useful. Surgical repair may be needed if reflux does not respond to medical treatment. The management of infants with prolonged apneas or ALTEs is discussed in the following section.

SUDDEN INFANT DEATH SYNDROME

And it came to pass . . . that this woman was delivered also: and we were together; there was no stranger with us in the house, save we two in the house. And this woman's child died in the night; because she overlaid it.

1 KINGS 3:18–19

SIDS, defined as unexpected sudden death in an infant in whom no cause of death is apparent despite thorough investigation, has occurred for thousands of years. With a prevalence of 1 per 1000 live births, it is a major cause of death in the first year of life.[37] The highest frequency of SIDS cases is found among infants 2 to 4 months of age, and 90% of cases occur in infants younger than 6 months.

Clinical Features

The clinical features of the syndrome are straightforward. In the usual case, an apparently healthy infant, assumed to be asleep, is found dead in a crib. In fewer than 10% of SIDS cases, apneas were observed before death.

Risk factors for SIDS are shown in Table 14–3.[37,38] One of the most important risk factors for SIDS is sleeping position. The risk of SIDS is three to seven times greater in infants who sleep in a prone position compared to infants who sleep in other positions, even after adjusting for confounding variables.[39–41] After ages 5 to 6 months, the infant is able to roll over easily and the risk associated with the prone position is probably negligible.[37] The low rate of SIDS in many Asian populations may be a result of a cultural preference for the supine sleeping position.[38]

Table 14–3. Risk Factors for Sudden Infant Death Syndrome (SIDS)[43]

Young, unwed mother
Maternal smoking or substance abuse
Maternal depression
Short interpregnancy interval
Low socioeconomic status
Deficient prenatal care
One or more siblings with SIDS or near-miss SIDS
Preterm birth
Low birth weight
Formula feeding
Prolonged apneas
Excessively warm sleeping environment
Prone sleeping position

Source: Adapted from Guntheroth and Spiers,[43] Table 1, with permission.

Biologic Basis

Although bradycardia may occur during apneas, the proximate cause of death is most likely to be respiratory arrest (apnea) rather than cardiac arrest.[42] The factors responsible for the respiratory arrest remain uncertain. Maternal overlaying, suspected as a cause for centuries, became a less plausible explanation as separate sleeping areas for infants became commonplace. In the 19th and early 20th centuries, SIDS was thought to be caused by status thymicolymphaticus—airway blockage by an enlarged thymus gland combined with a hypoplastic aorta.[37,43] More recent theories have emphasized the possible roles of impaired arousal mechanisms, infection, inflammation, and accidental or intentional suffocation.

The occurrence of apneas in normal infants suggests that an abnormal or absent arousal response to apnea, rather than apnea itself, leads to SIDS. Arousal responses to apnea are underdeveloped or absent before birth because breathing does not occur in utero; delayed or abnormal development of arousal responses could therefore contribute to SIDS, and the maturation of arousal responses could account for the decline in the occurrence of SIDS after the first few months of life.[44] Although defective arousal cannot explain the low rate of SIDS during the first few weeks of life, the newborn gasping mechanism may provide protection during this period.[45]

In some infants who die while sleeping in the prone position, accidental suffocation may be the cause of death. Death from suffocation may occur when infants are placed on waterbeds or cushions, or on hard surfaces (e.g., table tops) that compress the infant's nose.[46,47] Intentional suffocation (homicide) must be considered, especially if SIDS occurred in one or more older siblings or if the deceased infant has life insurance.

Infections may trigger SIDS in infants at risk. Many SIDS victims have evidence of minor viral respiratory infections, and the risk of SIDS is twice as high in the winter as in the summer, possibly because of the seasonal increase in viral infections.[48–50] Increased deep sleep as a result of high levels of interleukin-1 is one mechanism by which infection might affect the likelihood of arousal from apnea.[49]

Upper airway obstruction could also be a contributor in some cases. Families who have at least one family member with a diagnosis of OSA are more likely to have an infant die of SIDS than families with no history of OSA.[51]

Diagnosis

Because SIDS is a diagnosis of exclusion, postmortem examination and death scene evaluation are required to rule out infection, toxic ingestion, or cardiac, congenital, or metabolic defects. Retinal hemorrhages, intracranial bleeding, and long-bone fractures suggest prior abuse and should raise the suspicion of homicide.

Prevention

In populations in which the prone sleeping position is common, interventions to discourage its use are an effective means of prevention. For example, the rate of SIDS was reduced significantly in England, the Netherlands, and Washington state after public education programs to promote the supine sleeping position.[37,52,53] Increased publicity concerning the risks of the prone sleeping position probably contributed to a decline in the prevalence of prone sleeping in the United States from 74% in 1992 to about 25% to 30% in 1995,[54] and it may have accounted for a large part of the decline in the US rate of SIDS from 1.5 in 1000 to 1.0 in 1000 live births that occurred between the mid-1980s and 1995.[37] Despite parental concerns about choking, the risk of aspiration for normal infants is not increased in the supine position.

Although bed sharing appears to increase the risk of SIDS, infants who sleep in their own bed in the same room as one or more parents have a reduced risk of SIDS, suggesting that widespread adoption of room-sharing by parents and infants, at least for the first few months after birth, might reduce SIDS mortality.[55]

Interventions to reduce teen-age pregnancies, reduce maternal alcohol and illicit drug abuse during pregnancy, improve prenatal and postnatal care, and improve maternal education and nutrition would probably lead to a decline in the incidence of SIDS.[56] Unfortunately, however, social programs designed to achieve these goals generally have not been successful.

MANAGEMENT OF INFANTS WITH APPARENT LIFE-THREATENING EVENTS

Among infants with ALTEs, about 30% to 50% have additional episodes of prolonged apnea and 3% to 7% die of SIDS.[57] Long-term follow-up of infants who survive suggests that subsequent development is for the most part unremarkable.[58] In an attempt to reduce the risk of SIDS, infants in the United States with this diagnosis generally are sent home on monitors, which usually assess chest or abdominal movement as well as ECG or heart rate. In the United States, although convincing data that such monitoring reduces the rate of SIDS are lacking, most parents of infants who have had prolonged apneas prefer to have home monitoring, and the associated reduction in parental anxiety is one positive result. Parental training in cardiopulmonary resuscitation techniques may also allay anxiety. The presence of the monitors also may reduce the frequency of emergency room visits and hospital admissions of an infant whose family is hypersensitive to apneas.[59]

CENTRAL ALVEOLAR HYPOVENTILATION SYNDROME

Central alveolar hypoventilation (CAH) syndrome, often called Ondine's curse, is a rare disorder of automatic ventilation that may be congenital or acquired. In Germanic mythology, Ondine, a water nymph, married a mortal man, Hans, who promised never to marry a mortal woman. However, when Ondine returned to the sea, Hans broke his promise. In retaliation, Ondine placed a curse on him that required him to remember to breathe. When he finally fell asleep, he was unable to breathe and died. In the play, *Ondine*, published by Giraudoux in 1939, Hans explains how hard it is to live with his curse: "A single moment of inattention, and I forget to breathe. He died, they will say, because it was a nuisance to breathe . . ."

Clinical Features

CAH causes prolonged apneas and periods of hypoventilation that are worse during sleep. In severe cases, cyanosis and apneas are apparent on the first day of life and require mechanical ventilation. Less severe cases may present in infancy with episodic cyanosis and apnea, or cyanosis during feedings. Recurrent aspiration, difficulties with swallowing, and seizures resulting from hypoxia may complicate the picture and obscure the diagnosis.

In mild cases, the diagnosis is not made until childhood or even adulthood, when a relatively minor respiratory infection leads to respiratory failure with hypercapnia, which then persists after resolution of the infection. Acquired cases may present at any age and may manifest as insomnia, apneas observed by others, sleepiness, or daytime hypercapnia.

Biologic Basis

The cause of CAH is dysfunction of respiratory control systems that leads to a reduced or absent response to hypercapnia. Responsiveness to hypoxia is generally much less affected, and voluntary and nonchemical systems for control of breathing are usually normal, although persons with CAH may not develop respiratory discomfort or a subjective sense of shortness of breath.[60]

The disturbance may occur as a result of chemoreceptor dysfunction, brainstem dysfunction, or high cervical cord dysfunction. Disorders that have been associated with infantile CAH include ganglioneuroblastoma, Hirschsprung's disease, Leigh's syndrome, and familial dysautonomia. The association with Hirschsprung's disease and with neuroblastoma suggest that a defect in neural crest cell growth, migration, or differentiation is the cause in most congenital cases.[61] Acquired CAH is usually caused by medullary stroke or tumor, or by lesions of the upper cervical spinal cord.

Diagnosis

In cases presenting in infancy, infections, cardiac defects, toxic ingestions, metabolic disorders, and neuromuscular diseases need to be considered. For acquired cases, brainstem MRI is indicated to determine whether a structural lesion is present.

To confirm the diagnosis, sleep studies, ABGs, and evaluation of hypoxic and hypercapnic ventilatory responsiveness are useful. Polysomnography, most informative when combined with PCO_2 monitoring, usually shows slow, shallow respiratory efforts with progressive hypoxemia and hypercapnia during NREM sleep. Although ventilation may increase in some patients when $PaCO_2$ reaches 70 to 90 mm Hg,[62] severely affected patients continue to hypoventilate with $PaCO_2$ values of more than 100 mm Hg, and the associated acidosis and hypoxemia may lead to dangerous ventricular arrhythmias. Arousal eventually occurs, followed by increased ventilation, a decrease in $PaCO_2$, a return to sleep, and repetition of the cycle. Unlike hypoventilation caused by neuromuscular disease, which is usually worse in REM sleep (see Hypoventilation Associated with Neuromuscular Disorders, below), hypoventilation with CAH is usually worse in NREM sleep because of greater dependence of respiration on chemical drive. Assessment of ventilatory responsiveness to hypercapnic challenge during sleep, a technically difficult procedure, is sometimes necessary for definitive diagnosis.[62]

Management

CAH in infants usually does not improve and, without treatment, may lead to pulmonary hypertension, right heart failure, and death during sleep. Severely affected infants require ventilatory assistance during both sleep and wakefulness. When assisted ventilation is not required during wakefulness, nocturnal nasal positive-pressure ventilation may be an acceptable solution.[63,64] Some children use a negative-pressure cuirass ventilator, which can be combined with nasal CPAP if upper airway collapse is a problem. If these approaches do not succeed, ventilation via tracheostomy or diaphragm pacing with implantable phrenic nerve stimulators are other options.[65] Mild to moderate developmental delay and learning disabilities are usually present in long-term survivors, and some develop pulmonary hypertension and right heart failure as a consequence of chronic hypercapnia.[66]

OBESITY-HYPOVENTILATION SYNDROME

Osler noted that "an extraordinary phenomenon in excessively fat young persons is an uncontrollable tendency to sleep—like the fat boy in Pickwick."[67] Although Joe, the fat boy of Dickens' *The Pickwick Papers,* is the most famous literary case of OSA combined with the obesity-hypoventilation syndrome, Denys, the tyrant of Heraclea, was an even more striking example. According to Athenaeus of ancient Greece, he was so enormous that he was in constant danger of suffocation; most of the time he was in a stupor or asleep. To keep him from falling asleep when he was sitting, needles were put in the backs of his chairs.[68]

Clinical Features

Obesity-hypoventilation syndrome, often referred to as pickwickian syndrome, is characterized by obesity, hypercapnia, and hypersomnolence. Obesity is often severe; most patients are more than 100% above their ideal body weight. OSA with associated snoring is almost always present and is usually severe; it is the apnea and associated arousals, rather than hypercapnia, that accounts for most of the sleepiness.

Biologic Basis

Hypoventilation in this syndrome is caused by the excess adipose tissue. The large abdomen pushes the diaphragm rostrally while chest obesity restricts chest wall motion; the result is reduced tidal volume and reduced functional residual capacity. The large mechanical load on the respiratory muscles increases the work of breathing, particularly when the patient is in a horizontal position, and can lead to respiratory muscle fatigue.

Hypoventilation is worse during sleep for at least three reasons:
1. The patient is horizontal; consequently, there is a lower functional

residual capacity and higher impedance to diaphragm descent.
2. Upper airway narrowing during sleep leads to even greater work of breathing.
3. Ventilatory responsiveness is reduced during sleep compared to wakefulness. As the night progresses, hypoventilation may become more severe because of progressive respiratory muscle fatigue and acidosis-induced respiratory muscle dysfunction.

Although the mechanical load of adipose tissue is the major cause of respiratory failure in obesity-hypoventilation syndrome, the absence of hypercapnia in many massively obese patients indicates that other factors play a role as well. The distribution of obesity is significant: Patients with upper body obesity are at greater risk for this syndrome. Reduced ventilatory responsiveness to hypercapnia or hypoxemia is also a suspected contributor; however, patients often have normal ventilatory responses after weight loss, suggesting that reduced ventilatory response is a functional adaptation to excess weight. Furthermore, hypercapnia that is worse during sleep combined with compensatory elevations of blood bicarbonate, rather than an intrinsic defect in ventilatory responsiveness, could account for hypoventilation and flattening of the ventilatory response curve.[69] Respiratory muscle weakness and fatigue may also be contributing factors, as there is evidence that respiratory muscle strength tends to be lower in patients with obesity-hypoventilation syndrome than in equally obese patients without obesity-hypoventilation.[70]

Diagnosis

Obesity-hypoventilation syndrome is a consideration in any obese person with daytime hypercapnia. Differential diagnosis includes chronic obstructive pulmonary disease, neuromuscular disease, disorders of the chest wall, myxedema, and central alveolar hypoventilation. Obesity-hypoventilation syndrome also should be considered in obese patients with OSA who are found on polysomnographic studies to have hypoxemia during nonapneic periods. Diagnostic considerations in such patients include the other causes of hypoventilation mentioned above as well as pneumonia, atelectasis, pulmonary congestion, and other causes of ventilation-perfusion mismatch.

For patients with suspected obesity-hypoventilation syndrome, evaluation may include pulmonary function tests in the sitting and supine positions, ABGs, chest radiographs, thyroid function tests, polysomnography with monitoring of P_{CO_2}, and assessment of ventilatory responsiveness to hypoxia and hypercapnia.

Management

The goal of management is to reduce the work of breathing, especially during sleep. Weight loss is the most important measure. Nasal CPAP, effective for coexisting OSA, also reduces the work of breathing by reducing upper airway resistance and by providing positive-pressure support. Nasal CPAP can be used while the patient loses weight or if attempts at weight loss are unsuccessful. Although BPAP provides little inspiratory assistance in patients with obesity-hypoventilation syndrome because of the large load on inspiratory muscles, some patients, for uncertain reasons, respond better to BPAP than to CPAP. Tracheostomy also reduces the work of breathing and is an effective, albeit drastic, treatment. Although progesterone 20 mg three times per day reduces hypercapnia and hypoxemia to a mild degree, its clinical benefits, in my experience, are minimal, and it can cause sexual dysfunction in men.

HYPOVENTILATION ASSOCIATED WITH NEUROMUSCULAR DISORDERS

Any neuromuscular disorder that affects respiratory muscles can cause hypoventilation, and respiratory failure is a common cause of death in patients with progressive neuromuscular disorders. Although the epidemic of poliomyelitis in the 1940s and 1950s focused increased attention on the respiratory problems of neuromuscular diseases, sleep-related breathing disorders that were not immediately life-threatening were largely ignored until the 1980s. Since then, the scope of the problem has become more apparent; in one series of patients from a neuromuscular disease clinic, sleep-related breathing disturbances occurred in 42%.[71]

Clinical Features

Sleep-related complaints in patients with hypoventilation associated with neuromuscular disorders may include excessive sleepiness, apneas observed by others, inability to sleep supine, breathlessness in the prone or supine positions, waking up gasping or fighting for breath, and morning headaches. Examination features indicating increased risk for hypoventilation include proximal weakness, scoliosis, and reduced breath sounds. Dysarthria and dysphagia resulting from bulbar muscle dysfunction suggest that upper airway obstruction may occur during sleep, contributing to sleep-related hypoventilation.

A sleep-related breathing disturbance leading to disrupted sleep, morning headaches, and daytime sleepiness occasionally is the presenting manifestation of amyotrophic lateral sclerosis or a static myopathy, as in the following case.

CASE HISTORY: Case 14–2

A 60-year-old woman was admitted with respiratory failure. She had gained approximately 20 lb over the previous 3 to 4 years, and for about 1 year she had difficulty getting out of a low chair and standing up from a squatting position. Three months before admission, she experienced daytime sleepiness that worsened progressively, and she felt short of breath when climbing stairs. The day before admission, ankle edema and profound somnolence developed.

She took no medications. Her brother had "weakness" as a child. Examination revealed left convex kyphoscoliosis, reduced breath sounds at the left lung base, and moderate symmetric proximal weakness. ABGs revealed a Po_2 of 35 mm Hg, a Pco_2 of 77 mm Hg, and a pH of 7.34. After hospitalization and diuretic treatment, her oxygenation during wakefulness improved. During the first 6 hours of a polysomnogram, her oxyhemoglobin saturation (SaO_2) was 92% while awake and sitting, 89% while awake and supine, 80% to 86% while in NREM sleep, and 58% to 70% while in REM sleep. During the last 2 hours of the polysomnogram, the SaO_2 gradually declined to 35% to 50%. Although oxygenation during sleep did not improve with nasal CPAP on a subsequent study, it was almost normal with nasal BPAP combined with 1 L/min of supplemental oxygen. A quadriceps muscle biopsy revealed findings consistent with nemaline myopathy. Three weeks later, her sleepiness and ankle edema had improved; neurological examination was unchanged. ABGs showed a Po_2 of 74 mm Hg, a Pco_2 of 54 mm Hg, and a pH of 7.41.

In this patient with congenital nemaline myopathy, nocturnal hypoventilation was probably precipitated by weight gain. ABGs and polysomnography revealed hypoventilation and hypoxia that were severe in REM sleep and that responded to BPAP.

Biologic Basis

In neuromuscular diseases associated with hypoventilation (Table 14–4), weakness of the diaphragm and chest wall muscles reduces the forced vital capacity and alters chest mechanics, which leads to less efficient breathing and increased work of breathing. In this setting, several pathophysiologic mechanisms contribute to even greater impairments of ventilation during sleep. First, minute ventilation normally decreases with the onset of sleep. Second, diaphragmatic workload increases in NREM sleep and increases even further in REM sleep; as a result, patients with marginal respiratory reserve while awake may be unable to perform the increased respiratory work. Third, acces-

Table 14–4. **Neuromuscular Diseases Commonly Associated with Nocturnal Hypoventilation or Sleep Apnea**

Class of Disorder	Examples
Motor neuron diseases and motor neuropathies	Amyotrophic lateral sclerosis, poliomyelitis, syringomyelia, Guillain-Barré syndrome, chronic inflammatory demyelinating polyneuropathy, hereditary sensorimotor neuropathies, phrenic nerve palsies, Werdnig-Hoffmann disease
Neuromuscular junction disorders	Myasthenia gravis
Muscle diseases	Congenital myopathies, Duchenne's muscular dystrophy, myotonic dystrophy, limb-girdle dystrophy, polymyositis, acid maltase deficiency

sory muscles of respiration are inhibited during REM sleep, requiring a weakened diaphragm to perform almost all of the work of breathing. Finally, weakness of upper airway muscles may reduce airway diameter, leading to increased airway resistance with increased load on the respiratory muscles if airway closure is incomplete, or to OSA if airway closure is complete. The increased workload during sleep may lead to fatigue of the respiratory muscles, producing hypercapnia, acidosis-induced muscle dysfunction, and progressively worsening hypoventilation. Severe hypoxemia during the latter portion of the night is often the result. Respiratory muscle fatigue is thought to be an especially important factor in the hypoxemia that often develops during sleep, especially during REM sleep, in patients with myasthenia gravis.[72]

Muscle diseases that produce weakness early in life, such as Duchenne's muscular dystrophy, often lead to kyphoscoliosis, which further compromises breathing. In patients with poliomyelitis, nocturnal alveolar hypoventilation may develop during the initial paralytic episode or as part of a functional deterioration—the *post-polio syndrome*—years or decades after the initial illness. Cervical spinal cord injuries that affect respiratory neurons also may cause hypoventilation during sleep.[73]

Factors other than weakness and fatigue may contribute to impaired breathing in some neuromuscular diseases. In myopathies, abnormal afferent input to the brain caused by dysfunction of muscle spindles may contribute to abnormal respiratory patterns. In myotonic dystrophy, abnormalities in respiratory and hypnogenic neurons in the brainstem, in addition to respiratory muscle weakness, may contribute to breathing disorders during sleep.[74,75] Although sleepiness is a common complaint in myotonic dystrophy, some patients who are sleepy do not have a significant sleep-related breathing disturbance, suggesting a CNS cause of sleepiness.[74–77]

Diagnosis

In any patient with a neuromuscular disorder affecting pulmonary function who begins to experience insomnia, the clinician should consider the possibility of nocturnal hypoventilation. However, anxiety, contractures, pain caused by polyneuropathies or muscle cramps, and immobility caused by muscle weakness often contribute to sleep disturbance, and in some patients they are more significant contributors to sleep complaints than breathing disturbances.

For patients with suspected respiratory dysfunction, hand-held spirometers can be used to assess tidal volume and vital capacity during examination. A marked difference in vital capacity between sitting and supine positions suggests diaphragmatic weakness, which can be further assessed with fluoroscopy or diaphragmatic EMG. Pulmonary function tests and ABGs are helpful for assessing the severity of hypoventilation during wakefulness and determining whether intrinsic bronchopulmonary disease may be contributing to hypoventilation. EMG, nerve conduction studies, and muscle biopsy may be necessary if the cause of the neuromuscular disease is uncertain.

Polysomnographic tests should be performed if nocturnal hypoventilation is a consideration. If P_{CO_2} monitoring is included with polysomnography, diagnosis is usually straightforward, based on elevations of P_{CO_2} that are most pronounced during REM sleep. In the absence of P_{CO_2} monitoring, a gradual decline in SaO_2 during REM sleep, a substantially lower baseline SaO_2 during sleep compared to wakefulness, and a failure of SaO_2 to return to baseline between apneas, if they occur, provide inferential evidence of hypoventilation.

Management

The goal of management is to reduce the load on the respiratory muscles, which leads to better sleep and reduces the patient's risk of dangerous cardiac arrhythmias and cardiopulmonary complications of hypoventilation. Improved nocturnal ventilation with less hypercapnia, fatigue, and acidosis can also lead to improved daytime respiratory function with decreased waking $PaCO_2$, and occasionally to reduced need for assisted ventilation during wakefulness.[78]

For obese patients, weight loss reduces the load on the diaphragm and other inspiratory muscles. Alcohol, sedative drugs, and other medications, which may contribute to sleep disturbance and cause depression of breathing during sleep, should be avoided. Myasthenic patients may benefit from bedtime or nocturnal doses of pyridostigmine.

Although mild alveolar hypoventilation often can be treated with nocturnal oxygen alone, additional treatment is required for more severe disease. With CPAP, airway resistance and work of breathing are reduced, but CPAP may cause problems if expiratory muscle weakness is present, because the expiratory muscles may not be strong enough to overcome the increased load associated with positive expiratory pressure. In such cases, BPAP can be used with relatively low expiratory pressure to reduce expiratory work, and with relatively high inspiratory pressure to provide inspiratory assistance. However, the pressure settings used for BPAP and CPAP must be chosen carefully because both can cause overinflation of the lungs, thereby shortening the inspiratory muscles and making them mechanically less efficient. Thus, regular reassessments of pressure settings may be needed for patients with progressive diseases who are using CPAP or BPAP. A rocking bed or a pneumobelt, which uses gravity to encourage descent and subsequent ascent of the diaphragm, is helpful in some cases.

Patients who do not obtain relief with CPAP or BPAP may benefit from intermittent positive-pressure ventilation administered by nasal mask.[62] Despite the potential for leaks around the mask or via the mouth, nasal ventilation is usually effective and well tolerated, avoids the complications of tracheostomy, and improves sleep quality.[79-82] In patients with amyotrophic lateral sclerosis, BPAP followed when necessary by nasal ventilation can improve quality of life for months or years, particularly when respiratory failure is the initial manifestation of the disease.[83]

Although negative-pressure poncho- or cuirass-type ventilators are often used for patients with static neuromuscular diseases, such as congenital myopathies and poliomyelitis, the negative intrathoracic pressure and the lack of coordination between ventilator action and upper airway dilator muscle activation can induce partial or complete airway obstruction. In such cases, nasal CPAP or BPAP can be added to the negative-pressure ventilation system, but if nasal masks are to be used, it is often simpler to switch to nasal positive-pressure ventilation.

Tracheostomy alone is usually not beneficial if obstructive apneas are not present, but it may markedly improve daytime symptoms when combined with nocturnal positive-pressure ventilation. Diaphragmatic pacing is occasionally an option for patients who cannot maintain sufficient minute ventilation during wakefulness and who require 24-hour ventilation because of neurogenic respiratory failure, but it does not help persons with diseases of muscle or the neuromuscular junction, and as with negative-pressure ventilators, it often induces upper airway obstruction and OSA.

SUMMARY

CSA is a condition in which absence of respiratory effort leads to cessation of breathing. CSA has a variety of causes, and the clinical presentation of the syndrome may differ depending on the cause and the age of the patient. Although the breathing disturbance is caused by a metabolic, cardiac, neurological, or pulmonary disorder that leads to aberrant function of systems involved in respiratory control, respiratory function during wakefulness is usually normal. Four major features of the metabolic system for respiratory control determine whether central apneas are present: (1) sleep-wake state, (2) $Paco_2$, (3) Pao_2, and (4) responsiveness to ABG changes. As most patients with CSA do not have apnea while awake, the changes in breathing that occur during sleep account for the development of central sleep apnea. A decrease in $Paco_2$ during sleep and the destabilizing influence of hypoxia are the causes of CSA and associated insomnia in persons sleeping at high altitude. With congestive heart failure, poor left ventricular function leads to decreased cardiac output and to increased circulation time. As a result, chemoreceptors receive delayed information regarding gas exchange and acid-base balance, so the reduction in ventilation that occurs in response to hypocapnia is delayed, leading to further hypocapnia and eventually to apnea if the apneic threshold is reached. Abnormalities of the neural elements of the metabolic system for breathing control—chemoreceptors, autonomic pathways, medulla, descending motor pathways, motoneurons, neuromuscular junction, or respiratory muscles—can lead to CSA and other abnormal breathing patterns during sleep. Upper airway obstruction can lead to CSA, particularly in patients who also have OSA. In such cases, airway muscle relaxation during expiration leads to airway occlusion, and upper airway afferents activated by the occlusion trigger reflex inhibition of inspiration. Incomplete development of the metabolic control system for breathing is the cause of central apneas in premature infants. Treatment of CSA is most successful if it is directed at the underlying cause.

REFERENCES

1. Gastaut, HV, et al: Polygraphic study of the periodic diurnal and nocturnal (hypnic and respiratory) manifestations of the Pickwick syndrome. Brain Res 2:167, 1966.
2. Major, RH: Classic descriptions of disease. Charles C Thomas, Springfield, Ill, 1965, p 548.
3. Cheyne, JA: A case of apoplexy, in which the fleshy part of the heart was converted into fat. Dublin Hosp Reports 2:216, 1818.
4. Stokes, W: The disease of the heart and the aorta. Hodges and Smith, Dublin, 1854, pp 302–340.
5. Plum, F, and Swanson, AG: Abnormalities in central regulation of respiration in acute and convalescent poliomyelitis. Arch Neurol Psychiatry 80:267, 1958.
6. Dempsey, JA, and Skatrud, JB: A sleep-induced apneic threshold and its consequences. Am Rev Respir Dis 133:1163, 1986.
7. Xie, A, et al: Hypocapnia and increased ventilatory responsiveness in patients with idiopathic central sleep apnea. Am J Respir Crit Care Med 152:1950, 1995.
8. Weil, JV: Sleep at high altitude. In Kryger, MH, et al (eds): Principles and Practice of Sleep Medicine, ed 2. WB Saunders, Philadelphia, 1994, pp 224–230.
9. Coote, JH, et al: Sleep of Andean high altitude natives. Eur J Appl Physiol 64:178, 1992.
10. Hall, MJ, et al: Cycle length of periodic breathing in patients with and without heart failure. Am J Respir Crit Care Med 154:376, 1996.
11. Nathan, PW: The descending respiratory pathway in man. J Neurol Neurosurg Psychiatry 26:487, 1963.
12. Bullock, R, et al: Isolated central respiratory failure due to syringomyelia and Arnold-Chiari malformation. Br Med J 297:1448, 1988.
13. Khoo, MCK, et al: Dynamics of periodic breathing and arousal from sleep during sleep at extreme altitude. Respir Physiol 103:33, 1996.

14. Bradley, TD, et al: Clinical and physiologic heterogeneity of the central sleep apnea syndrome. Am Rev Respir Dis 134:217, 1986.
15. Kelly, D, et al: The care of infants with near-miss sudden infant death syndrome. Pediatrics 61:511, 1978.
16. Stein, IM, and Shannon, DC: The pediatric pneumogram: A new method for detecting and quantitating apnea in infants. Pediatrics 55:599, 1975.
17. Kahn, A, et al: Polysomnographic studies of infants who subsequently died of sudden infant death syndrome. Pediatrics 82:721, 1988.
18. Guilleminault, C: Sleep apnea in the full-term infant. In Guilleminault C (ed): Sleep and Its Disorders in Children. Raven, New York, 1987, pp 195–211.
19. Southall, DP: Role of apnea in the sudden infant death syndrome: A personal view. Pediatrics 81:73, 1988.
20. Oren, J, et al: Identification of a high-risk group for sudden infant death syndrome among infants who were resuscitated for sleep apnea. Pediatrics 77:495, 1986.
21. Kahn, A, et al: Polysomnnographic studies and home monitoring of siblings of SIDS victims and of infants with no family history of sudden infant death. Eur J Pediatr 145:351, 1986.
22. Issa, FG, and Sullivan, CE: Reversal of central sleep apnea using nasal CPAP. Chest 90:165, 1986.
23. Naughton, MT, et al: Effects of nasal CPAP on sympathetic activity in patients with heart failure and central sleep apnea. Am J Respir Crit Care Med 152:473, 1995.
24. Takasaki, Y, et al: Effect of nasal continuous positive airway pressure on sleep apnea in congestive heart failure. Am Rev Respir Dis 140:1578, 1989.
25. Bradley, TD, et al: Cardiac output response to continuous positive airway pressure in congestive heart failure. Am Rev Respir Dis 145:377, 1992.
26. Hagenah, G, et al: Nachtliche Sauerstoffgabe und Herzrhythmusstorungen wahrend Cheyne-Stokes-Atmung bei Patienten mit Herzinsuffizienz. Z Kardiol 85:435, 1996.
27. Villiger, PM, et al: Beneficial effect of inhaled CO_2 in a patient with non-obstructive sleep apnoea. J Neurol 241:45, 1993.
28. Steens, RD, et al: Effect of inhaled 3% CO_2 on Cheyne-Stokes respiration in congestive heart failure. Sleep 17:61, 1994.
29. White, DP, et al: Central sleep apnea: Improvement with acetazolamide therapy. Arch Intern Med 142:1816, 1982.
30. DeBacker, WA, et al: Central apnea index decreases after prolonged treatment with acetazolamide. Am J Respir Crit Care Med 151:87, 1995.
31. Hackett, PH, et al: Respiratory stimulants and sleep periodic breathing at high altitude: Almitrine versus acetazolamide. Am Rev Respir Dis 135:896, 1987.
32. Sutton, JR, et al: Effect of acetazolamide on hypoxemia during sleep at high altitude. N Engl J Med 301:1329, 1979.
33. Coote, JH: Sleep at high altitude. In Cooper, R (ed): Sleep. Chapman & Hill, London, 1994, pp 243–264.
34. Nicholson, AN, et al: Altitude insomnia: Studies during an expedition to the Himalayas. Sleep 11:354, 1988.
35. Bonnet, MH, et al: The effect of triazolam on arousal and respiration in central sleep apnea patients. Sleep 13:31, 1990.
36. Roloff, DW, and Aldrich, MS: Sleep disorders and airway obstruction in newborns and infants. Otolaryngol Clin North Am 23:639, 1990.
37. Guntheroth, WG, and Spiers, PS: Sudden infant death syndrome. In Gilman, S, et al (eds): Neurobase, ed 3. Arbor Publishing, La Jolla, Calif, 1997.
38. Davies, DP: Cot death in Hong Kong: A rare problem? Lancet 2:1346, 1985.
39. Brooke, H, et al: Case-control study of sudden infant death syndrome in Scotland, 1992–5. BMJ 314:1516, 1997.
40. Dwyer, T, et al: Prospective cohort study of prone sleeping position and sudden infant death syndrome. Lancet 337:1244, 1991.
41. Taylor, JA, et al: Prone sleep position and the sudden infant death syndrome in King County, Washington: A case-control study. J Pediatr 128:626, 1996.
42. Rognum, TO, and Saugstad, OD: Hypoxanthine levels in vitreous humor: Evidence of hypoxia in most infants who died of sudden infant death syndrome. Pediatrics 87:306, 1991.
43. Guntheroth, WG: The thymus, suffocation, and the sudden infant death syndrome: Social agenda or hubris? Perspect Biol Med 37:2, 1993.
44. Kinney, HC, et al: The neuropathology of the sudden infant death syndrome: A review. J Neuropathol Exp Neurol 51:115, 1992.
45. Guntheroth, WG, and Kawabori, I: Hypoxic apnea and gasping. J Clin Invest 56:1371, 1975.
46. Bass, M, et al: Death scene investigation in sudden infant death. N Engl J Med 315:100, 1986.
47. Kemp, JS, and Thach, BT: Sudden death in infants sleeping on polystyrene-filled cushions. New Engl J Med 324:1858, 1991.
48. Guntheroth, WG, et al: The role of respiratory infection in intrathoracic petechiae: Implication for SIDS. Am J Dis Child 134:364, 1980.
49. Guntheroth, WG: Interleukin-1 as intermediary causing prolonged sleep apnea and SIDS during respiratory infections. Med Hypotheses 28:121, 1989.
50. Forsyth, KD, et al: Lung immunoglobulins in the sudden infant death syndrome. Br Med J 298:23, 1989.
51. Tishler, PV, et al: The association of sudden unexpected infant death with obstructive sleep apnea. Am J Respir Crit Care Med 153:1857, 1996.
52. Willinger, M, et al: Infant sleep position and risk for sudden infant death syndrome: Report of meeting held January 13 and 14, 1994, National Institutes of Health, Bethesda, Md. Pediatrics 93:814, 1994.
53. Spiers, PS, and Guntheroth, WG: Recommendations to avoid the prone sleeping position and recent statistics for sudden infant death syndrome

54. Hunt, CE: Prone sleeping in healthy infants and victims of sudden infant death syndrome. J Pediatr 128:594, 1996.
55. Scragg, RKR, et al, and the New Zealand Cot Death Study Group: Infant room-sharing and prone sleep position in sudden infant death syndrome. Lancet 347:7, 1996.
56. Mitchell, EA, et al: Four modifiable and other risk factors for cot death: The New Zealand study. J Paediatr Child Health (suppl)28:S3, 1992.
57. Guntheroth, WG: Crib death: Sudden infant death syndrome, ed 3. Futura, Mt. Kisco, NY, 1995.
58. Kahn, A, et al: Long-term development of children monitored as infants for an apparent life-threatening event during sleep: A 10-year follow-up study. Pediatrics 83:668, 1989.
59. Sivan, Y, et al: Home monitoring for infants at high risk for the sudden infant death syndrome. Isr J Med Sci 33:45, 1997.
60. Shea, SA, et al: Respiratory sensations in subjects who lack a ventilatory response to CO_2. Respir Physiol 93:203, 1993.
61. Stovroff, M, et al: The complete spectrum of neurocristopathy in an infant with congenital hypoventilation, Hirschsprung's disease, and neuroblastoma. J Pediatr Surg 30:1218, 1995.
62. Guilleminault, C: Central alveolar hypoventilation. In Gilman, S, et al (eds): Neurobase, ed 3. Arbor, La Jolla, Calif, 1997.
63. Claman, DM, et al: Nocturnal noninvasive positive pressure ventilatory assistance. Chest 110:1581, 1996.
64. Ellis, ER, et al: Treatment of alveolar hypoventilation in a 6 year old girl with intermittent positive pressure ventilation through a nose mask. Am Rev Respir Dis 136:188, 1987.
65. Flageole, H, et al: Diaphragmatic pacing in children with congenital central alveolar hypoventilation syndrome. Surgery 118:25, 1995.
66. Oren, J, et al: Long-term follow-up of children with congenital central hypoventilation syndrome. Pediatrics 80:375, 1987.
67. Osler, W: The Principles and Practice of Medicine, ed 8. Appleton, New York, 1918.
68. Gould, GM, and Pyle, WL: Anomalies and curiosities of medicine: Being an encyclopedic collection of rare and extraordinary cases, and of the most striking instances of abnormality in all branches of medicine and surgery, derived from an exhaustive research of medical literature from its origin to the present day, abstracted, classified, annotated, and indexed. WB Saunders, Philadelphia, 1897, p 356.
69. Martin, TJ, and Sanders, MH: Chronic alveolar hypoventilation: A review for the clinician. Sleep 18:617, 1995.
70. Rochester, DF, and Enson, Y: Current concepts in the pathogenesis of the obesity-hypoventilation syndrome: Mechanical and circulatory factors. Am J Med 57:402, 1974.
71. Labanowski, M, et al: Sleep and neuromuscular disease: Frequency of sleep-disordered breathing in a neuromuscular disease clinic population. Neurology 47:1173, 1996.
72. Quera-Salva, MA, et al: Breathing disorders during sleep in myasthenia gravis. Ann Neurol 31:86, 1992.
73. McEvoy, RD, et al: Sleep apnoea in patients with quadriplegia. Thorax 50:613, 1995.
74. Guilleminault, C, et al: Respiratory and hemodynamic study during wakefulness and sleep in myotonic dystrophy. Sleep 1:19, 1978.
75. Hansotia, P, and Frens, D: Hypersomnia associated with alveolar hypoventilation in myotonic dystrophy. Neurology 31:1336, 1981.
76. Coccagna, G, et al: Alveolar hypoventilation and hypersomnia in myotonic dystrophy. J Neurol Neurosurg Psychiatry 38:977, 1975.
77. Van der Meche, FGA, et al: Daytime sleep in myotonic dystrophy is not caused by sleep apnoea. J Neurol Neurosurg Psychiatry 57:626, 1994.
78. Braun, SR, et al: Intermittent negative pressure ventilation in the treatment of respiratory failure in progressive neuromuscular disease. Neurology 37:1874, 1987.
79. Ellis, ER, et al: Treatment of respiratory failure during sleep in patients with neuromuscular disease: Positive pressure ventilation through a nose mask. Am Rev Respir Dis 135:148, 1987.
80. Guilleminault, C, and Kowall, J: Central sleep apnea in adults. In Thorpy, MJ (ed): Handbook of Sleep Disorders. Marcel Dekker, New York, 1990, pp 337–351.
81. Hill, NS, et al: Efficacy of nocturnal nasal ventilation in patients with restrictive thoracic disease. Am Rev Respir Dis 145:365, 1992.
82. Hill, NS, et al: Sleep-disordered breathing in patients with Duchenne muscular dystrophy using negative pressure ventilators. Chest 102:1656, 1992.
83. Sivak, ED, and Streib, EW: Management of hypoventilation in motor neuron disease presenting with respiratory insufficiency. Ann Neurol 7:188, 1980.

Chapter 15

PARASOMNIAS

AROUSAL DISORDERS
Clinical Features
Biologic Basis
Differential Diagnosis
Diagnostic Evaluation
Management
SLEEP-WAKE TRANSITION DISORDERS
Rhythmic Movement Disorder
Sleep Starts
Sleeptalking
Nocturnal Leg Cramps
PARASOMNIAS RELATED TO REM SLEEP
Sleep Paralysis
REM Sleep Behavior Disorder
REM Sleep–Related Sinus Arrest
Impaired Nocturnal Penile Tumescence
Sleep-Related Painful Erections
Nightmares
OTHER PARASOMNIAS
Bruxism
Enuresis
Nocturnal Dissociative Disorder
Sudden Unexplained Nocturnal Death

I have seen her rise from her bed, throw her nightgown upon her, unlock her closet, take forth paper, fold it, write upon't, read it, afterwards seal it, and again return to bed; yet all this while in a most fast sleep.

WM. SHAKESPEARE, MACBETH V.I.10–13

Parasomnias are disorders in which undesirable physical and mental phenomena occur mainly or exclusively during sleep.[1] They differ from the dyssomnias in that insomnia and daytime sleepiness are not the primary complaints; rather, the complaint is related to the undesirable phenomena. Central nervous system (CNS) activation, with skeletal muscle activity and signs of autonomic arousal, occurs with many of the parasomnias. Physical phenomena may include visceral contractions, jerks or twitches of the extremities, thrashing, shouts and screams, posturing, walking or running, and complex automatisms. Mental phenomena may include thoughts, dreams, images, and emotions.

The definition of parasomnias encompasses a broad range of physical and mental symptoms. Some parasomnias, such as nightmares, are so common that almost everyone has experienced them; others, such as hypnogenic paroxysmal dystonia, are exceedingly rare. The symptoms of some parasomnias, such as sleeptalking and sleep starts, are undesirable for some persons but of no concern to others. Other symptoms, however, such as the complex behaviors that accompany rapid eye movement (REM) sleep behavior disorder (RBD), are almost always considered undesirable. Violent motor activity may accompany some parasomnias, whereas others (e.g., enuresis) have no skeletal muscle disturbance.

The phenomena that occur with parasomnias often are paroxysmal events and thus share similarities with daytime "spells" that can tax the diagnostic acumen of physicians. Although the history provides crucial information for determining the cause of spells, paroxysmal events during sleep present a special challenge because patients, who are usually asleep when the event begins, are rarely able to give a complete history of the episode. The clinician must, therefore, rely on the history from the bed partner or parent because the subjective aspects of the events, so important in the diagnosis of waking paroxysmal events, are less available for analysis. Furthermore, parents or bed partners may not observe the beginning of an event, or they may be too upset or frightened by the event to allow accurate observation. Although assessment of level of consciousness and responsiveness is an important tool for determining the etiology of some phenomena, this assessment is more difficult when the patient has just aroused from sleep.

Parasomnias may be classified in a variety of ways, none of which is entirely satisfactory. In the organization of this chapter, I have followed the approach adopted by the *International Classification of Sleep Disorders*,[1] which classifies parasomnias into four categories according to the predominant sleep stage in which they occur: (1) arousal disorders, which usually occur during deep non–rapid eye movement (NREM) sleep; 2) sleep-wake transition disorders; 3) parasomnias usually associated with REM sleep; and (4) other parasomnias, which do not show a clear relation to a particular stage of sleep (see Table 7–2). This chapter covers all of the major parasomnias except parasomnias associated with breathing disorders, which are discussed in Chapters 13 and 14, and nocturnal paroxysmal dystonia, which is discussed in Chapter 20. Because its manifestations develop during nocturnal wakeful periods, nocturnal dissociative disorder is more properly a psychiatric disorder, not a true parasomnia. It is included in this chapter, however, because it is sometimes a consideration in the differential diagnosis of parasomnias.

AROUSAL DISORDERS

Confusional arousals, sleepwalking, and sleep terrors, the arousal disorders, are grouped together based on the hypothesis that disordered arousal mechanisms, leading to an impaired ability to arouse fully from slow-wave sleep (SWS), are responsible for the phenomena.[2] As a result, an internal or external stimulus that would normally produce a full awakening or a brief arousal leads to a prolonged state in between wakefulness and sleep. Although sleep terrors were described in reports of "pavor nocturnus" in children and "incubus" in adults, they were not clearly differentiated from nightmares until 1949.[3] Gastaut and colleagues[4,5] performed the first polygraphic recordings of sleep terrors in the 1960s.

Clinical Features

Although the arousal disorders are divided into three types (confusional arousals, sleepwalking, and sleep terrors), they exist on a continuum: *Confusional arousals* are associated with low levels of motor, emotional, and autonomic activation; *sleepwalking* is associated with more complex motor activity but with little autonomic or emotional activation; and *sleep terrors* are associated with pronounced autonomic and emotional activation and with varying degrees of motor activity. In a given person, episodes may be associated with motor, emotional, and autonomic activation of variable severity, leading to considerable clinical overlap among the three types. For example, patients with sleep terrors also may have confusional arousals, episodes of talking or moaning with mild tachycardia, episodes of agitated behavior without leaving the bed, or screams during sleep with minimal subsequent movements.

CONFUSIONAL AROUSALS

Confusional arousals, recognized at least since the 19th century, are characterized by a sudden arousal from sleep associated with complex behaviors without full alertness. During the arousal, patients are disoriented, confused, and slow to speak or respond. Aimless or inappropriate behavior may be accompanied by moaning and incoherent or nonsensical vocalizations, and some patients become agitated or violent, especially during attempts to awaken them. Although the episodes usually last just a few seconds or minutes, occasionally they may last for 10 minutes or more, during which it is difficult or impossible to arouse the patient. Although most common during the first third of the night, they may also occur later in the night and after awakening from long naps. In the morning, the patient is usually amnesic regarding the episode.

Almost all children have confusional arousals on occasion. The episodes tend to become less frequent with age, but some young adults have episodes as often as several times per week.

SLEEPWALKING

Sleepwalking (somnambulism) is characterized by complex automatisms during sleep typified by getting out of bed and walking. Both Hippocrates and Aristotle discussed somnambulism, and Galen was said to have walked in his sleep. Lady Macbeth is the most famous literary example of a somnambulist. Sleepwalking occurs in 10% to 30% of children, most commonly between the ages of 4 and 6 years. It occurs occasionally in 3% to 4% of adults, but the proportion of adults who sleepwalk at least weekly is only about 4 in 1000.[6] Episodes usually become less frequent during and after adolescence. Although childhood onset is the norm, some adult sleepwalkers were apparently unaffected as children.

Sleepwalking episodes may last a few minutes or as long as 30 minutes. A variety of behaviors may accompany sleepwalking: Some patients leave or attempt to leave the home through a door or window, whereas others eat or drink.[7] Patients do not always get out of bed and walk but may sit or stand in bed or fumble with clothing. Although sleepwalkers usually respond inappropriately or not at all to voices, they sometimes follow instructions to return to bed.

Injury during the episode is the most serious complication of sleepwalking. Patients may trip over objects or fall down stairs, resulting in fractures and other serious injuries. In a series of 100 cases of adults with nocturnal injuries, sleep terrors and sleepwalking were responsible for more than half of the cases.[8] Patients who eat during sleepwalking episodes have secondary complications of weight gain and tooth decay. Occasional sleepwalkers engage in violent behavior, sometimes directed at the person trying to awaken them, and sleepwalking has been used successfully as a defense against a charge of murder.[9]

SLEEP TERRORS

Sleep terrors are episodes of agitation and apparent fear or terror arising abruptly from sleep during which the patient is unresponsive or only partially responsive and inconsolable. These episodes are most prevalent among children aged 4 to 10 years; only about 1% to 2% of children between ages 6 and 14 have sleep terrors at least once per week. Onset of sleep terrors tends to occur in childhood and to decrease in frequency and severity during adolescence; however, onset can occur in adulthood.

Prototypical episodes are characterized by screaming and agitated behavior with attempts to leave the bed or the room. Autonomic manifestations may include tachycardia with rates of up to 160/min, flushing, mydriasis, and sweating. Extreme agitation may lead to injuries: Patients may jump from windows or fall down stairs and may react violently to attempts to restrain them. In one series of adults with sleep terrors, 18% had left their homes during an episode.[10] After several minutes, the patient quiets and returns to sleep. Morning recall is usually absent, although some patients have a vague recollection of the behavior or of a terrifying image or situation, such as suffocation, burial, impending death, or monsters. Occasional patients recall more elaborate dreamlike experiences. A case of sleep terrors that began in childhood and continued into middle age is presented below.

CASE HISTORY: Case 15–1

A 34-year-old woman complained of episodic nocturnal screaming. The episodes, which began at about age 7 when her parents were in the midst of a divorce, occurred nightly at first, but when she became an adult, they occurred one to three times per month, usually between 12:30 and 2 AM. Although her roommate described loud screams, agitation, and inconsolability lasting several minutes, she had no recall of the episodes. She did not get out of bed during the episodes and had not suffered any injuries. She was embarrassed by the episodes and worried that one would occur if she stayed at the house of a friend or relative.

She went to bed at about midnight and awakened at 6 AM during the week and at about 10 AM on weekends. She took no medications, although she had been treated for depression once in the past. Her father had nightmares and violent dreams after being a prisoner of war. Examination was normal.

She was given a diagnosis of sleep terrors, and because of her concerns that episodes might occur during overnight visits, treatment with imipramine 20 mg at bedtime was recommended for those nights. Over the next 3 years, she had no episodes on the nights that she took imipramine, but she continued to have about two episodes per month on other nights.

Biologic Basis

Although neuropathologic evidence of an abnormality of arousal systems is lacking, substantial inferential evidence supports the hypothesis that these syndromes are associated with incomplete cortical activation in response to an arousal stimulus. First, the appearance of the electroencephalogram (EEG) during the episode, which may show drowsy patterns, nonreactive alpha waves, synchronous or asynchronous theta or delta waves, or combinations of these patterns, is consistent with the idea that cortical activation is incomplete (Fig. 15–1). Second, episodes can be precipitated by attempts to awaken persons from SWS and are most likely to occur with spontaneous arousals from SWS, the stage in which depth of sleep is greatest and from which full awakening is most difficult. (Less commonly, episodes can occur during lighter stages of NREM sleep.[11]) Third, CNS depressants, which impair arousal mechanisms, and sleep deprivation also can precipitate confusional arousals. Fourth, bladder distention and loud noises frequently trigger episodes, and sleep disorders that cause arousal during sleep, such as obstructive sleep apnea (OSA), can precipitate episodes. Fifth, the disorders are most common during childhood, when the amount of SWS is at its peak.

Although the neurological basis for the presumed defect in arousal mechanisms is unknown, the prevalence is increased in first-degree relatives of persons with

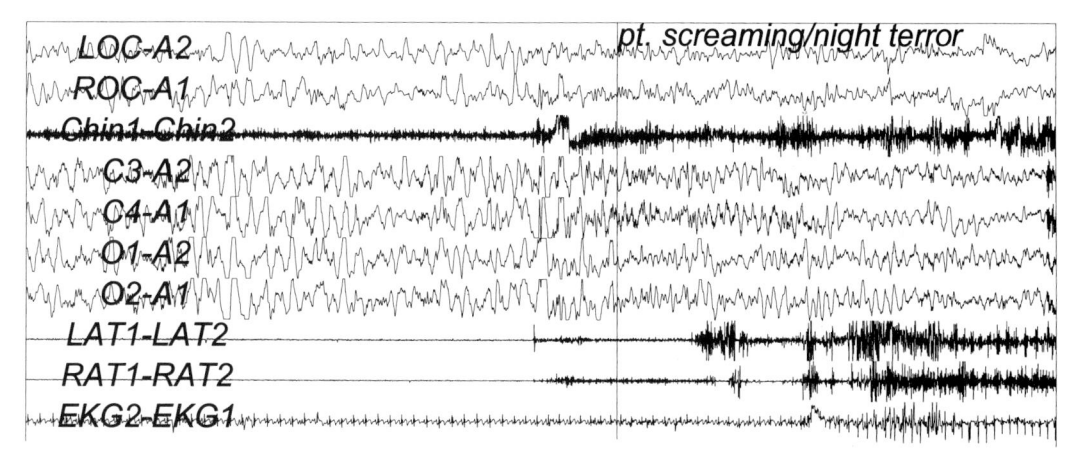

Figure 15–1. One-minute segment of a polysomnogram in a 9-year-old girl with sleep terrors. The segment shows a sudden arousal from slow-wave sleep, with screaming, agitation, and mild tachycardia. The EEG during the arousal shows a mixed frequency pattern composed mainly of theta and delta activity. The EEG electrodes for this figure and for Figures 15–2, 15–4, and 15–7 (A1, A2, C3, C4, F3, F4, F7, F8, P3, P4, T3, T4, T5, T6, O1, O2) were placed in accordance with the International 10-20 system. Other abbreviations: LOC, left outer canthus; ROC, right outer canthus; Chin1–Chin2, submental EMG; LAT1–LAT2, left lower leg extensor compartment surface EMG; RAT1–RAT2 right lower leg extensor compartment surface EMG; EKG2–EKG1, electrocardiogram.

arousal disorders,[12] and twin studies suggest that a genetic effect is probably a substantial contributor.[6,13] In adults and children with arousal disorders, symptoms often increase with sleep deprivation and with acute sleep-wake phase shifts, as occurs with jet lag or shift work.

In some patients with sleep terrors, such as children with violent and abusive families or adults with post-traumatic stress disorder, psychological factors are important, perhaps because they increase the intensity of emotion during sleep and increase the likelihood that emotional stimuli will trigger an arousal. Apart from such cases, however, sleep terrors that begin in childhood are usually not associated with any particular psychopathology, although they can be precipitated in susceptible children by emotional or physical stress. Sleep terrors that develop in late childhood or adolescence, or that persist into adulthood, may be more often associated with psychopathology, but many adults with this disorder have no overt psychiatric condition.[10,14,15]

Differential Diagnosis

The differential diagnosis of confusional arousals includes sleeptalking, RBD, partial complex seizures, and somnambulism. The patient with confusional arousals does not get out of bed, as do those with somnambulism; does not become violent and does not appear to act out dreams, as do those with RBD; has less autonomic arousal compared to those with sleep terrors; and does not behave in a stereotyped fashion, as do patients with nocturnal complex partial seizures.

As with confusional arousals, the diagnosis of sleepwalking frequently can be made on the basis of the history. Episodes of complex motor activity that began in childhood, that occur in the first third of the night, that are associated with amnesia, and that are not accompanied by stereotyped behaviors or tonic postures are almost always the result of an arousal disorder.

For patients with atypical presentations, other possible diagnoses must be considered (Table 15–1). RBD is associated with dream recall; the vague recollections that some patients with sleepwalking have are usually fragmentary images or sensations without the unfolding story that characterizes dream mentation. Nocturnal complex partial seizures should be considered in patients with somnambulism who have poor or no response to medications, have atypical or stereotyped behaviors during the episodes, have onset in adulthood, or have multiple nightly episodes. The absence of abnormal EEG activity during the episode does not exclude the possibility of epilepsy; some patients with nocturnal complex partial seizures have ictal discharges that are not apparent on surface EEG (see Chap. 20).[16,17]

The differential diagnosis of sleep terrors includes nightmares, RBD, nocturnal

Table 15–1 **Causes of Episodic Nocturnal Complex Motor Activity**

Disorder	Associated Stage of Sleep or Wakefulness
Sleepwalking	NREM sleep, especially slow-wave sleep
Sleep terrors	NREM sleep, especially slow-wave sleep
Partial seizures	Usually NREM sleep
Panic disorder	NREM sleep
REM sleep behavior disorder	REM sleep
Nightmares	REM sleep
Nocturnal delirium	Wakefulness
Nocturnal dissociative disorder	Wakefulness
Malingering	Wakefulness

delirium, sleep-related epilepsy, and other sleep disorders that produce anxiety, including panic disorder, OSA, and nocturnal cardiac ischemia. Sleep terrors usually occur within the first few hours of sleep and arise out of SWS, whereas nightmares are REM-sleep phenomena that are more common during the middle or latter half of the night. Rapid return to full alertness and vivid dream recall help to distinguish nightmares from sleep terrors (Table 15–2). RBD may be associated with violent behavior, running, or screaming; observers often report that patients seem to be "acting out" their dreams. Episodes tend to occur later in the night and are often associated with dream recall but with little or no autonomic activation. In some patients, however, behavior may be similar to that observed with sleep terrors, and some patients have an "overlap" syndrome, with elements of both sleep terrors and RBD.[18]

Generally, sleep terrors can be distinguished from epileptic seizures based on clinical features. Sleep terrors tend to last longer than complex partial seizures and are associated with more pronounced autonomic activation, whereas stereotyped motor behavior and oral automatisms, such as chewing, swallowing, and salivation, are rare. However, the distinction cannot always be made on the basis of the history: Nocturnal complex partial seizures may be associated with a fearful appearance, screaming, running, tachycardia, and vague, frightening perceptions. In nocturnal dissociative disorder the patient is alert and the behavior is purposeful and more complex.

Diagnostic Evaluation

In most cases of confusional arousals and many cases of sleepwalking and sleep terrors, no diagnostic studies are needed. For example, the diagnosis of an arousal disorder is generally secure in otherwise normal children or young adults with typical sleepwalking behaviors or sleep terrors occurring during the first third of the night. If the diagnosis is not apparent from the history, the clinician must decide whether to begin empiric treatment or to obtain diagnostic studies. The need for accurate diagnosis and effective treatment is increased if atypical features are present, such as adult onset, stereotyped behaviors, or frequent occurrence in the second half of the night, or if the patient has suffered injuries during nocturnal episodes or describes behavior suggesting the possibility of injuries.

If diagnostic studies are required, one or two nights of video-EEG polysomnography (VPSG) are usually informative, especially if the episodes in question occur almost every night. This procedure combines video-EEG monitoring with standard polysomnographic recording and thereby adds information unobtainable with standard polysomnography (i.e., behavioral events can be reviewed in detail, and EEG activity can be monitored to a much greater extent).[19] Systems with the capability to display EEG data at 30 mm/s or with

Table 15–2 **Comparison of Clinical Features of Nightmares, Sleep Terrors, and REM Sleep Behavior Disorder**

	Nightmares	Sleep Terrors	REM-Sleep Behavior Disorder
Behavior	None, moaning, or brief cries	Screaming, agitation, escape behavior	Talking, dream-enacting behavior
Mental status on awakening	Alert	Confused	Variable
Recall	Good dream recall	Amnesia	Variable dream recall
Autonomic activation	Moderate	High	Low
Timing	Often in second half of night	First third of night	Often in second half of night

comparable resolution on computer monitors, are the most useful because abnormal epileptiform activity may not be identifiable at lower resolution. With sophisticated systems, dozens of channels are available for EEG data, either superimposed or adjacent to one or more images of the patient, with capabilities for digital or analog storage and subsequent playback.

The cost of VPSG recording is greater than that of conventional PSG, largely because of the need for continuous observation of behavior by technologists or trained observers. Because of the higher cost, VPSG should be used only when there is a reasonable likelihood of obtaining diagnostically useful information. If the behaviors in question occur at least several times per week, one or two nights of recording will usually permit observation of at least one of the events, leading to a definite diagnosis. In one series of more than 100 adults and children with suspected parasomnias, VPSG provided diagnostically useful information in 65%.[19]

VPSG is most useful if a sleepwalking or sleep terror episode is recorded; high-voltage EEG slow-wave activity may be seen immediately before the episode, and the EEG during an episode, if it is not obscured by movement artifact, often shows regular, rhythmic, hypersynchronous delta or theta activity. Even if the behaviors do not occur, the study may be helpful. For example, confusional arousals may occur that support the presence of an arousal disorder, or the EEG may show runs of hypersynchronous delta waves during arousals from stage 3–4 sleep, a finding that is more common in persons with arousal disorders than in normal subjects (Fig. 15–2).[20] In contrast, psy-

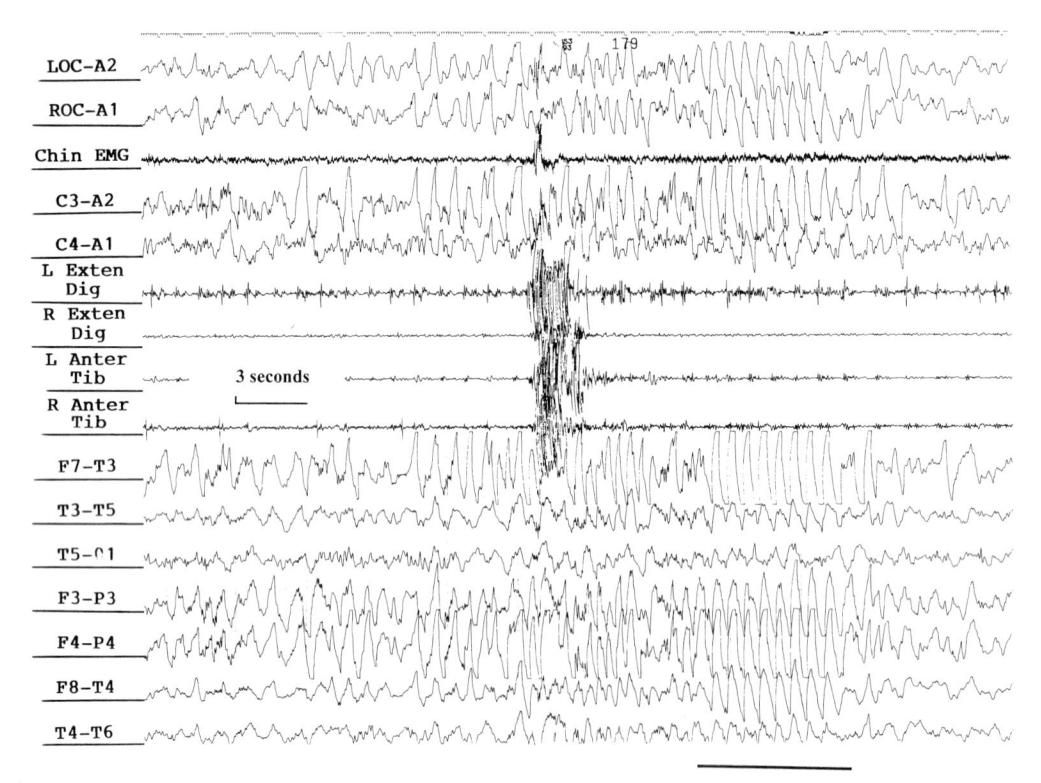

Figure 15–2. Forty-second segment of a polysomnogram in a 19-year-old woman with sleep terrors. Approximately halfway through the segment, there is a sudden arousal from slow-wave sleep, with high-amplitude muscle activity recorded from the extremities. Several seconds later, hypersynchronous delta activity (underlined) is recorded from EEG derivations, with maximal amplitude anteriorly. For abbreviations, see Figure 15–1. Other abbreviations: L, left; R, right; Exten dig, extensor forearm compartment surface EMG; Anter Tib, lower leg extensor compartment surface EMG.

chogenic dissociative episodes are associated with a waking EEG pattern, whereas RBD is associated with EEG features of REM sleep. If the behavior is highly stereotyped or is associated with tonic postures or clonic movements that suggest the possibility of nocturnal seizures, but no surface EEG abnormalities are evident during the episodes, long-term video-EEG monitoring in specialized epilepsy laboratories may be required (see Chap. 20).

Management

Arousal disorders tend to improve with age, perhaps because the amount of time spent in slow-wave sleep decreases with age, and they often resolve before adolescence. To reduce the frequency of episodes, patients should maintain regular sleep habits, avoid sleep deprivation, and minimize the use of CNS depressants. Because the disorder is benign, no other treatment is necessary. Parents may be reassured to know that the frequency of episodes tends to decrease with age.

For patients with infrequent sleepwalking episodes, safety measures and reassurance may be sufficient. Useful safety measures include having the patient sleep in a ground-floor bedroom, installing window locks, and removing sharp objects and toys on the bedroom floor. If the sleepwalking behavior is excessively disruptive to the family, or if injuries have occurred or appear likely to occur, bedtime doses of imipramine 20 to 100 mg, diazepam 2 to 5 mg, or clonazepam 0.5 to 2 mg are usually effective. A 6-month course of medication, followed by gradual withdrawal, is a typical program. If stress or psychopathology appear to be significant factors, counseling or psychiatric evaluation may be useful. Hypnosis or other behavioral treatments may be helpful for some patients.[21]

The approach to management of sleep terrors is similar to that used with sleepwalking, with emphasis on safety measures and reassurance that the episodes tend to improve with time. Patients who leave the bed and run or leave the house during the episodes are at increased risk for injuries. Medication doses are similar to those used for sleepwalking.

SLEEP-WAKE TRANSITION DISORDERS

The sleep-wake transition disorders—rhythmic movement disorder, sleep starts, sleeptalking, and nocturnal leg cramps—occur primarily during transitions between sleep and wakefulness.[1] Apart from leg cramps, the phenomena associated with these disorders can occur in otherwise healthy persons and thus may be considered normal events unless they occur with such frequency or severity that they disrupt sleep or cause anxiety, discomfort, or injury.

Rhythmic Movement Disorder

Rhythmic movement disorder, described in 1905 as *jactatio capitis nocturna* and as *rhythmie du sommeil*,[22,23] is also called head banging, head rolling, body rocking, and body rolling depending on the type of movement. It is characterized by stereotyped repetitive movements, usually of the head and neck but sometimes of the trunk, at a rate of 0.5 to 2 Hz. One typical pattern consists of repetitive banging of the head into the pillow while the patient is prone. Other patients rock back and forth on hands and knees or bang the forehead or occiput into the headboard or wall. Leg or arm banging and rolling head or body movements may occur, and repetitive humming or chanting may accompany the movements. In most patients the disorder causes no daytime symptoms, although it can be upsetting to family members and embarrassing to older children and adults. Occasional patients bang their heads hard enough to produces bruises with callus formation. Subdural hematoma and carotid dissection are extremely rare complications.

Episodes, which can last between a few seconds and 20 to 30 minutes, usually occur during quiet, relaxed wakefulness, drowsy wakefulness, stage 1 sleep, with arousals, and, rarely, in REM sleep.[24] The EEG patterns accompanying the movements are consistent with sleep-wake stage.

Rhythmic movements occur in most infants, but they are much less common by age 4 (Fig. 15–3).[25] Occasionally, symptoms persist into adolescence or adulthood. Because it is so common, rhythmic movement disorder is a common incidental finding in infants who have sleep recordings for other reasons. Other causes in infants of rhythmic muscle activity recorded on a polysomnogram include bruxism, thumb-sucking, and sucking on a pacifier.

The cause of the disorder is unknown; a relaxing effect induced by vestibular stimulation may contribute. Episodes are more common in patients who are mentally retarded.

Although in most cases the diagnosis is clear from the clinical description, occasionally the history may suggest the possibility of seizures, particularly in a patient with mental retardation and a history of seizures, and in rare cases epileptic seizures may be manifest as rocking movements. If the description is not clear or if epilepsy is suspected, VPSG may be useful. Episodes of rhythmic movements are not associated with EEG abnormalities, and they show little change in the frequency of movements over the course of an episode. In contrast, seizures often show an initial increase in the amplitude and frequency of movements, followed by a slowing in frequency in the second half of the episode.

Infants and toddlers usually do not require treatment, and parents should be reassured that the disorder usually resolves with time. Padding the headboard and sides of the crib may reduce bruising or other injuries. For patients with severe head banging that causes injuries, treatment other than safety measures is difficult; behavior therapy or a benzodiazepine is occasionally useful.

Sleep Starts

Sleep starts, described more than 100 years ago,[26] are myoclonic jerks occurring at the transition from wakefulness to sleep. Synonyms include sleep jerks, hypnic jerks, hypnagogic jerks, predormital myoclonus, and sleep-onset myoclonus.

Irregular low-amplitude jerks and twitches occur in almost everyone at sleep onset, but they are usually not perceived because they begin after the onset of sleep. More vigorous myoclonic jerks that produce flexion of the trunk or gross movements of the extremities are more likely to cause awakenings and are therefore more likely to be remembered. A sensation of falling, a sensory flash, a fragmentary dreamlike image, or a grunt or other

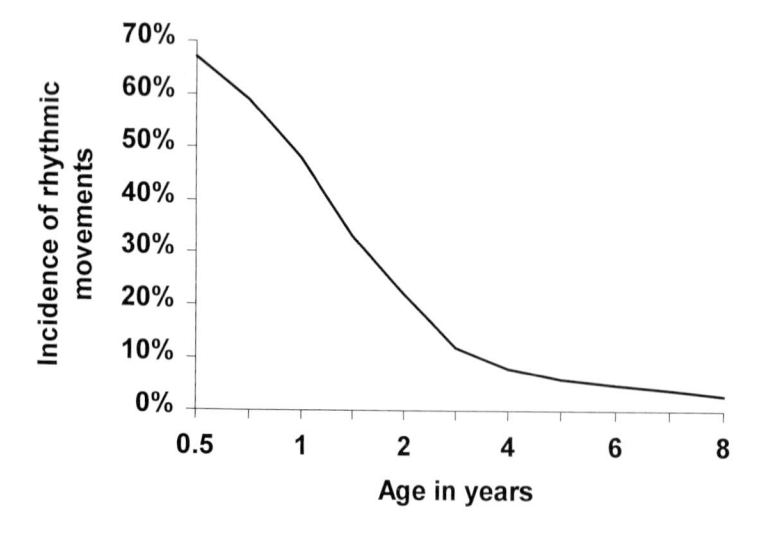

Figure 15–3. Incidence of rhythmic movements at bedtime during infancy and childhood. (Based in part on data from Klackenberg.[25])

vocalization sometimes accompanies the movement. Sleep starts usually occur while the EEG is showing drowsy patterns or stage 1 sleep, and a brief arousal may follow.

The cause is unknown. The more vigorous jerks of the whole body may be increased after caffeine intake, after physical exercise the previous day, or with stress.[27] Differential diagnosis includes periodic leg movements, myoclonic seizures, and excessive startle responses. If myoclonic seizures are suspected (see Chap. 20), an EEG is indicated.

Sleep starts are normal and for the most part do not require treatment other than reassurance. For the rare patient who develops sleep-onset insomnia as a result of anxiety associated with the episodes, relaxation therapy or other approaches to relieve anxiety are usually successful.

Sleeptalking

Sleeptalking occurs during sleep or during brief arousals from sleep. It usually consists of just a few words with minimal emotional content; some patients, however, may speak a few sentences and occasionally speech consists of loud or angry comments. Sleeptalking is common: About 10% of children talk during sleep every night or almost every night, and in a series of 600 nights of sleep studies with a range of subjects, sleeptalking occurred during 13% of the nights.[28]

About 80% to 90% of sleep-talking episodes occur during arousals from NREM sleep; the remainder occur in REM sleep. Episodes may occur at any time of night. Talking in NREM sleep, but not talking in REM sleep, is usually associated with an arousal or a gross body movement, or both. Episodes in NREM sleep tend to consist of fewer words and to have less affect than speech during REM sleep. If awakened after talking during NREM sleep, patients may report mundane thoughts or no recall of any mental activity. Episodes occurring during NREM or REM sleep are often accompanied by mentation appropriate to the speech, but subjects rarely recall the content of speech even if awakened immediately afterward.

The cause of sleeptalking is unknown. Children of parents with arousal disorders appear to be more likely to talk during sleep, suggesting a genetic contribution.[29] In predisposed persons, sleeptalking may be precipitated by sleep deprivation, stress, fever, and by arousals from any cause, including position changes, apneas, confusional arousals, nightmares, sleepwalking, or night terrors. Sleeptalking is common with RBD.

The differential diagnosis of nighttime talking includes talking during awakenings, talking during confusional arousals or sleep terrors, RBD, nocturnal dissociative disorder, and talking during arousals from apneas. For most patients, no evaluation is required. If the history suggests the possibility of RBD, OSA, or sleep terrors, polysomnography may be helpful. Psychiatric evaluation is indicated if nocturnal dissociative disorder is suspected.

Treatment is rarely, if ever, necessary unless talking is caused by one of the other disorders listed above. Stress reduction therapy may be helpful if sleeptalking is bothersome to the bed partner.

Nocturnal Leg Cramps

Nocturnal leg cramps affect up to one-sixth of the population and are especially common in older people. The cramp, a painful muscle contraction that usually involves the calf or foot and is associated with visible bulging and palpable tautness of the muscle, may occur during sleep or during arousals from sleep. The cramp may resolve spontaneously within a few seconds, or it may last for a few minutes or, rarely, as long as half an hour. Attempts to stretch the muscle by extension of the toes or foot may relieve the cramp. Leg cramps may also develop during the day, and some patients have them only during wakefulness. Suspected predisposing factors include prior exercise, fluid and electrolyte disturbances, pregnancy, diabetes mellitus, and nicotine and caffeine use. The course of the disorder is variable with frequent spontaneous remissions.

Although the biologic basis of nocturnal leg cramps is unknown, familial forms have been described, and disturbances of calcium metabolism and other metabolic disturbances are suspected contributors.[30–34] Nocturnal leg cramps occur with increased frequency in patients with myotonia congenita, stiff-man syndrome, myokymia, McArdle's disease, hypoparathyroidism, hypothyroidism, lead poisoning, and renal failure.[34,35]

Palpable tightness of the muscle helps to differentiate nocturnal leg cramps from restless legs syndrome, painful neuropathies, and claudication. Patients may obtain relief by dorsiflexion of the toes or foot during an episode. For patients who require pharmacological treatment, quinine (325 mg at bedtime) is usually the preferred treatment. If quinine is ineffective or contraindicated, vitamin E (400 IU), verapamil (120 mg), diphenhydramine (25 to 50 mg), or procainamide (250 mg) is sometimes effective.

PARASOMNIAS RELATED TO REM SLEEP

In the parasomnias related to REM sleep, features of REM sleep are disturbed and lead to or exacerbate the parasomnia. For example, sleep paralysis develops when the muscle atonia of REM sleep overlaps with wakefulness. Sleep-related painful penile erections and impaired sleep-related penile erections are related to the penile tumescence that normally accompanies REM sleep in men, and the pathophysiology of REM-sleep–related sinus arrest is probably related to alterations in autonomic nervous system activity during REM sleep. In RBD, the normal muscle atonia of REM sleep is disrupted and patients are able to act out their dreams.

Sleep Paralysis

Because sleep paralysis is such a common feature of narcolepsy, clinical features are described in Chapter 10. In the absence of narcolepsy, sleep paralysis is often called *isolated sleep paralysis*. The symptom is most common among adolescents, occurring at least occasionally in up to 15%. About 50% of adults have had at least one attack, but only about 3% to 6% have frequent attacks; in rare instances, isolated sleep paralysis occurs in several members of a family.[1,36–38]

In susceptible persons, isolated sleep paralysis is most likely to occur after a period of sleep disruption or a change in sleep schedule,[36] situations that increase the likelihood that naps will include REM sleep. Other factors that disrupt normal sleep patterns also can precipitate episodes.

Sleep paralysis is presumed to be a manifestation of REM-sleep intrusion into wakefulness. Transitions from sleep paralysis into REM sleep are common, and recordings during sleep paralysis show muscle atonia similar to the atonia of REM sleep and cataplexy.

Clinically, sleep paralysis should be distinguished from fatigue upon awakening. Some persons report that they feel so tired upon awakening that they are "unable to move." This symptom, which is not uncommon in depression and chronic fatigue syndrome, is caused by impaired motivation and is not a true paralysis. It also differs from sleep paralysis in that it may last for hours, and small body movements can be performed even though the patient may report being unable to get out of bed. Because persons with sleep paralysis are unable to lift even a finger, asking about the ability to "lift a finger" usually distinguishes sleep paralysis from morning fatigue.

Diagnostic considerations in persons with sleep paralysis include isolated or familial sleep paralysis, narcolepsy, idiopathic hypersomnia, irregular sleep-wake pattern, and disrupted sleep from any cause. The absence of daytime sleepiness distinguishes isolated sleep paralysis from narcolepsy and idiopathic hypersomnia. A review of the sleep history and the sleep-wake schedule usually helps to distinguish patients with irregular schedules and disrupted sleep from those with isolated sleep paralysis.

Diagnostic tests are unnecessary in most cases. If an irregular schedule is the suspected cause, a sleep log may be helpful. If daytime sleepiness is a symptom, or if narcolepsy or another sleep disorder is suspected, a polysomnogram and a Multiple Sleep Latency Test are usually indicated.

Treatment is not necessary except in the occasional patient with severe anxiety or insomnia as a result of the episodes, who may benefit from a benzodiazepine or a tricyclic antidepressant that suppresses REM sleep.

REM Sleep Behavior Disorder

RBD is associated with dream-enacting behaviors and loss or impairment of normal muscle atonia during REM sleep. The earliest descriptions of RBD were probably those from reports of cases of nocturnal hallucinations associated with lesions of the cerebral peduncles, a syndrome known as peduncular hallucinosis. Although isolated cases of increased EMG activity and complex behaviors during REM sleep were reported in the 1970s, a series of patients reported in the 1980s led to the recognition that RBD is a major cause of behavioral disturbance during sleep, especially in older persons and in those with degenerative diseases of the CNS.[39] For unknown reasons, the disorder is more common in men than in women by a ratio of between 2:1 and 5:1.

CLINICAL FEATURES

The presenting complaint may be from the patient, who may report "acting out dreams," incurring injuries during sleep, or having interrupted sleep; or it may come from the spouse or bed partner, who may describe shouting and violent behavior. Episodes, which may occur several times nightly or once every few weeks, usually last from a few seconds to 20 to 30 minutes. During the episodes, behavior may be limited to talking, laughing, and waving of the arms, or it may be so violent (e.g., yelling, cursing, punching, kicking, jumping out of bed) that patients or bed partners suffer cuts or bruises. Some patients have a "Jekyll and Hyde" syndrome with quiet, peaceable behavior during the day that contrasts with the swearing and violence that accompanies REM sleep.

In severe cases, fractures or subdural hematoma may occur[40]; in a series of patients with sleep-related injury, RBD was the cause in more than one-third.[8] Some patients devise elaborate means to prevent injury, such as tethering themselves to the bed or sleeping in a room without furniture.[41]

The timing and duration of episodes parallel the distribution of REM sleep across the night. Some patients have complex behaviors during virtually every REM period, while others have them no more than once per week. Behaviors often parallel dream content, as in the following case.

CASE HISTORY: Case 15–2

A 65-year-old retired grocer complained of episodic violent behavior during sleep. The episodes, characterized by sudden violent movements and talking or shouting, began at about age 55 and initially occurred about once every 2 months, later increasing to several times per week. He often jumped or dived out of bed and seemed to be fighting with someone. Although mild-mannered during the day, he often cursed during the episodes. He was amnesic regarding some activities; others were accompanied by dreamlike mentation. For example, he dreamed that a cat was on the bed, and as he yanked at the cat to pull it off the bed, he pulled his wife's hair. On another occasion, he dreamed he was riding a bicycle through an apple orchard on his way to a picnic; he saw a tree limb coming at him and dove off the bicycle, simultaneously diving out of bed. During a dream involving military activities, he suddenly jumped up in bed and saluted. Once, he dreamed that a crippled man in a wheelchair came up to the edge of a swimming pool and fell in; as he dived into the swimming pool to rescue the man, he simultaneously dived out of bed and fractured his wrist on the floor.

He had regular sleep habits, believed that he slept well, and felt refreshed in the morning. He did not sleepwalk or have sleep terrors as a child. Neurological examination was normal. A polysomnogram demonstrated increased muscle tone during REM sleep with frequent, vigorous movements. On one occasion during REM sleep, he appeared to be taking off his coat, then said, "Oh... don't want to race, eh?," and kicked violently several times.

When he took clonazepam 0.5 mg at bedtime, violent movements were completely eliminated, although he still had occasional talking during sleep. Four years later, clonazepam was discontinued on a trial basis, but the violent behaviors returned within a few nights. Clonazepam was resumed, and 9 years after initial evaluation, he continued to do well. Neurological examination remained normal.

Some patients with RBD report that their dreams have become more aggressive and action-filled, although such reports may reflect more frequent awakenings from such dreams rather than an actual change in dream content. Patients can usually be awakened fairly easily from an episode and do not have the prolonged confusion or difficulty arousing that characterizes sleep terror.

The disorder may begin fairly abruptly, or it may have a prolonged prodrome of progressively increasing nocturnal movements, vocalizations, and disturbing dreams. Tricyclic antidepressants and other medications with anticholinergic properties may exacerbate RBD, probably because muscle atonia during REM sleep is mediated at least in part through cholinergic systems.

RBD is primarily a disease of older persons; in the majority of patients, onset of symptoms occurs after age 50. Once established, the disorder is usually chronic, and patients rarely if ever experience spontaneous remission. A self-limited form of the disorder may occur after treatment with tricyclic antidepressants, monoamine oxidase inhibitors, or fluoxetine, or during withdrawal from alcohol.[41] Excessive caffeine intake may precipitate RBD.[42]

On polysomnography, the atonia that normally accompanies REM sleep is disrupted by periods of sustained increased tone, increased phasic muscle activity, or both. Simple as well as complex coordinated movements of the extremities occur during REM sleep, whereas periodic and aperiodic movements of the extremities may occur during NREM sleep (Fig. 15–4). Some patients have nocturnal events that include features of sleep terrors and RBD.[18]

BIOLOGIC BASIS

Normal REM sleep is associated with essentially complete atonia of all skeletal muscles except the diaphragm, interrupted by irregular asynchronous muscle twitches. The cause of RBD is presumed to be disruption of the neurological systems responsible for initiating and maintaining atonia during REM sleep.

In 1965, Jouvet and Delorme[43] reported that bilateral pontine lesions in the area adjacent to the locus ceruleus in cats eliminated or attenuated the atonia of REM sleep, and cats with such lesions were able to walk and perform complex movements during REM sleep. Subsequent animal studies demonstrated that the behavioral manifestations that accompany the state of REM sleep without muscle atonia depend on the size and site of the pontine lesions.[44] Cats with small bilateral lesions of the pontine tegmentum have increased proximal limb and head movements, but they do not display coordinated behavior. Cats with larger lesions that include regions projecting to the superior colliculus raise their heads, engage in apparent orienting behavior, and attempt to stand. Cats with even larger lesions that extend rostrally and ventrally into the midbrain have violent movements resembling attack behavior. Thus, the loss of muscle atonia during REM sleep is not in itself sufficient to produce behaviors; disinhibition of neurons involved in initiation of coordinated behaviors must also occur. The basis of human RBD is probably similar: Treatment with clonazepam, which usually attenuates or eliminates the behaviors, does not abolish the increased muscle tone during REM sleep. It appears, therefore, that the activation of locomotor drive responds to treatment, but the defect in muscle atonia does not.

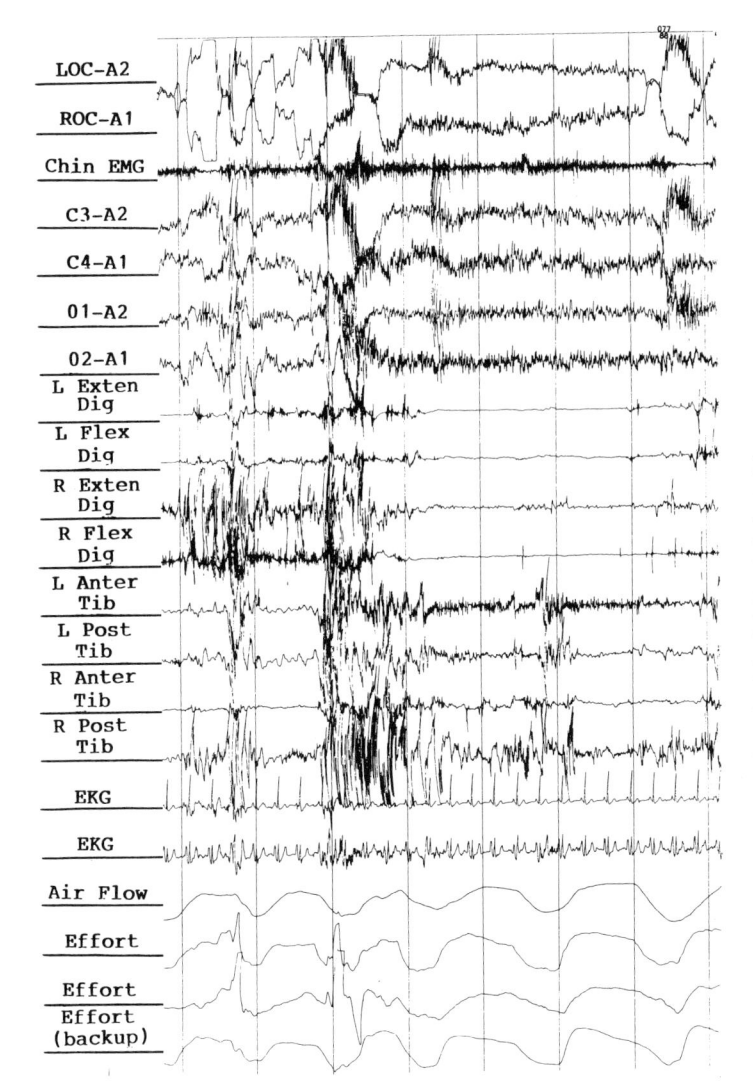

Figure 15–4. Twenty-two–second segment of a polysomnogram in a 68-year-old woman with REM-sleep behavior disorder that developed after a small left hemisphere subcortical stroke. Vertical lines are 3 seconds apart. Vigorous muscle activity is present asynchronously from all extremities, and increased muscle tone is apparent in the Chin EMG channel for most of the segment, with only two brief periods of atonia. Despite the vigorous movements, the heart rate is fairly constant. Rapid eye movements are apparent during the first 6 seconds and the last 3 seconds. For abbreviations, see Figure 15–1. Other abbreviations: Flex dig, flexor forearm compartment surface EMG; Post tib, posterior lower leg surface EMG; Air flow, airflow at the nose and mouth; Effort, thoracic and abdominal respiratory effort.

Although some patients have an underlying neurological disorder[45–49] (Table 15–3) and others have incidental magnetic resonance imaging (MRI) abnormalities, more than half have no abnormality on neurological examination and no identifiable structural brain lesion. In some patients, parkinsonism or dementia develops years after the onset of RBD, suggesting that for them, RBD is the first symptom of their degenerative disorder.[48] The RBD also may occur in patients with multiple sclerosis, cerebrovascular disease, and brainstem neoplasms. Some patients re-

Table 15–3 **Neurological Disorders Associated with Increased Prevalence of REM Sleep Behavior Disorder**[45–47]

Parkinson's disease
Lewy body disease
Multiple system atrophy
Alzheimer's disease
Olivopontocerebellar degeneration
Striatonigral degeneration
Multi-infarct dementia
Narcolepsy

port that the disorder began after apparently uncomplicated major surgery, presumably as a result of microembolization to the brain. These clinical findings suggest that pontine abnormalities may not be a necessary prerequisite for RBD: Dysfunction of centers and pathways projecting to pontine neurons involved in generation and maintenance of muscle atonia during REM sleep may also lead to RBD.

DIAGNOSIS

The differential diagnosis includes sleepwalking, sleep terrors, nocturnal seizures, delirium, postoperative psychosis, nocturnal confusional states resulting from dementia or other causes, nocturnal dissociative disorder, post-traumatic stress disorder, nocturnal panic disorder, and malingering. However, reports of dream mentation and dream-enacting behaviors are uncommon with any of these disorders, and a description of recurrent, nonstereotyped episodes of apparent dream-enacting behavior in an older person is sufficient to make the diagnosis in many cases. The recurrence of episodes several times throughout the night at approximately 90-minute intervals, and rapid return to full alertness upon awakening, are useful distinguishing features.

If the diagnosis cannot be made clinically, polysomnographic studies are required. Electromyographic monitoring of all four extremities helps to identify increased phasic muscle activity. Detailed observation by technologists and review of audiotaped and videotaped behavior by the clinician are also important aspects of diagnosis. If partial seizures are a diagnostic consideration, an expanded EEG montage is helpful (see Chap. 20).

Diagnostic criteria include a history of complex nocturnal behavior along with increased chin EMG or excessive phasic activity recorded from chin EMG or limb EMG electrodes. Although one night of monitoring is often sufficient, two are sometimes required; REM-sleep abnormalities are usually present even when no unusual behavior occurs. If focal abnormalities are present on neurological examination, brain computerized tomography or MRI may be indicated.

MANAGEMENT

Remissions rarely, if ever, occur in cases unrelated to medications or substance abuse. Fortunately, treatment with clonazepam 0.5 to 2 mg at bedtime is effective in 80% to 90% of patients. (For unknown reasons, clonazepam appears to be more effective than other benzodiazepines.) Although clonazepam reduces or eliminates the pathological behavior to minor movements and vocalizations, incomplete atonia persists during REM sleep. Tolerance occasionally develops, and daytime sedation and memory disturbance are side effects that limit the value of clonazepam therapy, particularly in patients with dementing disorders. If clonazepam is ineffective or cannot be used because of side effects, imipramine, levodopa-carbidopa, diazepam, temazepam, clonidine, or carbamazepine are occasionally beneficial.

REM Sleep–Related Sinus Arrest

Patients with this disorder have episodes of sinus arrest during REM sleep, with asystolic intervals of up to 9 seconds. Faintness, light-headedness, and blurred vision may develop during nocturnal awakenings, and syncope may occur if patients arise from bed during the night. Some patients report daytime symptoms of chest discomfort and intermittent palpitations.

The cause of the disorder is unknown; the few reported cases have occurred in otherwise healthy young adults.[50,51] As cardiac tests are usually normal, the basis is presumed to be transient autonomic dysfunction. For patients with prolonged asystolic periods, pacemaker treatment may be indicated.

Impaired Nocturnal Penile Tumescence

Although impaired nocturnal penile tumescence (NPT) is classified as a parasomnia, it produces no sleep-related symptoms; its importance lies in its usefulness for assessing impotence. Ohlmeyer and colleagues[52] noted periodic erections during sleep more than 50 years ago, and after the discovery of REM-sleep periods, Fisher and associates[53] noted a close relation between REM-sleep periods and erections.

Episodes of NPT are present at birth during REM sleep and increase in frequency and duration with puberty, perhaps because of changes in testosterone secretion.[54,55] After puberty, there is an age-related decline in total tumescence time but little or no decline in maximum circumference increase and number of erections (Fig. 15–5).[55]

BIOLOGIC BASIS

Nocturnal erections are produced by increases in penile blood flow and increases in bulbocavernosus and ischiocavernosus muscle activity, which occur during the transition from NREM to REM sleep (Fig. 15–6).[56,57] Tumescence may begin a few minutes before or after the onset of muscle atonia; similarly, detumescence may precede the end of the REM period or may follow the transition back to NREM sleep. Although the precise nature of the initiating events is unknown, increases in testosterone secretion during transitions from NREM sleep to REM sleep may contribute.[58]

During the initial phase of the erection, arterial blood flow increases into the corpora cavernosa. As intracavernous pressure increases as a result of increased blood volume, cavernous flow decreases until an equilibrium is reached and intracavernous pressure and cavernous flow stabilize. Subsequently, ischiocavernosus muscle contractions force blood into the cavernous system. During the final detumescent phase, there is a gradual return of cavernous blood flow and penile blood volume to baseline levels.

CAUSES

Small vessel arterial disease, caused by hypertension, diabetes mellitus, and tobacco use, is the most common cause of organic impotence. Autonomic dysfunction accompanying hypertension and diabetes mellitus probably also contributes to erectile dysfunction. In severe cases of diabetes mellitus or hypertension, NPT is completely or almost completely absent, with increases in penile circumference during REM sleep of less than 10 mm and often less than 5 mm. Less severe arterial insufficiency may lead to slowly evolving erections of long duration in which expansion of the base of the penis is substantially greater than expansion of the tip. Rapid

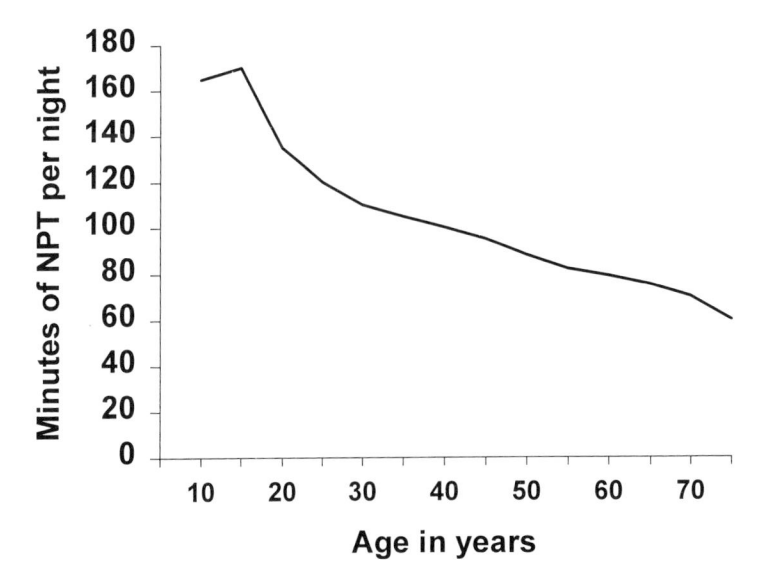

Figure 15–5. Minutes per night of nocturnal penile tumescence as a function of age. (Adapted from Karacan et al.,[55] with permission.)

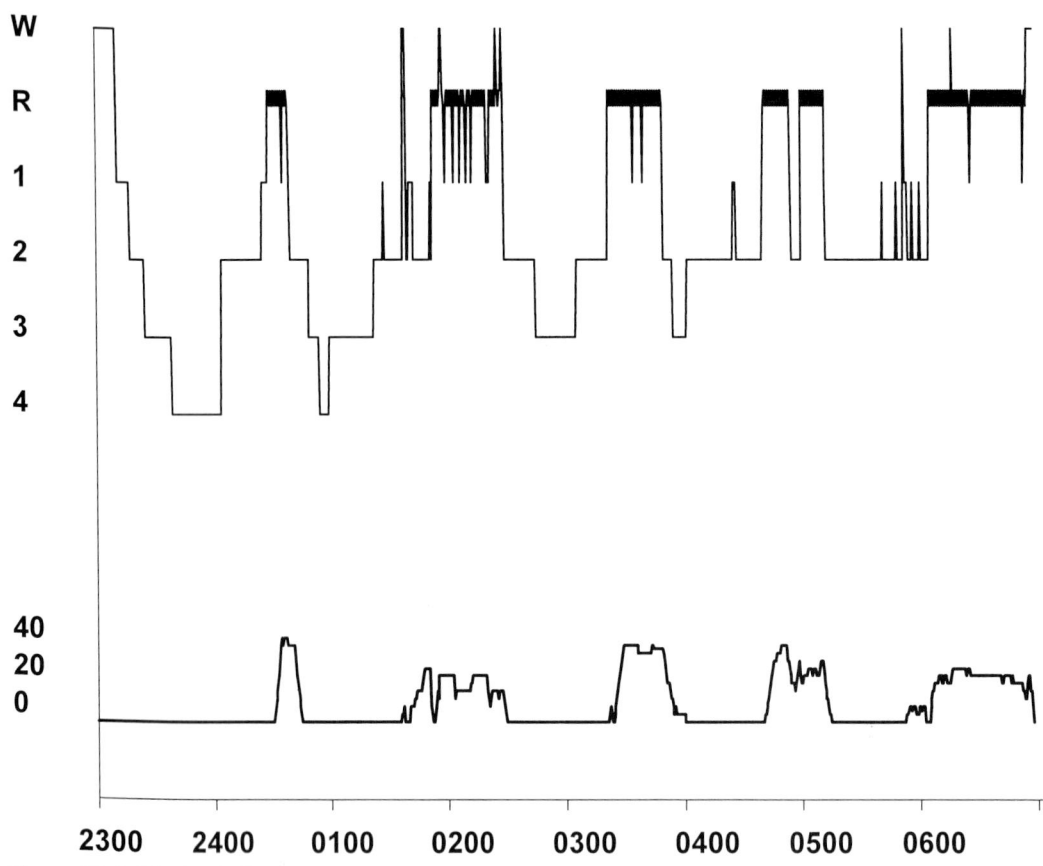

Figure 15–6. Idealized diagram of nocturnal penile tumescence (*lower tracing*) and sleep stage (*upper tracing*) during a night of sleep. The y-axis shows NREM sleep stages 1 to 4, REM sleep (R), wakefulness (W), and the increase in penile circumference in millimeters. Tumescence occurs in association with each REM period, but the amount and duration of tumescence varies considerably. The second and fifth tumescence periods begin a few minutes before the onset of the corresponding REM period.

loss of a full erection during REM sleep may be caused by abnormal venous outflow, and irregular detumescence may occur in patients with autonomic dysfunction. With spinal cord injuries, NPT may be absent or there may be brief erections during sleep as a result of spinal reflexes. Erectile impairment with abnormal NPT is also common in patients with end-stage renal disease, alcoholism, and endocrinopathies associated with subnormal testosterone or supranormal prolactin levels.

NPT can be affected by disorders that do not always cause impotence, such as major depression. The mechanism by which depression changes NPT is unknown; neuroendocrine changes accompanying depression may contribute.[59,60]

Patients with OSA also may have abnormal NPT, perhaps because a high proportion of these patients have hypertension, or because some men with OSA have reduced testosterone levels.[61] Drugs that may affect NPT findings include α-adrenergic receptor agonists, β-adrenergic receptor antagonists, agents with anticholinergic effects, and antiandrogenic agents.

DIAGNOSTIC USE OF NOCTURNAL PENILE TUMESCENCE

With the development of surgical implants for the treatment of impotence, the need for a biological marker for organic impotence increased. Karacan[62] suggested the use of NPT testing to evaluate the causes of impotence, based on the idea that nocturnal erections would be preserved in psychogenic impotence and impaired in organic impotence.

Although this idea has intuitive appeal, it is based on a fundamental dichotomy between organic and psychological processes that is not justified, as both play a role in many cases of impotence. Furthermore, erections during REM sleep probably are initiated by brainstem processes that are responsible for the generation of REM sleep; lesions of the nervous system above the brainstem can lead to organic impotence, while NPT remains normal or near-normal. Nonetheless, documentation of a rigid, normal-sized erection that lasts for several minutes during REM sleep provides clear evidence that the neurovascular processes responsible for erections are functional. Conversely, complete absence of tumescence during repeated episodes of REM sleep provides strong evidence that the neurovascular processes are impaired. Interpretation of NPT results is more difficult when partial erections or erections of brief duration occur.

NPT studies with sleep-state monitoring were performed frequently in sleep laboratories in the 1970s and 1980s as part of the evaluation of impotence, along with medical and sexual histories, physical and genital examinations, and psychiatric assessment. Such studies are superior to studies without sleep-state monitoring, such as snap gauges that pop open when an erection occurs, because the latter do not permit one to determine whether the patient has entered REM sleep. However, the use of NPT studies declined in the 1990s mainly because the widespread use of injections of papaverine and other vasoactive substances to treat impotence led to a decline in the use of surgical implants. Because the injections are associated with low morbidity, the need to document an organic abnormality of erectile function became less critical.

NOCTURNAL PENILE TUMESCENCE RECORDINGS

For recording sleep-related erections, strain gauges are placed around the penis. The electrical impedance of the gauges is a function of penile circumference. Two gauges, one at the penile base and the other at the coronal sulcus, are used to increase sensitivity to erectile anomalies. Each erectile episode is assessed for magnitude and duration.[63]

Measures of penile rigidity during maximal tumescence are generally included in the assessment because tumescence alone is not sufficient for intercourse: Vaginal penetration also requires adequate rigidity. Rigidity is usually assessed by determining the resistance to lateral buckling; if buckling occurs when a lateral force of less 500 g is applied, rigidity is probably insufficient to achieve vaginal penetration.[64]

A photograph of the erection is usually obtained to document abnormalities in penile size and shape, and the patient is asked to rate the adequacy of the erection. Patient ratings of apparently normal erections as inadequate may assist with diagnosis of sexual dysfunction.

Sleep-Related Painful Erections

In this rare condition, erections occurring during sleep are painful, disrupt REM sleep, and awaken the patient, but daytime erections are not painful. Nightly pain in severe cases may lead to insomnia. The cause is unknown; physical and genital examinations are usually normal, and psychiatric problems appear to be no more frequent than in the general population. Treatment is usually unsatisfactory; however, some patients improve over time.

Nightmares

O God, I could be bounded in a nut shell and count myself a king of infinite space, were it not that I have bad dreams.

WM. SHAKESPEARE, HAMLET II.II.397–398

CLINICAL FEATURES

Almost everyone has had nightmares—bad dreams—at some time. They are most common between ages 3 and 6, and although they decrease with age, about 10% of teenagers have nightmares at least once per month, and up to 3% of adults have them weekly.[65,66] Nightmares occur equally in boys and girls; however, about two-thirds of adults who present to clinicians with nightmares are women.

During a nightmare, some persons appear to be quietly asleep, whereas others talk or moan. Some persons awaken suddenly with anxiety and vivid immediate recall of the horrifying or frightening aspects of the dream, which, combined with a rapid heart rate and other somatic manifestations of anxiety, may make it difficult to return to sleep. Others recall the nightmare in the morning; in such cases, the autonomic activation may be less. Patients with nightmares that occur nightly may have difficulty falling asleep because of fear that nightmares will occur.

PSYCHOBIOLOGIC BASIS

Stress often precipitates nightmares; 40% of college students living near San Francisco had nightmares after the 1989 earthquake.[67] Medications, such as levodopa and β-adrenergic receptor blockers, may induce nightmares, and nightmares may also occur after withdrawal of major tranquilizers, tricyclic antidepressants, and monoamine oxidase inhibitors, presumably because of the rebound increase in REM sleep that may accompany drug withdrawal. A similar effect on REM sleep may account for the nightmares that occur with alcohol, benzodiazepine, and barbiturate withdrawal. The combination of stress and medications probably accounts for the postoperative nightmares that occur in about one-third of patients who have undergone major surgery.[68] Disorders or conditions that produce arousals and awakenings during REM sleep, such as sleep apnea, may also lead to nightmares.

Although nightmares that occur only occasionally are usually caused by stress or anxiety, chronic nightmares are usually caused by major psychological disturbances. Schizotypal personality, borderline personality, or schizophrenia is present in up to half of adults with frequent nightmares; in patients with psychotic disorders, nightmares may increase in frequency just before a psychotic break.[69] Repetitive nightmares, the content of which reflects the stressful event, are characteristic features of post-traumatic stress disorder (see Chap. 16). Vulnerable, sensitive persons appear to be more prone to nightmares than thick-skinned types.[66,70]

DIAGNOSIS

Differential diagnosis includes sleep terrors, RBD, sleep paralysis with hypnagogic hallucinations (see Chap. 10), and nocturnal panic disorder (see Chap. 16). Full alertness and immediate recall of a frightening dream are characteristic of nightmares, unlike the amnesia and confusion that typify sleep terrors, although some patients describe overlapping features. Sweating and rapid heart rate and respiration are common with nightmares, but these autonomic changes are less pronounced than those that accompany sleep terrors.

RBD is associated with dream-enacting behaviors, whereas nightmares are usually associated with either no behavior or only with moaning or brief cries. Although patients with sleep apnea or sleep paralysis may describe awakenings with feelings of suffocation, vivid recall of a dream story is rarely a feature.

Psychiatric evaluation should be obtained in young persons who report an increasing frequency of nightmares without obvious cause, as the increase may be the first indication of a psychotic illness.[71] Psychiatric assessment also is indicated for many patients with chronic nightmares that are not related to medication use, especially if post-traumatic stress disorder is a diagnostic possibility. Polysomnography is rarely required unless the history suggests RBD or an overlap between nightmares and sleep terrors.

MANAGEMENT

The first step in management is to address any underlying psychiatric illness, medical disease, or medication that is contributing to the nightmares. Behavioral approaches such as hypnosis, desensitization, and dream rehearsal may be helpful. Antidepressants or cyproheptadine may be helpful in some patients. Nighttime doses of antipsychotic medications are often useful in patients with nightmares associated with psychotic illnesses.

OTHER PARASOMNIAS

The other parasomnias constitute a heterogeneous group of disorders whose phenomena of concern are not closely linked to a sleep stage or to sleep-wake transitions.

Sleep Bruxism

CLINICAL FEATURES

Sleep bruxism refers to grinding or clenching of the teeth during sleep. Although dentists have been aware of the damaging effects of bruxism on teeth and supporting structures for more than 100 years, the differences between bruxism during sleep and bruxism during wakefulness were not recognized until the 1960s.

Sleep bruxism, which occurs at all ages and affects men and women equally, is characterized by rhythmic, chewing movements and, less commonly, by prolonged contractions of the jaw muscles lasting from several seconds to as long as several minutes. Bruxism occurs during arousals from all sleep stages, including REM sleep, and clicking or grating sounds accompany the grinding in about 20% of episodes. The severity of bruxism typically fluctuates from night to night and from week to week. Although sleep bruxism probably occurs in almost everyone at some point during the life span, the severity of sleep bruxism is highly variable. Significant tooth wear, muscular pain, or temporomandibular changes occur in about 5% to 10% of the population; another 10% to 20% have minor tooth wear.[72-74]

Patients may present for evaluation because of daytime symptoms, because the sounds disturb the sleep of bed partners, or because bruxism has caused significant damage to teeth, dental restorations, or supporting structures. Although most patients are unaware of sleep bruxism and do not have daytime symptoms, some patients have morning jaw or facial pain; others have tenderness or tightness of the temporalis muscle or of the temporomandibular joint. Periodontal pain and muscle contraction headaches are complaints in some patients with sleep bruxism. Although the contribution of sleep bruxism to these daytime symptoms is not always clear, in one study, jaw discomfort was a complaint in 6 of 18 bruxers compared to 0 of 18 control subjects.[75] Many patients report that bruxism or symptoms associated with bruxism, such as morning jaw pain, increase with stress. Although craniomandibular disorders are often attributed to bruxism, these disorders are more strongly associated with systemic joint laxity than with bruxism.[76]

The damage produced by bruxism is a function of the intensity, frequency, direction, and duration of bruxing. For unknown reasons, tooth damage appears to be more common in male bruxers than female bruxers.[77] Sleep bruxism is often more destructive than bruxism that occurs during wakefulness because the grinding forces are lateral. On physical examination, tooth wear is the most striking finding, most commonly on the incisal edges of the anterior teeth and on the cusps of the posterior teeth. Severe bruxism during sleep may also damage supporting structures and lead to spacing problems and food impaction.[77] Destruction of crowns and other dental repairs, masticatory muscle hypertrophy, limitation of jaw movement, and an overclosed or short-face appearance may occur in severe cases.[78-80] Although bruxism alone does not cause periodontal disease, it may be an aggravating factor and may increase the rate of bone loss.[79]

BIOLOGIC BASIS

The cause of sleep bruxism is unknown, although children of persons with sleep bruxism are at increased risk. Although stress is probably a precipitating or aggravating factor in most patients, psychological problems are not more frequent than in the general population.[81,82] Malocclusion and other dental and anatomic anomalies may contribute in some patients. Although reduced inhibition of motor generators may be a factor in institutionalized mentally retarded patients, in whom the prevalence of tooth wear from sleep bruxism approaches 50%,[83] most patients with sleep bruxism do not have motor system abnormalities.[84] Other possible contributors include OSA and such medications as amphetamines, levodopa, and phenothiazines.[72,78]

DIAGNOSIS

Oral examination, combined with a history of audible grinding at night, is generally sufficient for diagnosis. Smooth, shiny occlusal facets of the incisal surfaces of the maxillary canines are usually the first indication of tooth wear. Masseter muscle hypertrophy is suggestive of bruxism, but morning temporal headaches and temporomandibular joint pain are less specific for the disorder.

Polysomnographic monitoring is not required. If performed, episodes of rhythmic masseter and temporalis muscle activity during sleep are typical (Fig. 15-7); their absence does not exclude bruxism, as it may not occur every night. Rhythmic jaw movements that can accompany partial complex seizures may occasionally be confused with bruxism. VPSG or daytime EEG, or both, may be useful if a seizure disorder is under consideration (see Chap. 20).

MANAGEMENT

Tooth damage can usually be prevented with a dental guard, which reduces symptoms in 80% to 90% of patients but does not eliminate bruxing. Counseling on ways to reduce stress may help some patients; other treatments are of limited value. Diazepam (5 mg at bedtime), which may be helpful during acute exacerbations, is not helpful for long-term treatment. Nocturnal biofeedback provides little benefit. Bruxism in children often becomes insignificant by ages 9 to 12 and generally does not require treatment.

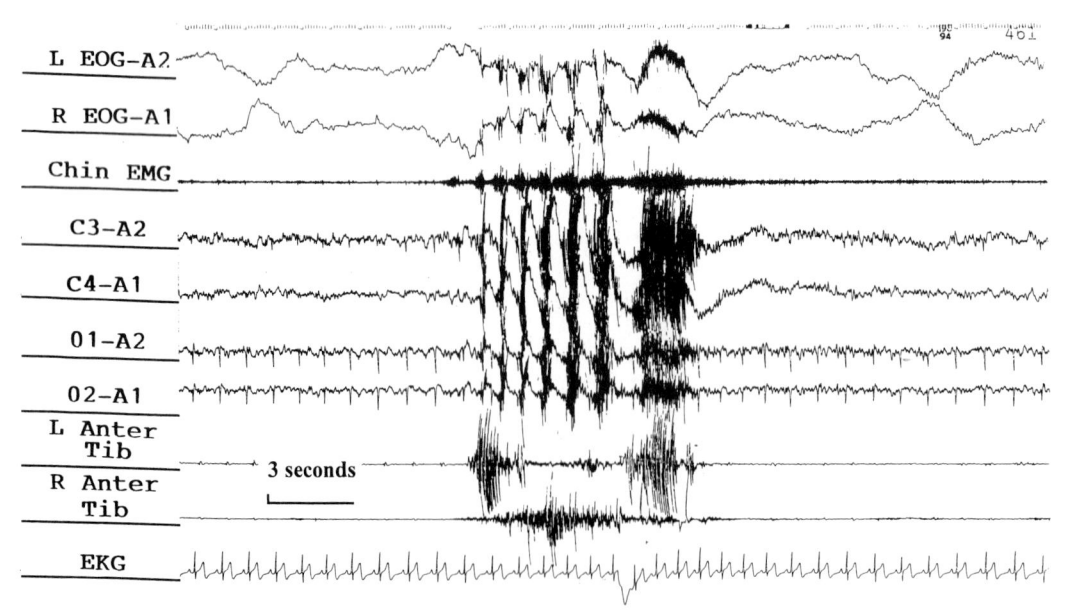

Figure 15-7. Bruxism during stage 1 sleep in a 48-year-old man undergoing evaluation for suspected obstructive sleep apnea. Rhythmic bursts of muscle activity are apparent in the first 7 channels. EOG, electro-oculogram. For other abbreviations, see Figures 15-1 and 15-2.

Sleep Enuresis

Recurrent involuntary bedwetting that occurs beyond the age of expected nocturnal bladder control is called sleep enuresis. First described more than 3000 years ago, sleep enuresis is a distressing problem for many children and adults. Embarrassment, shame, and fears of sleeping over at friends' houses or going on overnight trips are common, and enuresis often becomes a focus of family conflict.

Although early sleep studies suggested that enuresis occurred as a dream equivalent or that it was one of the disorders of arousal, with enuresis occurring during SWS, later studies demonstrated that episodes occur throughout the night in all sleep stages.

CLINICAL FEATURES

Although the prevalence of bedwetting at various ages is similar in many societies, the age of expected bladder control varies across cultures. In the United States, sleep enuresis is typically diagnosed after age 6 in boys and after age 5 in girls. While most children achieve bladder control by age 4, a substantial minority have episodes up to age 6, and about 1% to 3% still have episodes by age 12 (Fig. 15–8).[25,85,86] Because boys achieve nocturnal continence later than girls, bedwetting affects boys more than girls at all ages. After age 5, sleep enuresis disappears spontaneously at a rate of about 15% per year. In young adults, the prevalence of sleep enuresis is about 1%. Daytime wetting occurs in about 15% to 20% of nocturnal enuretics, mostly in those under age 6.[87]

Primary sleep enuresis, which refers to bedwetting in children who have never attained bladder control, accounts for 75% to 80% of cases. Secondary sleep enuresis, in which the child has been free of sleep enuresis for at least 3 months and then resumes bedwetting on a regular basis, is less common in the first decade, although 50% of cases are secondary by age 12. In children and adults with secondary sleep enuresis, psychological disorders or genitourinary pathology are more often involved than in those with primary sleep enuresis.

BIOLOGIC BASIS

The etiology of primary sleep enuresis is presumed to be psychosocial, genetic, maturational, endocrinological, or a combination of these. Only a small minority have anomalies of the genitourinary system, such as malformation of the posterior urethral valves or renal insufficiency. Renal medullary infarction probably accounts for the high incidence of enuresis in children with sickle cell anemia.[88] Sleep patterns between enuretic and nonenuretic children are similar, suggesting that sleep disorders are not the cause in most cases,

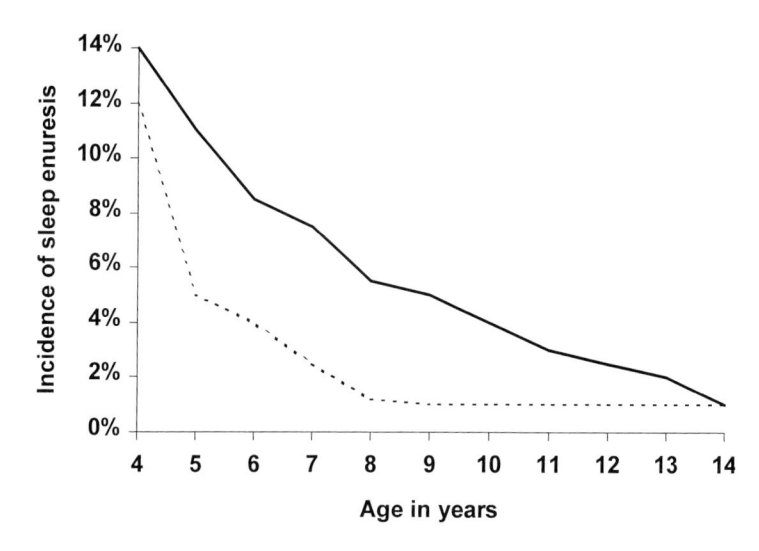

Figure 15–8. Incidence of sleep enuresis during childhood for boys (solid line) and girls (dotted line). Between ages 5 and 12, the incidence is much higher in boys than in girls. (Based in part on data from Klackenberg.[25])

although OSA may contribute in some. Nocturnal seizures, diabetes mellitus, and diabetes insipidus also may cause nocturnal incontinence.

Psychosocial factors play a role in most enuretics: Marital discord, parental separation, and the birth of a sibling may precipitate enuresis in some children. Children with sleep enuresis have suffered emotional trauma more often than children without enuresis, and sexual abuse may be a factor in some cases. Poor toilet training and inadequate parenting skills, along with limited expectations concerning toilet training, may contribute to the higher rate of sleep enuresis in lower socioeconomic groups and in institutionalized children. Evidence for a genetic component comes from family studies: A child's risk of sleep enuresis is increased if one parent was enuretic, and increased even further if both were enuretic.[87,89] However, these findings may result from expectations on the part of the parents or from other psychological, environmental, or cultural factors as well as from genetic influences.

A maturational component is suggested by studies indicating that enuretic children tend to have lower birth weight, shorter stature, younger bone age, delayed developmental milestones, and later onset of puberty.[90,91] On average, children with sleep enuresis have smaller functional bladder capacities than children without enuresis, but the contribution of this difference to enuresis is probably small. Some enuretic children do not show the normal nighttime peak in antidiuretic hormone (ADH) levels and have unusually high nocturnal urine output, suggesting that ADH metabolism may be abnormal in these patients.[92]

DIAGNOSIS

Diagnosis is usually based on history, exploration of family dynamics, physical examination, and urinalysis. The possible contributions of diabetes mellitus, diabetes insipidus, OSA, sickle cell anemia, and epilepsy should be considered when obtaining the history. For children with primary sleep enuresis and normal physical examination, normal urinalysis, and no daytime voiding symptoms, additional laboratory investigations are not required. Primary sleep enuresis complicated by urinary tract infections or daytime voiding dysfunction requires a more extensive evaluation for possible genitourinary anomalies, as does secondary enuresis; intravenous pyelogram, vesical sphincter electromyography, cystometry, and cystoscopy are often indicated. Psychiatric evaluation is often helpful in cases of secondary enuresis. Repeated urinary tract infections along with enuresis raise the possibility of sexual abuse, particularly in girls.

MANAGEMENT

For primary sleep enuresis, education, empathy, and patience may be all that is required. Since the annual spontaneous cure rate is substantial,[93] almost any approach will work, given enough time, because of the natural history of the disorder.

Because parental expectations are often a major contributor, especially among parents who themselves were enuretic, and because sleep enuresis may become the focus of family conflict, effective management requires a good understanding of family dynamics. In most cases, parents should be encouraged to allow the child to assume control of the treatment program in order to reduce the impact of parental expectations.

Alarm systems are the most effective behavioral technique. A sensor is placed in the bed or on the bedclothes that activates an alarm when wet. When the alarm sounds, the child awakens, stops voiding, turns off the alarm, and goes to the bathroom. By pavlovian conditioning, the child gradually learns to awaken before the alarm. Alarms can be combined with rewards, imaging techniques, and allowing the child to assume control of cleaning up after an episode. About 65% to 80% of patients have success with the alarm; however, about 10% to 40% of these patients relapse after the alarm is removed.[94] Bladder retention training, in which the child deliberately delays using the bathroom despite the urge to do so, is sometimes used in conjunction with alarms. Hypnotherapy may be successful in some cases.[87]

Although urine alarms have higher success rates than medications, drug therapy is often prescribed.[86,87] Desmopressin (DDAVP) (25 to 75 mg daily) reduces urine production and has few side effects, but less than one-third of patients are free of symptoms for more than 2 weeks.[95] Imipramine (25 to 75 mg at bedtime), with a success rate of about 25%, is only marginally superior to the approximate 15% annual spontaneous cure rate, and relapses are common.[86,96] Oxybutynin chloride, a spasmolytic, is rarely of benefit.

Nocturnal Dissociative Disorder

In dissociative disorder, recognized for hundreds of years, conscious awareness becomes dissociated from behavior on a psychogenic basis, and patients engage in complex behaviors regarding which they are subsequently amnesic. In the 19th century, Charcot and Janet popularized the term *hysteria* and described the combination of dissociated awareness and hysterical sensorimotor symptoms as *la grande hysterie*. Freud believed that the amnesia and dissociated awareness were caused by repression, while hysterical sensorimotor symptoms were a result of conversion of unacceptable wishes into bodily symptoms. In the *Diagnostic and Statistical Manual of Mental Disorders, Edition 4, Revised*, conversion disorder is included as one of the somatoform disorders, separate from the dissociative disorders, in which the predominant symptom is a disruption of consciousness, identity, memory, or perception. If the disturbance affects identity primarily, it may lead to depersonalization disorder or to multiple personality disorder. If it affects memory primarily, the diagnosis may be dissociative amnesia (psychogenic amnesia) or dissociative fugue (psychogenic fugue).[97]

Sleep is sometimes the portal by which patients enter or leave dissociative states. William James, in 1890, described the case of Rev. Bourne of Rhode Island, who vanished from his home. After living for 2 months as A.J. Brown, a shopkeeper in Norristown, Pennsylvania, he awoke one night "in a fright and called in the people of the house to tell him where he was. He said that his name was Ansel Bourne, that he was entirely ignorant of Norristown, that he knew nothing of shopkeeping, and that the last thing he remembered—it seemed only yesterday—was drawing the money from the bank, etc., in Providence."[98]

Although prolonged nocturnal fugues still occur in some patients, brief episodes are more common in patients presenting with this disorder in the 1990s. Patients are often young women with borderline personality disorder, anxiety disorder, or other psychiatric conditions who sometimes appear, during nocturnal episodes, to be reenacting previous assaults by others on them, sometimes accompanied by self-mutilating behavior and injuries.[99] They may be amnesic regarding the episode, or they may report dreamlike mentation. Poor psychosocial adjustment, substance abuse, suicide attempts, and a reported history of physical and sexual abuse beginning in childhood are often part of the clinical picture. Most patients have daytime as well as nighttime dissociative episodes.

The cause of the disorder is presumed to be the underlying psychiatric condition. Psychic trauma, physical and sexual abuse, and personality disorder are potential contributors.

Psychiatric evaluation is the key to diagnosis. VPSG, indicated if RBD, partial seizures, sleepwalking, or sleep terrors are suspected, almost invariably demonstrates EEG patterns of wakefulness preceding and during the episodes.[99,100] Nocturnal dissociative disorder is distinguished from malingering and factitious disorders by its unconscious motivation.

The underlying psychiatric disorder is the primary focus of treatment. Unfortunately, the disorder often lasts for years, with a poor response to psychotherapy and psychoactive medications.

Sudden Unexplained Nocturnal Death

Except for SIDS, nocturnal spells culminating in death fortunately are rare. The disorder of sudden unexplained nocturnal death (SUND) in adults, first described in the Philippines in the early twentieth century, appears to affect primarily young and middle-aged men from Southeast Asian countries, including the Philippines, Japan, Thailand, Vietnam, Cambodia, and Laos.[101-103] The syndrome assumed increasing importance in the United States after the influx of Southeast Asians in the 1970s and 1980s.

The annual incidence per 100,000 person in affected populations ranges from 25 in northeastern Thailand to 574 among Laotian and Cambodian refugees living in Thailand.[104,105] The syndrome is one of the leading causes of death in young and middle-aged Southeast Asian men, and in the early 1980s, it was a common cause of death among Southeast Asian refugees emigrating to the United States.[106] Migration from place of birth appears to increase the risk, particularly within the first 2 years after entry into the foreign country. For example, the incidence of SUND among Thai men working in Singapore is four times higher than that of Thai men living in rural Thailand.

Affected persons are usually previously healthy, although some have had episodes resembling sleep terrors in the weeks preceding death. Spells usually occur 3 to 4 hours after sleep onset, with choking, gasping, groaning, screaming, or labored breathing, followed a few seconds or minutes later by collapse and death.[107]

The cause of SUND is unknown, although in all cases of successful resuscitation or monitored death, ventricular fibrillation was the cardiac rhythm.[108] Excessive sympathetic discharge during REM sleep, congenital cardiac conduction abnormalities, and stress may contribute, and deficiencies of thiamine and potassium have also been implicated, suggesting that improved nutrition may be a means of prevention.[104,109-111]

For persons who have been successfully resuscitated, cardiologic evaluation is indicated, and treatment may include antiarrhythmics or implantable defibrillators along with measures to reduce stress.

often associated with skeletal muscle activity and autonomic arousal, occur mainly or exclusively during sleep. Some parasomnias, such as nightmares, are so common that almost everyone has experienced them, whereas others are exceedingly rare.

Parasomnias are classified according to the predominant sleep stage in which they occur. The arousal disorders—confusional arousals, sleepwalking, and sleep terrors—are grouped together based on the hypothesis that impaired ability to arouse fully from SWS is responsible for the symptoms. The sleep-wake transition disorders—rhythmic movement disorder, sleep starts, sleeptalking, and nocturnal leg cramps—occur primarily during transitions between sleep and wakefulness. In the parasomnias related to REM sleep, features of REM sleep are disturbed and lead to or exacerbate the parasomnia. For example, in RBD, the normal muscle atonia of REM sleep is disrupted, and patients are able to act out their dreams. The last group of parasomnias, which include sleep bruxism and sleep enuresis, is a heterogeneous group of disorders in which the phenomena of concern are not closely linked to a sleep stage or to sleep-wake transitions.

The phenomena that occur with parasomnias often are paroxysmal events and thus share similarities with daytime spells. Although the history provides crucial information for determining the cause of spells, paroxysmal events during sleep present a special challenge because patients, who are usually asleep when the event begins, are rarely able to give an accurate description of the episode. When the diagnosis cannot be made clinically, VPSG recordings allow behavioral analysis of episodes along with assessment of EEG and other physiologic measures. Management is based on the prognosis of the disorder, the degree of sleep disruption, the psychosocial consequences, and the potential for injury.

SUMMARY

Parasomnias are disorders in which undesirable physical and mental phenomena,

REFERENCES

1. American Sleep Disorders Association: International Classification of Sleep Disorders, Re-

1. vised: Diagnostic and Coding Manual. American Sleep Disorders Association, Rochester, Mn, 1997.
2. Broughton, RJ: Sleep disorders: Disorders of arousal? Science 159:1070, 1968.
3. Jones, E: On the nightmare. Hogarth, London, 1949.
4. Gastaut, H, and Broughton, R: A clinical and polygraphic study of episodic phenomena during sleep. Recent Adv Biol Psychiatry 7:197, 1965.
5. Gastaut, H, et al: Etude electro-clinique de terreurs nocturnes et diurnes concomitantes d'un reve ou d'une idee obsedante chez un nevrose. Rev Neurol 107:277, 1962.
6. Hublin, C, et al: Prevalence and genetics of sleepwalking: A population-based twin study. Neurology 48:177, 1997.
7. Schenck, CH, et al: Sleep-related eating disorders: Polysomnographic correlates of a heterogeneous syndrome distinct from daytime eating disorders. Sleep 14:419, 1991.
8. Schenck, CH, et al: A polysomnographic and clinical report on sleep-related injury in 100 adult patients. Am J Psychiatry 146:1166, 1989.
9. Broughton, R, et al: Homicidal somnambulism: A case report. Sleep 17:253, 1994.
10. Kales, JD, et al: Night terrors: Clinical characteristics and personality patterns. Arch Gen Psychiatry 37:1413, 1980.
11. Naylor, MW, and Aldrich, MS: The distribution of confusional arousals across sleep stages and time of night in children and adolescents with sleep terrors. Sleep Res 20:308, 1991.
12. Hallstrom, T: Night terror in adults through three generations. Acta Psychiatr Scand 48:350, 1972.
13. Bakwin, HI: Sleepwalking in twins. Lancet 2:446, 1970.
14. Kales, A, et al: Hereditary factors in sleepwalking and night terrors. Br J Psychiatry 137:111, 1980.
15. Llorente, MD, et al: Night terrors in adults: Phenomenology and relationship to psychopathology. J Clin Psychiatry 53:392, 1992.
16. Maselli, RA, et al: Episodic nocturnal wanderings in non-epileptic young patients. Sleep 11:156, 1988.
17. Pedley, TA, and Guilleminault, C: Episodic nocturnal wanderings responsive to anticonvulsant drug therapy. Ann Neurol 2:30, 1977.
18. Hurwitz, TD, et al: Sleepwalking—sleep terrors—REM sleep behavior disorder: Overlapping parasomnias. Sleep Res 20:260, 1991.
19. Aldrich, MS, and Jahnke, B: Diagnostic value of video-EEG polysomnography. Neurology 41:1060, 1991.
20. Blatt, I, et al: The value of sleep recording in evaluating somnambulism in young adults. Electroencephalogr Clin Neurophysiol 78:407, 1991.
21. Lask, B: Novel and non-toxic treatment for night terrors. Br Med J 297:592, 1988.
22. Zappert, J: Über nächtliche Kopfbewegungen bei Kindern (jactatio capitis nocturna). Jahrb Kinderheilkd 62:70, 1905.
23. Cruchet, R: Tics et sommeil. Presse Med 13:33, 1905.
24. Gagnon, P, and De Koninck, J: Repetitive head movements during REM sleep. Biol Psychiatry 20:176, 1985.
25. Klackenberg, G: Incidence of parasomnias in children in a general population. In Guilleminault, C (ed). Sleep and Its Disorders in Children. Raven, New York, 1987, pp 99–113.
26. Mitchell, SW: Some disorders of sleep. Int J Med Sci 100:109, 1890.
27. Thorpy, MJ: Sleep starts. In Gilman, S, et al (eds): Neurobase, ed 3. Arbor, La Jolla, Calif, 1997.
28. Rechtschaffen, A, et al: Patterns of sleep talking Arch Gen Psychiatry 7:418, 1962.
29. Abe, K, et al: Sleepwalking and recurrent sleeptalking in children of childhood sleepwalkers. Am J Psychiatry 141:800, 1984.
30. Cutler, P: Cramps in the legs and feet. JAMA 252:98, 1984.
31. Weiner, IH, and Weiner, HL: Nocturnal leg muscle cramps. JAMA 244:2332, 1980.
32. Ricker, K, and Moxley, RT: Autosomal dominant cramping disease. Arch Neurol 47:810, 1990.
33. Jacobsen, JH, et al: Familial nocturnal cramping. Sleep 9:54, 1986.
34. Lazaro, RP, et al: Familial cramps and muscle pain. Arch Neurol 38:22, 1981.
35. Whitely, AM: Cramps, stiffness and restless legs. Practitioner 226:1085, 1982.
36. Takeuchi, T, et al: Isolated sleep paralysis elicited by sleep interruption. Sleep 15:217, 1992.
37. Fukuda, K, et al: High prevalence of isolated sleep paralysis: Kanashibari phenomenon in Japan. Sleep 10:279, 1987.
38. Roth, B, et al: Familial sleep paralysis. Arch Suiss Neurol Neurochir Psychiatry 102:321, 1968.
39. Schenck, CH, et al: Chronic behavioral disorders of human REM sleep: A new category of parasomnia. Sleep 9:293, 1986.
40. Dyken, ME, et al: Violent sleep-related behavior leading to subdural hemorrhage. Arch Neurol 52:318, 1995.
41. Mahowald, MW, and Schenck, CH: REM sleep behavior disorder. In Kryger, MH, et al (eds): Principles and Practice of Sleep Medicine, ed 2. WB Saunders, Philadelphia, 1994, pp 574–588.
42. Stolz, SE, and Aldrich, MS: REM sleep behavior disorder associated with caffeine abuse. Sleep Res 20:341, 1991.
43. Jouvet, M, and Delorme, F: Locus coeruleus et sommeil paradoxal. C R Soc Biol 159:895, 1965.
44. Hendricks, JC, et al: Different behaviors during paradoxical sleep without atonia depend on pontine lesion site. Brain Res 239:81, 1982.
45. Tison, F, et al: REM sleep behaviour disorder as the presenting symptom of multiple system atrophy. J Neurol Neurosurg Psychiatry 58:379, 1995.
46. Uchiyama, M, et al: Incidental Lewy body disease in a patient with REM sleep behavior disorder. Neurology 45:709, 1995.
47. Wright, BA, et al: Shy-Drager syndrome presenting as a REM behavioral disorder. J Geriatr Psychiatry Neurol 3:110, 1990.

48. Schenck, CH, et al: Delayed emergence of a parkinsonian disorder in 38% of 29 older men initially diagnosed with idiopathic rapid eye movement sleep behavior disorder. Neurology 46:388, 1996.
49. Schenck, CH, and Mahowald, MW: Motor dyscontrol in narcolepsy: Rapid-eye-movement (REM) sleep without atonia and REM sleep behavior disorder. Ann Neurol 32:3, 1992.
50. Rattenborg, NC, et al: REM sleep-related asystole associated with unusual polysomnographic features: A case history. Sleep Res 24:324, 1995.
51. Guilleminault, C, et al: Sinus arrest during REM sleep in young adults. N Engl J Med 311:1006, 1984.
52. Ohlmeyer, P, et al: Periodische vorgange im schaf pflug. Arch Ges Physiol 248:559, 1944.
53. Fisher, C, et al: Cycle of penile erection synchronous with dreaming (REM) sleep: Preliminary report. Arch Gen Psychiatry 12:29, 1965.
54. Karacan, I: Nocturnal penile tumescence as a biologic marker in assessing erectile dysfunction. Psychosomatics 23:349, 1982.
55. Karacan, I, et al: Aging and sexual dysfunction in man: Contributions of the sleep laboratory. In Chase, MH, and Weitzman, ED (eds): Sleep Disorders: Basic and Clinical Research. Spectrum, New York, 1983, pp 347–362 and 503–513.
56. Karacan, I, et al: Erectile mechanisms in man. Science 220:1080, 1983.
57. Schmidt, HS, and Schmidt, MH: Current physiology of erectile mechanisms: Basis for five phases of penile erection. Sleep Res 20:52, 1991.
58. Roffwarg, HP, et al: Plasma testosterone and sleep: Relationship to sleep stage variables. Psychosom Med 44:73, 1982.
59. Thase, ME, et al: Nocturnal penile tumescence in depressed men. Am J Psychiatry 144:89, 1987.
60. Thase, ME, et al: Nocturnal penile tumescence is diminished in depressed men. Biol Psychiatry 24:33, 1988.
61. Grunstein, RR: Metabolic aspects of sleep apnea. Sleep (10 suppl)19:S218, 1996.
62. Karacan, I: Clinical value of nocturnal erection in the prognosis and diagnosis of impotence. Med Aspects Human Sex 4:27, 1970.
63. Ware, JC, and Hirshkowitz, M: Monitoring penile erections during sleep. In Kryger, MH, et al (eds): Principles and Practice of Sleep Medicine. WB Saunders, Philadelphia, 1994, pp 967–977.
64. Karacan, I, et al: Measurement of pressure necessary for vaginal penetration. Sleep Res 14:269,1985.
65. Wood, J, and Bootzin, R: The prevalence of nightmares and their independence from anxiety. J Abnorm Psychol 99:64, 1990.
66. Hartmann, E: The Nightmare. Basic Books, NY, 1984.
67. Wood, J, et al: Effects of the 1989 San Francisco earthquake on frequency and content of nightmares. J Abnorm Psychol 101:219, 1992.
68. Brimacombe, J, and Mcfie, AG: Peri-operative nightmares in surgical patients. Anesthesiology 48:527, 1993.
69. Cartwright, R: Nightmares. In Gilman, S, et al (eds): Neurobase, ed 3. Arbor, La Jolla, Calif, 1997.
70. Berquier, A, and Ashton, R: Characteristics of the frequent nightmare sufferer. J Abnorm Psychol 101:246, 1992.
71. Hartmann, E: Nightmares. In Kryger, MH, et al (eds): Principles and Practice of Sleep Medicine, ed 2. WB Saunders, Philadelphia, 1994, pp 407–410.
72. Rugh, JD, and Harlan, J: Nocturnal bruxism and temporomandibular disorders. Adv Neurol 49:329, 1988.
73. Glaros, AG: Incidence of diurnal and nocturnal bruxism. J Prosthet Dent 45:545, 1981.
74. Hartmann, E: Bruxism. In Kryger, MH, et al (eds): Principles and Practice of Sleep Medicine, ed 2. WB Saunders, Philadelphia, 1994, pp 598–601.
75. Lavigne, GJ, et al: Sleep bruxism: Validity of clinical research diagnostic criteria in a controlled polysomnographic study. J Dent Res 75:546, 1996.
76. Westling, L: Temporomandibular joint dysfunction and systemic joint laxity. Swed Dent J Suppl 81:1, 1992.
77. Seligman, DA, et al: The prevalence of dental attrition and its association with factors of age, gender, occlusion, and TMJ symptomatology. J Dent Res 67:1323, 1988.
78. Glaros, AG, and Rao, SM: Bruxism: A critical review. Psychol Bull 34:767, 1977.
79. Glaros, AG, and Rao, SM: Effects of bruxism: A review of the literature. J Prosthet Dent 38:149, 1977.
80. Clark, GT, et al: Nocturnal masseter muscle activity and the symptoms of masticatory dysfunction. J Oral Rehabil 8:279, 1981.
81. Harness, DM, and Peltier, B: Comparison of MMPI scores with self-report of sleep disturbance and bruxism in the facial pain population. Cranio 10:70, 1992.
82. Fischer, WF, and O'Toole, ET: Personality characteristics of chronic bruxers. Behav Med 19:82, 1993.
83. Richmond, G, et al: Survey of bruxism in an institutionalized mentally retarded population. Am J Ment Defic 88:418, 1984.
84. Dyken, ME, and Rodnitzky, RL: Periodic, aperiodic, and rhythmic motor disorders of sleep. Neurology (suppl 6)42:68, 1992.
85. Schmitt, BD: Nocturnal enuresis. Prim Care 11:485, 1984.
86. Friman, PC, and Warzak, WJ: Nocturnal enuresis: A prevalent, persistent, yet curable parasomnia. Pediatrician 17:38, 1990.
87. Rushton, HG: Nocturnal enuresis: Epidemiology, evaluation, and currently available treatment options. J Pediatr 114:691, 1989.
88. Readett, DR, et al: Nocturnal enuresis in sickle cell haemoglobinopathies. Arch Dis Child 65:290, 1990.
89. Bakwin, H: The genetics of enuresis. In Kolvin,

I, et al (eds): Bladder Control and Enuresis. Heinemann Medical, London, 1973, pp 39–46.
90. Jarvelin, MR, et al: Aetiological and precipitating factors for childhood enuresis. Acta Paediatr Scand 80:361, 1991.
91. Mimouni, M, et al: Retarded skeletal maturation in children with primary enuresis. Eur J Pediatr 144:234, 1985.
92. Nergaard, JP, et al: Nocturnal enuresis: An approach to treatment based on pathogenesis. J Pediatr 114:705, 1989.
93. Forsythe, WI, and Redmond, A: Enuresis and spontaneous cure rate, study of 1129 enuretics. Arch Dis Child 49:259, 1974.
94. Maizels, M, and Rosenbaum, D: Successful treatment of nocturnal enuresis: A practical approach. Prim Care 12:621, 1985.
95. Moffatt, MEK, et al: Desmopressin acetate and nocturnal enuresis: How much do we know? Pediatrics 92:420, 1993.
96. Fritz, GK, et al: Plasma levels and efficacy of imipramine treatment for enuresis. J Am Acad Child Adolesc Psychiatry 33:60, 1994.
97. American Psychiatric Association: Diagnostic and Statistical Manual of Mental Disorders, ed 4. American Psychiatric Association, Washington, DC, 1994.
98. James, W: The Principles of Psychology. H Holt, New York, 1890.
99. Schenck, CH, et al: Dissociative disorders presenting as somnambulism: Polysomnographic, video and clinical documentation (8 cases). Dissociation 2:194, 1989.
100. Rice, E, and Fisher, C: Fugue states in sleep and wakefulness: A psychophysiological study. J Nerv Ment Dis 163:79, 1976.
101. Baron, RC, et al: Sudden death among Southeast Asian refugees. JAMA 250:2947, 1983.
102. Aponte, GE: The enigma of "bangungut." Ann Intern Med 52:1258, 1960.
103. Sugai, M: A pathological study on sudden and unexpected death, especially on the cardiac death autopsied by medical examiners in Tokyo. Acta Pathol Jpn (suppl)9:723,1959.
104. Munger, RG, and Booton, EA: Thiamine and sudden death of South-East Asian refugees. Lancet 335:1154, 1990.
105. Tatsanavivat, P, et al: Sudden and unexplained deaths in sleep (Laitai) of young men in rural northeastern Thailand. Int J Epidemiol 21:904, 1992.
106. Parrish, RG, et al: Sudden unexplained death syndrome in Southeast Asian refugees: A review of CDC surveillance. MMWR 36(no. 1SS):43SS, 1987.
107. Otto, CM, et al: Ventricular fibrillation causes sudden death in Southeast Asian immigrants. Ann Intern Med 100:45, 1984.
108. Pressman, MR, et al: Polysomnographic and electrocardiographic findings in a sudden unexplained nocturnal death syndrome (SUNDS) survivor. Sleep Res 22:313, 1993.
109. Feest, TG, and Wrong, O: Potassium deficiency and sudden unexplained nocturnal death. Lancet 338:1406, 1991.
110. Munger, RG, et al: Prolonged QT interval and risk of sudden death of Southeast Asian men. Lancet 338:280, 1991.
111. Kirschner, RH, et al: The cardiac pathology of sudden, unexplained nocturnal death in Southeast Asian refugees. JAMA 256:2700, 1986.

Chapter 16

PSYCHIATRIC DISORDERS AND SLEEP

SLEEP DISORDERS ASSOCIATED
  WITH SCHIZOPHRENIA
Clinical Features
Psychobiologic Basis
Diagnosis
Management
SLEEP DISTURBANCE ASSOCIATED
  WITH MOOD DISORDERS
Major Depressive Episodes
Manic and Hypomanic Episodes
Dysthymic Disorder
Seasonal Affective Disorder
Bereavement
SLEEP DISORDERS ASSOCIATED WITH
  ANXIETY DISORDERS
Generalized Anxiety Disorder
Panic Disorder
Post-traumatic Stress Disorder
Obsessive-Compulsive Disorder
SLEEP DISTURBANCE ASSOCIATED
  WITH EATING DISORDERS
SLEEP DISTURBANCE ASSOCIATED
  WITH ALCOHOL USE
  AND ALCOHOLISM
Effects of Alcohol on Alertness
Effects of Alcohol on Sleep
Alcohol-Dependent Sleep Disorder
Alcoholism
SLEEP DISTURBANCE ASSOCIATED
  WITH OTHER SUBSTANCE-RELATED
  DISORDERS

Psychiatric disorders are common in patients with sleep complaints, particularly those with insomnia. About one-third of insomniacs evaluated at sleep centers have a primary psychiatric diagnosis and another 40% have a secondary psychiatric diagnosis.[1] According to community-based surveys, as many as 40% of those who complain of insomnia have a psychiatric illness, usually a mood disorder or an anxiety disorder.[2,3] In addition, mood disorders subsequently develop in many patients who initially present with insomnia, suggesting that insomnia can be an initial symptom of depression or that the irritability, anxiety, and dysphoria that often accompany insomnia increase the likelihood that depression will develop.[2,4] Whether treatment of insomnia can prevent depression is uncertain.

Just as psychiatric disorders are common in patients with sleep complaints, sleep disturbance is common in patients with psychiatric problems. Psychiatric disorders may cause sleep problems directly, and the medications used to treat psychiatric illness can exacerbate the sleep disturbance or can cause new sleep disorders.

SLEEP DISORDERS ASSOCIATED WITH SCHIZOPHRENIA

Schizophrenia, a chronic disorder with intermittent exacerbations and a prevalence of about 1% in the general population, is one of the two major causes of chronic psychosis, the other being bipolar disorder.[5] An association between disordered sleep and psychosis has been recognized for many centuries and a relation between dreams and psychosis has been a source of speculation since ancient times. The discovery of rapid eye movement (REM) sleep led to the idea that the resemblance between psychotic hallucinations and dreams was a result of REM sleep processes affecting or intruding into wakefulness.[6] Although subsequent research did not support this proposal, the concept led to a number of studies of sleep disturbance in persons with psychosis.

Clinical Features

Clinical and sleep-related features of schizophrenia are shown in Table 16–1. For many schizophrenics, sleep disturbance—particularly insomnia—is a chronic problem. Inability to fall asleep, frequent awakenings, and a sense that sleep is not restorative are typical complaints, and although early morning awakening is unusual in schizophrenia, it may occur in patients with schizoaffective disorder. Some patients sleep late; others with severe insomnia take daytime naps. Highly disorganized patients often have highly disorganized sleep-wake patterns: Sleep episodes are of variable length and can occur at almost any time of the day or night. Complaints of increased need for sleep, common in medicated patients, are unusual in untreated patients.

During acute psychotic episodes, insomnia usually becomes much worse, characterized by reduced sleep time and decreased sleep continuity, and some patients complain of nocturnal hallucinations and frightening dreams that are difficult to distinguish from reality.

Table 16–1. **Clinical and Sleep-Related Aspects of Schizophrenia**

Disorganized speech and behavior
Delusions and hallucinations
Flat affect
Impoverished speech
Decrease in goal-directed activities
Sleep
  Insomnia
  Disorganized sleep-wake patterns
  REM sleep abnormalities and reduced slow-wave sleep sometimes present

Psychobiologic Basis

The causes of insomnia in schizophrenia include anxiety that accompanies the delusions and hallucinations, as well as behavioral disorganization with loss of scheduled activities. Behavioral disorganization may also lead to delayed sleep-phase syndrome and poor sleep hygiene, which sometimes contribute to sleep problems.

Changes in REM sleep and slow-wave sleep (SWS) are common in schizophrenia, and although the impact of these changes on sleep symptoms in schizophrenia is uncertain, they may provide clues to the neurobiology of the disorder. Although the amount of REM sleep is usually not increased, short REM sleep latency is common in patients with schizophrenia and schizoaffective disorder, as it is in patients with mood disorders (see below).[7] Short REM sleep latency, along with increased numbers of rapid eye movements per minute of REM sleep (*REM density*), appears to be more common in patients with prominent negative symptoms of apathy and social withdrawal and may be caused by increased cholinergic activity, as it correlates with levels of an abnormal plasma cholinesterase isozyme in schizophrenics.[8,9] Overactivity of dopamine or norepinephrine, or both, caused by schizophrenia may prevent an increase in total REM sleep time that would be expected with cholinergic overactivity or supersensitivity.

Reductions in SWS, present in many schizophrenics,[10] appear to be more common in those with poor intellectual performance or a structural brain abnormality.[11] Possible causes of reduced amounts of SWS include dysfunction of the prefrontal cortex or increased pressure for REM sleep that shortens the first non–rapid eye movement (NREM) sleep period.[12]

Diagnosis

Sleep disturbance in persons with psychosis may be caused by the underlying disorder, medications, or dyssomnias that interact with the psychiatric condition. For example, when poor sleep hygiene is a major contributor to sleep complaints, the reason for poor sleep hygiene is often the behavioral disorganization and social isolation resulting from schizophrenia.

Although excessive sleepiness is often caused by medications, obstructive sleep apnea (OSA) should be considered in patients who snore loudly or have other risk factors for sleep apnea. In schizophrenic patients with OSA, somnolence and associated cognitive impairment may suggest psychomotor retardation. Nocturnal polysomnography, indicated if sleep apnea is a possible diagnosis, can be performed successfully on most psychotic patients as long as they receive appropriate explanations and care. In patients who are taking phenothiazines or other dopamine antagonists, motor restlessness (akathisia) is common and may occur at night. Differentiation of restless legs syndrome from akathisia caused by dopamine antagonists is discussed in Chapter 11. In rare patients with narcolepsy, nocturnal hallucinations lead to misdiagnosis of narcolepsy as schizophrenia (see Chap. 10).

Management

Unmedicated patients with psychosis and insomnia can usually be treated successfully with antipsychotic medications, both because of their antipsychotic effects and because of their sedative properties. If sleep disturbance is prominent, a large portion of the daily dose of these medications can be given at bedtime. Chlorpromazine or thioridazine are useful because they have sedating effects in addition to the dopamine-blocking properties that are the presumed basis for their antipsychotic effect. Chlorpromazine 100 mg at bedtime can be combined with less sedating agents, such as haloperidol in the daytime. Although sleep continuity and architecture usually improve after treatment, antipsychotic medications may cause increased daytime drowsiness and increased nocturnal sleep time. Antipsychotic medications that block dopamine receptors may cause leg cramps that contribute to disrupted sleep.

Although behavioral treatment of insomnia and instruction in sleep hygiene may be helpful for schizophrenic patients who complain of insomnia and are not in the midst of an acute episode, in my experience most of these patients do not follow such recommendations. Hypnotics should be used cautiously and only after consultation with the psychiatrist responsible for treatment of the psychosis.

SLEEP DISTURBANCE ASSOCIATED WITH MOOD DISORDERS

In the second year of the reign of Nebuchadnez'zar, Nebuchadnez'zar had dreams; and his spirit was troubled, and his sleep left him.
DANIEL 2:1–20

Mood disorders associated with disturbed sleep include unipolar depression, bipolar disorder, dysthymic disorder, and seasonal affective disorder. Although bereavement is not classified as a mood disorder, it is discussed in this section because symptoms may be similar to those seen with mood disorders.

Unipolar depression is associated with episodes of major depression, whereas *bipolar disorder* causes major depressive episodes as well as manic episodes, hypomanic episodes, and mixed episodes that have features of mania and depression. Episodes of major depression and mania, usually of limited duration, cause psychological and often physical distress, as well as impaired social and occupational functioning. Hypomanic episodes are less intense than manic episodes and cause little or no functional impairment. With unipolar depression and bipolar disorder, recovery is usually complete or nearly complete between major depressive episodes. In contrast, *dysthymic disorder* is characterized by chronic but less severe depressive symptoms.

Major depression is common: About 10% to 25% of women and 5% to 12% of men have at least one episode of major depression during their lifetimes. Bipolar disorder has a lifetime prevalence of 1% to 2% and affects men and women equally.

Table 16–2. **Clinical and Sleep-Related Aspects of Major Depressive Episodes**

Depressed mood
Loss of interest in activities
Weight change
Psychomotor agitation or retardation
Fatigue
Feelings of guilt or worthlessness
Impaired concentration
Recurrent thoughts of suicide or death
Sleep
  Difficulty falling asleep, reduced sleep time, early morning awakening
  Reduced amounts of SWS
  Increased amounts of stage 1 sleep
  Frequent awakenings, decreased sleep efficiency
  Short REM sleep latency, high REM density, increased amount of REM sleep

These disorders are not benign: Suicide is the cause of death in 10% to 15% of patients with episodes of major depression.[10]

Major Depressive Episodes

Clinical features of major depressive episodes, as well as the changes in sleep that occur in at least 85% to 90% of adult patients with unipolar depression, are shown in Table 16–2. Early morning awakening with inability to return to sleep, which occurs in most patients, is one of the diagnostic criteria for major depressive episodes[5] and may be accompanied by difficulty falling asleep, frequent awakenings, and vivid, disturbing dreams. If psychotic features accompany the depressive episode, sleep disturbance is usually severe.

Sleep studies during major depressive episodes usually demonstrate short, shallow, and fragmented sleep (see Table 16–2). The severity of sleep disturbance is a function of the age of the patient and the type and severity of depression, with the greatest sleep disruption occurring in older patients (Fig. 16–1) and in those with negative life events or inadequate social support.[13,14] Although sleep continuity tends to improve during remission, insomnia may persist for months or years after the mood disturbance has subsided.[15,16] On the other hand, interrupted sleep and early morning awakening are much less prominent in adolescents with major depression.[13,17] In fact, young depressives, when given the opportunity to sleep longer, tend to do so,[18] and they often have delayed sleep phase syndrome rather than early morning awakening.

Although many patients with bipolar disorder have sleep disturbance and sleep complaints during major depressive episodes similar to those seen in patients with unipolar depression, others complain

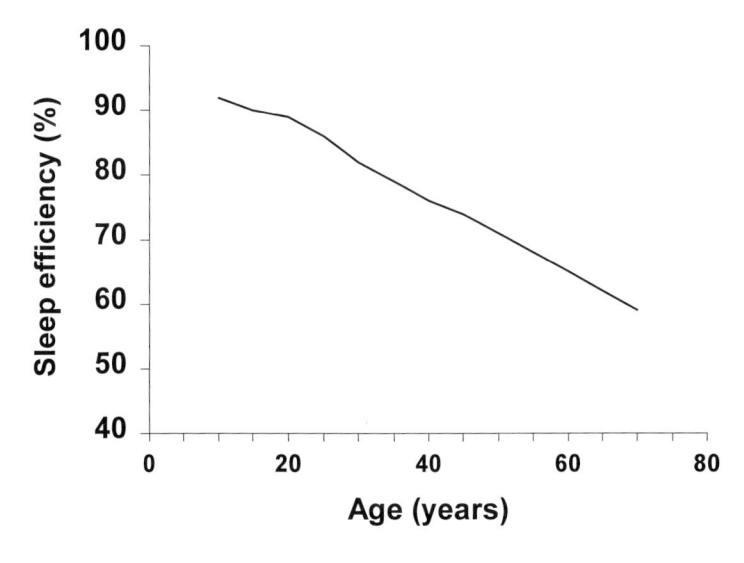

Figure 16–1. Sleep efficiency (sleep time divided by time in bed) during major depressive episodes as a function of age. Time awake at night is highest in elderly depressives and lowest in young depressives. (Based on data from Knowles and MacLean.[13])

more of hypersomnia.[19] Such patients may describe long nocturnal sleep periods, difficulty awakening in the morning, and increased daytime sleep. They may also report problems with daytime fatigue and an increased desire for daytime sleep, but these complaints are usually not accompanied by a tendency to fall asleep in inappropriate settings and Multiple Sleep Latency Tests are generally normal.[20]

PSYCHOBIOLOGIC BASIS

The basis for sleep abnormalities in affective disorders is not known for certain. A disturbance of REM sleep—early onset of REM sleep and increased REM density—was first observed during major depressive episodes by Kupfer and Foster.[21] Reduced latency to REM sleep can occur at the beginning of daytime naps as well as at the beginning of nocturnal sleep.[22] The REM sleep abnormalities, which are usually most apparent during the early phase of the depressive episode, are age dependent: They are most frequent in older patients with severe depression and least frequent in adolescent depressives (Fig. 16–2).

Of the usual hypotheses for these REM sleep abnormalities (Table 16–3), current evidence favors cholinergic overactivity, as effects of cholinergic agents on sleep and REM sleep appear to be increased in depressives and their first-degree relatives.[23–28] Reductions in the amount of SWS usually accompany the REM sleep changes, at least in part because the reduced REM sleep latency leads to a shorter initial period of NREM sleep. Although REM density tends to return to normal levels during remission, reduced REM sleep latency and reductions in SWS often persist during remission, even in patients who no longer have sleep complaints.[15,16]

Anxiety and obsessional thoughts contribute to insomnia in many patients, and elevated nocturnal cortisol levels and other neuroendocrine abnormalities may also play a role. The association of insomnia with an increased risk for subsequent development of depression and mania suggests that disturbed sleep may contribute to the development of these two conditions. On the other hand, sleep itself appears to have a depressant effect in some persons with mood disorders, although the mechanism by which this effect occurs is unknown. Because sleep can have a depressant effect, it is not surprising that sleep deprivation can have an antidepressant effect, which is sometimes used for therapeutic purposes (see Management below).

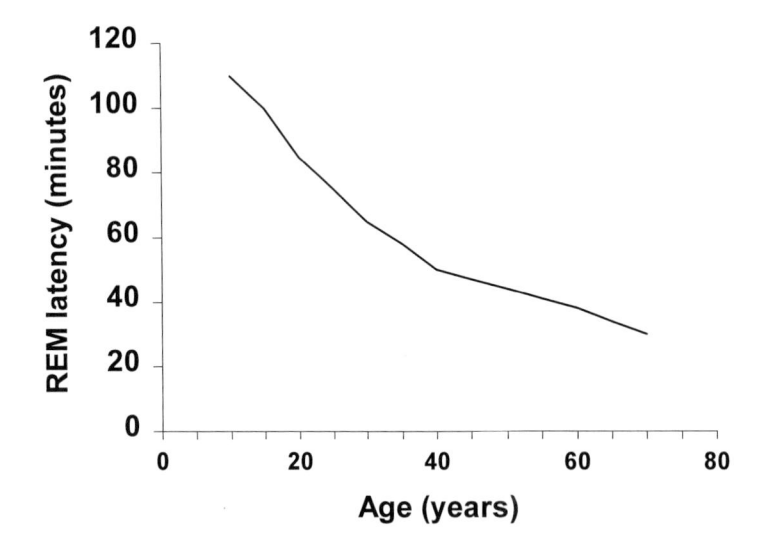

Figure 16–2. Latency to REM sleep as a function of age during major depressive episodes. Short REM latency is most prominent in elderly depressives, whereas REM latency is normal or nearly normal in adolescent depressives. (Based on data from Knowles and MacLean.[13])

Table 16–3. **Possible Causes of REM Sleep Abnormalities in Major Depression**

Relative increase in central cholinergic activity compared to monoaminergic activity
Increased REM sleep pressure
Deficiency in the homeostatic sleep process
Phase-advance of the sleep cycle

DIAGNOSIS

Major depression should be considered as a cause or contributor to sleep problems in almost all patients who present with sleep complaints, particularly in those with a chief complaint of insomnia and in those whose sleep problems began at about the same time as a mood disturbance. Bipolar disorder or seasonal affective disorder may cause cyclic variation of sleep disturbance; in some patients, insomnia is the principal symptom of mania or hypomania. In older adults, insomnia with early morning awakening is particularly suggestive of a mood disorder, whereas in adolescents and young adults with depression, difficult awakening in the morning with delayed sleep phase syndrome is a more common manifestation. Although many patients with sleep disturbance resulting from a mood disorder admit having mood changes, others deny or minimize mood disturbance, or attribute it to the effects of insomnia.

The contribution of depression to sleep complaints is sometimes difficult to separate from the contribution of intrinsic or extrinsic sleep disorders. If anxiety is a prominent symptom of affective illness, the sleep complaint and clinical presentation may suggest that disordered sleep is the result of an anxiety disorder. Depressive symptoms may be secondary to intrinsic sleep disorders, such as sleep apnea and narcolepsy. Likewise, some sleep disorders may be secondary to other psychiatric and medical illnesses, such as alcoholism, degenerative neurological conditions, and stroke. Other potential contributors to insomnia are discussed in Chapter 9.

For a time after the original report by Kupfer and Foster, it appeared that REM sleep changes were specific for major depression and could be used for diagnostic purposes. However, similar REM sleep changes have been found with a number of other psychiatric disorders as well (Table 16–4).[29,30] Thus, although sleep studies during major depression are of great interest to researchers, their value for diagnosis is limited because the changes in REM sleep and SWS, although prominent with mood disorders, can occur with other psychiatric disorders and do not help much to differentiate subtypes of depression. In patients undergoing sleep studies for other reasons, short REM sleep latency, high REM density, or early morning awakening during polysomnography should alert the clinician to the possibility of an affective disorder.

MANAGEMENT

Treatment of sleep disturbance in mood disorders is best accomplished by treatment of the underlying mood disorder with medication or psychotherapy, or both. Patients with mild depressive symptoms often respond to psychotherapy without requiring medications; however, patients with moderate or severe symptoms may require medications in addition to or instead of psychotherapy. Patients with psychotic symptoms may require

Table 16–4. **Conditions Associated with Short REM Sleep Latency**

Narcolepsy
Circadian rhythm sleep disorders
Prolonged bedrest
REM sleep deprivation and withdrawal of REM-suppressing medications
Untreated sleep apnea and first night of CPAP use for sleep apnea
Acute and chronic schizophrenia; schizoaffective disorders
Mania, hypomania
Alcoholism with depression
Alcoholism in recovery
Substance abuse disorders

antipsychotic as well as antidepressant medications, and electroconvulsive therapy may be necessary in suicidal patients with severe depression because antidepressant medications are seldom effective during the first 3 weeks of treatment.

Total sleep deprivation for one night produces short-term improvements in mood in about 60% of unipolar and bipolar depressives, presumably those for whom sleep acts as a depressant. Beneficial effects are also apparent in some depressed patients after selective REM sleep deprivation or partial sleep deprivation.[31,32] In addition, sleep deprivation can accelerate the response to antidepressant medication in severe depression when a rapid clinical response is needed.[33] Patients with marked diurnal mood variations and those with bipolar depression are more likely to respond to sleep deprivation.[33,34] Unfortunately, the benefits of sleep deprivation are transient: More than 80% relapse after one night of sleep,[35] and relapse also may occur with daytime naps, independent of the amount of REM sleep and delta sleep obtained during the naps.[36,37]

For patients with hypersomnia, selective serotonin reuptake inhibitors (SSRIs) are useful because of their low incidence of sedative side effects. Monoamine oxidase inhibitors also can be used, but they can lead to hypertensive crisis if used in combination with sympathomimetic drugs or tyramine-containing foods.

As the mood episode remits, sleep problems usually improve. Antidepressant medications, indicated for most patients with major depressive episodes caused by unipolar depression, are generally effective for depressed mood. SSRIs—fluoxetine, paroxetine, sertraline, and venlafaxine—are commonly used for depression because they have fewer side effects than the tricyclic antidepressants; however, those with activating effects may not help much in cases involving insomnia. In such cases, addition of a bedtime dose of a sedating antidepressant, such as trazodone 50 to 100 mg or a tricyclic compound such as amitriptyline 50 to 100 mg, may be helpful. Nefazodone, a less activating SSRI, may be preferable when insomnia is a prominent complaint.[38] Side effects of tricyclic antidepressants include dry mouth, urinary retention, constipation, weight gain, orthostatic hypotension, blurred vision, and confusion. Although the SSRIs usually have fewer side effects than tricyclic compounds, some patients complain of drowsiness, anxiety, tremor, or nausea.

Most antidepressants increase levels of monoaminergic transmission and therefore have pronounced effects on sleep architecture. Monoamine oxidase inhibitors can lead to almost total suppression of REM sleep. Tricyclic antidepressants and SSRIs increase REM latency and decrease the amount of REM sleep; these effects may continue for months to years in patients who take tricyclics.[39] REM sleep rebound, however, may occur for days to weeks after discontinuation of these drugs. Some antidepressants, such as nefazodone and trimipramine, do not suppress REM sleep,[40] indicating that REM sleep suppression is not required for their efficacy.

If sleep does not improve as other symptoms of depression improve, the patient should be assessed for other causes of sleep disturbance. Although symptomatic treatment aimed only at the sleep difficulty is unlikely to succeed, many patients with mood disorders also have elements of psychophysiologic insomnia and poor sleep hygiene and may benefit from behavioral treatments and sleep hygiene instruction (see Chap. 9).

Manic and Hypomanic Episodes

During manic episodes (Table 16-5), patients often describe a decreased need for sleep with little or no subjective daytime fatigue despite much less sleep than usual; some feel that they need no sleep at all. The time to fall asleep may be increased, and periods of sleep, interrupted by frequent awakenings, are often less than 6 hours, sometimes less than 3 hours. With hypomania, changes in sleep are similar, but less pronounced. The REM sleep changes seen with major depressive episodes—short REM latency and increased REM density—occur in some patients with mania, but not in others.[41,42] Some pa-

Table 16–5. **Clinical and Sleep-Related Aspects of Manic Episodes**

Elevated or irritable mood
Grandiosity
Talkativeness
Racing thoughts
Distractibility
Increased goal-directed activity
Excessive involvement in pleasurable activities with potential for negative consequences
Sleep
  Decreased sleep need
  Decreased amounts of sleep
  REM sleep alterations sometimes present

---

tients with bipolar disorder alternate between periods of depression with increased sleep, and periods of mania with reduced sleep.

Mood-stabilizing agents, including lithium carbonate, carbamazepine, and sodium valproate, are sedating and are therefore useful for managing insomnia associated with bipolar illness. During acute manic episodes, antipsychotics also may be required; hypnotics are generally ineffective.[43] Occasional bipolar patients may switch out of a depressed phase following sleep deprivation or switch into hypomania or mania, suggesting that sleep loss may sometimes trigger mania and that measures to reduce sleep loss might reduce the frequency of manic episodes in some patients.[44]

Dysthymic Disorder

Dysthymic disorder is a common disorder, with a lifetime prevalence of about 3% to 6%, characterized by chronic depression with mood disturbance that is less severe than occurs with major depressive episodes (Table 16–6). Insomnia is common, although it is usually less severe than the insomnia associated with acute psychosis or major depression. Abnormalities of REM sleep also are much less pronounced in patients with dysthymia than in patients with major depression.[45]

Treatments of dysthymic disorder are similar to those used for major depression. Although insomnia may respond to treatment of the underlying disorder, poor sleep hygiene and psychophysiologic aspects, which are often present, may require behavioral treatments (see Chap. 9).

Seasonal Affective Disorder

Seasonal affective disorder (SAD) (see Table 16–6), more common in women than men, is characterized by recurrent episodes of depression that usually occur in the fall or winter.[46] During the episodes, patients may go to bed earlier, get up

Table 16–6. **Clinical and Sleep-Related Aspects of Dysthymia, Seasonal Affective Disorder, and Bereavement**

Disorder	Clinical Aspects	Sleep Aspects
Dysthymia	Chronic depressive symptoms for at least 2 years with impaired functioning but without major depression	Insomnia, usually less severe than occurs with major depressive episodes
Seasonal affective disorder	Mood disorder with a regular temporal relationship to a particular time of year	Increased sleep amounts Restless sleep Daytime drowsiness
Bereavement	Depressed mood in reaction to the death of a loved one	Insomnia, sometimes with excessive daytime sleepiness Minimal or no REM sleep alterations

later, or both. They also may report restless sleep and daytime drowsiness. Other symptoms include decreased desire for physical activity, reduced libido, overeating and carbohydrate craving, and weight gain.

Decreased light exposure contributes to SAD, and symptoms usually begin during the period of reduced daylight in the fall and winter. The link to light exposure appears to be the basis for the higher prevalence of the disorder in extreme latitudes: The prevalence is about 1% to 2% in Florida, while in New Hampshire, it is about 10%; in Norway, the disorder is more common north of the Arctic Circle than south of it. The pathophysiology is uncertain, although seasonal and light-dependent variations in melatonin secretion or serotonin metabolism may contribute.

Light therapy, usually more effective in the morning than in the evening, can improve symptoms in about 60% of patients, particularly those with hypersomnia and hyperphagia.[33,47,48] Improvements in mood generally occur after several days of treatment, but relapse can occur after 2 to 3 days without light.

The beneficial effects, proportional to the intensity and duration of light exposure,[33] are caused by retinal stimulation rather than skin exposure. The middle range of visible wavelengths appears optimal; ultraviolet light, with its potential for retinal damage, is probably not necessary. The usual recommendations are for 30 minutes of light exposure daily at 10,000 lux, which can be obtained with the use of several 40-watt fluorescent light bulbs placed 2 to 3 feet in front of the face. Two-hour exposures at 2,500 lux produce similar results. Patients with more severe depressive symptoms may require simultaneous treatment with activating antidepressants, such as SSRIs. Stimulants are sometimes used in refractory cases.

Side effects of light therapy include headache and eye irritation, and in patients with bipolar disorder, light treatment may have the potential to trigger mania. Light therapy is contraindicated in patients with retinopathies, and should be administered cautiously in patients who are taking photosensitizing medications such as tricyclic antidepressants and lithium carbonate.

Bereavement

When the king heard this news, he was astounded and badly shaken. He took to his bed and became sick from grief. . . . He lay there for many days, because deep grief continually gripped him, and he concluded that he was dying. So he called all his friends and said to them, "Sleep departs from my eyes and I am downhearted with worry."

1 MACCABEES 6:8–11

Insomnia is common with bereavement—the loss by death of a person to whom the patient was emotionally attached—and it may be accompanied by daytime sleepiness (see Table 16–6). Insomnia may be severe and can last for several months as the person experiences denial, anger, and depression in response to the loss. During sleepless intervals, the survivor may be preoccupied with the loss and may experience feelings of guilt for having survived and about issues that were unresolved before the death occurred. Sleep disturbance may be accompanied by changes in appetite and emotional lability, but feelings of worthlessness and psychomotor retardation are uncommon unless bereavement is complicated by the development of major depression. Similar sleep problems may develop with other forms of loss, such as divorce.[49]

Sleep studies following bereavement sometimes show increased phasic activity during REM sleep, but in uncomplicated bereavement, they do not show the other changes typical of major depression.[50]

With uncomplicated bereavement, sleep disturbance usually improves after several weeks or months as the patient works through the grieving process. If insomnia is severe, a short course of treatment with a hypnotic may be useful. With pathological grief reactions, however, sleep disturbance may persist for years; in such cases, psychotherapy is the primary treatment, supplemented by antidepressant medications if major depression is present.

SLEEP DISORDERS ASSOCIATED WITH ANXIETY DISORDERS

Anxiety is a major contributor to insomnia, not only in patients with anxiety disorders but also in many patients with psychosis, mood disorders, medical illnesses, and acute stress. Among the anxiety disorders, sleep problems are common in patients with generalized anxiety disorder and panic disorder, almost universal in patients with post-traumatic stress disorder (PTSD), and uncommon in patients with simple phobias.

Generalized Anxiety Disorder

Generalized anxiety disorder is characterized by chronic and persistent anxiety with its associated symptoms (Table 16–7). In severe cases, the anxiety interferes with social or occupational functioning. Problems with insomnia are common; its severity usually depends on the severity of anxiety, and because of the chronic course, additional sleep problems may develop because of poor sleep hygiene, negative conditioning, or hypnotic dependence (see Chap. 9). Sleep studies may show nonspecific features of insomnia,

Table 16–7. **Clinical and Sleep-Related Aspects of Anxiety Disorders**

Anxiety Disorder	Clinical Aspects	Sleep Aspects
Generalized anxiety disorder	Poorly controlled anxiety with distress or impaired social functioning Fatigability Poor concentration Hypervigilance Irritability Muscle tension	Long sleep latency Early-morning awakening Reduced sleep amount
Panic disorder	Recurrent panic attacks with pounding heart, sweating, trembling, shortness of breath, choking, chest pain, nausea, dizziness, derealization, fear of dying, paresthesias, chills, flushing Agoraphobia	Delayed sleep onset Panic attacks from sleep, usually from stage 2 sleep
Post-traumatic stress disorder	Fear, helplessness, or horror in response to exposure to a traumatic event Recurrent distressing recollections of the event Avoidance of stimuli associated with the event Increased arousal	Insomnia Recurrent nightmares
Obsessive-compulsive disorder	Obsessions (intrusive thoughts, impulses, or images that are disturbing although they are recognized as irrational) or compulsions (repetitive behaviors that the individual is driven to perform) that cause distress or interfere with social functioning	Frequent awakenings Reduced sleep efficiency

such as prolonged sleep latency, early morning awakening, and decreased total sleep time,[51-53] but changes in REM sleep patterns typical of major depression are uncommon.

Diagnosis is usually based on clinical features, supplemented if necessary by a sleep log; sleep laboratory studies are rarely necessary. The diagnosis of generalized anxiety disorder as the cause of disrupted sleep should be considered if anxiety occurs during a range of daytime activities as well as at night. If anxiety is mainly or exclusively focused on sleep, the diagnosis may be psychophysiologic insomnia (see Chap. 9).

Benzodiazepines are commonly used for the symptomatic treatment of daytime anxiety, and a dose at bedtime is useful for associated insomnia. If chronic use—more than 1 month—is considered, the clinician must weigh the benefits of the anxiolytic effects against the potential for dependence and side effects. Patients should be advised not to increase the dose without authorization. In my experience, however, many patients with insomnia caused by anxiety disorders respond well to hypnotics and remain on stable doses for many years. As with all of the anxiety disorders, behavioral and relaxation therapies may also lead to improved sleep, and in early cases, psychotherapy directed at the psychosocial causes of the anxiety is often effective. Antidepressants are helpful if a mood disorder is also present.

Panic Disorder

Patients with panic disorder suffer from sudden attacks of extreme anxiety accompanied by a number of psychological and somatic symptoms (see Table 16–7). *Agoraphobia*—the fear of being in a place from which escape might not be possible in the event of a panic attack—commonly develops.

As in other anxiety disorders, insomnia with impaired sleep continuity is common.[54-56] Most patients have had at least one panic attack during sleep, up to one-third have recurrent nocturnal panic attacks,[57] and occasional patients with panic disorder have attacks only at night. Beginning with an abrupt arousal, usually during stage 2 sleep or during the transition from stage 2 to SWS,[55,58] the sensations during episodes are similar to those that accompany daytime attacks. Less commonly, episodes occur during REM sleep or just after sleep onset.[58] In some patients, fear of having nocturnal panic attacks may contribute to difficulty falling asleep.

Differential diagnosis may include sleep terrors, nightmares, partial seizures, nocturnal angina, or arrhythmias. Recall of the episode and the absence of associated dream mentation distinguish panic attacks from nightmares and sleep terrors. Choking sensations and feelings of breathlessness may suggest the possibility of sleep apnea or gastroesophageal reflux. In such cases, the occurrence of daytime panic episodes is a helpful diagnostic feature.

Patients with panic attacks during sleep usually respond to antidepressants; therapy to prevent daytime panic attacks may include desensitization or other behavioral treatments, benzodiazepines, or antidepressants. The approach to chronic insomnia in patients with panic disorder is similar to that used in other patients with chronic insomnia (see Chap. 9).

Post-traumatic Stress Disorder

With PTSD, patients reexperience one or more traumatic events in intrusive recollections (sometimes called flashbacks) or in recurrent nightmares, or both (see Table 16–7). Typical causes include psychic or physical trauma associated with combat or torture, concentration camp experiences, and childhood sexual, physical, or emotional abuse. The disorder is often chronic, and sleep disturbance can continue for as long as 50 years after the trauma.[59]

PTSD is one of the major causes of chronic nightmares in adults. The content of the nightmares may include reenactments of the traumatic events or, in a symbolic or transformed version, may include violence and mayhem that does not relate directly to the previous trauma. Although the nightmares are usually the chief sleep-related complaint, some patients complain mainly of sleep disruption associated with awakening from nightmares or of difficulty falling asleep because of fear that nightmares will occur. Just as dreams can occur in NREM as well as in REM sleep, nightmares and arousals with anxiety may occur during both NREM and REM sleep.[60,61]

The traumatic event and the associated heightened arousal and anxiety are the major causes of sleep disturbance. Periodic leg movements also contribute to sleep disruption in many patients with PTSD.[62] PTSD does not develop in every person who experiences extreme trauma: Although the other factors that lead to the disorder have not been established with certainty, patients who are unable to work through the emotional responses to such events appear to be at higher risk for the development of PTSD. Ordinarily, dreams themselves appear to contribute to the process of "working through" the trauma; in PTSD, the repetitive dreams with their failure to evolve over time may reflect a failure of this process.[63]

The diagnosis of PTSD should be considered in any patient with chronic nightmares, particularly when the nightmares are repetitive and relate to previous trauma. Stereotyped features and good recall of the dream content help to differentiate nightmares associated with PTSD from nocturnal panic attacks and sleep terrors.

Psychotropic medications that suppress REM sleep are often useful in the treatment of PTSD, probably at least in part because the reduced time in REM sleep leads to fewer opportunities for nightmares.[60] Psychotherapy, including cognitive behavioral therapy and desensitization, may also be helpful.

Obsessive-Compulsive Disorder

Patients with obsessive-compulsive disorder suffer from obsessions and compulsions (see Table 16–7). The compulsive behaviors may be performed in response to an obsession in an attempt to reduce distress or prevent a dreaded event.

Impaired sleep continuity and reduced sleep efficiency are common.[64,65] Obsessional thoughts that occur while the patient is in bed may make it difficult to fall asleep or to return to sleep after awakenings during the night. Similarly, compulsions to bathe, wash hands, or check door locks repeatedly may interfere with the process of preparing to sleep. Treatment includes behavioral and pharmacological approaches; clomipramine, fluvoxamine, and fluoxetine are among the drugs used to treat this disorder.

SLEEP DISTURBANCE ASSOCIATED WITH EATING DISORDERS

Eating disorders occur predominantly in adolescent girls and young women. The two major symptom complexes, anorexia nervosa and bulimia nervosa, may occur separately or in combination. *Anorexia nervosa* is characterized by intense fear of gaining weight, a body weight less than 85% of expected, severe food restriction, amenorrhea, and altered body weight perception. *Bulimia nervosa* patients binge on food and believe that they have lost control over their eating. After consuming large amounts of food quickly, they induce vomiting or engage in other inappropriate behaviors to minimize the impact of the food intake on body weight. Excessive exercise, fasting, and inappropriate use of laxatives, diuretics, and enemas often accompany both disorders.

Some patients with eating disorders restrict their time in bed in order to allow more time for exercise. Others, particularly those with anorexia nervosa who are severely underweight, experience insomnia. Patients with bulimia often stay up late into the night during eating binges and sometimes have episodes of eating during sleepwalking (see Chap. 15).[66] Eating binges may be followed by periods of increased sleep. Treatment of sleep disturbance is secondary to treatment of the eating disorder, which may involve psychotherapy, medications, and, in severe cases of anorexia, parenteral nutrition.

SLEEP DISTURBANCE ASSOCIATED WITH ALCOHOL USE AND ALCOHOLISM

The sleep-promoting effects and the sleep-disrupting effects of alcohol probably have been known since prehistoric times. The first polygraphic recordings of the effects of alcohol on sleep were performed in the 1960s.[67]

Effects of Alcohol on Alertness

Alcohol has both stimulating and sedating effects. Stimulating effects generally are produced at low doses and as blood alcohol levels are rising; conversely, sedative effects occur at high doses and as blood levels are falling.[68,69] Although sedating effects are dose dependent, they occur even with low doses of alcohol: two to three drinks can lead to increased sleepiness for several hours.[70,71] Furthermore, the sedating effect of alcohol is greater in subjects who are sleepy than in subjects who are alert and well-rested (Fig. 16–3).[68,70–73] Sleepiness associated with alcohol use, which leads to impaired performance in motor vehicle driving simulators,[74,75] is probably more dangerous than sleepiness from other causes because of the impaired judgment that accompanies alcohol use.

Effects of Alcohol on Sleep

The effects of alcohol on sleep are a function of the dose and the interval between alcohol consumption and bedtime. Alcohol consumed at bedtime may interfere with sleep onset because of its stimulating effects, whereas alcohol consumed 3 to 4 hours before bed can lead to sleepiness and promote sleep onset. Low to moderate doses of alcohol (one to three drinks) consumed more than 5 hours before bedtime may cause sleepiness in the evening, but they have little effect on nighttime sleep. On the other hand, evening ingestion of larger amounts of alcohol (four to eight drinks) also promotes sleep onset, but it leads to restless sleep during the latter part of the night and to sleepiness and fatigue the next day.[76] Sleep changes induced by alcohol on one night tend to be less apparent on subsequent nights, suggesting that some tolerance to the effects of alcohol occurs.[77,78]

Dose-related differences in effects on sleep may be related to short-term withdrawal effects or to hangover effects. Low doses have mostly sedative effects that increase sleep time; higher doses produce not only sedative effects, but also short-term withdrawal effects or other toxic effects associated with increased sympathetic activity. Higher doses, therefore,

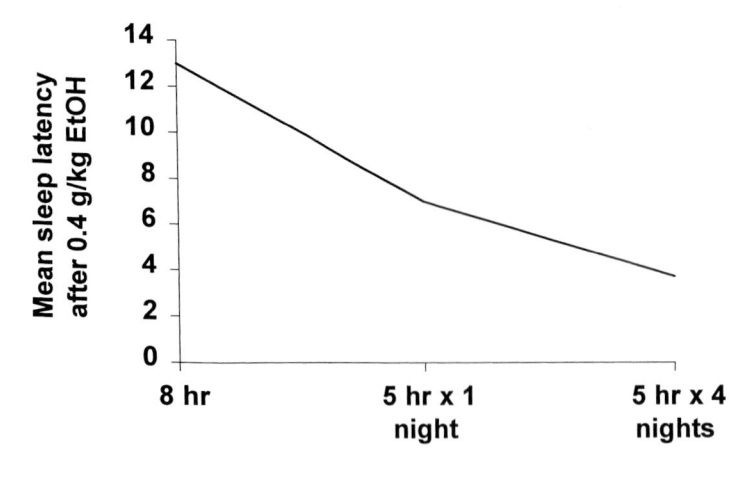

Figure 16–3. Effects of reduced amounts of nighttime sleep on the sedating effects of a daytime dose of alcohol. After 8 hours in bed, the mean sleep latency (in minutes) after a moderate dose of alcohol (EtOH) is within the normal range. Sleep latency is reduced after one night of 5 hours in bed and further reduced after four nights of 5 hours in bed. (Based on data from Zwyghuizen-Doorenbos et al.[70])

lead to sleep disruption—especially during the second half of the night and especially in older persons, who have less robust sleep.[79]

Alcohol-Dependent Sleep Disorder

Many insomniacs discover that alcohol can help them fall asleep both because of its sedative effect and because of its relaxing effect, which may reduce anxiety about sleep difficulties. Of 590 patients seen at a sleep center who complained of insomnia, 10% used alcohol at least once a month to help them sleep (Aldrich, unpublished data).

Alcohol-dependent sleep disorder is characterized by regular use of alcohol to promote sleep and the psychological dependence that accompanies such use. Alcohol-dependent sleep disorder should be considered in any patient with insomnia who uses alcohol regularly in the evening or at night. The alcohol use may be the major factor responsible for the insomnia; more often, however, other causes are also present (see Chap. 9).

Although alcohol is often effective initially as a hypnotic because of its sedative effect, tolerance develops, and persons who use alcohol to promote sleep may gradually increase their evening consumption. Subsequent sleep is often less restful than usual and associated with fragmented dreams, sweating, and nocturnal awakenings in the latter portion of the night. Daytime sleepiness and fatigue caused by alcohol-induced sleep disruption may be attributed by the insomniac to inability to fall asleep; the patient may then use alcohol again in the evening to facilitate sleep onset.

Although insomnia may not be chronic initially, use of alcohol to treat insomnia may make insomnia a regular occurrence, and attempts to refrain from alcohol can lead to severe insomnia, at least for the first few nights (see also Chap. 9).

In patients with light to moderate alcohol use, sleepiness caused by alcohol-related sleep disruption usually resolves after 2 to 3 days of abstinence. In alcohol-dependent sleep disorder, abstinence is the best treatment, but it may be difficult to convince patients that alcohol is the cause of insomnia, and if they do follow a recommendation for abstinence, insomnia may become more severe for several weeks. Counseling, support, and encouragement that insomnia is likely to improve may help the patient to resist the temptation to resume drinking in order to deal with the insomnia. As with any insomnia, behavioral techniques and sleep hygiene may be helpful (see Chap. 9). Although low doses of sedating tricyclic antidepressants or antihistamines may be useful during the first several weeks, hypnotics generally should be avoided because of the potential for dependence and the likelihood of cross-tolerance. Sometimes, discontinuation of alcohol unmasks a significant depression.

Alcoholism

CLINICAL FEATURES

Insomnia is common among alcoholics, and because the first few days of a drinking episode are often associated with less time awake at night, alcoholics may believe that sleep improves with drinking. This initial subjective improvement in sleep, combined with insomnia during abstinent intervals, may contribute to relapse. With continued drinking, sleep disturbance and subjective sleep quality often become worse. Some alcoholics describe sleepiness and fatigue associated with heavy alcohol use; others report intermittent sleepiness during periods of abstinence.

During alcohol withdrawal, particularly during the first week, severe insomnia is common.[80,81] The suppression of REM sleep that accompanies heavy drinking can lead to reduced REM sleep latency and increased amounts of REM sleep during withdrawal.[80] In patients with delirium tremens, sleep episodes may consist almost entirely of brief periods of REM sleep, sometimes with impaired muscle atonia that leads to a self-limited form of REM sleep behavior disorder (see Chap.

15).[80,82,83] Sleep disturbance during the second week of withdrawal generally is less severe, and some alcoholics have deep, restful sleep after the initial withdrawal symptoms subside.[84]

During periods of abstinence, many alcoholics complain of interrupted sleep and a sense that sleep is not restorative. Although sleep quality often improves gradually, it may remain abnormal (i.e., reduced amounts of SWS and long sleep latencies) for as long as 2 years after the cessation of drinking.[85,86] Sleep apnea, common in older alcoholics, contributes in some patients to the subjective sense of poor sleep.[87] Periodic limb movements during sleep also may contribute.[88]

The risk of resuming alcohol abuse correlates with the severity of sleep disturbance during abstinence,[89] and REM sleep and SWS measures at the time of treatment can be used to predict relapse risk.[90,91] Whether relapse can be predicted more accurately if based on sleep findings than if based on other clinical data is as yet unknown.

PSYCHOBIOLOGIC BASIS

Alcohol produces sleepiness because of its sedative effects and because the sleep disruption that occurs with alcohol use makes sleep less restful. Although SWS usually increases with consumption of moderate or high doses of alcohol at bedtime,[78,92] the increase is probably not an indication of an increase in the homeostatically regulated restorative process that underlies natural SWS; it is more likely to be caused by cortical dysfunction, with an increase in EEG slow waves resulting from toxic effects of alcohol or its metabolites. Inhibition of REM sleep by alcohol may contribute to the loss of the restorative quality of sleep that occurs with heavy drinking,[93,94] while late-night sleep disruption probably contributes to increased dream recall.

Alcohol relaxes upper airway dilator muscles, which increases the likelihood of OSA in susceptible persons, and it depresses arousal responses, which tends to increase the duration of apneas and associated hypoxemia (effects of alcohol on sleep apnea are discussed in more detail in Chap. 13). The increased frequency and duration of obstructive apneas lead to increased sleep disruption and more pronounced daytime sleepiness.

The prolonged insomnia and reduced amounts of SWS that characterize the sleep of abstinent alcoholics are probably caused, at least partly, by toxic effects of chronic alcohol abuse.[89,95,96] Secondary depression, present in 25% to 50% of alcoholics, other psychological disturbances, and gastrointestinal and other systemic symptoms contribute in some patients. Although a short REM sleep latency may be caused by primary or secondary depression, similar changes in REM sleep occur during withdrawal in alcoholics without depression.[97]

DIAGNOSIS

In alcoholics with insomnia, the major differential diagnosis is often depression, which may be secondary to alcoholism or may be the primary problem that led to or contributed to alcohol abuse. A history of major depressive episodes antedating the onset of heavy alcohol use suggests that unipolar depression or bipolar disorder is present. Sleepiness may be caused by the sedating effects of alcohol, to the sleep-disrupting effects of alcohol, to the systemic symptoms related to alcohol use, or to alcohol-induced OSA.

MANAGEMENT

If the underlying alcoholism is not addressed, treatment of sleep disturbance in alcoholics is rarely successful. For heavy alcohol users, recommendations to stop drinking usually are insufficient. For patients who are motivated to deal with their alcohol problem, referral to an alcohol treatment program is a better approach. Insomnia may respond to behavioral treatments or to low-dose antidepressants. Benzodiazepines should be avoided unless employed as part of the treatment for alcoholism.

SLEEP DISTURBANCE ASSOCIATED WITH OTHER SUBSTANCE-RELATED DISORDERS

In addition to alcohol, most of the common drugs of abuse—amphetamines, caffeine, cannabis, cocaine, nicotine, opioids and sedatives—affect sleep. Substance-related disorders include substance dependence, abuse, intoxication, and withdrawal, as well as substance-induced dementia. Furthermore, substance abusers often have co-existing primary or secondary psychiatric disorders that contribute to sleep difficulties.

The type of sleep disturbance varies depending on the substance being abused (Table 16–8). Amphetamine and cocaine abusers may stay awake for days during drug binges, and then "crash" and become hypersomnolent during withdrawal, with increased sleep and an increased proportion of REM sleep lasting 1 to 2 weeks. Opiate addicts, on the other hand, often have severe insomnia during withdrawal.

Caffeine can cause sleep disturbance, although its effects vary considerably across individuals: 300 mg of caffeine (about three to four cups of coffee), consumed at or close to bedtime, can disrupt sleep, especially in older persons. Persons who drink excessive amounts of caffeinated beverages (e.g., 10 to 20 cups of coffee per day) often do not realize that caffeine is causing or contributing to insomnia. Many such persons drank coffee heavily as young adults without experiencing sleep problems. With increasing age, they become more sensitive to the sleep-disturbing effects of caffeine.

For abusers of illegal drugs, initial treatment may include detoxification, counseling, and behavioral treatments. Associated anxiety or depression should also be addressed; if medications such as antidepressants or antipsychotics are clinically indicated, agents with greater sedative effects may be used in patients with significant sleep disruption. Behavioral treatments for insomnia may be helpful, but hypnotics and other potentially addictive substances should generally be avoided. Although sleep may improve somewhat after the acute period of intoxication or withdrawal, subjective sleep difficulties may persist for months, and sometimes permanently.

SUMMARY

Psychiatric disorders, especially mood disorders and anxiety disorders, are common in patients with sleep complaints, particularly in those with insomnia. In addition, many patients who present with insomnia later develop a mood disorder, suggesting

Table 16–8. **Effects of Drugs of Abuse on Sleep**

Drug	Acute Effects on Sleep	Withdrawal Effects on Sleep
Amphetamines and cocaine	↓ sleepiness, ↓ sleep, ↓ REM sleep	↑ sleep, ↑ REM sleep
Caffeine*	↑ sleep latency, ↓ sleep	
Cannabis	↑ sleepiness, ↓ REM sleep	↑ REM sleep
Opioids	↓ sleep, ↓ SW sleep ↓ REM sleep	Yawning, restless sleep, insomnia
Barbiturates	↓ REM sleep	Insomnia, ↑ REM sleep
Nicotine	Minor ↓ sleep efficiency	

*Caffeine content: coffee (8-ounce cup) = 60–120 mg; caffeinated soft drinks (12 ounces) = 30–60 mg.

that insomnia can be an initial symptom of depression. Psychiatric disorders are common in patients with sleep complaints, and the converse is also true: Psychiatric disorders may cause sleep problems directly, and the medications used to treat psychiatric illness can exacerbate the sleep disturbance or can cause new sleep disorders.

For many schizophrenics, insomnia is a chronic problem. Highly disorganized patients often have highly disorganized sleep-wake patterns, and during acute psychotic episodes, insomnia usually becomes much worse. With major depression, early morning awakening occurs in most patients. REM sleep abnormalities are common, particularly in older patients, perhaps caused by cholinergic overactivity. Manic patients often describe a decreased need for sleep, with little or no subjective daytime fatigue despite much less sleep than usual; some feel that they need no sleep at all. On the other hand, hypersomnia is common with seasonal affective disorder and with the depressed phase of bipolar disorder. Insomnia often accompanies bereavement; however, the REM sleep abnormalities characteristic of major depression are usually absent. Up to one-third of patients with panic disorder have recurrent nocturnal panic attacks. In adults, PTSD is one of the major causes of chronic nightmares.

Alcohol and other substances of abuse also affect sleep. Alcohol-dependent sleep disorder is characterized by regular use of alcohol to promote sleep, resulting in psychological dependence. Among alcoholics, insomnia is common, particularly during periods of abstinence, and severe sleep disturbance may accompany abuse and withdrawal from stimulants and opiates.

Causes of sleep problems in patients with psychiatric disorders include the underlying psychiatric disorder, medications, and coexisting dyssomnias and parasomnias. Although management of sleep disturbance associated with psychiatric disorders is based fundamentally on treatment of the psychiatric condition, therapies directed at coexisting disorders may lead to substantial improvement in symptoms.

REFERENCES

1. Buysse, DJ, et al: Clinical diagnoses in 216 insomnia patients using the International Classification of Sleep Disorders (ICSD), DSM-IV and ICD-10 categories: A report from the APA/NIMH DSM-IV Field Trial. Sleep 17:630, 1994.
2. Breslau, N, et al: Sleep disturbance and psychiatric disorders: A longitudinal study of young adults. Biol Psychiatry 39:411, 1996.
3. Ford, DE, and Kamerow, DB: Epidemiologic study of sleep disturbances and psychiatric disorders: An opportunity for prevention? JAMA 262:1479, 1989.
4. Vollrath, M, et al: The Zurich study: VIII. Insomnia: Association with depression, anxiety, somatic syndromes, and course of insomnia. Eur Arch Psychiatry Neurol Sci 239:113, 1989.
5. American Psychiatric Association: Diagnostic and Statistical Manual of Mental Disorders, ed 4, revised. Washington, DC, American Psychiatric Association, 1991.
6. Fisher, C, and Dement, WC: Studies in the psychopathology of sleep and dreams. Am J Psychiatry 119:1160, 1963.
7. Benson, KL, and Zarcone, VP Jr: Rapid eye movement sleep eye movements in schizophrenia and depression. Arch Gen Psychiatry 50:474, 1993.
8. Keshavan, MS, et al: Plasma cholinesterase isozymes and REM latency in schizophrenia. Psychiatry Res 43:23, 1992.
9. Tandon, R, et al: Association between abnormal REM sleep and negative symptoms in schizophrenia. Psychiatry Res 27:359, 1988.
10. Benca, RM: Sleep in psychiatric disorders. Neurol Clin North Am 14:739, 1996.
11. Keshavan, MS, et al: EEG sleep and cerebral morphology in functional psychoses: A preliminary study with computed tomography. Psychiatry Res 39:293, 1991.
12. Keshavan, MS, et al: Electroencephalographic sleep in schizophrenia: A critical review. Compr Psychiatry 31:34, 1990.
13. Knowles, JB, and MacLean, AW: Age-related changes in sleep in depressed and healthy subjects: A meta-analysis. Neuropsychopharmacology 3:251, 1990.
14. Dew, MA, et al: Psychosocial correlates and sequelae of electroencephalographic sleep in healthy elders. J Gerontol 49:P8, 1994.
15. Thase, ME, et al: Polysomnographic studies of unmedicated depressed men before and after cognitive behavioral therapy. Am J Psychiatry 151:1615, 1994.
16. Giles, DE, et al: Prospective assessment of electroencephalographic sleep in remitted major depression. Psychiatry Res 46:269, 1993.
17. Goetz, RR, et al: Electroencephalographic sleep of adolescents with major depression and normal controls. Arch Gen Psychiatry 44:61, 1987.
18. Hawkins, DR, et al: Extended sleep (hypersomnia) in young depressed patients. Am J Psychiatry 142:905, 1985.

19. Kupfer, DJ, et al: Hypersomnia in manic-depressive disease (a preliminary report). Dis Nerv Syst 33:720, 1972.
20. Nofzinger, EA, et al: Hypersomnia in bipolar depression: A comparison with narcolepsy using the Multiple Sleep Latency Test. Am J Psychiatry 148:1177, 1991.
21. Kupfer, DJ, and Foster, G: Interval between onset of sleep and rapid eye movement as an indicator of depression. Lancet 2:684, 1972.
22. Kupfer, DJ, et al: REM sleep, naps, and depression. Psychiatry Res 5:195, 1981.
23. Janowsky, DS, et al: A cholinergic-adrenergic hypothesis of mania and depression. Lancet 2:632, 1972.
24. Sitaram, N, et al: Cholinergic regulation of mood and REM sleep: Potential model and marker of vulnerability to affective disorder. Am J Psychiatry 139:571, 1982.
25. Schreiber, W, et al: Cholinergic REM sleep induction test in subjects at high risk for psychiatric disorders. Biol Psychiatry 32:79, 1992.
26. Riemann, D, et al: Sleep in depression: The influence of age, gender and diagnostic subtype on baseline sleep and the cholinergic REM induction test with RS 86. Eur Arch Psychiatry Clin Neurosci 243:279, 1994.
27. Gillin, JC, et al: The cholinergic rapid eye movement induction test with arecoline in depression. Arch Gen Psychiatry 48:264, 1991.
28. Dahl, RE, et al: Cholinergic REM induction test with arecoline in depressed children. Psychiatry Res 51:269, 1994.
29. Campbell, SS, and Gillin, JC: Sleep measures in depression: How sensitive? How specific? Psychiatr Ann 17:647, 1987.
30. Tandon, R, et al: Electroencephalographic sleep abnormalities in schizophrenia. Arch Gen Psychiatry 49:185, 1992.
31. Giedke, H, et al: The timing of partial sleep deprivation in depression. J Affect Disord 25:117, 1992.
32. Leibenluft, E, and Wehr, TA: Is sleep deprivation useful in the treatment of depression? Am J Psychiatry 149:159, 1992.
33. Wehr, T: Manipulations of sleep and phototherapy: Nonpharmacologic alternatives in the treatment of depression. Clin Neuropharmacol 13:S54, 1990.
34. Szuba, MP, et al: Effects of partial sleep deprivation on the diurnal variation of mood and motor activity in major depression. Biol Psychiatry 30:817, 1991.
35. Wu, JC, and Bunney, WE: The biological basis of an antidepressant response to sleep deprivation and relapse: Review and hypothesis. Am J Psychiatry 147:14, 1990.
36. Wiegand, M, et al: Effect of morning and afternoon naps on mood after total sleep deprivation in patients with major depression. Biol Psychiatry 33:467, 1993.
37. Riemann, D, et al: Naps after total sleep deprivation in depressed patients: Are they depressogenic? Psychiatry Res 49:109, 1993.
38. Gillin, JC, et al: A comparison of nefazodone and fluoxetine on mood and on objective, subjective and clinical-related measures of sleep in depressed patients: A double-blind, 8-week clinical trial. J Clin Psychiatry 58:185, 1997.
39. Kupfer, DJ, et al: Persistent effects of antidepressants: EEG sleep studies in depressed patients during maintenance treatment. Biol Psychiatry 35:781, 1994.
40. Vogel, GW, et al: Drug effects on REM sleep and on endogenous depression. Neurosci Biobehav Rev 14:49, 1990.
41. Linkowski, P, et al: Sleep during mania in manic-depressive males. Eur Arch Psychiatr Neurol Sci 235:339, 1986.
42. Hudson, JI, et al: Polysomnographic characteristics of young manic patients: Comparison with unipolar depressed patients and normal control subjects. Arch Gen Psychiatry 49:378, 1992.
43. Douglass, AB: Sleep disorders associated with mental disorders. In Gilman, S, et al (eds): Neurobase, ed 3. Arbor, La Jolla, Calif, 1997.
44. Wehr, TA: Sleep-loss as a possible mediator of diverse causes of mania. Br J Psychiatry 159:576, 1991.
45. Arriaga, F, et al: The sleep of dysthymic patients: A comparison with normal controls. Biol Psychiatry 27:649, 1990.
46. Rosenthal, NE, et al: Seasonal affective disorder: A description of the syndrome and preliminary findings with light therapy. Arch Gen Psychiatry 41:72, 1984.
47. Kuhs, H, and Tolle, R: Sleep deprivation therapy. Biol Psychiatry 29:1129, 1991.
48. Avery, DH, et al: Morning or evening bright light treatment of winter depression? The significance of hypersomnia. Biol Psychiatry 29:117, 1991.
49. Cartwright, RD: REM sleep during and after mood-disturbing events. Arch Gen Psychiatry 40:197, 1983.
50. Reynolds, CF III, et al: Sleep after spousal bereavement: A study of recovery from stress. Biol Psychiatry 34:791, 1993.
51. Saletu, B, et al: Insomnia in generalized anxiety disorder: Polysomnographic, psychometric and clinical investigations before, during and after therapy with a long- versus a short-half-life benzodiazepine (quazepam versus triazolam). Neuropsychobiology 29:69, 1994.
52. Reynolds, CF III, et al: EEG sleep in outpatients with generalized anxiety: A preliminary comparison with depressed outpatients. Psychiatry Res 8:81, 1983.
53. Papadimitriou, GN, et al: Sleep EEG recordings in generalized anxiety disorder with significant depression. J Affect Disord 15:113, 1988.
54. Stein, MB, et al: Sleep in nondepressed patients with panic disorder: II. Polysomnographic assessment of sleep architecture and sleep continuity. J Affect Disord 28:1, 1993.
55. Mellman, TA, and Uhde, TW: Electroencephalographic sleep in panic disorder. Arch Gen Psychiatry 46:178, 1989.
56. Lauer, CJ, et al: Panic disorder and major depression: A comparative electroencephalographic sleep study. Psychiatry Res 44:41, 1992.

57. Mellman, TA, and Uhde, TW: Sleep panic attacks: New clinical findings and theoretical implications. Am J Psychiatry 146:1204, 1989.
58. Hauri, PJ, et al: Sleep in patients with spontaneous panic attacks. Sleep 12:323, 1989.
59. Rosen, J, et al: Sleep disturbances in survivors of the Nazi Holocaust. Am J Psychiatry 148:62, 1991.
60. Ross, RJ, et al: Rapid eye movement sleep disturbance in posttraumatic stress disorder. Biol Psychiatry 35:195, 1994.
61. van der Kolk, B, et al: Nightmares and trauma: A comparison of nightmares after combat with lifelong nightmares in veterans. Am J Psychiatry 141:187, 1984.
62. Brown, TM, and Boudewyns, PA: Periodic limb movements of sleep in combat veterans with posttraumatic stress disorder. J Trauma Stress 9:129, 1996.
63. Kramer, M, et al: Nightmares in Vietnam veterans. J Am Acad Psychoanal 15:67, 1987.
64. Insel, TR, et al: The sleep of patients with obsessive-compulsive disorder. Arch Gen Psychiatry 39:1372, 1982.
65. Hohagen, F, et al: Sleep EEG of patients with obsessive-compulsive disorder. Eur Arch Psychiatry Clin Neurosci 243:273, 1994.
66. Schenck, CH, et al: Sleep-related eating disorders: Polysomnographic correlates of a heterogeneous syndrome distinct from daytime eating disorders. Sleep 14:419, 1991.
67. Yules, RB, et al: The effect of ethyl alcohol on man's electroencephalographic sleep cycle. Neurophysiology 20:109, 1966.
68. O'Boyle, DJ, et al: Effects of alcohol, at two times of day, on EEG-derived indices of physiological arousal. Electroencephalogr Clin Neurophysiol 95:97, 1995.
69. Petrucelli, N, et al: The biphasic effects of ethanol on sleep latency. Sleep Res 23:75, 1994.
70. Zwyghuizen-Doorenbos, A, et al: Increased daytime sleepiness enhances ethanol's sedative effects. Neuropsychopharmacology 1:279, 1988.
71. Roehrs, T, et al: Sedative effects and plasma concentrations following single doses of triazolam, diphenhydramine, ethanol and placebo. Sleep 16:301, 1993.
72. Zwyghuizen-Doorenbos, A, et al: Individual differences in the sedating effects of ethanol. Alcohol Clin Exp Res 14:400, 1990.
73. Roehrs, T, et al: Sleep extension, enhanced alertness and the sedating effects of ethanol. Pharmacol Biochem Behav 34:321, 1989.
74. Roehrs, T, et al: Sleepiness and ethanol effects on simulated driving. Alcohol Clin Exp Res 18:154, 1994.
75. Krull, KR, et al: Simple reaction time event-related potentials: Effects of alcohol and sleep deprivation. Alcohol Clin Exp Res 17:771, 1993.
76. Williams, DL, et al: Dose-response effects of ethanol on the sleep of young women. J Stud Alcohol 44:515, 1983.
77. Rundell, OH, et al: Alcohol and sleep in young adults. Psychopharmacologia 26:201, 1972.
78. Prinz, PN, et al: Effect of alcohol on sleep and nighttime plasma growth hormone and cortisol concentrations. J Clin Endocrinol Metab 51:759, 1980.
79. Block, AJ, et al: Effect of alcohol ingestion on breathing and oxygenation during sleep: Analysis of the influence of age and sex. Am J Med 80:595, 1986.
80. Johnson, L, et al: Sleep during alcohol intake and withdrawal in the chronic alcoholic. Arch Gen Psychiatry 22:406, 1970.
81. Thompson, PM, et al: Polygraphic sleep measures differentiate alcoholics and stimulant abusers during short-term abstinence. Biol Psychiatry 38:831, 1995.
82. Tachibana, M, et al: A sleep study of acute psychotic states due to alcohol and meprobamate addiction. Adv Sleep Res 2:177, 1977.
83. Kotorii, T, et al: The sleep pattern of chronic alcoholics during the alcohol withdrawal period. Folia Psychiatr Neurol Jpn 34:89, 1980.
84. Gross, M, et al: Experimental study of sleep in chronic alcoholics before, during and after four days of heavy drinking, with a non-drinking comparison. Ann NY Acad Sci 215:254, 1973.
85. Williams, HL, and Rundell, OH Jr: Altered sleep physiology in chronic alcoholics: Reversal with abstinence. Alcohol Clin Exp Res 5:318, 1981.
86. Snyder, S, and Karacan, I: Sleep patterns of sober chronic alcoholics. Neuropsychobiology 13:97, 1985.
87. Aldrich, MS, et al: Sleep disordered breathing in alcoholics: Association with age. Alcohol Clin Exp Res 17:1179, 1993.
88. Aldrich, MS, and Shipley, JE: Alcohol use and periodic limb movements of sleep: Alcohol Clin Exp Res 17:192, 1993.
89. Allen, RP, et al: Slow wave sleep: A predictor of individual differences in response to drinking? Biol Psychiatry 15:345, 1980.
90. Gillin, JC, et al: Increased pressure for rapid eye movement sleep at a time of hospital admission predicts relapse in nondepressed patients with primary alcoholism at three month follow-up. Arch Gen Psychiatry 51:189, 1994.
91. Aldrich, MS, et al: Slow wave sleep decrement and relapse tendency in alcoholics in treatment. Sleep Res 23:185, 1994.
92. Dijk, DJ, et al: The effects of ethanol on human sleep: EEG power spectra differ from those of benzodiazepine receptor agonists. Neuropsychopharmacology 7:225, 1992.
93. Knowles, JB, et al: Effects of alcohol on REM sleep. Q J Stud Alcohol 29:342, 1968.
94. Gresham, SC, et al: Alcohol and caffeine: Effect on inferred visual dreaming. Science 140:1226, 1963.
95. Gross, MM, and Best, S: Behavioral concomitants of the relationship between baseline slow wave sleep and carry-over of tolerance and dependence in alcoholics. Adv Exp Med Biol 59:633, 1975.
96. Ishibashi, M, et al: Cerebral atrophy and slow wave sleep of abstinent chronic alcoholics. Drug Alcohol Depend 19:325, 1987.
97. Clark, CP, et al: The relationship of sleep abnormalities to short-term sobriety in primary alcoholics with secondary depression. Sleep Res 25:155, 1996.

Chapter 17

MEDICAL CAUSES OF DISORDERED SLEEP

SLEEP DISTURBANCE IN ISCHEMIC
  HEART DISEASE
Clinical Features
Biologic Basis
Diagnosis
Management
SLEEP DISTURBANCE IN CHRONIC
  LUNG DISEASES
Clinical Features
Biologic Basis
Diagnosis
Management
SLEEP DISTURBANCE AND
  GASTROINTESTINAL DISEASE
Gastroesophageal Reflux Disease
Peptic Ulcer Disease
Intestinal Disorders
SLEEP DISTURBANCE AND
  RHEUMATOLOGIC DISORDERS
Arthritis
Fibromyalgia
SLEEP DISTURBANCE AND CHRONIC
  RENAL FAILURE
Clinical Features
Biologic Basis
Diagnosis
Management

Medical diseases are frequent causes of sleep disturbance, especially in older people, who are more likely to have serious medical problems. Among patients who are severely ill, such as those in intensive care units and those recovering from major surgery, sleep is often severely reduced and fragmented.

The temporal association between increased sleep disturbance and exacerbation of the underlying medical disease is the key to diagnosis of sleep disorders caused by medical disease. Although symptoms of the medical disorder (e.g., pain, dyspnea, nocturia, impaired mobility) are common contributors to disturbed sleep in patients with serious medical illness, anxiety and depression concerning the loss of function and the potential for disability or death are often even more significant. In addition, medications used to treat medical conditions can cause or exacerbate insomnia or daytime sleepiness.

Because the pathophysiology of many medical disorders is altered during non—rapid eye movement (NREM) and rapid eye movement (REM) sleep, and because nocturnal symptoms and their management often differ from daytime symptoms and their management, a comprehensive review of sleep disturbance in medical diseases would require a large textbook. Major textbooks of internal medicine, often more than 2000 pages in length, focus largely on daytime manifestations of disease; a similar volume covering disease manifestations during the third of our lives spent asleep would probably require 1000 pages, far beyond the scope of this work.

SLEEP DISTURBANCE IN ISCHEMIC HEART DISEASE

Clinical Features

Ischemic heart disease is caused by disease of the coronary arteries, leading to inadequate blood flow to the heart. Impairment of heart muscle function resulting from ischemia or other causes can lead to congestive heart failure (CHF). Sleep-related complaints in patients with ischemic heart disease and CHF are shown in Table 17–1. Nocturnal angina, usually similar to the patient's daytime angina, occurs in 10% to 15% of patients with ischemic heart disease, more often in REM than NREM sleep.[1,2] Anxiety associated with angina and palpitations may make it difficult to return to sleep.

In patients with CHF, inability to lie down or sleep in a horizontal position, awakening with shortness of breath and gasping (*paroxysmal nocturnal dyspnea*), and daytime sleepiness are common complaints.[3] Many patients report that they must sleep in a semirecumbent position or in a chair. The bed partner may note periods of apnea during sleep that are accompanied by snoring or regular variations in the degree of chest wall motion, or both. Polysomnography in patients with ischemic heart disease often reveals central as well as obstructive apneas, and complaints related to sleep apnea sometimes are the initial presentation of heart failure.[4] Insomnia and daytime sleepiness tend to become worse as heart disease progresses.

Biologic Basis

In patients with ischemic heart disease, nocturnal angina is often caused by obstructive sleep apnea (OSA), which causes hypoxemia and changes in hemodynamics.[5,6] In one study, 9 of 10 patients with nocturnal angina had OSA; angina diminished during continuous positive airway pressure (CPAP) treatment.[7] Prinzmetal's angina, also called variant angina, is another cause of nocturnal angina. In this disorder, angina typically occurs at rest, not with effort, and is often worst at night. Vasospasm is thought to contribute to its pathogenesis and about 25% of patients with Prinzmetal's angina have other symptoms suggestive of abnormal vasomotor tone, such as migraine or Raynaud's phenomenon.

Cheyne-Stokes breathing is a major cause of frequent awakenings, paroxysmal nocturnal dyspnea, and daytime sleepiness (Fig. 17–1). Although periodic breathing was recognized by Hippocrates more than 2000 years ago, the commonly used eponym derives from two Irish physicians: Cheyne,[8] who described the breathing pattern in 1818, including a description of apnea accompanying the pattern; and Stokes, who elaborated on it in 1854 (see Chap. 14).[9] This breathing pattern occurs during NREM sleep in as many as 50% of patients with severe CHF; it also may be apparent during REM sleep and wakefulness in patients with end-stage heart disease.[10,11] Even patients with stable CHF have a high rate of sleep-disordered breathing.[11]

A patient's risk of Cheyne-Stokes breathing and associated central sleep apnea (CSA) depends in part on the severity of his or her left heart failure, as indicated by the degree of left ventricular dysfunction, the circulation time, and the size of the left atrium.[11,12] Hypoxemia during the apneas may further impair left ventricular function, which in turn leads to more severe apnea.

In patients with heart disease, awakening from a central apnea associated with hypoxemia is one of the major causes of paroxysmal nocturnal dyspnea. Dyspnea is caused by hypoxemia and, in patients with left heart failure, by pulmonary edema along with its associated increased activity of intrapulmonary receptors. Ventricular and supraventricular arrhythmias may cause shortness of breath, palpitations, and chest pain.

Table 17–1. **Sleep-Related Symptoms Associated with Ischemic Heart Disease and Congestive Heart Failure**

Angina
Palpitations
Shortness of breath
Orthopnea
Insomnia
Excessive daytime sleepiness
Apneas during sleep witnessed by others
Nocturia

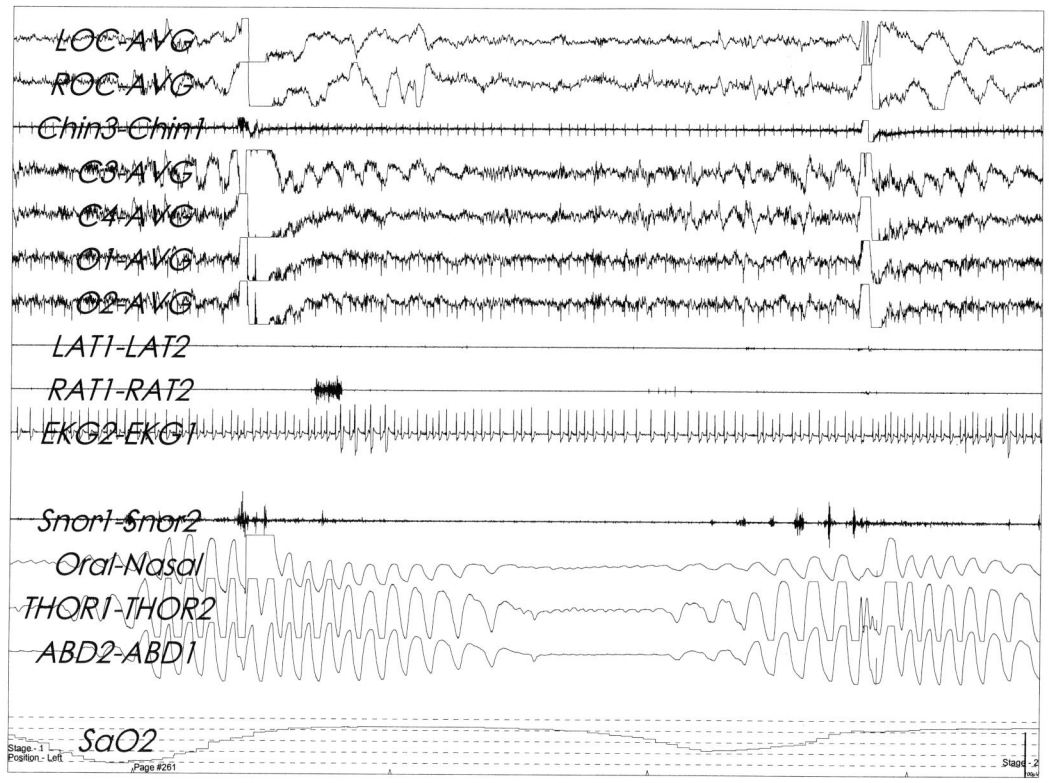

Figure 17–1. Two-minute epoch of NREM sleep from a polysomnogram in a 67-year-old man with atrial fibrillation, congestive heart failure, and a history of snoring and awakening gasping for breath. Cheyne-Stokes breathing is apparent, with regular increases and decreases in the frequency and amplitude of respiratory efforts, as well as a central apnea. Snoring sounds suggest that an obstructive component is also present. The EEG electrodes for the figures in this chapter (A1, A2, C3, C4, O1, O2) were placed in accordance with the International 10-20 system. Other abbreviations: LOC, left outer canthus; ROC, right outer canthus; AVG, average reference; Chin3–Chin1, submental EMG; LAT1–LAT2 and RAT1–RAT2, surface EMG electrodes over the left and right anterior tibialis; EKG2–EKG1, electrocardiogram; Snor1–Snor2, snoring sound; Oral-Nasal, oral-nasal airflow; THOR1–THOR2, thoracic wall motion; ABD2–ABD1, abdominal wall motion; SaO2, pulse oximetry by finger probe.

Left heart failure also leads to fluid overload, and the resulting airway edema may narrow the airway lumen and precipitate or exacerbate OSA, which may then increase left ventricular work during sleep and lead to further cardiac dysfunction and sleep disruption. The recumbent position promotes resorption of pedal edema, which leads to increased blood pressure, increased left ventricular preload, and increased cardiac output. The increased renal perfusion resulting from increased cardiac output often causes nocturia, as a result of which patients may awaken several times each night.

Anxiety and depression, common in patients with ischemic heart disease, also contribute to sleep disturbance in many patients. Finally, medications used for the treatment of ischemic heart disease may contribute to sleep disturbance. Diuretic medications often worsen nocturia, and β-adrenergic antagonists may cause nightmares. The need to awaken to take nitrates or inotropic agents on a regular schedule may also interrupt sleep.

Diagnosis

In patients who complain of nocturnal dyspnea and chest pain, the differential diagnosis includes panic disorder and gastroesophageal reflux disease (GERD). A therapeutic trial of an antireflux medication is sometimes helpful diagnostically (see below). Features of nocturnal panic attacks are discussed in Chapter 16. For patients with known cardiac disease, nocturnal chest pain that is similar to daytime angina is usually a result of cardiac ischemia.

CSA or OSA should be considered in all patients with ischemic heart disease who have sleep complaints. A history of snoring, combined with obesity and evidence of a narrow airway, is suggestive of OSA. Apnea observed by others, accompanied by regular waxing and waning of chest wall motion, is suggestive of Cheyne-Stokes breathing with CSA, especially if the patient does not snore. Concomitant paroxysmal nocturnal dyspnea also suggests that Cheyne-Stokes breathing is occurring. However, because clinical differentiation between CSA and OSA is unreliable in patients with heart disease, and because both central and obstructive components may be present in a given patient (Fig. 17–2), sleep studies should be performed if either type of apnea is suspected. Echocardiographic or nuclear medicine studies to assess left ventricular function may be of value if CHF is suspected.

On polysomnography, Cheyne-Stokes breathing with CSA is distinguished from other forms of CSA by the presence of a waxing and waning amplitude of airflow and breathing effort before and after apneas, and by the occurrence of arousals at the apogee of the breathing cycle rather than at the end of apnea (see Fig. 14–3). A sinusoidal pattern of highly regular variations in oxyhemoglobin saturation (SaO_2) and a peak in SaO_2 of 97% to 98%, higher than the usual daytime SaO_2, are also typical of Cheyne-Stokes breathing. If upper airway obstruction contributes to sleep apnea, mixed apneas are often seen (Fig. 17–3).

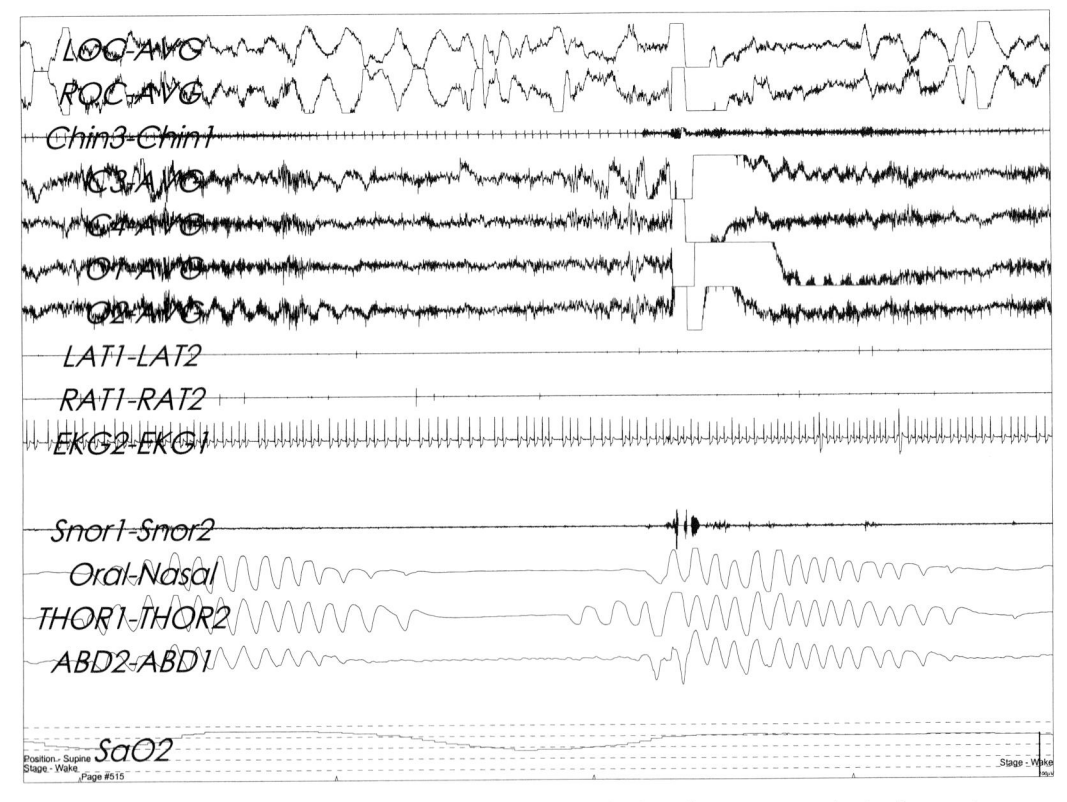

Figure 17–2. Same patient as in Figure 17–1. Two-minute epoch of a polysomnogram. A mixed apnea is apparent in the setting of Cheyne-Stokes breathing. For abbreviations, see Figure 17–1.

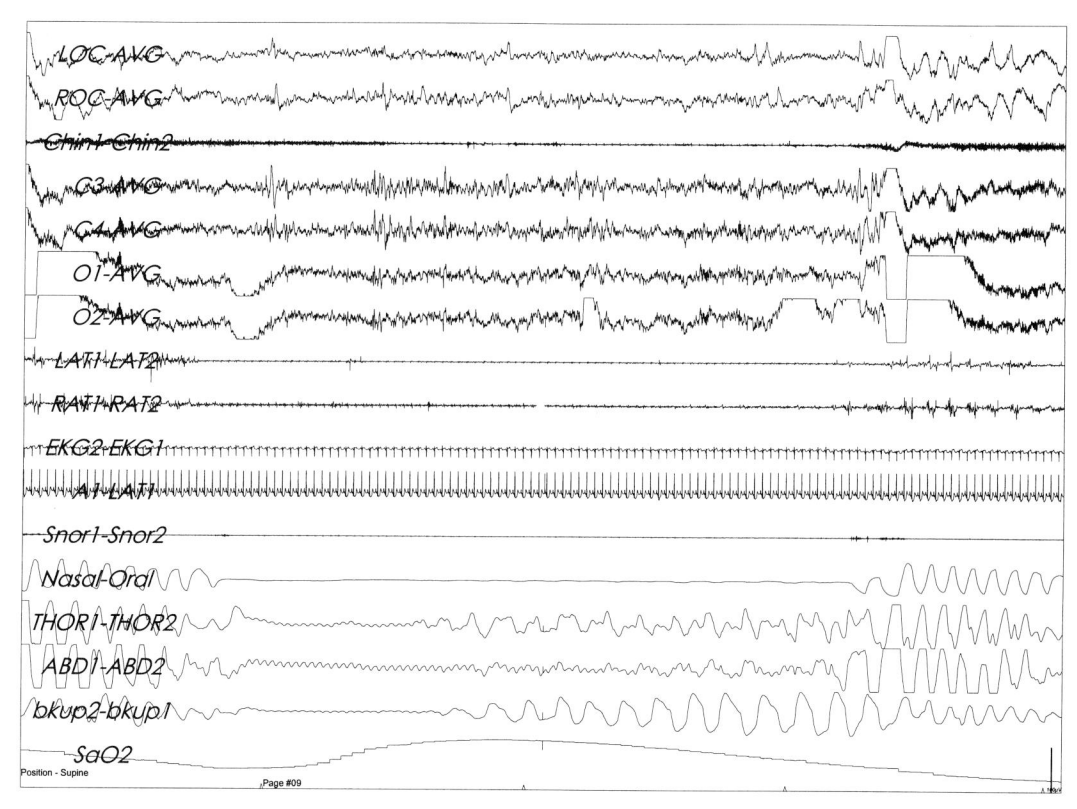

Figure 17–3. Two-minute epoch of a polysomnogram in a 69-year-old man with congestive heart failure and apneas observed by his wife. A 75-second mixed apnea is shown; the initial central component lasts approximately 20 seconds and is followed by a 55-second obstructive component. For abbreviations, see Figure 17–1. Other abbreviations: bkup2–bkup1, backup respiratory effort monitor placed between the thoracic and abdominal sensors.

Although obstructive apneas can be differentiated readily from central apneas by the occurrence of respiratory effort during the apnea, it is often difficult to determine whether hypopneas are caused by obstruction or a mild form of Cheyne-Stokes breathing. Obstructive hypopneas are usually more frequent and longer in duration during REM sleep, whereas central hypopneas are more frequent during NREM sleep because of their dependence on chemical stimuli (see Chap. 3). The timing of associated arousals may help to distinguish the two as well: Arousals resulting from obstructive hypopneas usually occur after a crescendo of breathing effort over a few breaths, with little change in air flow; arousals from central hypopneas are preceded by or accompanied by gradually increasing effort and airflow. The use of an esophageal catheter to measure breathing effort more precisely (see Chap. 13) also is useful if breathing patterns are ambiguous and if management depends on precise characterization of the breathing disturbance.

Sleep apnea is less likely, though still possibly present, in the absence of complaints of daytime sleepiness. Early morning awakening or frequent awakenings with inability to return to sleep are suggestive of anxiety or depression, particularly if other psychiatric symptoms are present (see Chap. 16).

Management

In patients with CHF and Cheyne-Stokes breathing, optimal treatment of heart disease is the most effective treatment for sleep disturbance. Increased cardiac output usually improves sleep quality by reducing the amount of Cheyne-Stokes breathing and associated hypoxemia, and by eliminating or reducing airway edema.[13] Metabolic alkalosis, which may occur with diuretics and which potentiates Cheyne-Stokes breathing, should be avoided. In patients with end-stage heart disease, cardiac transplant is probably the most effective treatment for heart failure and associated CSA or OSA. Nocturnal oxygen (0.5 to 2 L/min) may reduce Cheyne-Stokes breathing, improve oxygenation, and reduce arousals,[9] but whether it has long-term beneficial effects on cardiac and daytime function is unknown.

When insomnia is the major complaint associated with Cheyne-Stokes breathing, benzodiazepines may reduce arousals and thereby reduce ventilatory instability associated with frequent state changes; however, these medications must be used with caution because an obstructive component of apnea, if present, may become worse. Acetazolamide (250 mg at bedtime), which causes urinary excretion of bicarbonate and leads to metabolic acidosis, reduces the ventilatory response to hypocapnia, which can then lead to a reduction in CSA, a decrease in associated arousals, and improved sleep.[14,15] However, sleep disruption associated with nocturia may partially offset the benefits of improved breathing.

For patients with OSA, treatment with nasal CPAP may lead to improved left ventricular function.[16] Nasal CPAP also is sometimes useful for patients with CSA: in such cases, it may reduce the numbers of apneas and hypopneas, improve oxygenation and sleep architecture, and reduce dyspnea and daytime fatigue.[17] Some patients with CHF and central apneas also have improved left ventricular ejection fraction with nasal CPAP, probably because of (1) increased residual volume with increased body stores of CO_2 and O_2, (2) reductions in preload or afterload,[18] or (3) increased respiratory muscle strength.[19] On the other hand, because CPAP causes increased systemic vascular resistance and decreased cardiac output in some patients with CHF,[20] it should not be prescribed for patients with severe CHF without an initial assessment of the cardiac response.

SLEEP DISTURBANCE IN CHRONIC LUNG DISEASES

Chronic lung diseases are common medical disorders that cause cough, fatigue, and shortness of breath. In the 17th century, Floyer[21] noted an association between asthma and sleep; two centuries later, Salter[22] wrote that "spasm of the bronchial tubes is more prone to occur during . . . sleep than during the waking hours." These early clinical impressions were eventually confirmed by studies showing that expiratory flow rates are decreased during sleep in patients with asthma.[23] Furthermore, sleep studies performed after the development of oximeters demonstrated that hypoxemia is usually worse during sleep than during wakefulness in patients with chronic lung disease, and that hypoxemia may be severe during REM sleep in patients with near-normal oxygenation during wakefulness.[24-27] Sleep has a major impact on morbidity and mortality in patients with chronic lung diseases; in asthmatics, for example, more than two-thirds of deaths and three-fourths of respiratory arrests occur between midnight and 6 AM.[28,29]

Clinical Features

Difficulty falling asleep, frequent awakenings, and a sense that sleep is not restorative are problems for 40% to 50% of patients with chronic lung disorders. Wheezing, coughing, and shortness of breath often occur during nocturnal awakenings, especially during exacerbations of the lung disorder. Three-fourths of asthmatic patients report having nocturnal awakenings caused by asthma at least once per week and almost 40% report such awakenings every night.[30] Nightmares, probably related at least in part to breathing problems, are an additional concern for some patients.[31] Daytime fatigue, also a common problem, may be caused by diminished lung capacity, sleep disturbance, or other factors; daytime sleepiness, however, is a complaint in only about 10%.[32]

Biologic Basis

BASIS FOR HYPOXEMIA DURING SLEEP

In patients with chronic lung diseases, including chronic obstructive pulmonary disease (COPD), asthma, cystic fibrosis, and interstitial lung disease, several factors contribute to nocturnal hypoxemia (Table 17–2). As a result, Sao_2 is usually reduced by a few percent in NREM sleep and reduced even further in REM sleep. Patients with hypercapnia during wakefulness, obesity, erythrocythemia, pulmonary hypertension, and cor pulmonale usually have the most severe hypoxemia, in part because of reduced oxygen reserves in the lungs.[33] Clinically, the daytime features that best predict hypoxemia during sleep include low waking Sao_2, hypoventilation, and ventilation-perfusion mismatch.[34]

SLEEP DISTURBANCE

Sleep disturbance in patients with chronic lung diseases may be caused by a variety of factors. Reduced minute ventilation and impaired gas exchange lead to hypoxemia, hypercapnia, and increased respiratory effort, all of which can cause arousals and sleep disruption (see Chap. 3). Inflammation of the airways with increased secretions leads to bronchospasm and cough, which may prolong the duration of awakenings and make a return to sleep more difficult. Sleep disruption in obese hypercapnic patients is often related to severe hypoxemia during REM sleep, whereas sleep disturbance in thin patients is more often related to airway inflammation and cough. Stimulating agents used to treat abnormal breathing, including theophylline and adrenergic bronchodilators, also may contribute to sleep disruption, especially if they are used in the evening or at night.[35,36]

CSA induced by hypoxemia also may cause sleep disturbance, similar to the mechanism whereby CSA is induced at high altitudes (see Chap. 14).

NOCTURNAL ASTHMA ATTACKS

Although sleep is associated with increased airway resistance in patients with asthma, the occurrence of nocturnal bronchospastic attacks appears to be related to circadian variation in bronchial reactivity and hormonal secretion, rather than to time of night and sleep stage.[37–42] For patients who are heavily dependent on inhalers that contain adrenergic bronchodilators, the short duration of action of the medication, usually 2 to 3 hours, may promote early-morning "rebound" asthma.

Perhaps because of altered diaphragm mechanics, asthmatic patients tend to have low esophageal sphincter pressures, which tends to promote GERD. Airway irritation caused by aspiration of gastric contents after reflux may therefore contribute to nocturnal asthma, particularly if the asthmatic episodes occur while the patient is supine.[43] Even in the absence of acid, gastroesophageal reflux may induce bronchoconstriction via a vagally mediated reflex arc from the esophagus to the lung.[44]

Diagnosis

Although coughing at night is usually caused by excess secretions, GERD should be considered if cough continues despite aggressive pulmonary hygiene. Improvement in nocturnal asthma after effective treatment of GERD also suggests that the

Table 17–2. **Contributors to Nocturnal Hypoxemia in Chronic Lung Diseases**

Disease-related factors leading to increased minute ventilation requirements
  Abnormal alveolar gas exchange
  Increased physiologic dead space
  Abnormal diaphragm mechanics
Sleep-related factors
  Decreased ventilation
  Retained secretions
  Altered pulmonary mechanics associated with the horizontal position
  Increased upper airway resistance
REM-sleep related factors
  Decreased accessory muscle activity

reflux was contributing to the nocturnal asthma. Irregular sleep hours and excessively long times in bed may be partly responsible for sleep complaints in some patients.

Daytime sleepiness may be a result of sleep disruption caused by bronchodilators or by either OSA or CSA. Although home oximetry may be sufficient if nocturnal hypoxemia is the major question, polysomnography is indicated if sleep apnea is suspected. For patients with apnea observed by others, OSA is likely if there is a history of snoring or if the patient is obese and has a crowded upper airway. CSA is a consideration in patients with co-existing CHF, or in thin patients with daytime hypoxemia.

Lung disease can be assessed with spirometry and arterial blood gases. A chest radiograph, electrocardiogram, and cardiac imaging studies may be useful in selected patients.

Management

Bronchodilator therapy and good pulmonary hygiene to improve clearance of secretions are the first steps for most patients. Bedtime use of salmeterol, a long-acting bronchodilator, may prevent early morning exacerbation of asthma in some patients. Corticosteroids, often used during exacerbations of asthma and COPD, may improve sleep continuity during the episodes. Evening theophylline use, which may help sleep by reducing bronchospasm, can also cause sleep disruption via its stimulating effects. Respiratory stimulants (e.g., medroxyprogesterone, almitrine) may improve oxygenation, but it is uncertain whether they have beneficial effects on sleep.[33]

For a patient with overt or incipient hypercapnic respiratory failure, nocturnal respiratory support by nasal intermittent positive-pressure ventilation or bilevel positive airway pressure (BPAP) provides three benefits: (1) it prevents sleep-related hypoxemia; (2) it allows adequate, restful sleep; and (3) it may relieve or prevent chronic respiratory muscle fatigue.[44] Good sleep hygiene is another important aspect of management: Alcohol, benzodiazepines, and other hypnotics may worsen hypoventilation and should be avoided in most patients.

Supplemental oxygen, the mainstay of treatment of nocturnal hypoxemia caused by chronic lung disease, improves SaO_2 in sleep, prolongs survival, and improves the quality of life in patients with hypoxemia caused by COPD, although it is unknown whether the benefit derives from nocturnal effects or daytime effects, or both.[33,45] Although oxygen therapy can cause hypoventilation and acidosis in severely hypercapnic patients, most patients with COPD, even those with waking hypercapnia, tolerate low flow rates of oxygen (1 to 3 L/min) without developing respiratory failure. Nocturnal oxygen therapy may prevent progression of pulmonary hypertension in patients who do not have severe hypoxemia, but it does not improve pulmonary hypertension in the most severely affected patients.[45,46]

When OSA accompanies COPD or asthma, nasal CPAP is usually the best therapy; in addition to eliminating apnea, it may improve lung function in such patients.[47] Its value in patients who do not have OSA is doubtful. Transtracheal oxygen, an option for patients with COPD and OSA who cannot tolerate CPAP, improves SaO_2 but has little effect on sleep disruption and daytime sleepiness.[48]

For asthma exacerbated by GERD, vigorous antireflux regimens may reduce the number of nocturnal asthma attacks and decrease bronchial hyper-reactivity.[49,50]

SLEEP DISTURBANCE AND GASTROINTESTINAL DISEASE

Indigestion was probably recognized as a cause of sleep disturbance even before the time of Aristotle, who believed that sleep "arises from the evaporation attendant upon the process of nutrition" and that "a person awakes from sleep when digestion is completed." In addition to indigestion, nocturnal symptoms in patients with gastrointestinal disease may include chest and abdominal pain, coughing and choking, and diarrhea.

Gastroesophageal Reflux Disease

CLINICAL FEATURES

Heartburn—an epigastric or substernal burning sensation—is the principal symptom of GERD. More than 40% of adults have heartburn at least once per month and 7% have it daily. Although heartburn is often considered a benign condition, repeated episodes can lead to dysplasia of the esophageal mucosa, pulmonary fibrosis, and bronchiectasis.

Nocturnal heartburn may occur nightly, or it may develop only after late-evening consumption of alcohol, caffeine, or large meals. Evening tobacco use can also exacerbate heartburn, as can certain foods, such as peppermint. Nocturnal choking and coughing, awakening with a bitter taste in the mouth, and hoarseness or laryngitis in the morning are other sleep-related manifestations of GERD that can accompany heartburn or occur independently. The sleep disturbance caused by these symptoms can lead to daytime fatigue and sleepiness.

In infants, severe reflux may contribute to arousals from sleep and to CSA or OSA, particularly if refluxed material ascends to the larynx.[51-54] In addition, prolonged or frequent episodes of reflux may contribute to dysphagia, emesis, colic, lung disease, and failure to thrive.

BIOLOGIC BASIS

Reflux of stomach contents can occur only with relaxation of the lower esophageal sphincter, a region of tonically elevated pressure that separates the stomach from the esophagus. Reflux is more likely if gastric emptying is delayed or if lower esophageal sphincter function is abnormal. Once reflux occurs, the severity of esophageal inflammation is a function of the acid content of the refluxed material and the time required to neutralize it by swallowed saliva and to remove it by peristaltic sequences associated with swallows. Thus, abnormal esophageal peristalsis or impaired neutralization may contribute to esophageal inflammation.

Table 17–3. **Effects of Sleep on Factors Involved in Gastroesophageal Reflux**[60-63]

Slower gastric emptying
Transient relaxation of the lower esophageal sphincter during arousals
Decreased numbers of swallows
Decreased saliva production
Horizontal position

Sleep has a number of deleterious effects on the factors involved in reflux (Table 17–3). In addition, in patients with OSA, high negative intrathoracic and intrapharyngeal pressures generated during obstructive episodes may suck gastric contents through the lower esophageal sphincter, thereby contributing to reflux.

Reflux during sleep can lead to arousal even in the absence of esophageal inflammation, particularly when large volumes of refluxate are produced.[55,56]

DIAGNOSIS

In addition to GERD, the differential diagnosis of nocturnal chest pain includes angina caused by coronary artery disease, peptic ulcer disease, aerophagia, and panic disorder. The association with precipitating factors and relief by antacids or histamine H_2-receptor blockers are helpful diagnostic features. A number of conditions besides reflux can lead to nocturnal choking or gasping episodes (Table 17–4); GERD should be considered when

Table 17–4. **Differential Diagnosis of Nocturnal Gasping or Choking**

Cardiac arrhythmias
Paroxysmal nocturnal dyspnea
Orthopnea
Obstructive sleep apnea
Central sleep apnea
Gastroesophageal reflux
Bronchospasm
Panic attacks
Laryngospasm

other reflux symptoms are present or when sleeping in the sitting position improves symptoms.

A trial of antireflux therapy is indicated for most patients with suspected nocturnal reflux before further diagnostic evaluation is undertaken. For patients who do not respond and those with atypical symptoms, continuous ambulatory esophageal pH monitoring, which permits correlation of symptoms with esophageal pH, is often helpful. A pH less than 4 for more than 5% of the night, or more than 6% for infants, usually indicates significant reflux.[51] If OSA is suspected as a contributing factor, esophageal pH monitoring can be performed during polysomnography. Radiographic barium studies, endoscopy, esophageal manometry, and acid-perfusion studies are useful in selected cases.

MANAGEMENT

Dietary changes with reduced evening meal size and fat intake, elimination of bedtime snacks, and reduction or elimination of evening caffeine, tobacco, and alcohol use often reduce nighttime symptoms. Elevation of the head of the bed with blocks or with a foam wedge allows gravity to assist with acid clearance. Obese patients should be encouraged to lose weight. Hypnotics should not be prescribed because the reduction in arousals caused by hypnotics leads to fewer swallows and consequently to slower clearance of acid from the esophagus.

Antacids provide short-term relief; for long-term treatment, bedtime doses of H_2-receptor antagonists, which reduce acid secretion by blocking the histamine receptor, are effective. Omeprazole, which inhibits the H^+-K^+-adenosine triphosphatase (ATPase) pump, provides nearly complete acid suppression in refractory cases. Although nocturnal symptoms usually respond to treatment, relapse after discontinuation of therapy is common. In patients with reflux and OSA, treatment with nasal CPAP may reduce the frequency and duration of reflux episodes during sleep.[57]

Peptic Ulcer Disease

With peptic ulcer disease, inadequate gastric mucosal defenses or excess levels of acid and pepsin, or both, lead to ulceration of the gastric or duodenal mucosa. Epigastric pain relieved by food or antacids is characteristic. Nocturnal epigastric pain occurs in about one-third of patients with gastric ulcers and in about two-thirds of those with duodenal ulcers.[58,59]

In patients with active duodenal ulcer disease, increased vagal tone and perhaps other factors lead to acid secretion that is higher than normal during the day and at night.[60] Hypersecretion of gastric acid is particularly prominent during REM sleep and may contribute to the nocturnal pain.[61,62]

Although epigastric pain relieved by food suggests peptic ulcer disease, similar pain occurs in about one-third of patients with nonulcerous dyspepsia.[63] Reflux esophagitis also can cause epigastric pain, although substernal pain is more common. Pain associated with specific foods suggests the possibility of a food allergy.

If peptic ulcer disease is the suspected cause of nocturnal abdominal pain, evaluation may include esophagogastroduodenoscopy, radiographic barium studies, abdominal ultrasound, or abdominal computerized tomographic scanning.

Proton pump inhibitors and histamine H_2-receptor antagonists, the mainstays of treatment, promote healing of duodenal or gastric ulcers by reducing acid secretion. Bedtime doses help to prevent nocturnal reductions in pH. If *Helicobacter pylori* is present, treatment with antibiotics reduces the risk of relapse.

Intestinal Disorders

Nocturnal diarrhea, sometimes accompanied by fecal incontinence, may occur with inflammatory bowel disease and with diabetic autonomic neuropathy affecting the small and large bowel. On the other hand, with irritable bowel syndrome, the most common bowel motility disorder, abdominal pain and diarrhea are usually more frequent during the day than at night.

Treatment of nocturnal diarrhea associated with inflammatory bowel disease is directed at the underlying disorder. Although management of diabetic diarrhea and incontinence is often unsatisfactory, antidiarrheal agents, clonidine, and a somatostatin analogue are sometimes helpful.

SLEEP DISTURBANCE AND RHEUMATOLOGIC DISORDERS

Arthritis

CLINICAL FEATURES

Osteoarthritis, rheumatoid arthritis, and other connective tissue disorders, such as systemic lupus erythematosus, cause joint inflammation, which can cause pain, stiffness, and reduced mobility. Common sleep-related complaints include difficulty falling asleep, frequent awakenings, and the absence of a feeling of restoration in the morning. Although joint pain is sometimes a prominent complaint and can disrupt sleep when it occurs at bedtime or during nocturnal awakenings, stiffness during awakenings is more common and may cause patients to get up at night to walk and loosen up.[64]

Although daytime fatigue is common in patients with rheumatoid arthritis, often varying in severity according to the activity of the underlying disease, daytime sleepiness is unusual unless sleep apnea is present or sleep is severely disrupted for other reasons.

BIOLOGIC BASIS

Stiffness, pain, and immobility are probably the most common causes of sleep disturbance. Stiffness and decreased movement during brief arousals leads to increased discomfort at pressure points and therefore to more severe insomnia. In patients with rheumatoid arthritis, the degree of sleep disruption often varies with the severity of evening pain.[65]

OSA and periodic leg movements (PLMs) also contribute to sleep disruption in many patients. In patients with rheumatoid arthritis, the risk of OSA is increased if atlantoaxial subluxation or temporomandibular joint destruction lead to narrowing of the upper airway. Corticosteroids, commonly used in rheumatoid arthritis, increase fat deposition in the neck and thereby increase the risk of OSA. Brainstem compression caused by cervical spine disease can cause OSA or CSA, or both.[66] About one-half of patients with rheumatologic diseases who complain of fatigue or daytime sleepiness have PLMs during sleep and daytime symptoms suggestive of restless legs syndrome (RLS), a much higher proportion compared to that in the general population.[64] Associated psychiatric conditions, particularly depression, also contribute to insomnia in many patients with chronic painful conditions.

DIAGNOSIS

Diagnostic considerations for complaints of disrupted sleep include pain, stiffness, and reduced mobility; PLMs and RLS; OSA and CSA; depression; and medication effects. Recent worsening of pain and stiffness and inflamed joints on examination support the likelihood that sleep disturbance is related to disease activity. Depression should be considered when early morning awakening is prominent and when a complaint of increased pain is not accompanied by physical or laboratory evidence of increased disease activity. Weight gain associated with steroid use, combined with loud snoring or witnessed apnea, suggests the presence of OSA. Leg jerks during sleep suggest the presence of PLMs. A recent change in medications or the addition of psychoactive medications such as antidepressants or anxiolytics suggests that medication effects may be contributing to symptoms.

Because daytime fatigue is such a common problem for patients with active rheumatologic disease, it is important to distinguish fatigue alone from fatigue associated with sleepiness. Fatigue accompanied by drowsiness and a tendency to fall asleep in inactive situations is more suggestive of a sleep disorder than fatigue associated with muscle weakness, joint aching, or exertional dyspnea (see Chap. 8). When daytime sleepiness is present, considerations include sleep apnea, PLMs, medication effects, and poor sleep hygiene.

Laboratory studies or radiographs of affected joints may be indicated to assess disease activity. Sleep laboratory assessment is useful if OSA or PLMs are suspected or if objective confirmation of daytime sleepiness is needed.

MANAGEMENT

Nonsteroidal anti-inflammatory medications, the mainstays of treatment for nocturnal pain and stiffness, often lead to improved sleep. Appropriate treatment of associated sleep apnea, PLMs, and depression is also likely to improve sleep. The value of hypnotics is limited in these patients because insomnia is often chronic, and the joint disease may lead to an increased risk of falls and injury when sedative-hypnotics are ingested. Short-term use of hypnotics may be of value in selected patients who have acute insomnia associated with an exacerbation of their underlying disease.

Fibromyalgia

Fibromyalgia refers to a condition characterized by tenderness in specific areas ("trigger points") and diffuse musculo-

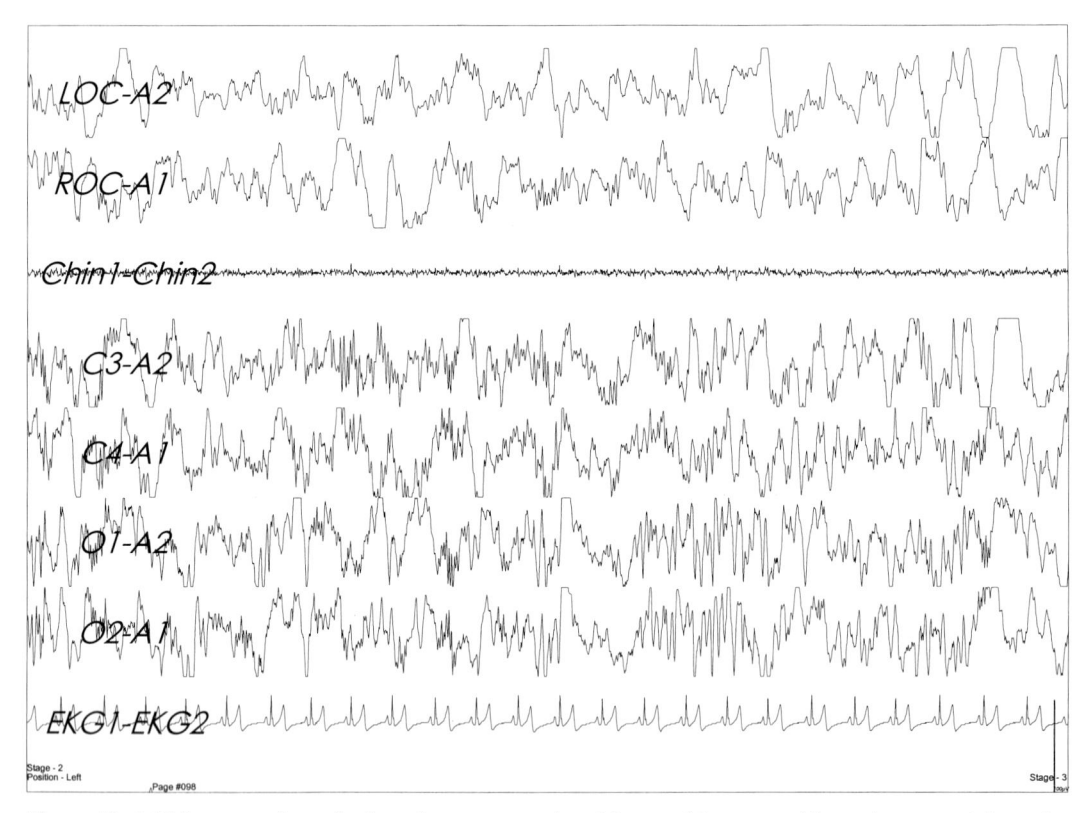

Figure 17-4. Thirty-second epoch of a polysomnogram in a 27-year-old woman with no sleep complaints who was recruited as a control subject for a research protocol. The figure shows slow-wave sleep with prominent alpha frequency intrusion in all EEG channels, more prominently from central derivations. Compare to Figure 17–5. The EEG electrodes (A1, A2, C3, C4, O1, O2) for this figure and Figure 17–5 were placed using the International 10-20 system. Other abbreviations: LOC, left outer canthus; ROC, right outer canthus; AVG, average reference; Chin2–Chin1, submental EMG; EKG1–EKG2, electrocardiogram.

skeletal pain, usually involving the back, neck, and shoulders. Fatigue, a prominent complaint, is often worse in the morning, and many patients report difficulty falling asleep, nocturnal awakenings with inability to return to sleep, and a failure of sleep to provide the usual sense of restoration.

The cause of fibromyalgia is uncertain; laboratory studies and joint radiographs are usually unremarkable. An association in many patients with chronic headaches, irritable bowel syndrome, temporomandibular joint dysfunction, and irritable bladder suggests a functional disturbance of autonomic function. The high number of complaints also suggests that, at least in some patients, fibromyalgia is a manifestation of a somatoform disorder. Depression or dysthymia probably contributes as well in many patients.

Electroencephalographic studies show that many fibromyalgia patients have prominent alpha-frequency activity during all stages of sleep, usually most prominent during stage 3–4 sleep (Figs. 17–4 and 17–5).[67,68] Although some investigators suggest that the pattern, referred to as *alpha-delta sleep or alpha intrusion into sleep*, indicates heightened arousal during sleep and is a marker of nonrestorative sleep,[69] the topography of the alpha activity, which usually has a central maximum, is quite different from the normal waking posterior-dominant alpha rhythm. The pattern was at one time thought to be specific for fibromyalgia, but it also can be seen in patients with other rheumatologic disorders and in up to 15% of asymptomatic persons (see Fig. 17–4).[70] Furthermore, in one study, alpha intrusion into delta sleep was similar between a group of

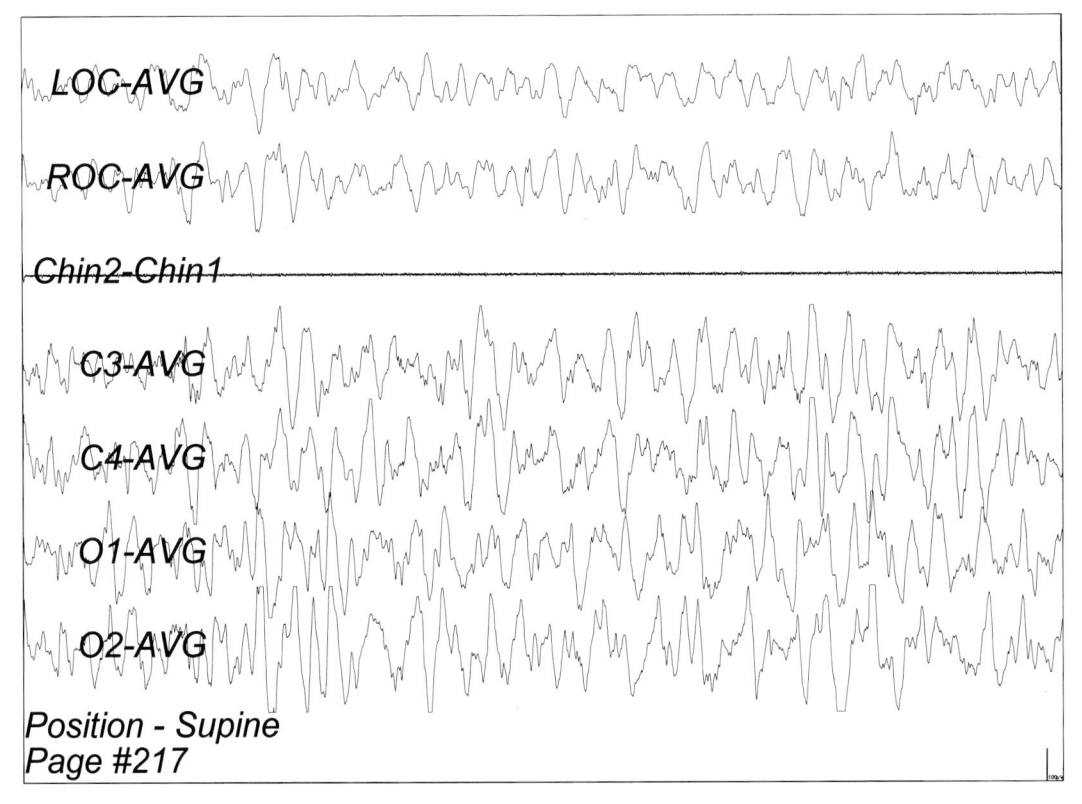

Figure 17–5. Thirty-second epoch of a polysomnogram showing normal slow-wave sleep, with minimal activity in the alpha frequency range. Compare to Figure 17–4, which shows alpha intrusion into slow wave sleep. Abbreviations: LOC, left outer canthus; ROC, right outer canthus; AVG, average reference; Chin2–Chin1, submental EMG; C3, C4, 01, 02, EEG electrodes.

patients with fibromyalgia and a group of normal controls.[71] The pathophysiologic basis for the pattern is currently unknown, and the link between the alpha-delta sleep pattern and complaints of nonrestorative sleep, daytime fatigue, and daytime sleepiness, if any, is obscure.

Treatment of sleep disturbance in patients with fibromyalgia should begin with good sleep hygiene (see Chap. 9). Amitriptyline (25 to 100 mg hs), nortriptyline, doxepin, fluoxetine, or sertraline may be used as well. If major depression is present, higher doses of antidepressants are usually required.

SLEEP DISTURBANCE AND CHRONIC RENAL FAILURE

Clinical Features

Chronic renal failure, usually caused by kidney damage resulting from inflammation, diabetes mellitus, or vascular disease, is a modern disorder. Before the development of effective dialysis in the mid-20th century and the subsequent use of renal transplantation, advanced kidney disease was uniformly fatal. Long-term survival is now possible with hemodialysis or peritoneal dialysis, and renal transplant offers many patients the opportunity to live without dialysis.

Sleep disturbance is a problem for most patients on either peritoneal dialysis or hemodialysis.[72–74] About 70% to 90% of patients report frequent nocturnal awakenings; difficulty falling asleep and daytime sleepiness are problems for about half of patients.[75–77] Most patients report little difference in sleep quality on nights after dialysis treatment compared to other nights.[73]

Sleep studies have shown reduced amounts of slow-wave sleep and poor sleep efficiency with increased awakenings.[74,78] PLMs and RLS are common among these patients, and sleep apnea occurs in 50% to 80%,[72,73,79,80] more often obstructive than central. Among patients with OSA, abdominal distention associated with peritoneal dialysis may lead to more severe hypoxemia associated with apneas; however, ventilation during sleep is rarely affected by peritoneal fluid in patients without apnea.[73]

Biologic Basis

Several factors contribute to sleep disturbance in patients with chronic renal failure. Anemia and iron deficiency, which develop in most patients with chronic renal failure, can cause or exacerbate RLS and PLMs.[76,81] CSA may be facilitated by reduced respiratory drive caused by chronic metabolic acidosis and resultant decreased P_{CO_2}; OSA risk is increased in patients with upper airway edema associated with fluid overload and in those with autonomic dysfunction that impairs respiratory and upper airway muscle activity and coordination. The composition of the dialysis fluid can also affect breathing during sleep. For example, CSA is more common after acetate dialysis than after bicarbonate dialysis.[82]

Bone pain and pruritus cause awakenings in some patients; in others, depression, side effects of medications, and poor sleep hygiene play a role. Toxic effects of "middle molecules" and removal of sleep-inducing cytokines by dialysis are suspected, but unproved, contributors to sleep disturbance.

Diagnosis

Determining the cause of sleep problems is often difficult in patients with chronic renal failure because of the many possible causes of sleep disturbance. Diagnostic considerations include CSA related to metabolic derangements or CHF; OSA related to upper airway edema or steroid use; PLMs and RLS; depression; and poor sleep hygiene. Recent weight gain associated with corticosteroid use after renal transplantation increases the likelihood of OSA. OSA or CSA may be responsible for sleep disturbance when it varies in severity depending on the patient's fluid status.

If sleep apnea is suspected, a polysomnogram is indicated; if possible, it should be performed when the patient's renal and other associated medical conditions are under optimal control.

Management

Patients with RLS and associated sleep disturbance can be treated with levodopa or dopamine agonists, opiates, or benzodiazepines (see Chap. 11), with appropriate modification of doses and dose intervals in accordance with the patient's impaired renal clearance. Treatment with recombinant human erythropoietin may be helpful if anemia is exacerbating sleep disturbance associated with RLS and PLMs.

Nasal CPAP can be used successfully by many patients with OSA who are on dialysis, although some patients on dialysis are reluctant to add another piece of medical technology to their lifestyle. Although CSA and OSA improve in some patients after renal transplantation, others develop OSA after transplantation because of the weight gain induced by corticosteroids. It is unknown whether treatment of sleep apnea leads to improved functional status or to improved survival in patients on dialysis.

Unfortunately, even with optimal management of sleep apnea, RLS, and PLMs, many patients still have problems with insomnia or daytime sleepiness. Adjustment of medications and treatment of nocturnal bone pain, depression, anxiety, and circadian rhythm disturbances may improve sleep in some of these patients.

SUMMARY

Medical diseases are frequent causes of sleep disturbance, especially for older people. The association between increased sleep disturbance and exacerbation of the underlying medical disease is the key to diagnosis. Although symptoms of the medical disorder (e.g., pain, dyspnea, nocturia, impaired mobility) are common contributors to disturbed sleep in patients with serious medical illness, anxiety and depression concerning the loss of function and the potential for disability or death often are even more significant. In addition, medications used to treat medical conditions can cause or exacerbate insomnia or daytime sleepiness.

Sleep-related complaints in patients with ischemic heart disease include nocturnal angina, often precipitated by OSA, as well as inability to lie down or sleep in a horizontal position, awakening with shortness of breath, and daytime sleepiness. A patient's risk for Cheyne-Stokes breathing and associated CSA, a major cause of frequent awakenings and daytime sleepiness, depends in part on the severity of his or her left heart failure.

Difficulty falling asleep, frequent awakenings, and a sense that sleep is nonrestorative are problems for 40% to 50% of patients with chronic lung disorders. Wheezing, coughing, and shortness of breath often occur during nocturnal awakenings, especially during exacerbations of the lung disorder. In patients with chronic lung diseases, hypoxemia is usually worse in sleep than in wakefulness, and hypoxemia may be severe during REM sleep, even in patients with near-normal oxygenation during wakefulness.

Sleep disturbance is common with several gastrointestinal, rheumatologic, and renal disorders. Nocturnal heartburn, common in patients with GERD, results from the deleterious effects of sleep on lower esophageal sphincter function and on mechanisms responsible for clearance of acid from the esophagus. Nocturnal epigastric pain also occurs in a substantial proportion of patients with peptic ulcer disease. In patients with rheumatologic disorders, stiffness, pain, and immobility are common causes of sleep disturbance. Sleep disturbance in patients on dialysis may be related to anemia and iron deficiency, which can cause or exacerbate RLS and PLMs; CSA, which may be facilitated by metabolic acidosis; or OSA, caused by upper airway edema associated with fluid overload.

Diagnosis of sleep disorders in patients with chronic medical diseases requires an understanding of the effects of the medical conditions on sleep and of the interactions between intrinsic sleep disorders and medical illness. In many patients, effective treatment of sleep disturbance can lead to improved quality of life.

REFERENCES

1. Lichstein, E, et al: Significance and treatment of nocturnal angina preceding myocardial infarction. Am Heart J 93:723, 1977.
2. Nowlin, JB, et al: The association of nocturnal angina pectoris with dreaming. Ann Intern Med 63:1040, 1965.
3. Harrison, TR, et al: Congestive heart failure: Cheyne-Stokes respiration as the cause of paroxysmal dyspnea at the onset of sleep. Arch Intern Med 53:891, 1934.
4. Baylor, P, et al: Cardiac failure presenting as sleep apnea: Elimination of apnea following medical management of cardiac failure. Chest 94:1298, 1988.
5. Liston, R, et al: Role of respiratory sleep disorders in the pathogenesis of nocturnal angina and arrhythmias. Postgrad Med J 70:275, 1994.
6. Loui, WS, et al: Obstructive sleep apnea manifesting as suspected angina: Report of three cases. Mayo Clin Proc 69:244, 1994.
7. Franklin, KA, et al: Sleep apnoea and nocturnal angina. Lancet 345:1085, 1995.
8. Cheyne, J: A case of apoplexy in which the fleshy part of the heart was converted into fat. Dublin Hosp Rep 2:216, 1818.
9. Stokes, W: The disease of the heart and the aorta. Hodges and Smith, Dublin, 1854, pp 302–340.
10. Hanly, PJ, et al: The effect of oxygen on respiration and sleep in patients with congestive heart failure. Ann Int Med 111:777, 1989.
11. Javaheri, S, et al: Occult sleep-disordered breathing in stable congestive heart failure. Ann Intern Med 122:487, 1995.
12. Andreas, S, et al: Periodic respiration in patients with heart failure. Clin Invest 71:281, 1993.
13. Walsh, JT, et al: Effects of captopril and oxygen on sleep apnoea in patients with mild to moderate congestive cardiac failure. Br Heart J 73:237, 1995.
14. Weil, J: Sleep at high altitudes. In Kryger, M, et al (eds): Principles and Practice of Sleep Medicine, ed 2. WB Saunders, Philadelphia, 1994, pp 224–230.
15. White, DP, et al: Central sleep apnea: Improvement with acetazolamide therapy. Arch Intern Med 142:1816, 1982.
16. Malone, S, et al: Obstructive sleep apnoea in patients with dilated cardiomyopathy: Effects of continuous positive airway pressure. Lancet 338:1480, 1991.
17. Takasaki, Y, et al: Effect of nasal continuous positive airway pressure on sleep apnea in congestive heart failure. Am Rev Respir Dis 140:1578, 1989.
18. Baratz, DM, et al: Effect of nasal continuous positive airway pressure on cardiac output and oxygen delivery in patients with congestive heart failure. Chest 102:1397, 1992.
19. Granton, JT, et al: CPAP improves inspiratory muscle strength in patients with heart failure and central sleep apnea. Am J Respir Crit Care Med 153:277, 1996.
20. Liston, R, et al: Haemodynamic effects of nasal continuous positive airway pressure in severe congestive heart failure. Eur Respir J 8:430, 1995.
21. Floyer, J: A treatise of the asthma. R. Wilkin and W. Inngs, London, 1698, pp 7–8.
22. Salter, HH: Asthma: Its pathology and treatment, ed 1 (American). Wood, New York, 1882, p 33.
23. Clark, TJH, and Hetzel, MR: Diurnal variation of asthma. Br J Dis Chest 71:87, 1977.
24. Koo, KW, et al: Arterial blood gases and pH during sleep in chronic obstructive lung disease. Am J Med 58:663, 1975.
25. Flick, MR, and Block, AJ: Continuous in-vivo monitoring of arterial oxygenation in chronic obstructive lung disease. Ann Intern Med 86:725, 1977.
26. Coccagna, G, and Lugaresi, E: Arterial blood gases and pulmonary and systemic arterial pressure during sleep in chronic obstructive pulmonary disease. Sleep 1:117, 1978.
27. Douglas, NJ, et al: Transient hypoxemia during sleep in chronic bronchitis and emphysema. Lancet 1:1, 1979.
28. Cochrane, GM, and Clark, TJH: A survey of asthma mortality in patients between ages 35 and 65 in the greater London hospitals in 1971. Thorax 30:300, 1975.
29. Hetzel, MR, et al: Asthma: Analysis of sudden deaths and ventilatory arrests in hospital. Br Med J 1:808, 1977.
30. Turner-Warwick, M: Epidemiology of nocturnal asthma. Am J Med (suppl 1B)85:6, 1988.
31. Klink, M, and Quan, SA: Prevalence of reported sleep disturbances in a general population and their relationship to obstructive airways diseases. Chest 91:540, 1987.
32. Klink, ME, et al: The relation of sleep complaints to respiratory symptoms in a general population. Chest 105:151, 1994.
33. Douglas, NJ, and Flenley, DC: Breathing during sleep in patients with obstructive lung disease. Am Rev Respir Dis 141:1055, 1990.
34. Fletcher, EC, et al: Nonapneic mechanisms of arterial oxygen desaturation during REM sleep. J Appl Physiol 54:632, 1983.
35. Avital, A, et al: Effect of theophylline on lung function tests, sleep quality, and nighttime Sa_{O_2} in children with cystic fibrosis. Am Rev Respir Dis 144:1245, 1991.
36. Berry, RB, et al: Effect of theophylline on sleep and sleep-disordered breathing in patients with COPD. Am Rev Respir Dis 143:245, 1991.

37. Montplaisir, J, et al: Nocturnal asthma: features of attacks, sleep and breathing patterns. Am Rev Respir Dis 125:18, 1982.
38. Cibella, F, et al: Ventilatory response to spontaneous resistive load variations during sleep. J Appl Physiol 76:2394, 1994.
39. Ballard, RD, et al: Effect of sleep on nocturnal bronchoconstriction and ventilatory patterns in asthmatics. J Appl Physiol 67:243, 1989.
40. Barnes, PJ, et al: Nocturnal asthma and changes in circulating epinephrine, histamine and cortisol. N Engl J Med 303:263, 1980.
41. Hetzel, MR, and Clark, TJH: Comparison of normal and asthmatic circadian rhythms in peak expiratory flow rate. Thorax 35:732, 1980.
42. Martin, RJ: Nocturnal asthma. Clin Chest Med 13:533, 1992.
43. Sontag, S, et al: Most asthmatics have gastroesophageal reflux with or without bronchodilator therapy. Gastroenterology 99:613, 1990.
44. Masa, JF, et al: Noninvasive positive pressure ventilation and not oxygen may prevent overt ventilatory failure in patients with chest wall diseases. Chest 112:207, 1997.
45. Nocturnal Oxygen Therapy Trial Group: Continuous or nocturnal oxygen therapy in hypoxemic chronic obstructive lung disease: A clinical trial. Ann Intern Med 93:391, 1980.
46. Fletcher, EC, et al: A double-blind trial of supplemental oxygen for sleep desaturation in patients with chronic obstructive pulmonary disease and a daytime PaO_2 above 60 mm Hg. Am Rev Respir Dis 145:1070, 1992.
47. Chan, CS, et al: Nocturnal asthma: Role of snoring and obstructive sleep apnea. Am Rev Respir Dis 137:1502, 1988.
48. Chauncey, JB, and Aldrich, MS: Preliminary findings in the treatment of obstructive sleep apnea with transtracheal oxygen. Sleep 13:167, 1990.
49. Harper, PC, et al: Antireflux treatment for asthma: Improvement in patients with associated gastroesophageal reflux. Arch Intern Med 197:56, 1987.
50. Ekstrom, T, et al: Effects of ranitidine treatment in patients with asthma and a history of gastroesophageal reflux: A double-blind crossover study. Thorax 44:19, 1989.
51. Sondheimer, JM, and Hoddes, E: Gastroesophageal reflux with drifting onset in infants: A phenomenon unique to sleep. J Pediatr Gastroenterol Nutr 15:418, 1992.
52. Kahn, A, et al: Arousals induced by proximal esophageal reflux in infants. Sleep 14:39, 1991.
53. Jolley, SG, et al: The risk of sudden infant death from gastroesophageal reflux. J Pediatr Surg 26:691, 1991.
54. Herbst, JJ, et al: Gastroesophageal reflux causing respiratory distress and apnea in newborn infants. J Pediatr 85:763, 1979.
55. Orr, WC, et al: Effect of sleep in swallowing, esophageal peristalsis, and acid clearance. Gastroenterology 86:814, 1984.
56. Orr, WC, et al: The effect of esophageal acid volume on arousals from sleep and acid clearance. Chest 99:351, 1991.
57. Kerr, P, et al: Nasal CPAP reduces gastroesophageal reflux in obstructive sleep apnea syndrome. Chest 101:1539, 1992.
58. Edwards, FC, and Coghill, NF: Clinical manifestations in patients with chronic atrophic gastritis, gastric ulcer and duodenal ulcer. Q J Med 37:337, 1968.
59. Earlam, R: A computerized questionnaire of duodenal ulcer symptoms. Gastroenterology 71:314, 1976.
60. Orr, WC, et al: Sleep patterns and gastric acid secretion in duodenal ulcer disease. Arch Intern Med 136:655, 1976.
61. Kales, A, and Kales, JD: Recent findings in the diagnosis and treatment of disturbed sleep. N Engl J Med 290:487, 1974.
62. Segawa, K, et al: The nocturnal intragastric pH in EEG sleep stages in peptic ulcer patients. Gastroenterol Jpn 12:1, 1977.
63. Aldrich, L: Sleep disorders associated with gastrointestinal disorders. In Gilman, S, et al (eds): Neurobase, ed 3. Arbor, La Jolla, Calif, 1997.
64. Mahowald, MW, et al: Sleep fragmentation in rheumatoid arthritis. Arthritis Rheum 32: 974, 1989.
65. Crosby, LJ: EEG sleep variables of rheumatoid arthritis patients. Arthritis Care Res 1:198, 1988.
66. Pepin, JL, et al: Sleep apnoea syndrome secondary to rheumatoid arthritis. Thorax 50:692, 1995.
67. Moldofsky, H: Sleep and musculoskeletal pain. Am J Med (suppl 3A)81:85, 1986.
68. Hauri, P, and Hawkins, DR: Alpha-delta sleep. Electroencephalogr Clin Neurophysiol 34:233, 1973.
69. Moldofsky, H, et al: Sleep and symptoms in fibrositis syndrome after a febrile illness. J Rheumatol 15:1701, 1988.
70. Lavigne, GJ, et al: Muscle pain, dyskinesia, and sleep. Can J Physiol Pharmacol 69:678, 1991.
71. Drewes, AM, et al: Alpha intrusion in fibromyalgia. J Musculoskel Pain 3/4:223, 1995.
72. Wadhwa, NK, and Mendelson, WB: A comparison of sleep-disordered respiration in ESRD patients receiving hemodialysis and peritoneal dialysis. Adv Perit Dial 8:195, 1992.
73. Wadhwa, NK, et al: Sleep related respiratory disorders in end-stage renal disease patients on peritoneal dialysis. Perit Dial Int 12:51, 1992.
74. Mendelson, WB, et al: Effects of hemodialysis on sleep apnea syndrome in end stage renal disease. Clin Nephrol 33:247, 1990.
75. Walker, S, et al: Sleep complaints are common in a dialysis unit. Am J Kidney Dis 26:751, 1995.
76. Stepanski, E, et al: Sleep disorders in patients on continuous ambulatory peritoneal dialysis. J Am Soc Nephrol 6:192, 1995.
77. Holley, JL, et al: A comparison of reported sleep disorders in patients on chronic hemodialysis and continuous peritoneal dialysis. Am J Kidney Dis 19:156, 1992.
78. Passouant, P, et al: Nocturnal sleep in chronic uraemic patients undergoing extrarenal detoxication. Electroencephalogr Clin Neurophysiol 25:91, 1968.

79. Hallett, M, et al: Sleep apnea in end-stage renal disease patients on hemodialysis and continuous ambulatory peritoneal dialysis. ASAIO J 41:M435, 1995.
80. Holley, JL, et al: Characterizing sleep disorders in chronic hemodialysis patients. ASAIO Trans 37:M456, 1991.
81. Winkelman, JW, et al: Restless legs syndrome in end-stage renal disease. Am J Kidney Dis 28:372, 1996.
82. Jean, G, et al: Sleep apnea incidence in maintenance hemodialysis patients: Influence of dialysate buffer. Nephron 71:138, 1995.

Chapter 18

SLEEP DISORDERS IN DEMENTIAS AND RELATED DEGENERATIVE DISEASES

ALZHEIMER'S DISEASE AND
  SLEEP DISTURBANCE
Clinical Features
Biologic Basis
Diagnosis
Management
PARKINSON'S DISEASE AND
  SLEEP DISTURBANCE
Clinical Features
Biologic Basis
Diagnosis
Management
PROGRESSIVE SUPRANUCLEAR PALSY
  AND SLEEP DISTURBANCE
Clinical Features
Biologic Basis
Diagnosis and Management
SLEEP DISTURBANCE IN DISORDERS
  WITH MULTISYSTEM DEGENERATION
Clinical Features
Biologic Basis
Diagnosis and Management

Degenerative disorders of the central nervous system (CNS) may disrupt sleep in a number of ways (Table 18–1), determined partly by the regions of the brain that degenerate. In some disorders, such as Parkinson's disease and progressive supranuclear palsy, sleep disturbance is a virtually universal symptom; in others, such as Huntington's disease, sleep complaints are less common.

Sleep disturbance is rarely the presenting complaint among patients with degenerative diseases, except for those with fatal familial insomnia (see Chap. 19). More often, the patient complains of disturbed sleep and is known to have a neurological disease, although a diagnosis may not be established. The history of the sleep problem and its relation to the neurological disorder provide the most useful information. Examination is helpful for evaluating the underlying disorder, but unless sleep apnea is suspected, it does not assist much in diagnosing the sleep disorder.

Although clinicians often tend to ignore or underestimate the impact of disturbed sleep because of the severity of the neurological problems, such as progressive dementia, seizures, ataxia, immobility, or paralysis, disturbed sleep can make life miserable for patients with degenerative disorders. Thus, accurate diagnosis and appropriate management of disturbed sleep can lead to significant improvements in quality of life and in some cases may facilitate management of other aspects of the disorder.

ALZHEIMER'S DISEASE AND SLEEP DISTURBANCE

Clinical Features

Alzheimer's disease (AD), which produces progressive deficits in memory and ulti-

Table 18–1. **Causes of Disordered Sleep in Degenerative Disorders**

Cause	Sleep-Related Effects
Degeneration of anterior hypothalamus, basal frontal lobes, dorsomedial thalamus	Insomnia
Degeneration of posterior hypothalamus, pontomesencephalic reticular formation, paramedian thalamic regions	Hypersomnolence
Degeneration of suprachiasmatic nucleus, retinohypothalamic tract, retina	Altered timing of sleep and wakefulness
Motor system dysfunction	Periodic and aperiodic movements during sleep, altered muscle tone, immobility, sleep apnea, REM sleep behavior disorder
Cognitive dysfunction	Nocturnal delirium (sundowning), hallucinations, violent behavior
Depression, anxiety, social isolation	Insomnia, early morning awakening
Medications	Insomnia, nightmares, nocturnal delirium, daytime sleepiness

mately in all cognitive abilities, is the most common cause of dementia and is becoming more prevalent as the number of older persons increases. Although AD cannot be diagnosed definitively before death, it is present in about 10% of persons aged 65 and in 47% of those over age 85; about 2% to 3% of persons over age 65 are given a new diagnosis of AD each year.[1] Dementia, often caused by AD, is the leading cause of institutionalization and affects, to some degree, nearly 90% of older persons in nursing homes.

Although the cognitive decline that occurs with AD is its most tragic aspect, sleep disruption can be a major problem. The following case provides an example of the impact of AD on sleep disturbance.

CASE HISTORY: Case 18–1

Memory difficulties began at age 77 for Mr. S, a retired engineer and widower who lived alone. Before that time, his sleep was disturbed only by the need to urinate several times each night because of benign prostatism. As his memory disturbance increased, he became confused during the night and sometimes, unable to find his way back to the bedroom, he wandered through the house. As the time spent awake with each nocturnal awakening increased, he became sleepy during the day and dozed frequently.

After he suffered a brief medical illness, his family realized that he could no longer live alone, and he moved in with his daughter. Although he remained affable during the day, at night he wandered and sometimes dressed and left the house, left the water running in the bath, or urinated on the floor. On one occasion he mixed lye into his coffee, but fortunately he did not drink it. He often became agitated and sometimes abusive when his daughter tried to help him back to bed. Sedatives and hypnotics prescribed by his physician increased his daytime somnolence and nocturnal confusion. Antidepressant and antipsychotic medications produced hallucinations and aggravated his bladder condition.

His daughter became progressively more exhausted as her sleep deteriorated because of her father's behavior. After she awoke one night to find the kitchen extremely hot and all the stove elements turned on, she arranged for him to live in a nursing home.

Sleep fragmentation, more severe in AD patients than in healthy older persons, tends to worsen as the disease progresses.[2,3] Patients with severe AD may have 25 or more awakenings each night, separated by brief periods of light, restless sleep. Sleep loss at night leads to daytime sleepiness and sleep episodes, which may exacerbate difficulties with memory and attention and disrupt social life for the family or caregiver. With severe sleep fragmentation, brief sleep episodes may be scattered throughout the 24-hour day. Although caregivers may report day-night reversal, with most sleep occurring during the day, a more or less random distribution of sleep throughout the 24-hour day is more common. The total amount of sleep during 24 hours is similar to the amount for older persons without dementia.

Nocturnal wandering with disorientation and confusion, which occurs in about one-quarter of demented persons living at home and in institutions, is a major problem for caregivers.[4-6] This syndrome, often called the sundown syndrome because it is more apparent during hours of darkness, may be associated with nocturnal talking or shouting, dressing or undressing, breaking household items, turning on kitchen appliances, leaving the house, and incontinence. Patients with the sundown syndrome usually are less disruptive during the day, in part because of daytime sleepiness and a tendency to doze. However, more time spent asleep during the day often leads to less sleep at night and to greater severity of "sundowning."[7]

Nocturnal wandering with associated confusion and agitation can have a tremendous impact on caregivers and family members, who may have to work or care for children during the day. Lack of adequate rest for the family and caregivers probably explains why the sundown syndrome often leads to institutionalization of demented patients.[8-10]

Some patients with AD exhibit rapid eye movement (REM) sleep behavior disorder (RBD). In addition, RBD appears to be present in most patients with Lewy body dementia, a more rapidly progressive dementia associated with attentional impairments, hallucinations, and prominent fluctuations in cognitive abilities.[11] Furthermore, in some patients, RBD is the initial manifestation of Lewy body dementia.[12]

Biologic Basis

Two major factors contribute to sleep disturbance in AD: primary degenerative changes of brain systems involved in sleep-wake regulation; and secondary disturbances of sleep resulting from immobility, discomfort, social isolation, and depression. Degeneration of noradrenergic, serotonergic, and cholinergic neurons, all of which are involved in sleep-wake control, probably contributes to sleep disruption, whereas degeneration of the suprachiasmatic nucleus may lead to circadian rhythm dysfunction.[13,14] Involvement of brainstem regions that modulate REM sleep probably accounts for the development of RBD. Although medications and depression contribute to sleep disturbance in many AD patients, sleep disturbance is common in unmedicated nondepressed patients.

Sleep apnea appears to be more common in AD than in controls. It was present in 43% of patients with probable AD in one series, and the severity of dementia correlated with the severity of apnea.[15] Nonetheless, sleep-disordered breathing tends to be mild with AD, even when the degree of sleep disturbance is severe. Periodic leg movements (PLMs), about as common in AD as in age-matched controls, are probably not a major contributor to sleep disturbance in most patients.

Although the cause of the sundown syndrome is not fully established, darkness appears to be a major contributor, probably in part because of loss of visual cues for orientation. In 1941, Cameron[16] reported that demented patients brought into a dark room during the daytime soon became agitated and confused, and sundowning is often worse in the winter, when the hours of darkness are increased. Many patients with AD have little exposure to direct sunlight, and those residing in nursing homes seldom are exposed to light levels above 100 lux, equivalent to low-moderate levels of indoor illumination.[17] Other factors that increase the likelihood of sundowning are shown in Table 18–2.

Diagnosis

Diagnosis of the sleep problem is based primarily on the history provided by the patient, spouse, and caretaker. A sleep problem that steadily worsens as disease progresses is probably a direct or indirect consequence of the disease, whereas a sleep problem that antedates the onset of neurological signs and symptoms is usually from other causes, although it may be exacerbated by the degenerative process. A sleep log maintained by the patient or caregiver for 1 to 2 weeks helps in determining the severity of the insomnia and disruptive nocturnal behavior, as well as their effects on the caregiver, although the number of awakenings recorded on the log is almost always an underestimation of the true number. The timing of nocturnal sleep and daytime naps should be recorded in the log along with the timing of medications and stimulants, such as caffeine. Mood should be assessed because depression is a common contributor to sleep disturbance in this population.

Polysomnography or actigraphy may be indicated in some cases, particularly if sleep apnea or a circadian rhythm disorder is suspected. If medication effects or medical problems such as thyroid dysfunction, urinary tract infections, lung infections, dehydration, or other metabolic derangements are suspected as contributors to sundowning, evaluation may include a complete blood count, serum chemistries, urinalysis, and thyroid function tests. The pattern of sleep disturbance also may suggest previously overlooked psychiatric or psychosocial problems.

Management

Although successful treatment of sleep disturbance usually requires determining the specific causes, some general rules apply. A regular schedule of bedtime and awakening may promote better sleep, and too much time in bed should be avoided because it may contribute to fragmented sleep. Morning and evening naps may exacerbate circadian rhythm disturbances and should probably be avoided in order to increase the amount of sleep at night; however, afternoon naps may help some patients obtain more total sleep.

Caffeine and alcohol late in the day should be eliminated, and psychoactive medications should be reduced or eliminated if possible. In institutional settings, staff members should minimize the number of times they awaken demented patients during the night. Although national guidelines require that institutions carry out this practice to prevent bedsores and contractures, such awakenings probably are of little value except in patients with paralysis, and they often lead to agitated behavior.[18,21] Nocturnal structured activities or occupational therapy, if available, may be helpful.

Although increased light exposure during the day may be helpful,[22] most studies of light therapy have used a light intensity equivalent to several hours outdoors on a partly sunny day; thus, lower intensity light, such as sitting in front of a window, may be insufficient, particularly in older patients with cataracts or other ocular pathology. Although I do not recommend melatonin to patients because the content

Table 18–2. **Contributors to Nocturnal Delirium (Sundowning) in Patients with Dementia**

Cognitive impairment
Reduced daytime light exposure
Change in living situation
Psychoactive medications
Medical illnesses
Dehydration
Malnutrition
Electrolyte disturbances
Infections

of commercially available formulations is not regulated, some patients appear to benefit from its use.[23]

Despite the application of the above measures, sundowning frequently persists, and families and other caregivers are often desperate for relief. In this setting, psychotropic medications are often used. Unfortunately, they usually do not help much and sometimes make the situation worse. Benzodiazepines frequently exacerbate nocturnal confusion, and their effects on memory are much more pronounced in demented persons than in cognitively intact persons. Even the shortest-acting benzodiazepines can produce next-day effects of sedation or memory disturbance in older people. Sedating tricyclic antidepressants usually have anticholinergic properties that may make cognitive disturbance worse. Some patients benefit from low doses of antipsychotic medications, such as thioridazine 10 to 25 mg, butyrophenone 0.5 to 1 mg, or clozapine 10 mg at bedtime, although the risk of falls caused by orthostatic hypotension may outweigh any benefit.[24] Regular or occasional use of nighttime caregivers may be an option for some families.

Although sleep apnea may be treated with nasal continuous positive airway pressure (CPAP), demented persons generally do not tolerate nasal CPAP and the complications associated with tracheostomy in demented persons usually outweigh the benefits.

PARKINSON'S DISEASE AND SLEEP DISTURBANCE

Parkinson's disease (PD), a common cause of neurological disability in older people, has an unmistakable clinical picture when fully developed: The patient has an expressionless face; infrequent blinking; a soft, monotonous voice; slow movements; few positional adjustments; a slow, shuffling gait, with small steps and a flexed posture; and a coarse resting tremor of the hands when the hands are immobile, which diminishes or disappears when the patient is in motion or completely relaxed.

PD usually begins after age 50 and progresses inexorably. Walking becomes difficult and finally impossible, speech becomes inaudible, and swallowing dysfunction may make eating impossible. Death usually occurs 10 to 25 years after the onset of the disease.

Clinical Features

In the first detailed description of the clinical syndrome, Parkinson,[25] in 1817, noted that ". . . sleep becomes much disturbed" and that in the final stages of the disease, there is ". . . constant sleepiness, with slight delirium, and other marks of extreme exhaustion . . ." Charcot,[26] describing a 53-year-old woman with PD, noted: "Her slumber is generally short . . . [with] an incessant need of change of posture." In advanced disease, Charcot observed: "This need of change of position is principally exhibited at night in bed . . . Half an hour, a quarter of an hour, has scarcely elapsed until they require to be turned again, and if . . . not . . . gratified they give vent to moans, which . . . testify to the intense uneasiness they experience."

About two-thirds of patients with PD have difficulty falling asleep and nearly 90% complain of frequent awakenings.[27] Patients are awake on average for 20% to 40% of the night and, even in recently diagnosed patients, sleep disturbance is greater than in controls.[28,29] The small postural adjustments during brief arousals, which reduce discomfort and enhance the continuity of sleep, are difficult because of bradykinesia. As a result, patients often lie awake and immobile for extended periods, and their inability to turn over in bed and to get out of bed to use the bathroom is especially bothersome. Early morning awakening, nightmares, nocturnal vocalizations, and jerking movements of the trunk or extremities are other common complaints.

In sleep studies, stage 1 sleep is usually increased while slow-wave sleep and REM sleep are reduced. Tremor disappears at the onset of sleep and generally does not cause sleep disturbance, although it may recur for brief intervals during arousals as well as in light non-rapid eye movement (NREM) sleep and occasionally in REM sleep. Although PLMs are common, in my experience they are only occasionally responsible for insomnia or daytime drowsiness. Fragmentary, irregular myoclonic twitches and jerks of the extremities, repetitive muscle contractions, and prolonged tonic contractions of limb extensor or flexor muscles may occur during NREM sleep and probably contribute to awakenings in some patients (Fig. 18–1). During REM sleep, prolonged elevations of muscle tone may occur as well as more complex movements and RBD (see Chap. 15), which in some cases precedes the onset of daytime PD symptoms of rigidity, tremor, and bradykinesia.[30]

Sleep apnea is common in persons with PD, although perhaps no more common than in age-matched controls. Patients with autonomic impairment are probably more likely to have sleep-disordered breathing, including central sleep apnea (CSA), obstructive sleep apnea (OSA), and nocturnal hypoventilation.[31]

With disease progression, sleep disturbance tends to increase and daytime drowsiness becomes more common. Sleep disruption is even more severe if dementia accompanies advanced disease.

Biologic Basis

Although loss of pigmented dopaminergic neurons of the substantia nigra and locus ceruleus (LC) is the most prominent pathologic abnormality in PD, abnormalities of the mesocorticolimbic dopamine system, as well as the mesostriatal system, occur and may contribute to sleep-wake disturbances.[32] Dopamine neurons of the ventral tegmental area contribute to electroencephalographic (EEG) desynchronization and behavioral arousal; abnormal activity of these neurons may contribute to impaired sleep-wake patterns. In addition, serotonergic neurons of the dorsal raphe, noradrenergic neurons of the LC, and cholinergic neurons of the pedunculopontine nucleus, all implicated in the regulation of REM and NREM sleep, are reduced in number in PD; these neurochemical changes may contribute to sleep-wake disturbance.[33,34]

Sleep apnea, circadian rhythm disturbances, RBD, or depression contribute to sleep disruption in many patients. Medications used for PD also affect sleep. Low

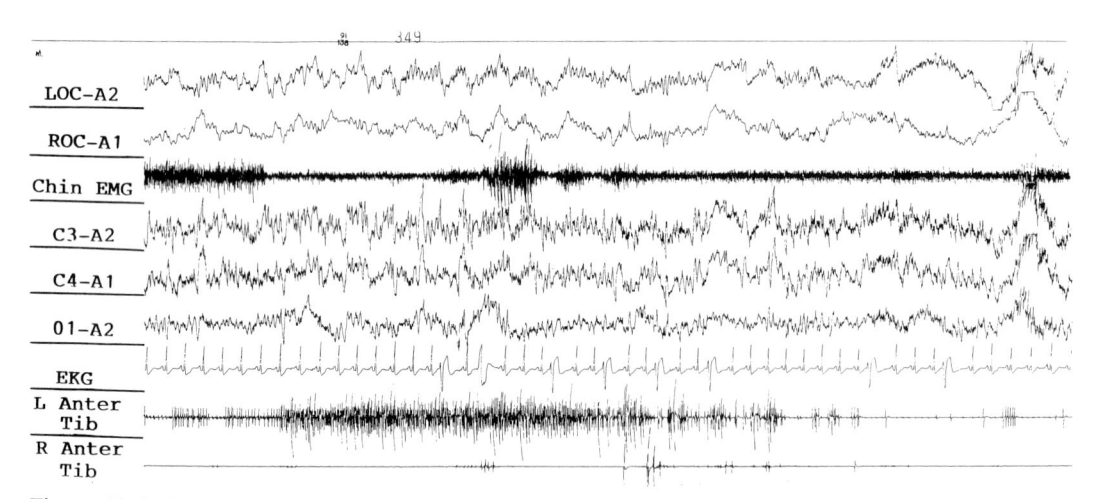

Figure 18–1. Forty-two–second segment of NREM sleep from a polysomnogram in a 72-year-old man with Parkinson's disease. A sustained muscle contraction lasting approximately 15 seconds is apparent in the left leg. The EEG electrodes for this figure and for Figure 18–3 (A1, A2, C3, C4, O1, O2) were placed in accordance with the International 10-20 system. Other abbreviations: LOC, left outer canthus; ROC, right outer canthus; Chin EMG, submental EMG; L Anter Tib and R Anter Tib, surface EMG electrodes over the left and right anterior tibialis, respectively; EKG, electrocardiogram.

doses of dopamine agonists have sedating effects, probably because of preferential stimulation of self-inhibitory dopamine autoreceptors; higher doses are activating and therefore can increase time awake at night and contribute to nightmares and hallucinations.[35] Although sleep latency may increase with moderate doses, the amount of nocturnal time awake and the number of awakenings usually decrease, presumably because the initial arousing effect of levodopa prolongs the onset of sleep, but sleep continuity is enhanced as the dopaminergic effect diminishes. Long-term use of levodopa sometimes leads to myoclonic jerks during drowsiness and sleep, perhaps a result of secondary dysregulation of serotonin activity.[36]

Abnormal tone in upper airway muscles, dyskinetic movements of glottic and supraglottic structures, respiratory muscle incoordination with decreased effective muscle strength, and abnormalities of respiratory drive all may contribute to the development of sleep apnea.[37,38]

Diagnosis

Diagnosis requires a history of the sleep problem, the sleep habits before the onset of PD, the schedule of medications, and the relation of sleep complaints to the development and progression of daytime symptoms. For patients with sleep-onset insomnia, diagnostic considerations include excessive evening doses of dopamine agonists, excessive daytime napping, anxiety, and restless legs syndrome (RLS). Early morning awakening, on the other hand, is often caused by depression or by lack of dopamine medication during the night, leading to bradykinesia and rigidity. Frequent awakenings may be related to breathing disturbances, PLMs, neurochemical changes, or the need to urinate. Causes of nocturnal confusion and hallucinations include dopaminergic or anticholinergic medications, sundowning, nightmares, or RBD. Daytime drowsiness may be a result of sleep disruption associated with sleep apnea or PLMs, sleep-wake schedule disturbances, neurochemical changes, or medication effects. Some patients become sleepy soon after taking small doses of dopamine agonists and may benefit from higher doses taken at less frequent intervals.

The examination sometimes provides evidence pointing to OSA (see Chap. 13). Patients may have signs of autonomic impairment (e.g., orthostatic hypotension, heart rate invariability with the Valsalva maneuver, absence of axillary sweat) resulting from anticholinergic medications or to underlying disease.

Polysomnography should be performed if there is a reasonable suspicion of sleep apnea, if violent behavior suggests RBD or other parasomnias, or if daytime sleepiness is a significant problem. If medications appear to be a major factor in the sleep disturbance, definite diagnosis may require two or more nights of recording under different treatment regimens. A Multiple Sleep Latency Test is sometimes helpful for determining the severity of daytime sleepiness and its circadian variation.

Management

When insomnia is the principle complaint, good sleep hygiene, as outlined in the section on AD (see above), is the first step. Psychosocial and behavioral factors and concurrent psychiatric disorders contributing to sleep disturbance should be addressed, and a portable commode placed at the bedside may help when nocturia is a major complaint. In advanced stages of the disease, it may be helpful for the spouse to sleep in a different bed or room for part or all of the night, or to arrange for occasional respites from care.

For the patient with insomnia who does not have nocturnal hallucinations, a small dose of dopaminergic medication at bedtime may be helpful, such as levodopa 100 mg combined with carbidopa 25 mg (Sinemet 25/100) in either the regular or controlled-release formulation. A second similar dose may be given at 2 or 3 AM if the patient awakens. Evening and nocturnal doses of levodopa or dopamine agonists are especially helpful for RLS and can facilitate sleep by improving nocturnal

Table 18–3. **Effects of Evening and Nighttime Dopaminergic Medications on Sleep in Parkinson's Disease**

Beneficial Effects	Harmful Effects
Sedation (low doses)	Insomnia (high doses)
Reduced periodic leg movements	Dyskinesias (high doses)
Reduced sensations of restless legs	Nightmares
Reduced nocturnal bradykinesia and stiffness	Hallucinations

mobility when sleep disruption is caused by stiffness and difficulty turning in bed.[39,40]

On the other hand, dopamine agonists given at bedtime may cause greater sleep disturbance (Table 18–3). Because the response to levodopa or dopamine agonists depends on the severity of the disease, the amount of medication, and the use of concurrent medications, clinical responses are highly variable. Trials of several medication regimens, along with a sleep diary to assess effectiveness, are often needed to determine the optimum management. The physician also must balance the effects on sleep of changes in medication dosage with the effects of such changes on daytime PD symptoms.

If insomnia does not respond to changes in the dosage and timing of dopamine agonists, small doses of sedating tricyclic antidepressants (e.g., amitriptyline 25 to 75 mg) may be helpful, and the anticholinergic effects may improve daytime PD symptoms. Insomnia caused by dyskinetic nocturnal movements may respond to reduced doses of dopamine agonists or to benzodiazepines, such as clonazepam or temazepam.

Patients with RBD secondary to PD, who have nocturnal vocalizations and disruptive behavior, usually respond well to clonazepam therapy, as do patients with idiopathic RBD. For demented patients with PD, nocturnal confusion and hallucinations are often so disruptive that only minimal doses of benzodiazepines and dopaminergic agonists can be used, and management of sleep disturbance and daytime symptoms becomes exceedingly difficult.

Despite the abnormal respiratory muscle activity observed in some patients with PD, dopaminergic agents are of little or no value for treating sleep-related respiratory disturbances. Nasal CPAP can be used effectively by most cognitively intact patients until the advanced stages of the disease. For patients with autonomic dysfunction and significant OSA, I recommend definitive treatment with tracheostomy or nasal CPAP because these patients appear to be more likely than others to have cardiac arrhythmias and to die suddenly during sleep. Other forms of upper airway surgery are rarely helpful.

PROGRESSIVE SUPRANUCLEAR PALSY AND SLEEP DISTURBANCE

Clinical Features

Progressive supranuclear palsy (PSP), much less common than AD or PD, is characterized by supranuclear gaze palsy, dementia, dysarthria, and axial rigidity. In its early stages, it resembles and is often mistaken for PD.

Insomnia, the most common sleep-related symptom of PSP, appears to be more severe than in most other degenerative neurological diseases.[41,42] With mild to moderate PSP, periods of uninterrupted sleep may last 1 to 2 hours, but as the disease progresses, the total amount of sleep decreases; on some nights there may be no sleep at all (Fig. 18–2).[42–45] Often there is significant sleep disturbance without any clinical complaint, perhaps reflecting the apathy that accompanies the disorder. Daytime sleepiness also tends to become more severe as the disorder progresses.[45] Although RBD occasionally develops in these patients, many have normal muscle

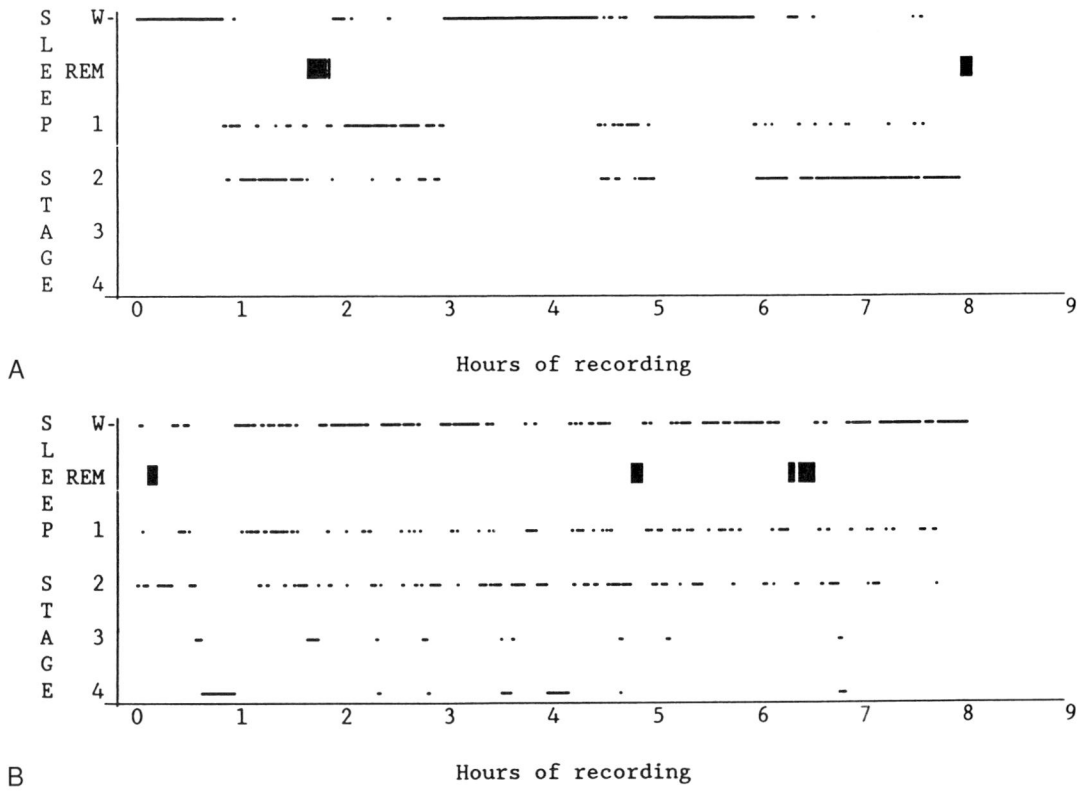

Figure 18–2. Hypnograms in a 72-year-old man with PSP of moderate severity (A) and a 67-year-old woman with severe PSP (B). The moderately affected man had a long sleep latency and two long awakenings during the middle of the night. The woman with advanced disease had severely fragmented sleep with more than 60 brief awakenings. (From Aldrich et al.,[42] p. 579, with permission.)

tone during sleep, even in the advanced stages of the disease. Sleep apnea and PLMs are usually only minor contributors to sleep disturbance.[42]

Rapid eye movements during REM sleep are abnormal, indicating that the gaze disturbances of PSP are not exclusively of supranuclear origin. In PSP of moderate severity, vertical eye movements are absent during REM and horizontal eye movements remain present, but they are slower than usual and reduced in number and amplitude.[41,43,44] Horizontal eye movements eventually disappear as the disease progresses.[45] Reduced phasic twitching during REM sleep in some patients suggests widespread dysfunction of generators of phasic activity of REM sleep.

EEG features of sleep are abnormal, with excessive slowing during all sleep stages. Increased fast activity and intrusion of alpha frequency activity are common during sleep, while sleep spindles, vertex sharp waves, and K-complexes are poorly formed and reduced in number. The combination of increased alpha activity during sleep; increased slow waves during waking; loss of spindles, K-complexes, and vertex waves; and absence of rapid eye movements leads eventually in some cases to almost complete inability to differentiate sleep-wake states.

Biologic Basis

Several brain regions involved in sleep regulation are affected in PSP, including the pontine tegmentum, periaqueductal gray, LC, pedunculopontine nuclei, and median thalamic nuclei.[46,47] Dopaminergic systems are also affected, and cholinergic neurons of the pedunculopontine nucleus—essential components of the brain systems that generate REM sleep—are markedly reduced in number.[47] Functional alterations of the thalamus associated with deafferentation may be responsible for the loss of K-complexes and sleep spindles.[46,48]

Diagnosis and Management

The approach to diagnosis and management of sleep disturbance in PSP is similar to that used for AD and PD. Immobility and dysarthria may make the prominent vocalizations and behaviors that usually accompany RBD or sundowning less prominent and thus harder to diagnose. Although PSP has motor signs similar to those of PD, levodopa and dopamine agonists are rarely helpful unless PLMs or symptoms of RLS are present.

SLEEP DISTURBANCE IN DISORDERS WITH MULTISYSTEM DEGENERATION

Clinical Features

A variety of neurological disorders cause degeneration of more than one brain system. *Multiple system atrophy* (MSA), also called Shy-Drager syndrome, is a progressive multisystem disorder characterized primarily by impaired autonomic function and extrapyramidal symptoms. In *olivopontocerebellar atrophy* (OPCA), which is similar to MSA, cerebellar degeneration is combined with cortical, autonomic, or extrapyramidal degeneration. This disorder occurs sporadically and also runs in families.

Sleep complaints associated with these disorders are common and may include insomnia with frequent awakenings, dream-enacting behaviors, irregular breathing and apneas, stridor, and daytime sleepiness.

Progressive reductions in REM sleep and SWS occur with both syndromes, and RBD, which occurs in at least two-thirds of patients with MSA, may develop 1 year or more before other clinical features emerge.[49–51] Similar changes in REM sleep accompany OPCA.[52] Motor activity is often prominent in NREM sleep as well (Fig. 18–3), and the combination of motor activity, loss of spindles during NREM sleep, and prominent alpha activity during sleep sometimes leads to almost complete loss of electrographic sleep-wake state differentiation.

Breathing abnormalities during sleep, more prominent with MSA than OPCA, include obstructive apneas, central apneas, arrhythmic respirations, and stridor caused by vocal cord paresis.[53–56] In contrast to patients with OSA alone, patients with MSA and OSA sometimes have decreased systemic blood pressure during sleep.[54,57]

Biologic Basis

Because these syndromes frequently affect the brainstem structures involved in sleep regulation—including the LC, the paramedian and dorsolateral brainstem reticular formation, and the nucleus tractus solitarius—it is not surprising that sleep maintenance is disrupted. The brainstem pathology also contributes to prominent abnormalities of REM sleep. Reduced daytime ventilatory responses to hypercapnia and hypoxia in some patients with MSA indicate impaired central control of breathing as a result of autonomic dysfunction and brainstem pathology.[55,58]

Diagnosis and Management

Polysomnography is indicated if concomitant OSA or RBD is suspected. Because the risk of serious consequences of disordered breathing, particularly sudden death during sleep, appears to be high in MSA patients, probably because of autonomic dysfunction,[59,60] tracheostomy or treatment with nasal CPAP should be considered in all patients with MSA and OSA.

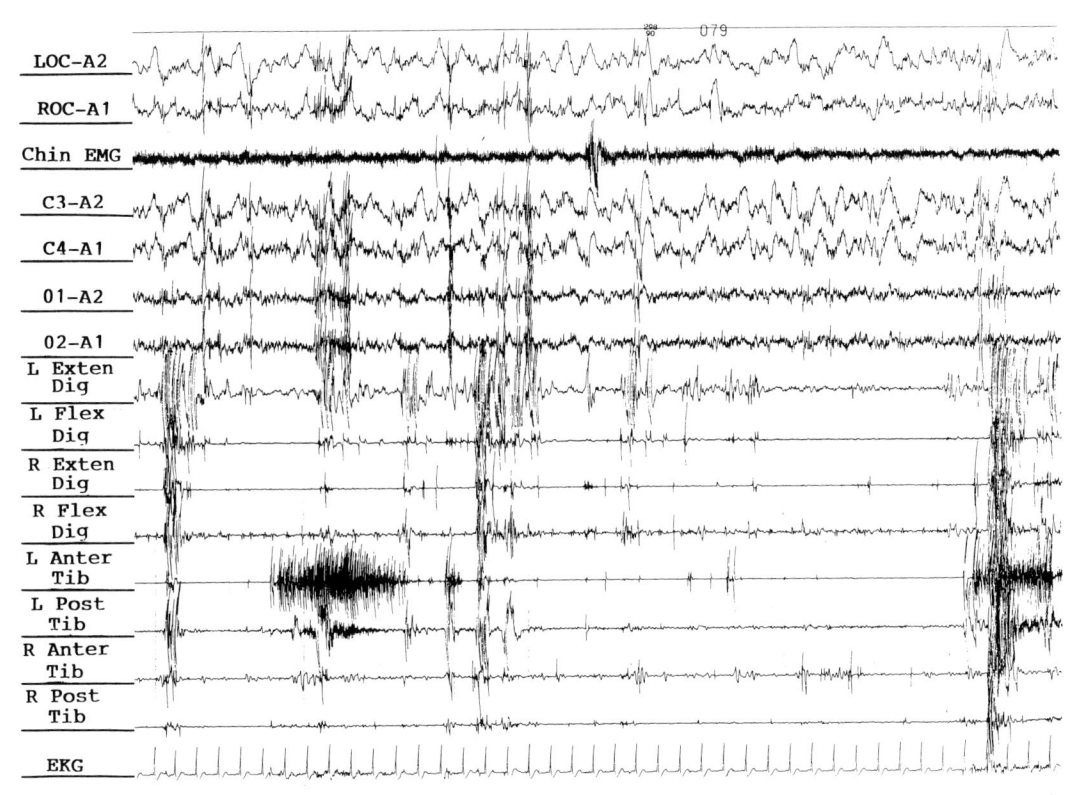

Figure 18 3. Thirty-eight–second segment of NREM sleep from a polysomnogram in a 74-year-old man with multiple system atrophy. Irregular asynchronous muscle activity is prominent. Exten Dig, extensor forearm compartment surface EMG; Flex Dig, flexor forearm compartment surface EMG. For other abbreviations, see Figure 18–1.

Patients with RBD may respond to clonazepam, at least initially.

SUMMARY

Degenerative CNS disorders may disrupt sleep in a number of ways, determined in part by the regions of the brain that degenerate. In some disorders, such as PD and PSP, sleep disturbance is almost universal. Sleep fragmentation with frequent awakenings during the night, the most common pattern of insomnia, may be caused not only by degenerative brain changes, but also by nocturnal pain, stiffness, immobility, or bladder symptoms. Daytime hypersomnolence can be caused by degenerative changes, medication effects, circadian rhythm disturbances, or sleep fragmentation associated with sleep apnea. The likelihood of OSA is increased if weakness, spasticity, or rigidity affect bulbar muscles; CSA or OSA, or both, may be present if respiratory control systems of the lower brainstem are involved. Parasomnias, such as RBD, are more likely to occur in patients with abnormal arousal patterns, fragmented sleep, or abnormal muscle tone.

Sleep disturbance is rarely the presenting complaint in patients with degenerative diseases; more often, these patients have disturbed sleep secondary to a known neurological disease. Although clinicians often underestimate the impact of disturbed sleep because of the severity of the neurological problems, disturbed sleep can make life miserable for patients with degenerative disorders. Accurate diagnosis and appropriate management of disturbed sleep can lead to significant improvements in quality of life and in some cases may facilitate management of other aspects of the disorder.

REFERENCES

1. van Duijn, CM: Epidemiology of the dementias: Recent developments and new approaches. J Neurol Neurosurg Psychiatry 60:478, 1996.
2. Bliwise, DL, et al: Observed sleep/wakefulness and severity of dementia in an Alzheimer's disease special care unit. J Gerontol Med Sci 50: M303, 1995.
3. Vitiello, MV, and Prinz, PN: Sleep/wake patterns and sleep disorders in Alzheimer's disease. In Thorpy, MJ (ed): Handbook of Sleep Disorders. Marcel Dekker, New York, 1990, pp 703–718.
4. Bliwise, DL: Dementia. In Kryger, MH, et al (eds): Principles and Practice of Sleep Medicine, ed 2. WB Saunders, Philadelphia, 1994, pp 790–800.
5. Bliwise, D, et al: Sundowning and rate of decline in mental function in Alzheimer's disease. Dementia 3:335, 1992.
6. Little, JT, et al: Sundown syndrome in severely demented patients with probable Alzheimer's disease. J Geriatr Psychiatry Neurol 8:103, 1995.
7. Bliwise, D, et al: Systematic 24-hr behavioral observations of sleep and wakefulness in a skilled care nursing facility. Psychol Aging 5:16, 1990.
8. Gallagher-Thompson, D, et al: The relations among caregiver stress, "sundowning" symptoms and cognitive decline in Alzheimer's disease. J Am Geriatr Soc 40:807, 1992.
9. Pollak, C, and Perlick, D: Sleep problems and institutionalization of the elderly. J Geriatr Psychiatry Neurol 4:204, 1991.
10. Pollak, C, et al: Sleep problems in the community elderly as predictors of death and nursing home placement. J Commun Health 15:123, 1990.
11. McKeith, LG, et al: Consensus guidelines for the clinical and pathologic diagnosis of dementia with Lewy bodies (DLB): Report of the consortium on DLB international workshop. Neurology 47:1113, 1996.
12. Turner, R, et al: Probable diffuse Lewy body disease presenting as REM sleep behavior disorder. Neurology 49:523, 1997.
13. Swaab, DF, et al: The suprachiasmatic nucleus of the human brain in relation to sex, age and senile dementia. Brain Res 342:37, 1985.
14. Goudsmit, E, et al: The supraoptic and paraventricular nuclei of the human hypothalamus in relation to sex, age and Alzheimer's disease. Neurobiol Aging 11:529, 1990.
15. Reynolds, CF 3d, et al: Sleep apnea in Alzheimer's dementia: Correlation with mental deterioration. J Clin Psychiatry 46:257, 1985.
16. Cameron, DE: Studies in senile nocturnal delirium. Psychiatric Q 14:47, 1941.
17. Campbell, SS, et al: Exposure to light in healthy elderly subjects and Alzheimer's patients. Physiol Behav 42:141, 1988.
18. Evans, LK: Sundown syndrome in institutionalized elderly. J Am Geriatr Soc 35:101, 1987.
19. Francis, J: Delirium in older patients. J Am Geriatr Soc 40:829, 1992.
20. Inouye, SK, and Charpentier, PA: Precipitating factors for delirium in hospitalized elderly persons. JAMA 275:852, 1996.
21. Cohen-Mansfield, J: The relationship between sleep disturbances and agitation in a nursing home. Int J Aging Health 2:42, 1990.
22. Mishima, K, et al: Morning bright light therapy for sleep and behavior disorders in elderly patients with dementia. Acta Psychiatr Scand 89:1, 1994.
23. Singer, C, et al: Physiologic melatonin administration and sleep-wake cycle in Alzheimer's disease: A pilot study. Sleep Res 23:84, 1994.
24. Rabey, JM, et al: Low-dose clozapine in the treatment of levodopa-induced mental disturbances in Parkinson's disease. Neurology 45:432, 1995.
25. Parkinson, J: Essay on the Shaking Palsy. Sherwood, Neely, and Jones, London, 1817, p 17.
26. Charcot, JM: Lectures on the Diseases of the Nervous System: Lecture V. Translation by G. Sigerson. The New Sydenham Society, London, 1877.
27. Factor, SA, et al: Sleep disorders and sleep effect in Parkinson's disease. Mov Disord 4:280, 1990.
28. Kales, A, et al: Sleep in patients with Parkinson's disease and normal subjects prior to and following levodopa administration. Clin Pharm Ther 12:397, 1971.
29. Fish, DR, et al: The effect of sleep on the dyskinetic movements of Parkinson's disease, Gilles de la Tourette syndrome, Huntington's disease, and torsion dystonia. Arch Neurol 48:210, 1991.
30. Schenck, CH, et al: Delayed emergence of a parkinsonian disorder in 38% of 29 older men initially diagnosed with idiopathic rapid eye movement sleep behavior disorder. Neurology 46:388, 1996.
31. Apps, MCP, et al: Respiration and sleep in Parkinson's disease. J Neurol Neurosurg Psychiatry 48:1240, 1985.
32. Javoy-Agid, F, and Agid, Y: Is the mesocortical dopaminergic system involved in Parkinson's disease? Neurology 30:1326, 1980.
33. Jellinger, K: Pathology of parkinsonism. In Fahn, S, et al (eds): Recent Developments in Parkinson's Disease. Raven, New York, pp 33–66.
34. Zweig, RM, et al: The pedunculopontine nucleus in Parkinson's disease. Ann Neurol 26:41, 1989.
35. Corsini, GU, et al: Evidence for dopamine receptors in the human brain mediating sedation and sleep. Life Sci 20:1613, 1977.
36. Klawans, HL, et al: Levodopa-induced myoclonus. Arch Neurol 32:331, 1975.
37. Vincken, WG, et al: Involvement of upper-airway muscles in extrapyramidal disorders: A cause of airflow limitation. New Engl J Med 311: 438, 1984.
38. Feinsilver, SH, et al: Respiration and sleep in Parkinson's disease. J Neurol Neurosurg Psychiatry 49:964, 1986.
39. Lang, AE, et al: Pergolide in late-stage Parkinson disease. Ann Neurol 12:243, 1982.

40. Askenasy, JJM, and Yahr, MD: Reversal of sleep disturbance in Parkinson's disease by antiparkinsonian therapy: A preliminary study. Neurology 35:527, 1985.
41. Gross, RA, et al: Sleep disturbances in progressive supranuclear palsy. Electroencephalogr Clin Neurophysiol 45:16, 1978.
42. Aldrich, MS, et al: Sleep abnormalities in progressive supranuclear palsy. Ann Neurol 25:577, 1989.
43. Perret, JL, and Jouvet, M: Etude du sommeil dans la paralysie supranucleaire progressive. Electroencephalogr Clin Neurophysiol 49:323, 1980.
44. Laffont, F, et al: Polygraphic sleep recordings in 9 cases of Steele-Richardson's disease. Rev Neurol 135:127, 1979.
45. Leygonie, F, et al: Troubles du sommeil dans la maladie de Steele-Richardson: Etude polygraphique de 3 cas. Rev Neurol (Paris) 2:125, 1976.
46. Steele, JC, et al: Progressive supranuclear palsy: A heterogeneous degeneration involving the brain stem, basal ganglia and cerebellum with vertical gaze and pseudobulbar palsy, nuchal dystonia, and dementia. Arch Neurol 10:335, 1964.
47. Zweig, RM, et al: Loss of pedunculopontine neurons in progressive supranuclear palsy. Ann Neurol 22:18, 1987.
48. Foster, NL, et al: Cerebral hypometabolism in progressive supranuclear palsy studied with positron emission tomography. Ann Neurol 24: 399, 1988.
49. Plazzi, G, et al: REM sleep behavior disorders in multiple system atrophy. Neurology 48:1094, 1997.
50. Neil, JF, et al: EEG sleep alterations in olivopontocerebellar degeneration. Neurology 30:660, 1980.
51. Cicirata, F, et al: Spindle and EEG sleep alterations in subjects affected by cortical cerebellar atrophy. Eur Neurol 26:120, 1987.
52. Quera-Salva, MA, and Guilleminault, C: Olivopontocerebellar degeneration, abnormal sleep, and REM sleep without atonia. Neurology 36: 576, 1986.
53. Guilleminault, C, et al: The impact of autonomic nervous system dysfunction on breathing during sleep. Sleep 4:263, 1981.
54. Guilleminault, C, et al: Sleep apnoea syndrome: States of sleep and autonomic dysfunction. J Neurol Neurosurg Psychiatry 40:718, 1977.
55. McNicholas, WT, et al: Abnormal respiratory pattern generation during sleep in patients with autonomic dysfunction. Am Rev Respir Dis 128: 429, 1983.
56. Castaigne, P, et al: Syndrome de Shy et Drager avec troubles du rhythme respiratoire et de la vigilance: A propos d'un cas anatomo-clinique. Rev Neurol (Paris) 133:455, 1977.
57. Martinelli, P, et al: Changes in systemic arterial pressure during sleep in Shy-Drager syndrome. Sleep 4:139, 1981.
58. Chokroverty, S, et al: Periodic respiration in erect posture in Shy-Drager syndrome. J Neurol Neurosurg Psychiatry 41:980, 1978.
59. Katayama, S, et al: Nocturnal sudden death in cases with spinocerebellar degeneration. Sleep Res 16:483, 1987.
60. Munschauer, FE, et al: Abnormal respiration and sudden death during sleep in multiple system atrophy with autonomic failure. Neurology 40: 677, 1990.

Chapter 19

DIENCEPHALIC AND BRAINSTEM SLEEP DISORDERS

ENCEPHALITIS LETHARGICA
Clinical Features
Biologic Basis
Diagnosis and Management
SLEEPING SICKNESS
Clinical Features
Biologic Basis
Diagnosis and Management
FATAL FAMILIAL INSOMNIA
Clinical Features
Biologic Basis
Diagnosis and Management
PRADER-WILLI SYNDROME
Clinical Features
Biologic Basis
Diagnosis and Management
KLEINE-LEVIN SYNDROME
Clinical Features
Biologic Basis
Diagnosis
Management
IDIOPATHIC RECURRING STUPOR AND OTHER DIENCEPHALIC AND BRAINSTEM DISTURBANCES
Thalamic Infarcts
Diencephalic Lesions
Brainstem Lesions

Epimenides, sent by his father into the field to look for a sheep, turned out of the road at mid-day and lay down in a certain cave and fell asleep. He slept there fifty-seven years, and after that, when awake, he went looking for the sheep, thinking that he had been taking a short nap.

EPIMENIDES II. DIOGENES LAERTIUS, CIRCA 200 AD

Disorders affecting the brainstem and the diencephalon, which includes the hypothalamus and the thalamus, can lead to profound alterations of sleep patterns. Although isolated cases of somnolence or insomnia associated with diencephalic lesions were recognized in the 19th century, the encephalitis lethargica epidemic of the early 20th century focused greater attention on these disorders. After von Economo[1] demonstrated in 1930 that insomnia occurred in patients with anterior hypothalamic damage caused by encephalitis lethargica, whereas hypersomnia was more likely to occur with posterior hypothalamic injury, Hess[2,3] in the 1940s showed that thalamic lesions also can lead to sleep-wake disturbances, and Cairns[4] in 1952 observed prolonged hypersomnia in patients with tumors in the area of the third ventricle. Several diencephalic syndromes associated with sleep-wake disturbances are reviewed in this chapter.

ENCEPHALITIS LETHARGICA

Encephalitis lethargica occurs sporadically and in epidemics. The most recent epidemic began in Vienna in 1916, spread through most of Europe, and reached North and Central America, Asia, and Africa before it subsided in about 1926. Of the more than 1 million persons affected, more than 33% died and 40% were left with permanent brain injury. Similar previous epidemics were given various names: "Schlafkrankheit" in Germany in

1580, Sydenham's coma fever in 1672, "febre lethargica" in 1695, "coma somnolentium" in 1780, "vertige paralysante" in 1887, and "nona" in Italy in 1890 (L. Selwa, Univ. of Michigan, 1997, personal communication). Although no major epidemics have occurred since the 1920s, encephalitis lethargica can occur sporadically. More than 30 cases have been reported since 1952.

Clinical Features

The disease can cause somnolence, insomnia, or both. In the early 20th century epidemic, initial symptoms often included fever, headache, joint pain, and somnolence, followed by weeks to months of fluctuating consciousness accompanied by ptosis, rigidity, and corticospinal tract signs. In other cases, insomnia was a prominent manifestation. After recovering from the acute illness, many patients were left with a chronic sleep disorder characterized by either excessive sleepiness and increased sleep or insomnia. More recently described sporadic cases have similar features. The following is a recent case (clinical and laboratory information kindly provided by L. Selwa, Univ. of Michigan, 1997).

CASE HISTORY: Case 19–1

An 11-year-old girl began to have fever of up to 105°F, arthralgias, and oral ulcers. During the next few weeks, she slept for 16 to 18 hours each day and exhibited dystonic postures, initially of the left arm but later of all four extremities. When aroused to eat, drink, and use the bathroom, she was able to interact normally. The somnolence continued for several weeks, improved, and then returned for weeks at a time during the next several months. Two months after the onset of the illness, dystonia and stiffness were severe: Her arms were flexed and her legs extended and turned inward. During the next 6 months, somnolence and dystonia improved, and she was able to resume schooling and horseback riding.

At the age of 14, she noticed the gradual onset of speech problems and a return of stiffness in her left arm. Her legs also tended to stiffen, she had difficulty riding, and she noted cramps in her hands when typing. Over the next several years, her speech and gait gradually became worse. At age 21, she slept 9 to 10 hours per day and had minimal daytime sleepiness. Examination showed normal intellect, severe dystonic dysarthria, dystonic postures of the extremities that were more severe on the left, brisk reflexes, and extensor plantar responses.

Magnetic resonance imaging (MRI) of the brain was normal, as were electroencephalogram (EEG), electromyogram (EMG), and nerve conduction velocities. Cerebrospinal fluid (CSF) was normal except for the presence of three oligoclonal bands. Antistreptolysin titers, thyroid function tests, ceruloplasmin, vitamin E, and very-long-chain fatty acids were normal. Dystonia improved but did not resolve with levodopa-carbidopa treatment.

Biologic Basis

The virus responsible for the disease, not yet identified, has a predilection for the mesencephalic tegmentum and hypothalamus.[1] Persons with posterior hypothalamic or tegmental lesions have hypersomnolence as a result of disruption of ascending arousal pathways. Those with anterior hypothalamic lesions have insomnia or hyposomnia because of disruption of somnogenic neurons of the basal forebrain and preoptic region.

Diagnosis and Management

Diagnosis is based on clinical features. CSF may show lymphocytosis and elevated protein during the acute phase; imaging studies are unremarkable. Management of the chronic phase is dependent on the symptoms. Stimulants may be helpful and should be used in accordance with the guidelines for treatment of narcolepsy (see Chap. 10).

SLEEPING SICKNESS

Sleeping sickness, endemic to West Africa for centuries, came to the attention of Europe and North America after descriptions by late 19th century European explorers.[5] It is the most common infectious cause of prolonged somnolence, with more than 20,000 cases recorded in Africa each year.[6]

Clinical Features

Sleeping sickness is caused by *Trypanosoma brucei gambiense* or *Trypanosoma brucei rhodesiense* infection of the central nervous system (CNS). Transmitted by the tsetse fly, the disease begins with a chancre and localized lymphadenopathy. Systemic spread leads to fevers, arthralgia, myalgia, anemia, hepatosplenomegaly, and cardiomyopathy. CNS invasion leads to meningoencephalitis with headache, tremor, dyskinesias, choreoathetosis, personality changes, and sleep disturbance.

Of these symptoms, sleep disturbance is a prominent feature. Initially, patients usually have drowsiness or increased daytime sleep that may be accompanied by nocturnal restlessness and delirium. As the disease progresses, sleep eventually becomes continuous.

The course of the illness varies. *T. brucei gambiense* infection often has an indolent course, with 1 to 3 years of intermittent symptoms before neurological involvement becomes prominent. *T. brucei rhodesiense* infection is usually fatal within a few months.

Biologic Basis

The tsetse fly acquires the parasite from infected animals and, after a developmental cycle in the fly, the trypanosomes are passed to humans or other animals. Intermittent parasitemia leads to multiorgan injury. Although only a few parasites enter the brain, breakdown of the blood-brain barrier leads to inflammation with infiltration of mononuclear cells.[7-9]

Although the cause of somnolence is not known with certainty, prostaglandin D2 (PGD_2), which has somnogenic effects (see Chap. 2), is increased in the CSF late in the course of the disease. In contrast, levels of interleukin-1 and PGE_2, which are not somnogenic, are not increased.[10] Increased levels of endotoxin also may contribute to symptoms.[11]

Diagnosis and Management

Demonstration of the presence of trypanosomes in the CSF is diagnostic, and accompanying lymphocytosis and pleocytosis are characteristic. CNS involvement can usually be prevented by early treatment of the infection with suramin, pentamidine, or diminazine. Even after CNS involvement, cure is often possible with early treatment using the same medications. Although symptoms improve in most patients who survive, insomnia may persist for months or years. Without drug therapy, sleeping sickness is always fatal.

FATAL FAMILIAL INSOMNIA

Clinical Features

Fatal familial insomnia, a rare familial disorder first reported in 1986 and now recognized in at least five kindreds, usually begins in middle age. Reduced sleep need and difficulty falling asleep are the initial symptoms.[12-14] Insomnia quickly becomes severe, and after the first few months, patients may sleep less than 2 hours each night.[15] Vivid dreams and dream-enacting behaviors occur during sleep or quiet wakefulness, or both.[16] During wakefulness, the level of vigilance may vary quickly from normal alertness to an oneiric state with dream-enacting behaviors.[16] Personality changes, depression, and memory dysfunction may accompany the insomnia, but a global decline in intellectual function does not occur.[16]

As the disease progresses, manifestations of autonomic disturbance become apparent, including increased sweating, salivation, and lacrimation; increased heart rate and blood pressure; abnormal sympathetic skin responses; increased plasma epinephrine and norepinephrine; and impotence in men.[17] Tachypnea during wakefulness, paradoxical breathing, loud snoring, and obstructive sleep apnea (OSA) may also occur.

Dysarthria and gait ataxia develop weeks to months after the onset of sleep disturbance and autonomic dysfunction. Late motor manifestations may include dysmetria, dysphagia, dystonic postures, hyperreflexia with extensor plantar responses, and spontaneous and evoked myoclonus. Smooth-pursuit eye movements may become impaired, and some patients have occasional generalized seizures.

The waking EEG changes progressively from normal alpha activity to low-voltage irregular slow activity, and the background EEG activity becomes gradually slower and less reactive.[18] Spindles, K-complexes, and other features of non–rapid eye movement (NREM) sleep are often completely absent; although periods of rapid eye movement (REM) sleep occur, they are associated with incomplete muscle atonia and dream-enacting behaviors, similar to what is seen in patients with REM sleep behavior disorder.[12,19] Late in the course, myoclonic jerks may be associated with periodic slow waves, similar to EEG findings in Creutzfeldt-Jakob disease.[19] Patients eventually become stuporous, with loss of sleep-wake differentiation. Coma and death occur 6 months to 3 years after the onset of symptoms.

Biologic Basis

The disorder is genetically based, with an autosomal-dominant pattern of inheritance caused by a point mutation at codon 178 of the prion protein gene.[13,20] In addition to the familial cases, sporadic cases have been reported, sometimes described as selective thalamic degeneration or Creutzfeldt-Jakob–type spongiform thalamic degeneration.[21–25] It appears that all affected persons have methionine at codon 129 of the mutant allele, which is the site of a common methionine/valine polymorphism.[26,27] Thus, current evidence suggests that the ^{129}Met, ^{178}Asn haplotype must be present for the disease to occur.[27] The prion fragments generated by treatment with proteinase K differ from the fragments generated from patients with Creutzfeldt-Jakob disease and other familial prion diseases.[28]

Severe spongiform degeneration of the anteroventral and dorsomedial thalamic nuclei is the most prominent pathological abnormality. Gliosis and spongiform changes may also occur in other thalamic nuclei, in the inferior olives, in the cerebellar cortex, and in deep layers of the cerebral cortex.[18] Consistent with the pathological changes, positron emission tomography (PET) studies show relatively isolated thalamic hypometabolism or a more diffuse hypometabolism that increases with disease progression.[29] The thalamic changes are most likely to be the cause of the sleep disturbance; presumably, the thalamocortical circuits involved in sleep initiation and maintenance are disrupted.

Diagnosis and Management

Diagnosis is based on the characteristic clinical features and the family history. Results of computerized tomography and MRI studies of the brain are usually normal. The early sleep abnormalities help to differentiate fatal familial insomnia from familial Creutzfeldt-Jakob disease, which also has the ^{178}Asn mutation of the prion protein gene but has ^{129}Val as well. Demonstration of the ^{178}Asn, ^{129}Met haplotype of the prion protein gene is diagnostic. Unfortunately, no treatment has been discovered.

PRADER-WILLI SYNDROME

Clinical Features

This genetic disorder, first described by Prader, Labhart, and Willi, is characterized by infantile hypotonia, mental retardation, and clinical manifestations of hypothalamic dysfunction, including growth hormone deficiency, hypogonadism, temperature instability, and hyperphagia with obesity.[30] Other features commonly present include short stature, small hands and feet, a narrow face with a small mouth, and kyphoscoliosis. Behavioral aspects may include temper tantrums and stealing or hoarding of food. With an incidence of 1 in 10,000 to 30,000 live births, Prader-Willi syndrome is one of the most common genetic causes of mental retardation.

Excessive sleepiness and increased sleep develop during childhood in most patients with Prader-Willi syndrome. Excessive daytime sleepiness occurs in more than 90% of adolescent and adult patients, and more than 25% sleep more than 10 hours at night.[31,32] Daytime sleep episodes may last anywhere from a few minutes to several hours.[31]

Although cataplexy-like episodes have been reported in some patients, it is often difficult to obtain an accurate history of weakness linked to emotion because of the retardation and hypotonia that are part of the syndrome. Thus, such episodes may not represent true cataplexy.

Biologic Basis

The syndrome is caused by the lack of a paternally inherited gene or genes on chromosome 15. An interstitial deletion (q11-q13) is present on the long arm of the paternally inherited chromosome 15 in 50% to 75% of cases. A similar maternally inherited deletion leads to *Angelman's syndrome,* characterized by mental retardation, microcephaly, and seizures. Thus, similar deletions have different phenotypic expression depending on the parent of origin, a phenomenon known as *imprinting.*[33] *Maternal uniparental disomy,* the inheritance of two copies of a chromosome from the mother and none from the father, occurs in 20% to 25% of Prader-Willi syndrome patients.

Sleep studies have demonstrated four major findings in Prader-Willi patients: (1) OSA, which is severe in some patients but mild or absent in others; (2) nonapneic episodes of hypoxemia that are most likely to be caused by hypoventilation (obesity-hypoventilation syndrome); (3) short sleep latencies on the Multiple Sleep Latency Test (MSLT); and (4) REM sleep abnormalities, with sleep-onset REM periods and fragmented REM sleep periods.[34–37]

The major cause of sleep-related breathing abnormalities in these patients appears to be obesity, although craniofacial anomalies may also contribute to a narrowed airway.[37] In some patients, sleepiness and sleep-onset REM periods are caused at least in part by obesity and OSA; however, they also occur in patients who do not have breathing disturbances, probably because of hypothalamic dysfunction.[38]

Diagnosis and Management

Diagnosis of Prader-Willi syndrome is based on clinical features and chromosomal analysis. Sleep studies are indicated in sleepy patients to assess the presence and severity of OSA and hypoventilation.

Weight loss, difficult to achieve in these patients, may improve nocturnal oxygenation. Although nasal continuous positive airway pressure (CPAP) or tracheostomy usually eliminates OSA, nocturnal oxygen desaturations may continue if hypoventilation is a factor. Although stimulants may be of value in some cases, they can aggravate behavioral disturbances.

KLEINE-LEVIN SYNDROME

Clinical Features

In this rare syndrome, recurrent episodes of prolonged sleep are associated with mental disturbance, hyperphagia, and sexual disinhibition.[39–45] The disorder usually begins in adolescence and is more common in boys than in girls by a ratio of 4:1.[44]

The onset is usually abrupt, and prolonged sleep, sometimes with vivid dreaming, usually lasts for several days, although it may occur for just a few hours or for several weeks. If awakened, patients are often irritable and may refuse to get out of bed. Confusion, a sense of unreality, and visual or auditory hallucinations may accompany the sleep disturbance, and a few nights of moderate insomnia, with depression or euphoria, may follow the hypersomnolence. Subsequent episodes recur at irregular intervals of approximately several months.

Abnormal eating during waking intervals is a striking feature of the symptomatic episodes. Although patients usually do not seek out food, they eat compulsively when food is available and may gain substantial amounts of weight. Other inappropriate behaviors may include public masturbation and indiscriminate sexual advances.

Neurological examination is normal between episodes, and although a variety of neurological findings have been reported during symptomatic periods, their significance is uncertain. Similarly, apart from mood disturbances immediately following episodes, psychiatric examination during asymptomatic intervals is usually unremarkable.

Incomplete forms, such as recurrent hypersomnia without an eating disorder, are more common than the complete syndrome. In one series of more than 6000 patients seen at a sleep center, there were no cases of the complete syndrome and only seven cases of the incomplete form (Aldrich, unpublished data).

Biologic Basis

The cause of the syndrome is unknown, although viral gastrointestinal or respiratory symptoms precede the initial episode in up to one-half of patients, suggesting that a virus may trigger the disorder.[44] Other possible precipitating factors include concussion, sunstroke, fever, and alcohol intoxication.[44]

Diencephalic disturbance is suspected because of the disturbance of vegetative functions and because some patients have symptoms suggestive of autonomic dysfunction, such as increased sweating and facial congestion.[46] In a few cases, postmortem examination showed focal encephalitic changes involving the hypothalamus or thalamus.[47,48] The mood disturbance that sometimes follows episodes and the presence of bipolar disorder in some family members suggest that Kleine-Levin syndrome occasionally may be a variant of bipolar disorder.[45,49]

Although neuroendocrine and CSF studies have sometimes demonstrated abnormalities, it is difficult to know whether these are primary changes or the result of sleep or mood alteration.[50–54]

Diagnosis

With the first episode of hypersomnia, before a pattern of recurrence is established, other diagnostic considerations include mass lesions of the diencephalon, intermittent obstruction of the third ventricle caused by colloid cysts or pediculated astrocytomas, encephalitis, head trauma, and thalamic stroke.

The differential diagnosis of recurrent hypersomnia includes diencephalic lesions, idiopathic recurring stupor, bipolar affective disorder, and psychogenic recurrent hypersomnia.[55] Recurrent hypersomnia, with or without mental and eating disturbances, may also occur in association with menstrual periods.[56] In contrast to Kleine-Levin patients, those with psychogenic recurrent hypersomnia spend many hours in bed and report increased sleep or increased need for sleep, but laboratory studies reveal no increase in sleep time; in some cases, the latter disorder may be a form of hysteria.[44,57]

Although no laboratory tests provide a definitive diagnosis of Kleine-Levin syndrome, CSF and brain imaging studies are often useful to help to rule out other diagnoses. The EEG is reported to show generalized slowing of background activity, which is not found in psychogenic states.[58] Prolonged day-night monitoring may help to document an actual increase in sleep time, while short sleep latencies on MSLT help to confirm objective sleepiness and can be compared to MSLT values during asymptomatic periods.[59]

Management

Effective treatments that terminate the episodes have not been identified, and symptomatic treatment with stimulants during episodes appears to be of little value. Lithium or carbamazepine may reduce the frequency of subsequent episodes, particularly when affective changes suggest that the condition may be a manifestation of bipolar disorder. Although some patients have repeated episodes for more than 20 years, in most patients the episodes become less frequent with time and eventually cease.[45]

IDIOPATHIC RECURRING STUPOR

Idiopathic recurring stupor consists of recurrent episodes, lasting for hours to days, of drowsiness or obtundation with confusion, dysarthria, and unsteadiness. The episodes begin with fatigue and drowsiness, followed in seconds to minutes by obtundation and in severe cases stupor or coma.[60-62] With mild attacks, patients may be ataxic and severely drowsy, or obtunded but arousable with slurred speech. During severe attacks, they appear to be asleep and cannot be aroused, although brainstem reflexes are intact. Recovery occurs after 2 to 72 hours; after the episodes, patients may be confused for several hours and amnesic regarding the episode, and they may complain of a headache. The disorder appears to be benign, and episodes tend to become less severe over time.

Although the cause of the disorder has not been identified, stuporous episodes are associated with marked increases in CSF levels of endozepine-4, an endogenous benzodiazepine receptor ligand.[61] Thus, there may be an underlying metabolic defect that leads to neurological symptoms after specific precipitating events, analogous to acute intermittent porphyria.

The rapid onset of an obtunded or stuporous state may suggest a mass lesion, a metabolic encephalopathy, or a sedative overdose. The key to diagnosis is the demonstration of characteristic EEG changes during the attack—diffuse, fast-frequency activity with prominent 10- to 14-Hz rhythms (Fig. 19–1)—and the reversibility of the clinical state and the EEG changes with flumazenil, a benzodiazepine receptor antagonist.

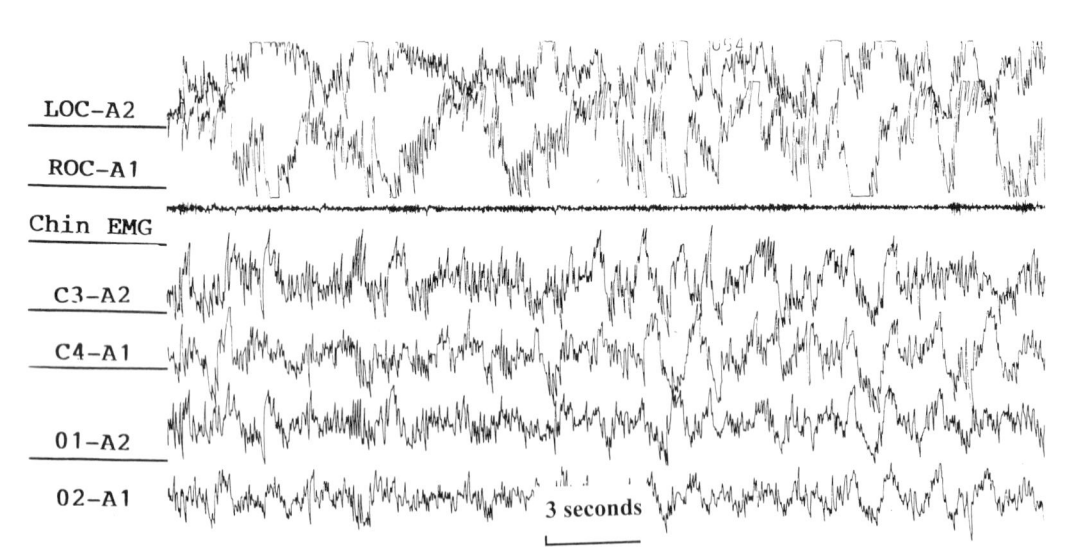

Figure 19–1. A 28-second segment of a polysomnogram in a 40-year-old woman with recurrent episodes of hypersomnolence and ataxia lasting up to 24 hours. Episodes occurred three or four times per year and were consistent with the diagnosis of idiopathic recurring stupor. The segment, recorded duirng an episode of hypersomnolence, shows abundant diffusely distributed 10- to 14-Hz activity intermixed with slower frequencies. The EEG electrodes (A1, A2, C3, C4, O1, O2) were placed in accordance with the International 10-20 system. Other abbreviations: LOC, left outer canthus; ROC, right outer canthus; Chin EMG, submental EMG.

OTHER DIENCEPHALIC AND BRAINSTEM DISTURBANCES

Thalamic Infarcts

Drowsiness and somnolence occur in patients with bilateral paramedian thalamic infarcts and in patients with "top-of-the-basilar" syndromes that involve the midbrain paramedian tegmentum, probably because of interruption of ascending noradrenergic pathways.[63-65] These patients may sleep up to 20 hours per day, but the amount of stage 3-4 sleep is often reduced, and most of the time asleep is spent in stage 1-2 NREM sleep, suggesting that the lesions disrupt the pathways required for slow-wave sleep as well as those involved in wakefulness.[65] Bilateral paramedian thalamic infarctions can also cause a syndrome of compulsive presleep behavior in which patients repeatedly get into bed and assume sleeping postures although they are not asleep and do not fall asleep quickly during MSLTs or EEGs.[66]

Diencephalic Lesions

Tumors of the diencephalon can lead to sleep-wake disturbances, which are often accompanied by diabetes insipidus and other endocrine and temperature disturbances if the rostral hypothalamus is affected. Craniopharyngiomas, benign tumors that arise from remnants of the hypophyseal duct, are often associated with insomnia or hypersomnia, and with other features of narcolepsy.[67-70] With astrocytomas involving the rostral midbrain and hypothalamus, daytime somnolence, nocturnal sleep disruption, and REM sleep abnormalities can occur, sometimes with cataplexy.[71,72] Cataplexy became almost continuous ("limp man syndrome") in a 36-year-old man with a rostral brainstem glioblastoma, who also experienced diplopia, weakness, gait instability, sleepiness, sleep paralysis, and hypnagogic hallucinations.[73]

A variety of diencephalic disorders other than tumors can cause sleep-wake disturbances. Hypothalamic disturbances are common in patients with CNS sarcoidosis and can lead to somnolence and sleep-onset REM periods.[74] Arteriovenous malformations adjacent to the third ventricle and brain radiation involving the hypothalamus can cause irregular sleep-wake patterns, sometimes with daytime sleepiness and REM sleep abnormalities (see Chap. 10).[75,76] In Whipple's disease, which affects the CNS, involvement of the hypothalamus is common and can cause hypersomnolence, occasionally as the presenting symptom.[77] Idiopathic hypothalamic syndromes, accompanied by such endocrine abnormalities as diabetes insipidus and hypothalamic hypothyroidism, may be associated with daytime sleepiness.[67,68,78]

Diencephalic lesions should be suspected and brain MRI should be obtained if sleep-wake disturbances are accompanied by endocrine abnormalities or temperature rhythm abnormalities, or in patients with other neurological disturbances that accompany the sleep-wake symptoms. Sleepiness sometimes responds to stimulants such as methylphenidate or dextroamphetamine.

Brainstem Lesions

Brainstem lesions can lead to daytime sleepiness by their effects on arousal functions of the reticular activating system and on breathing during sleep. The relative rarity of brainstem causes of excessive sleepiness probably reflects the nature of the reticular activating system: Lesions affecting arousal pathways are more likely to be lethal or to cause severe arousal deficits, such as coma and persistent vegetative state, than sleepiness. Narcolepsy with cataplexy may develop with pontine lymphoma or after anoxic injury to the pons.[79,80]

Dysfunction of central respiratory control neurons in the brainstem, particularly in the medulla, can lead to central sleep apnea (CSA) and periods of excessively slow, irregular breathing or excessively rapid breathing (Fig. 19-2).[81] OSA can occur if motor pathways or neurons involved in the coordination of upper airway muscles are affected, or if medullary lesions and lower cranial nerve dysfunction pro-

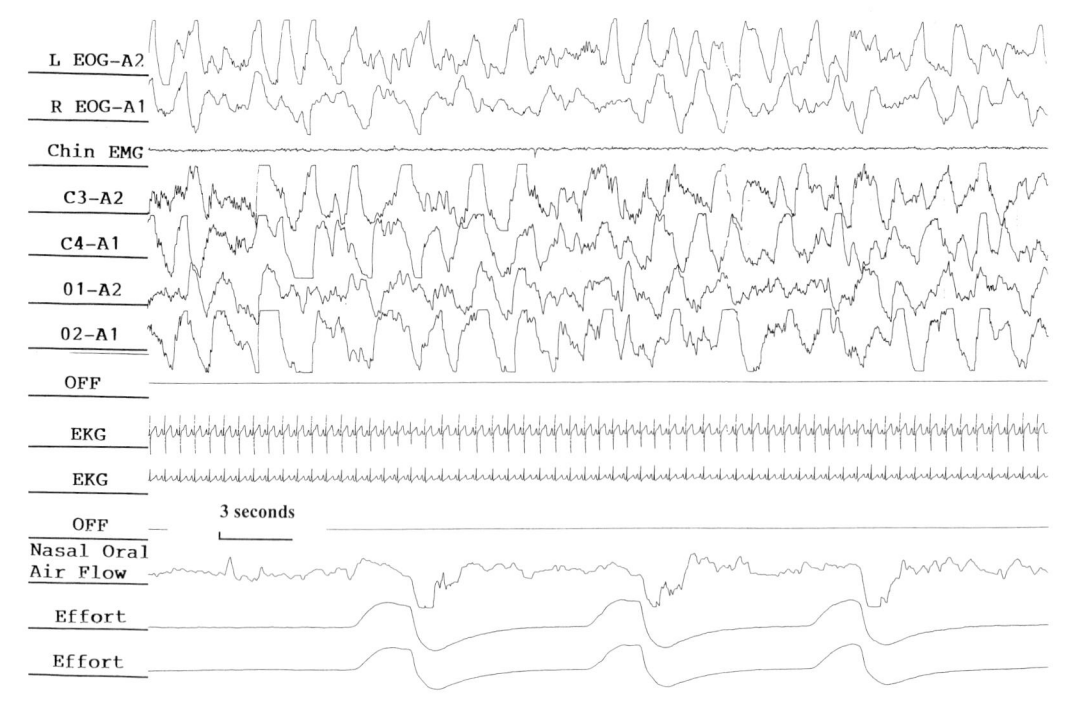

Figure 19–2. A 40-second segment of NREM stage 4 sleep during a polysomnogram in a 12-year-old girl with somnolence following a traumatic mid-brain injury. Between breaths, oxyhemoglobin saturation dropped from 95% to between 80% and 85%. The respiratory rate of 3 to 6 breaths/min during NREM sleep was a consequence of injury to neurons involved in respiratory control. During REM sleep, the respiratory rate was 6 to 10 breaths per minute and moderately irregular. The EEG electrodes (A1, A2, C3, C4, O1, O2) were placed in accordance with the International 10-20 system. EOG, left electro-oculogram; R EOG, right electro-oculogram; EKG, electrocardiogram; Effort, thoracoabdominal wall motion. For other abbreviations, see Figure 19–1.

duce pharyngeal or tongue weakness, leading to upper airway collapse during sleep. Potential causes include syringomyelia, syringobulbia, Arnold-Chiari malformation, cranial base abnormalities, and tumors.[82] In some patients, effects on medullary respiratory neurons may lead to pronounced hypoventilation during sleep, even when other significant neurological abnormalities are not apparent.

Brainstem disorders should be considered in any patient who presents with unexplained sleep-related hypoventilation, unusual respiratory patterns during sleep, or frank respiratory failure for which a cause is not apparent. In such cases, MRI of the brainstem and high cervical cord should be performed to assess for possible Arnold-Chiari malformation, syringomyelia or syringobulbia, or other structural lesions. However, CSA alone (i.e., without associated hypoventilation or abnormal respiratory patterns) is rarely, if ever, a result of brainstem disease, and if the neurological examination is normal, imaging studies are not required.

Management depends on the cause of the disorder. If structural lesions or anomalies that compress the medulla are the cause, then surgery to decompress the posterior fossa is indicated, and sleep-related respiratory failure may improve after decompressive surgery.[83] Nasal bilevel positive airway pressure or positive-pressure ventilation is required for some patients.

SUMMARY

Disorders affecting the diencephalon and brainstem can lead to profound alterations of sleep patterns. In encephalitis lethargica, presumed to be caused by a virus, insomnia is caused by anterior hypothalamic damage, whereas hypersomnia is more likely to be associated with posterior hypothalamic injury. Sleeping sickness, transmitted by the tsetse fly, is a result of trypanosomal invasion of the CNS; somnolence may be caused by increased levels of PGD_2.

Diencephalic disorders caused by genetic conditions also can lead to sleep-wake disturbances. Spongiform degeneration of the anteroventral and dorsomedial thalamic nuclei is the most prominent pathological abnormality of fatal familial insomnia, a disorder caused by a mutation of the prion protein gene. Symptoms include difficulty falling asleep, dream-enacting behaviors, personality changes, and manifestations of autonomic disturbance. In Prader-Willi syndrome, a genetic disorder characterized by hypotonia, hypogonadism, hyperphagia, and mental retardation, excessive sleepiness and REM-sleep abnormalities develop in most patients, caused by hypothalamic dysfunction or sleep-disordered breathing, or both.

Recurrent episodes of hypersomnolence occur in patients with Kleine-Levin syndrome and with idiopathic recurring stupor. The cause of Kleine-Levin syndrome, characterized by episodes of prolonged sleep associated with mental disturbance, hyperphagia, and sexual disinhibition, is unknown; episodes usually become less frequent with time. In idiopathic recurring stupor, recurrent episodes of drowsiness and obtundation are associated with confusion, dysarthria, unsteadiness, prominent 10- to 14-Hz EEG rhythms, and high CSF levels of an endogenous benzodiazepine receptor ligand.

Infarcts and mass lesions of the diencephalon can lead to sleep-wake disturbances, often accompanied by endocrine and temperature disturbances if the rostral hypothalamus is affected. Brainstem lesions can lead to daytime sleepiness as a result of their effects on arousal functions of the reticular activating system; dysfunction of respiratory neurons in the medulla can lead to CSA, OSA, and irregular breathing patterns during sleep.

REFERENCES

1. von Economo, C: Sleep as a problem of localization. J Nerv Ment Dis 71:249, 1930.
2. Hess, WR: Das Schlafsyndrom als Folge diencephaler Reizung. Helv Physiol Pharmacol Acta 2:305, 1944.
3. Hess, WR: Hernreizversuche uber den mechanismus des Schlafes. Arch Psychiatr Nervenkr 86:287, 1929.
4. Cairns, H: Disturbances of consciousness with lesions of the brain-stem and diencephalon. Brain 75:109, 1952.
5. Mott, FW: Histological observations on sleeping sickness and other trypanosome infections. Rep Sleeping Sickness Commiss R Soc 7:3, 1906.
6. Pentreath, VW: Sleeping sickness. In Gilman, S, et al (eds): Neurobase, ed 4. Arbor, La Jolla, Calif, 1997.
7. Adams, JH, et al: Human African trypanosomiasis (*T. b. gambiense*): A study of 16 fatal cases of sleeping sickness with some observations on acute arsenical encephalopathy. Neuropathol Appl Neurobiol 12:81, 1986.
8. Pentreath, VW: Neurobiology of sleeping sickness. Parasitol Today 5:215, 1989.
9. Hunter, CA, and Kennedy, PGE: Immunopathology in central nervous system human African trypanosomiasis. J Neuroimmunol 36:91, 1992.
10. Pentreath, VW, et al: The somnogenic T lymphocyte suppressor prostaglandin D2 is selectively elevated in the cerebrospinal fluid of advanced sleeping sickness patients. Trans R Soc Trop Med Hyg 84:795, 1990.
11. Pentreath, VW, et al: Endotoxins in the blood and cerebrospinal fluid of patients with African sleeping sickness. Parasitology 112:67, 1996.
12. Lugaresi, E, et al: Fatal familial insomnia and dysautonomia with selective degeneration of thalamic nuclei. New Engl J Med 315:997, 1986.
13. Medori, R, et al: Fatal familial insomnia is a prion disease with a mutation at codon 178 of the prion disease. N Engl J Med 326:444, 1992.
14. Medori, R, et al: Fatal familial insomnia: A second kindred with mutation of prion protein gene at codon 178. Neurology 42:669, 1992.
15. Rancurel, G, et al: Familial thalamic degeneration with fatal insomnia: Clinicopathological and polygraphic data on a French member of Lugaresi's Italian family. In Guilleminault, C, et al (eds): Fatal Familial Insomnia: Inherited Prion Diseases, Sleep, and the Thalamus. Raven, New York, 1994, pp 15–25.
16. Gallassi, R, et al: Fatal familial insomnia: Neuropsychological study of a disease with thalamic degeneration. Cortex 28:175, 1992.
17. Portaluppi, F, et al: Diurnal blood pressure variation and hormonal correlates in fatal familial insomnia. Hypertension 23:569, 1994.
18. Manetto, V, et al: Fatal familial insomnia: Clinical and pathological study of five new cases. Neurology 42:312, 1992.

19. Tinuper, P, et al: The thalamus participates in the regulation of the sleep-waking cycle: A clinico-pathological study in fatal familial thalamic degeneration. Electroencephalogr Clin Neurophysiol 73:117, 1989.
20. Petersen, RB, et al: Analysis of the prion gene in thalamic dementia. Neurology 42:1859, 1992.
21. Stern, K: Severe dementia associated with bilateral symmetrical degeneration of the thalamus. Brain 62:157, 1939.
22. Schulman, S: Bilateral symmetrical degeneration of the thalamus. J Neuropathol Exp Neurol 16:446, 1957.
23. Garcin, R, et al: Le syndrome de Creutzfeldt-Jakob et les syndromes corticotries du presenium (à l'occasion de 5 observations anatomocliniques). Rev Neurol 109:419, 1963.
24. Kornfeld, M, et al: Pure thalamic dementia with a single focus of spongiform change in cerebral cortex. Clin Neuropathol 13:77, 1994.
25. Mizusawa, H, et al: Degeneration of the thalamus and inferior olives associated with spongiform encephalopathy of the cerebral cortex. Clin Neuropathol 7:81, 1988.
26. Lugaresi, E, and Gambetti P: Fatal familial insomnia. In Gilman, S, et al (eds): Neurobase, ed 4. Arbor, La Jolla, Calif, 1997.
27. Goldfarb, LG, et al: Fatal familial insomnia and familial Creutzfeldt Jakob disease: Disease phenotype determined by a DNA polymorphism. Science 258:806, 1992.
28. Monari, L, et al: Fatal familial insomnia and familial Creutzfeldt-Jakob disease: Different prion proteins determined by a DNA polymorphism. Proc Natl Acad Sci 91:2839, 1994.
29. Perani, D, et al: [18F]FDG PET in fatal familial insomnia: The functional effects of thalamic lesions. Neurology 43:2565, 1993.
30. Prader, A, et al: Ein Syndrom von Adipositas, Kleinwuchs, Kryptorchismus und Oligophrenie nach Myotonieartigen Zustand in Neugeborenenalter. Schweiz Med Wochenschr 86:1260, 1956.
31. Clarke, DJ, et al: Adults with Prader-Willi syndrome: Abnormalities of sleep and behaviour. J R Soc Med 82:21, 1989.
32. Greenswag, LR: Adults with Prader-Willi syndrome: A survey of 232 cases. Dev Med Child Neurol 29:145, 1987.
33. Nicholls, RD: Genomic imprinting and candidate genes in the Prader-Willi and Angelman syndromes. Curr Opin Genet Dev 3:445, 1993.
34. Vela-Bueno, A, et al: Sleep in the Prader-Willi Syndrome. Arch Neurol 41:294, 1984.
35. Sforza, E, et al: Sleep and breathing abnormalities in a case of Prader-Willi syndrome: The effects of acute continuous positive airway pressure treatment. Acta Paediatr Scand 80:80, 1991.
36. Kaplan, J, et al: Sleep and breathing in patients with the Prader-Willi syndrome. Mayo Clin Proc 66:1124, 1991.
37. Hertz, G, et al: Sleep and breathing patterns in patients with Prader Willi syndrome (PWS): Effects of age and gender. Sleep 16:366, 1993.
38. Vgontzask, AN, et al: Prader-Willi syndrome: Effects of weight loss on sleep-disordered breathing, daytime sleepiness and REM sleep disturbance. Acta Pediatr 84:813, 1995.
39. Kleine, W: Periodische Schlafsucht. Mschr Psychiat Neurol 57:285, 1925.
40. Lewis, NDC: The psychosomatic approach to the problems of children under twelve years of age. Psychosom Rev 13:424, 1926.
41. Levin, M: Narcolepsy (Gelineau's syndrome) and other varieties of morbid somnolence. Arch Neurol Psychiatry 22:1172, 1929.
42. Levin, M: Periodic somnolence and morbid hunger: A new syndrome. Brain 59:494, 1936.
43. Critchley, M, and Hoffman, HL: The syndrome of periodic somnolence and morbid hunger (Kleine-Levin syndrome). Br Med J 1:137, 1942.
44. Billiard, M: Recurrent hypersomnia/Kleine-Levin syndrome. In Gilman, S (eds): Neurobase, ed 4. Arbor, La Jolla, Calif, 1997.
45. Critchley, M: Periodic hypersomnia and megaphagia in adolescent males. Brain 85:627, 1962.
46. Hegarty, A, and Merriam, AE: Autonomic events in Kleine-Levin syndrome. Am J Psychiatry 147:951, 1990.
47. Fenzil, F, et al: Clinical features of Kleine-Levin syndrome with localized encephalitis. Neuropediatrics 24:292, 1993.
48. Carpenter, S, et al: A pathological basis for Kleine-Levin syndrome. Arch Neurol 39:25, 1982.
49. Bonkalo, A: Hypersomnia: A discussion of psychiatric implications based on three cases. Br J Psychiatry 114:69, 1968.
50. Thompson, C, et al: Neuroendocrine rhythms in a patient with the Kleine-Levin syndrome. Br J Psychiatry 147:440, 1985.
51. Gadoth, N, et al: Episodic hormone secretion during sleep in Kleine-Levin syndrome: Evidence for hypothalamic dysfunction. Brain Dev 9:309, 1987.
52. Fernandez, JM, et al: Disturbed hypothalamic-pituitary axis in idiopathic recurring hypersomnia syndrome. Acta Neurol Scand 82:361, 1990.
53. Chesson, AL, and Levine, SN: Neuroendocrine evaluation in Kleine-Levin syndrome: Evidence of reduced dopaminergic tone during periods of hypersomnolence. Sleep 14:226, 1991.
54. Koerber, RK, et al: Increased cerebrospinal fluid 5-hydroxytryptamine and 5-hydroxyindoleacetic acid in Kleine-Levin syndrome. Neurology 34:1597, 1984.
55. Jeffries, JJ, and Lefebvre, A: Depression and mania associated with Kleine-Levin-Critchley syndrome. Can Psychiatr Assoc J 18:439, 1973.
56. Billiard, M, et al: A menstruation-linked periodic hypersomnia. Neurology 25:436, 1975.
57. Billiard, M, and Cadilhac, J: Les hypersomnies recurrentes. Rev Neurol (Paris) 144:249, 1988.
58. Thacore, VR, et al: The EEG in a case of periodic hypersomnia. EEG Clin Neurophysiol 27:605, 1969.
59. Manni, R, et al: Electrophysiological and immunogenetic findings in recurrent monosymptomatic-type hypersomnia: A study of two unrelated Italian cases. Acta Neurol Scand 88:293, 1993.

60. Lotz, BP, et al: Recurrent attacks of unconsciousness with diffuse EEG alpha activity. Sleep 16:671, 1993.
61. Rothstein, JD, et al: Endogenous benzodiazepine receptor ligands in idiopathic recurring stupor. Lancet 340:1002, 1992.
62. Tinuper, P, et al: Idiopathic recurring stupor: A case with possible involvement of the GABAergic system. Ann Neurol 31:502, 1992.
63. Castaigne, P, et al: Paramedian thalamic and midbrain infarcts: Clinical and neuropathological study. Ann Neurol 10:127, 1981.
64. Bogousslavsky, J, et al: Thalamic infarcts: Clinical syndromes, etiology, and prognosis. Neurology 38:837, 1988.
65. Bassetti, C, et al: Hypersomnia following paramedian thalamic stroke: A report of 12 patients. Ann Neurol 39:471, 1996.
66. Catsman-Berrevoets, CE, and von Harskamp, F: Compulsive pre-sleep behavior and apathy due to bilateral thalamic stroke: Response to bromocriptine. Neurology 38:647, 1988.
67. Aldrich, MS, and Naylor, MW: Narcolepsy associated with lesions of the diencephalon. Neurology 39:1505, 1989.
68. Schwartz, WJ, et al: Transient cataplexy after removal of a craniopharyngioma. Neurology 34:1372, 1984.
69. Palm, L, et al: Sleep and wakefulness after treatment for craniopharyngioma in childhood: Influence on the quality and maturation of sleep. Neuropediatrics 23:39, 1992.
70. Cohen, RA, and Albers, HE: Disruption of human circadian and cognitive regulation following a discrete hypothalamic lesion: A case study. Neurology 41:726, 1991.
71. Anderson, M, and Salmon, MV: Symptomatic cataplexy. J Neurol Neurosurg Psychiatry 40:186, 1977.
72. Haugh, RM, and Markesbery, WR: Hypothalamic astrocytoma: Syndrome of hyperphagia, obesity, and disturbances of behavior and endocrine and autonomic function. Arch Neurol 40:560, 1983.
73. Stahl, SM, et al: Continuous cataplexy in a patient with a midbrain tumor: The limp man syndrome. Neurology 30:1115, 1980.
74. Rubinstein, I, et al: Neurosarcoidosis associated with hypersomnolence treated with corticosteroids and brain irradiation. Chest 94:205, 1988.
75. Mechanik, JI, et al: Hypothalamic dysfunction following whole-brain irradiation. J Neurosurg 65:490, 1986.
76. Clavelou, P, et al: Narcolepsy associated with arteriovenous malformation of the diencephalon. Sleep 18:202, 1995.
77. Adams, M, et al: Whipple's disease confined to the central nervous system. Ann Neurol 21:104, 1987.
78. Gurewitz, R, et al: Recurrent hypothermia, hypersomnolence, central sleep apnea, hypodipsia, hypernatremia, hypothyroidism, hyperprolactinemia and growth hormone deficiency in a boy: Treatment with clomipramine. Acta Endocrinol 279:468, 1986.
79. Onofrj, M, et al: Narcolepsy associated with primary temporal lobe B-cells lymphoma in a HLA DR2 negative subject. J Neurol Neurosurg Psychiatry 55:852, 1992.
80. Rivera, VM, et al: Narcolepsy following cerebral hypoxic ischemia. Ann Neurol 19:505, 1986.
81. Roloff, DW, and Aldrich, MS: Sleep disorders and airway obstruction in neonates and infants. Otolaryngol Clin North Am 23:639, 1990.
82. Adelman, S, et al: Obstructive sleep apnea in association with posterior fossa neurologic disease. Arch Neurol 41:509, 1984.
83. Bullock, R, et al: Isolated central respiratory failure due to syringomyelia and Arnold-Chiari malformation. Br Med J 297:1448, 1988.

Chapter 20

SLEEP AND EPILEPSY

OVERVIEW OF EPILEPSY
Epilepsy and the Electroencephalogram
SLEEP AND EPILEPTIC MANIFESTATIONS
Epileptiform Activity During Sleep
Effects of Sleep on Seizure Frequency
Effects of Sleep Deprivation on Seizures and Epileptiform Activity
Effects of Sleep Disorders on Seizures
Effects of Seizures and Antiepileptic Medications on Sleep
DIAGNOSIS OF SEIZURES DURING SLEEP
Daytime Electroencephalography
Video-Electroencephalographic Polysomnography
Portable Electroencephalographic Recorders and Inpatient Monitoring
The Unexpected Electroencephalographic Finding During Polysomnography
GENERALIZED EPILEPSIES ASSOCIATED WITH SEIZURES DURING DROWSINESS, SLEEP, OR UPON AWAKENING
Generalized Epilepsy with Tonic-Clonic Seizures During Sleep
Epilepsy with Grand Mal Seizures on Awakening
Juvenile Myoclonic Epilepsy
Childhood Absence Epilepsy
Lennox-Gastaut Syndrome
PARTIAL EPILEPSIES ASSOCIATED WITH SEIZURES DURING SLEEP
Benign Childhood Epilepsy with Centrotemporal Spikes (Benign Rolandic Epilepsy)
Temporal Lobe Epilepsy
Frontal Lobe Epilepsy
Nocturnal Paroxysmal Dystonia
Autosomal-Dominant Nocturnal Frontal Lobe Epilepsy
Epilepsy with Continuous Spike Waves During Slow-Wave Sleep

Sleep is like epilepsy, and, in a sense, actually is a seizure of this sort. Accordingly, the beginning of this malady takes place with many during sleep, and their subsequent habitual seizures occur in sleep, not in waking hours.

ARISTOTLE, ON SLEEP AND SLEEPLESSNESS, 350 BC (TRANSLATED BY J.I. BEARE)

OVERVIEW OF EPILEPSY

Epilepsy, present in about 1 in 200 persons, is characterized by intermittent disruption of brain function—seizures—associated with bursts of abnormal brain electrical activity. The sometimes extraordinary features of epileptic seizures have intrigued physicians, philosophers, and writers for centuries, and the occurrence in many epileptics of seizures during sleep has been recognized since antiquity. Hippocrates described seizures during the night, and one of three main types of seizures recognized in the 1st and 2nd centuries AD was associated with sleepy states.[1] In the 19th century, Gowers[2] noted that more than 20% of epileptics have seizures exclusively at night, that 30% to 40% have seizures during sleep as well as during wakefulness, and that some patients have seizures almost exclusively in the morning.

Epileptic seizures are diverse; their classification is based on the clinical characteristics of the seizure and the appearance of the ictal and interictal EEG. There are two major categories: (1) *generalized seizures*, which affect both hemispheres of the brain from the beginning of the seizure, and (2) *partial seizures*, which affect only one part of the brain at the onset of the seizure.

There are several types of generalized seizures: absence, myoclonic, tonic, atonic, and tonic-clonic. *Absence (petit mal) seizures* cause brief episodes of staring with loss of awareness but with little or no alteration of postural tone. *Myoclonic seizures* are associated with brief single or multiple body jerks with little change in consciousness. *Tonic seizures* are characterized by stiffening, whereas *atonic seizures* are associated with sudden loss of muscle tone. *Tonic-clonic (grand mal) seizures* cause stiffening and violent jerking movements.

Partial seizures are classified as *simple* if consciousness is not affected or *complex* if consciousness is impaired. A simple partial seizure may progress to become a complex partial seizure, and either type of partial seizure may progress to become a generalized seizure, which is then called a partial seizure with secondary generalization.

Partial seizures cause an enormous variety of sensory, autonomic, and motor events depending on the location of the epileptic focus, ranging from twitches, lip smacking, focal jerks, and paresthesias, to elaborate hallucinations, vocalizations, and complex motor behavior. The type of phenomena that occur with the seizure depend on the location of the epileptic focus and its propagation pathway; for example, seizures beginning in the visual cortex of the occipital lobe may cause hallucinations of flashing lights or formed images, partial seizures beginning in the primary motor area cause jerking movements, and those beginning in the supplementary motor area may cause coordinated movements and dystonic postures.

Patients with epilepsy may have one or more types of seizures. About 40% to 50% of epileptics have generalized seizures as their primary seizure type, 30% to 40% have partial seizures primarily, and the remainder have partial seizures and generalized seizures.[3] In the Classification of Epilepsies and Epileptic Syndromes developed by the International League Against Epilepsy, the classification of epileptic syndromes is based on several features: (1) the types of seizures—generalized, partial, or both; (2) whether the disorder is idiopathic (primary) or symptomatic (secondary to an identifiable central nervous system [CNS] lesion); and (3) the age of onset of seizures.[4] Patients with epilepsy also may be grouped according to the portion of the brain affected. For example, patients with partial seizures that originate in the temporal lobe are said to have temporal lobe epilepsy. Major categories of epilepsies are shown in Table 20–1.

Epilepsy and the Electroencephalogram

The appearance on the electroencephalogram (EEG) of abnormal *epileptiform activity*, both during and between seizures, is

Table 20–1. **Classification of Sleep-Related Epilepsies within the International Classification of Epilepsies and Epileptic Syndromes**[4]

1. Localization-related (focal, partial)
   1.1 Idiopathic (primary)
       Benign childhood epilepsy with centrotemporal spikes
   1.2 Symptomatic (secondary)
       Temporal lobe epilepsies
       Frontal lobe epilepsies
       Occipital lobe epilepsies
   1.3 Cryptogenic
2. Generalized
   2.1 Idiopathic (primary)
       Benign neonatal familial convulsions
       Childhood absence epilepsy
       Juvenile myoclonic epilepsy (impulsive petit mal)
       Epilepsy with grand mal seizures on awakening
   2.2 Cryptogenic or symptomatic
       West's syndrome (infantile spasms)
       Lennox-Gastaut syndrome
3. Undetermined epilepsies
   3.1 With both generalized and focal seizures
       Epilepsy with continuous spike-waves during slow-wave sleep
4. Special syndromes
   Febrile convulsions
   Seizures caused by drugs or alcohol

one of the most striking and diagnostically useful features of epilepsy. Epileptiform activity, defined as "distinctive waves or complexes, distinguished from background [EEG] activity, and resembling those recorded in a proportion of human subjects suffering from epileptic disorders . . . ,"[5] include spikes, sharp waves, and spike-and-slow-wave discharges. Spikes and sharp waves are sharply contoured EEG transients that are clearly distinguished from background EEG activity; by convention, a spike has a duration of 20 to 70 msec and a sharp wave has a duration of 70 to 200 msec. The voltage is usually negative at the scalp, and the amplitude may vary from a few microvolts to hundreds of microvolts. Spikes and sharp waves in persons with epilepsy are often followed by a surface-negative slow wave; this combination is often referred to as a *spike-and-slow-wave complex*; other terms for this are sharp-and-slow-wave complex and spike-wave complex. A rapid series of spikes is referred to as a *polyspike*; if followed by a slow wave, it is called a *polyspike* and *wave*.

The overall appearance of the EEG also can be useful in the assessment of epileptic syndromes. In general, normal background EEG activity in between epileptiform discharges is suggestive of idiopathic epilepsy, whereas abnormal slowing of background activity is more typical of a symptomatic epilepsy.

The cellular basis for EEG spikes associated with focal epilepsies appears to be large excitatory postsynaptic potentials that lead to *paroxysmal depolarizing shifts* of the membrane potential.[6,7] The depolarization, which causes the spike, is followed by a period of hyperpolarization that causes the slow wave. The basis for the spike-and-slow-wave discharges that occur with generalized epilepsies is less certain, although they appear to be generated by interactions of thalamic and cortical neurons.

SLEEP AND EPILEPTIC MANIFESTATIONS

Sleep influences the expression of epileptic disorders in a number of ways:
1. It affects the amount and appearance of interictal epileptiform activity.[8]
2. Seizures in many epileptics are more frequent during sleep than during wakefulness.
3. In some epileptic disorders, sleep deprivation influences the likelihood of seizures and of recording epileptiform activity.
4. Disorders that cause sleep disturbance may affect seizure frequency.
5. Epilepsy can affect sleep continuity.
6. Antiepileptic medications can affect sleep and wakefulness.

Epileptiform Activity During Sleep

The changes in firing patterns of brainstem and diencephalic projection neurons that occur with sleep have pronounced effects on EEG activity and neuronal excitability (see Chaps. 1 and 2) that lead to changes in the frequency and appearance of epileptiform activity during sleep. In particular, epileptiform activity in many patients with epilepsy is more frequent in non–rapid eye movement (NREM) sleep than in wakefulness or rapid eye movement (REM) sleep.[9] In patients with generalized seizures, epileptiform discharges are sometimes facilitated by K-complexes. In addition, NREM sleep often facilitates the appearance of focal spikes in patients with partial seizures; for example, in benign childhood epilepsy with centrotemporal spikes (see Partial Epilepsies Associated with Seizures during Sleep below), spike frequency may be near zero during wakefulness but increase to 20 to 60 spikes/min during stage 1–2 sleep. Furthermore, spikes occur exclusively during sleep in up to one-third of patients with temporal lobe epilepsy.[10,11] During REM sleep, on the other hand, focal interictal discharges usually become less frequent and spatially more restricted, although amygdaloid and frontal lobe foci are activated during REM sleep in some patients.

In addition to changes in the amount of epileptiform activity, the morphology of interictal EEG activity may change during sleep because of altered excitability of thalamic and cortical neurons. In idiopathic generalized epilepsy with absence seizures, the three per second spike-wave discharges that are classically recorded during wakefulness become spatially fragmented and more irregular during NREM sleep, with polyspike-and-wave complexes of 2 to 5 Hz.[12] Similarly, in other patients with generalized seizures, focal or lateralized discharges, sometimes referred to as *fragmentary* or *larval* discharges, may be seen instead of generalized spike-and-slow-wave discharges. Thus, focal discharges seen only during sleep, with generalized discharges during wakefulness, do not necessarily indicate a focal source.

Effects of Sleep on Seizure Frequency

As with epileptiform discharges, seizures generally are more frequent in NREM sleep than in REM sleep. In many patients with sleep-related seizures, the seizures occur exclusively during NREM sleep. Neuronal synchronization, which occurs as a result of altered thalamic excitability and is reflected in the synchronous EEG activity of NREM sleep, appears to facilitate the spread of abnormal epileptic discharges through the brain and thereby increases the likelihood that seizures will occur. Thus, in temporal lobe epilepsy, partial seizures are more likely to become secondarily generalized during sleep than during wakefulness.[13] On the other hand, the desynchronized activity of REM sleep usually does not facilitate such spread.

Effects of Sleep Deprivation on Seizures and Epileptiform Activity

Sleep deprivation increases the likelihood of seizures in some persons, particularly in those with specific types of generalized epilepsy (see Generalized Seizures Associated with Seizures during Drowsiness, Sleep, or upon Awakening below), although the basis for these effects is unknown. In addition, in some patients with epilepsy, sleep deprivation increases the likelihood that epileptiform abnormalities will be observed, and consequently sleep deprivation is often used as an *activating procedure* for EEG recordings. In one study, EEG abnormalities, mainly during wakefulness, occurred in 41% of epileptic subjects after sleep deprivation; only 18% had abnormalities on a subsequent EEG performed after a normal night of sleep.[14] Consequently, the practice of obtaining EEGs after sleep deprivation may be helpful in: (1) patients with suspected epilepsy and normal EEGs, to increase the chance of detecting an interictal abnormality; (2) patients with known epilepsy and normal EEGs, to try to detect an interictal abnormality that may help determine the type of epilepsy; and (3) patients with known epilepsy and abnormal EEGs, to help to characterize the type of epilepsy.

Although sleep deprivation the night before a daytime EEG increases the likelihood that epileptiform abnormalities will be observed, this relationship does not necessarily mean that sleep deprivation affects neuronal excitability. The activating effects of drowsy wakefulness induced by sleep deprivation contribute to the increased likelihood of observing epileptiform abnormalities during wakefulness in some patients with epilepsy.[15] Furthermore, sleep-deprived patients are more likely to fall asleep during the EEG, which activates the EEG in many epileptics. Finally, EEG abnormalities in patients with epilepsy are transient events that may or may not be recorded on a given tracing, and a positive finding after sleep deprivation may be a result of day-to-day variability in the expression of EEG abnormalities. Although some investigators suggest that epileptiform activity is more likely with sleep that follows sleep deprivation than with sleep in well-rested persons, others found similar rates of epileptiform abnormalities in sleep-wake EEGs performed with and without prior sleep deprivation.[16,17]

In summary, although sleep deprivation increases the likelihood of observing EEG abnormalities in patients with epilepsy, sampling effects, amount of recorded sleep, and level of alertness during wakefulness probably account for most of the increased yield. Therefore, the evidence that sleep deprivation increases the frequency of epileptiform discharges, independent of level of alertness and sleep stage, is inconclusive. Daly[18] noted that "published reports [of sleep deprivation as an activating procedure] are a quagmire into which hasty and careless readers can quickly sink." Sleep deprivation does not cause epileptiform abnormalities in healthy persons.

Effects of Sleep Disorders on Seizures

Given the impact of sleep and sleep deprivation on seizures and epileptiform EEG activity, it is not surprising that sleep disruption associated with sleep disorders can affect seizure frequency in some epileptic patients. For example, in patients with epilepsy and obstructive sleep apnea (OSA), seizure control may improve after treatment of OSA, suggesting that sleep disruption, hypoxemia, or autonomic changes associated with sleep apnea lead to increased seizure frequency in some epileptics.[19–22] Sleep disturbance secondary to periodic limb movements (PLMs) and medical illnesses may also increase seizure frequency.

Effects of Seizures and Antiepileptic Medications on Sleep

Abnormal brain activity associated with epilepsy sometimes contributes to sleep disruption. In animal studies, seizures can cause prolonged reductions in the amount of REM sleep, and the sleep disruption associated with frequent nocturnal seizures may lead to daytime sleepiness in some patients. Furthermore, bursts of generalized spike-wave complexes sometimes may be associated with arousals and body movements that lead to sleep fragmentation and daytime sleepiness.[23,24]

Antiepileptic medications can affect sleep and wakefulness in at least three ways:

1. Most antiepileptic medications have sedative effects that promote drowsiness; somnolence, the most common side effect of antiepileptic drugs, may occur in the absence of other signs of drug toxicity, such as ataxia and nystagmus.[25] Although daytime sleepiness is particularly common with barbiturates and benzodiazepines, it also may occur with other antiepileptic medications.[26,27] Besides having sedative effects, antiepileptic medications may also increase the frequency of seizures in certain epileptic patients. For example, in patients with seizures facilitated by wakeful drowsiness, the sedative effects of these drugs may increase seizure frequency; conversely, dose reduction in these patients may lead to greater alertness and, paradoxically, to reduced seizure frequency.[28]
2. Some antiepileptic medications, such as sodium valproate, promote weight gain, which may precipitate or aggravate OSA, which in turn can increase the frequency of seizure episodes (see Effects of Sleep Disorders on Seizures above).
3. Antiepileptic medications promote better sleep in some epileptic patients because of their sedative effects or because they lead to less epileptiform activity and fewer seizures during sleep. Thus, in patients with frequent nocturnal seizures and in those with spike waves that induce arousals, these drugs may reduce paroxysmal events during sleep and improve daytime symptoms.

DIAGNOSIS OF SEIZURES DURING SLEEP

Because of the varied clinical manifestations of seizures and the many possible causes of episodic behaviors during sleep (Table 20–2), the approach to diagnosis of a suspected seizure during sleep depends on the clinical information and the type of epilepsy suspected to be responsible. For patients with a history of sudden episodes during sleep of stiffening and jerking accompanied by tongue biting and urinary incontinence, the clinical history is usually sufficient to make a diagnosis. Home videotapes of generalized seizures supplied by the patient or family also may be diagnostic. Even in such apparently

Table 20–2. **Differential Diagnosis of Episodic Nocturnal Behaviors**

Sleep terrors
Somnambulism
Confusional arousals
Bruxism
Panic attacks
Rhythmic movement disorder
REM sleep behavior disorder
Periodic limb movement disorder
Nocturnal dissociative episodes
Sleep enuresis
Epileptic seizures

straightforward cases of generalized tonic-clonic seizures, however, the possibility that the events are partial seizures with secondary generalization must be considered, as the distinction may be important for evaluation, prognosis, and treatment. In such cases, an EEG recording often helps determine the type of seizure disorder and the approach to management.

In patients who may have partial seizures during sleep without secondary generalization, diagnosis is more difficult, and EEG recordings are often required to help determine whether the episodes in question are epileptic in origin. The type of study obtained depends on the clinical setting and may include one or more of the following: daytime EEG, video-EEG polysomnography (VPSG), ambulatory EEG monitoring, or long-term inpatient or outpatient VPSG monitoring.

Daytime Electroencephalography

If epilepsy is suspected, a daytime EEG may be all that is required to confirm the presence of epileptiform abnormalities. Focal spikes over the anterior portions of the temporal lobes are strongly suggestive of epilepsy; they rarely occur in persons without seizures. Spikes in other regions are also associated with epilepsy, although less strongly. For example, approximately 90% of children with spikes over the anterior temporal regions have seizures, whereas only about 40% of children with central (rolandic) spikes have seizures.[29] Generalized spike-and-slow-wave discharges are also strongly suggestive of epilepsy, although they can occur in asymptomatic relatives of persons with some forms of familial epilepsy.

Interpretation of EEG recordings is complicated by the existence of a number of benign patterns that resemble the abnormal interictal activity characteristic of epilepsy, including benign epileptiform transients of sleep (BETS; also called small sharp spikes); 14- and 6-Hz positive bursts; 6-Hz spike and wave; and wicket spikes.[30,31] The electroencephalographer must be able to distinguish abnormal sharp waves and spikes from these benign epileptiform variants; from normal patterns such as mu rhythms, vertex sharp waves, and posterior occipital sharp transients (POSTs); and from artifacts such as lateral rectus electromyographic spikes and electrode "pops."

Because the occurrence of abnormal EEG activity is increased during sleep in many types of epilepsy, most EEG laboratories attempt to obtain recordings during sleep as well as during wakefulness. Sleep EEG recordings are especially useful in patients with seizures occurring exclusively during sleep, because EEGs during wakefulness are often normal. For example, in one series of EEGs that contained epileptiform activity, the abnormality was not apparent during wakefulness in 40% of the recordings.[32] Recording of stage 1 and stage 2 sleep is sufficient on routine EEG; recordings of stage 3–4 sleep rarely yield additional useful information.[33] Thus, all-night recordings are rarely required for the purpose of detecting interictal epileptiform activity. On the other hand, a substantial proportion of epileptics do not have abnormal epileptiform activity on an initial EEG, even if they fall asleep during the recording. As discussed above, a repeat EEG after sleep deprivation may increase the yield if the initial EEG is normal or shows only nonspecific abnormalities.[34]

In a patient with a history suggestive of seizures during sleep, an abnormal EEG with epileptiform patterns often allows the clinician to make a diagnosis of epilepsy with reasonable certainty. In many cases, however, the history is ambiguous or the EEG does not show epileptiform patterns, or both. In such cases, additional diagnostic studies are required.

Video-Electroencephalographic Polysomnography

VPSG, which combines standard polysomnographic recording with audiovisual recording and with more extensive EEG montages than are used in most polysomnographic studies, is more likely to detect epileptiform abnormalities compared to standard polysomnography. It is useful in patients who have normal or nonspecifically abnormal EEGs, despite frequent nocturnal spells that are suggestive of epilepsy, and in patients with epilepsy who are having frequent nocturnal behavioral episodes that differ clinically from their daytime seizures. In such cases, the simultaneous all-night recording and analysis of behavior, sleep stage, and EEG permits characterization and diagnosis of many nocturnal spells (see Chap. 15). In one study, VPSG was diagnostically useful in a majority of patients with frequent nocturnal episodes associated with simple or complex behaviors.[35] This type of recording is most useful in patients with episodes that occur at least several times per week;

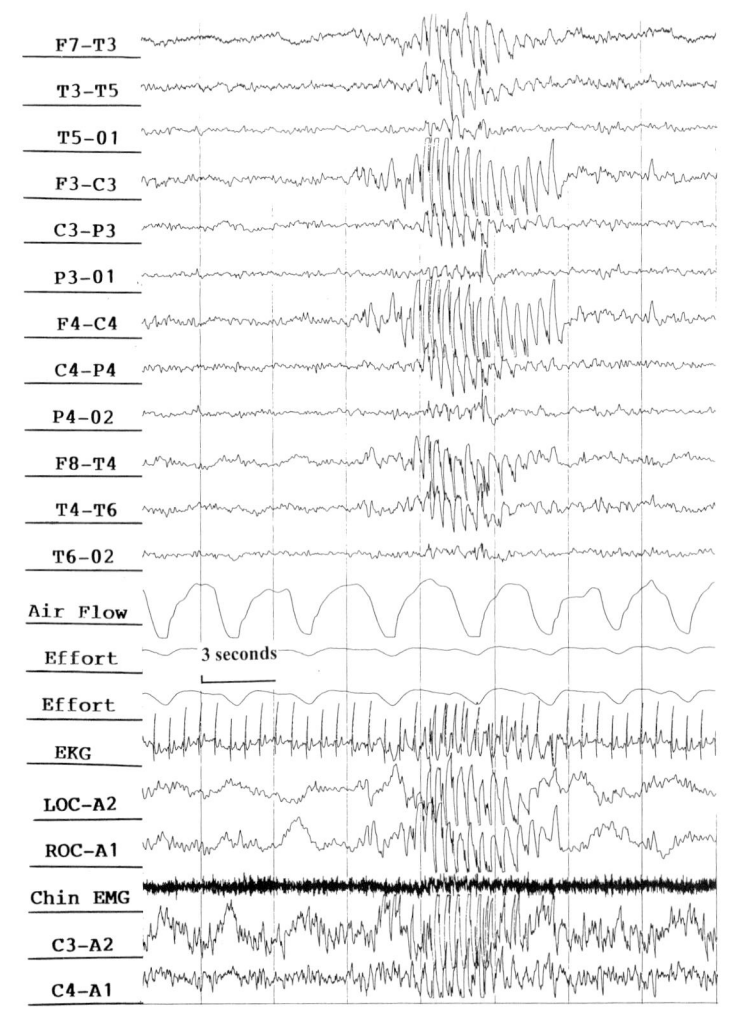

Figure 20–1. This 24-second polysomnographic sample, obtained from a 7-year-old boy with a history of generalized and partial seizures, illustrates a burst of high-amplitude 2- to 2.5-Hz spike waves during stage 2 sleep. Although the spike waves are readily identified in the lower EEG and EOG channels (channels 17, 18, 20, and 21), the frontal maximum, helpful for diagnosing the type of epilepsy, is apparent only with the addition of an extended EEG montage (channels 1 to 12). In all figures in this chapter, the EEG electrodes (A1, A2, F3, F4, F7, F8, Fp1, Fp2, C3, C4, T3, T4, T5, T6, P3, P4, O1, O2) were placed in accordance with the International 10-20 system. Other abbreviations: LOC and ROC, left and right outer canthus, respectively; Chin EMG, submental EMG; EKG, electrocardiogram; Effort, thoracoabdominal wall motion.

the likelihood of recording a spell is reduced in patients whose episodes are less frequent.

Portable Electroencephalographic Recorders and Inpatient Monitoring

Portable EEG recording devices, also called ambulatory EEG monitors, typically record 8 to 16 EEG channels.[36,37] Although they can be used to assist with the diagnosis of certain types of daytime spells, their usefulness for diagnosis of nocturnal spells is limited because the absence of visual analysis makes it difficult to characterize behaviors. In addition, artifacts are common because no technologist is present to detect recording problems. Systems that combine video recording with digital EEG monitoring, which has several advantages over analog monitoring, are becoming more widely available for home use and will probably be used increasingly in the future. Long-term continuous monitoring in specialized epilepsy inpatient units is required for definitive diagnosis in some patients.[36,38]

The Unexpected Electroencephalographic Finding During Polysomnography

Unexpected EEG transients are sometimes encountered during a polysomnographic recording performed for reasons

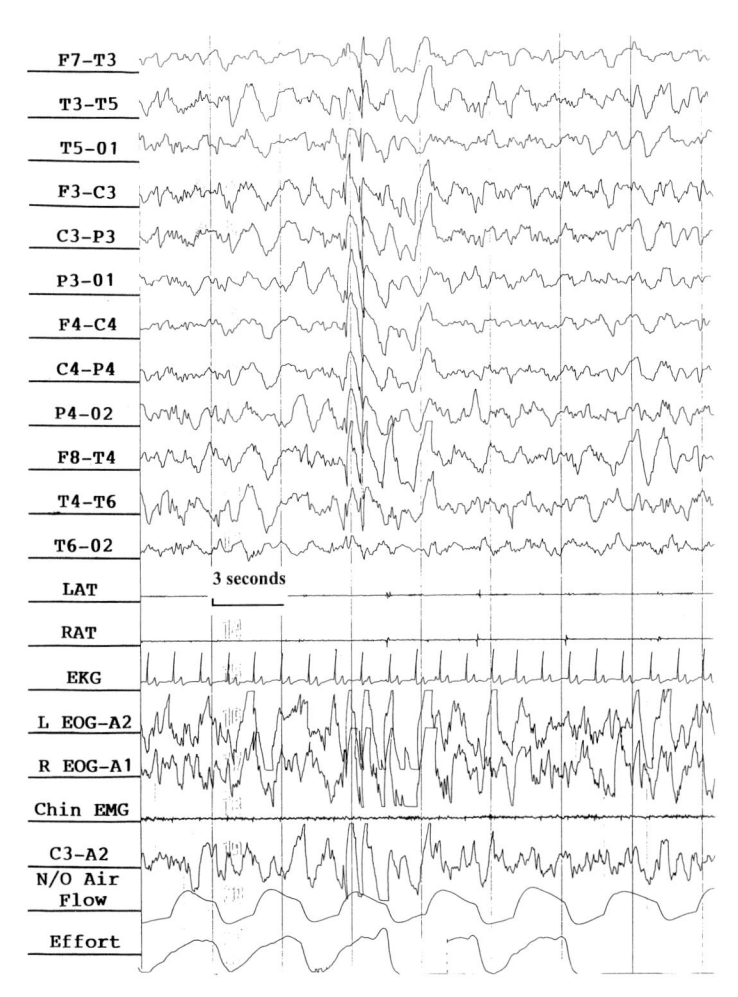

Figure 20–2. This 25-second polysomnographic sample, obtained from a 49-year-old woman with epilepsy, illustrates the difficulty of identifying epileptiform activity in the absence of an extended EEG montage. Although the spike waves during the 9th and 10th seconds of the sample produced deflections in the C3–A2 EEG channel (channel 19), as well as in the EOG channels, they probably would not have been identified as abnormal epileptiform activity if the recording had been restricted to the lower nine channels. The morphology and distribution of the burst of generalized spike waves, however, is readily apparent in the top 12 channels of EEG recording. (L EOG, left electro-oculogram; R EOG, right electro-oculogram; LAT and RAT, left and right anterior tibialis surface EMG, respectively; N/O, nasal/oral.) For other abbreviations, see Figure 20–1.

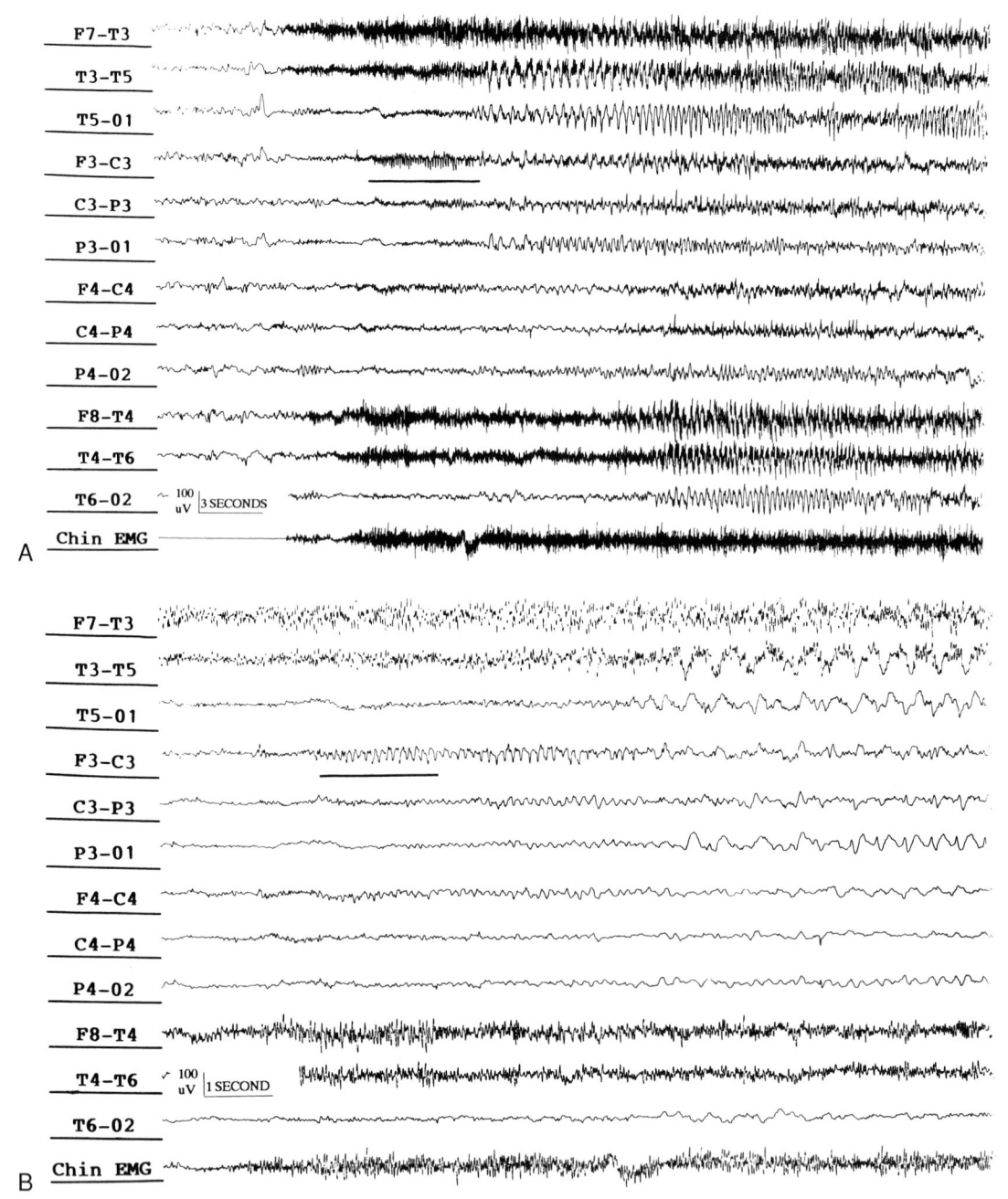

Figure 20–3. Portion of a polysomnogram at 10 mm/sec (*A*) and at 30 mm/sec (*B*) illustrating the onset of a partial seizure. Although rhythmic EEG activity is evident from most EEG derivations during the latter portion of the sample, the underlined activity appears to be muscle artifact at 10 mm/sec. At 30 mm/sec, however, it is apparent that the same underlined activity is the initial scalp manifestation of the ictal discharge. See Figure 20–1 for abbreviations. (From Aldrich and Jahnke,[35] p. 1063, with permission.)

unrelated to epilepsy. In such a situation, the clinician must determine whether such transients are abnormal and indicative or suggestive of epilepsy. First, however, the limitations of standard polysomnography in assessing EEG transients must be understood. For example, the spatial distribution of EEG abnormalities, which helps determine the type of epileptic syndrome, is difficult or impossible to characterize on

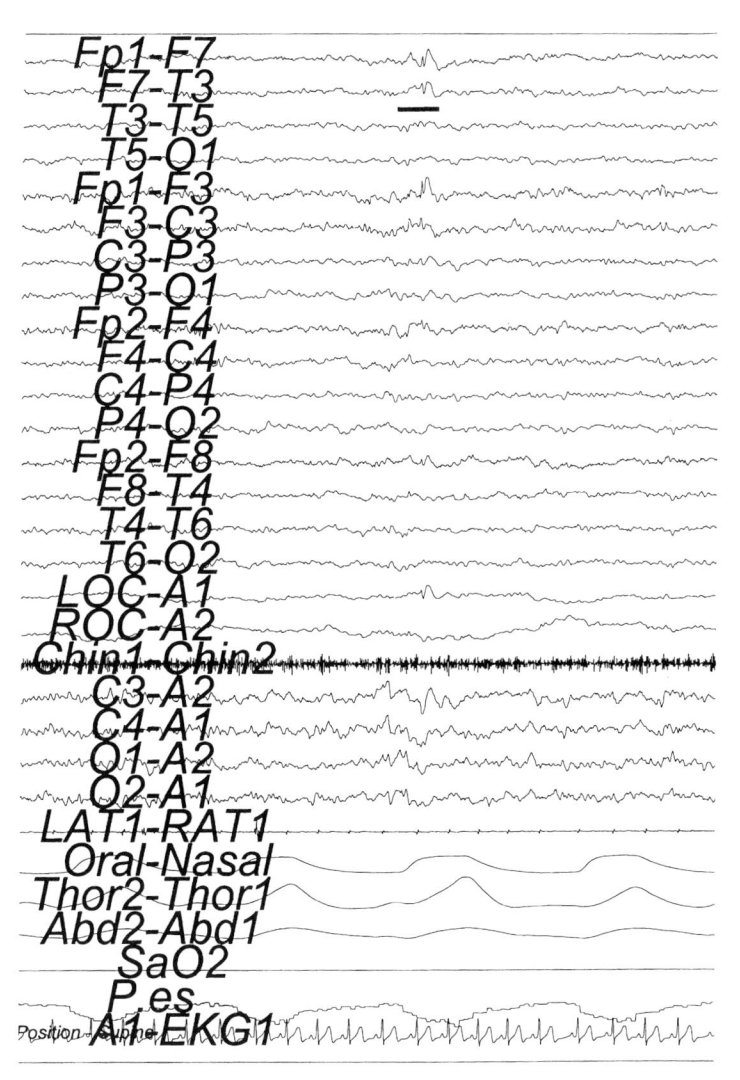

Figure 20–4. Recording from an 8-year-old boy with spells of screaming at night accompanied by bladder incontinence. A spike wave over the left frontal region (underlined) is apparent in the left EOG channel (LOC-A1) and in the extended EEG montage (channels 1 to 16), but it is undetectable in the EEG channels commonly used in standard polysomnography (C3–A2, C4–A1, O1–A2, O2–A1). Other abbreviations: Chin1–Chin2, submental EMG; LAT1–RAT1, left and right anterior tibialis surface EMG, respectively; Oral-Nasal, oral-nasal airflow; Thor2–Thor1, thoracic respiratory effort; Abd2–Abd1, abdominal respiratory effort; P.es, endoesophageal pressure; SaO$_2$, oxyhemoglobin saturation. For remaining abbreviations, see Figure 20–1.

the basis of a montage that includes only central and occipital leads (Fig. 20–1). In addition, the 10 mm/s paper speed of conventional polysomnography, combined with a limited montage, does not permit easy distinction of epileptiform abnormalities from artifacts and normal EEG features (Figs. 20–2 and 20–3). Although generalized spike waves at a rate of three per second may be identifiable, spikes over the central regions are difficult to identify, and spikes originating in the frontal or temporal lobes may be missed entirely (Fig. 20–4). If the polysomnographer does not have extensive training in EEG interpretation, an electroencephalographer should review the finding; if indicated, a daytime sleep-wake EEG or neurological consultation, or both, should be obtained.

GENERALIZED EPILEPSIES ASSOCIATED WITH SEIZURES DURING DROWSINESS, SLEEP, OR UPON AWAKENING

Generalized epilepsies include idiopathic syndromes (primary generalized epilepsies) and syndromes secondary to CNS lesions (symptomatic generalized epilepsies). Seizures associated with several of the generalized epilepsies are affected by sleep-wake state.

Generalized Epilepsy with Tonic-Clonic Seizures During Sleep

Of patients with generalized epilepsy who have tonic-clonic seizures, a substantial proportion have seizures exclusively during sleep, almost always during NREM sleep (Fig. 20–5).[39] The seizures usually begin before age 20; EEG abnormalities are present during NREM sleep in about 40%, but fewer than 10% show abnormalities during REM sleep or during wakefulness.[39]

Diagnosis is usually straightforward, based on the bed partner's description of stiffening and generalized jerking movements, which are often accompanied by enuresis and tongue-biting. A period of unresponsiveness usually follows the seizure, and many patients describe diffuse muscle aching the next morning. Seizures are easily controlled in most patients with valproate (20 to 60 mg/kg body weight), phenytoin (5 to 7.5 mg/kg), or carbamazepine (10 to 20 mg/kg). In most of these patients, seizures do not subsequently develop during wakefulness.

Epilepsy with Grand Mal Seizures on Awakening

With *epilepsy associated with grand mal seizures on awakening*, which usually begins in the second decade and accounts for about 2% to 4% of adult epilepsies, seizures occur within several minutes of awakening from nighttime sleep or from naps. Some patients have seizures only after awakenings; others have additional seizures during relaxed wakefulness or drowsiness in the evening. A few patients who initially have generalized tonic-clonic seizures only on awakening later also have seizures during sleep or during wakefulness and sleep. Most patients have no warning preceding the seizure, although some have a few myoclonic jerks just before loss of consciousness.

This syndrome may be caused by the same gene that is linked to juvenile myoclonic epilepsy (see below), as about 40% to 50% of these patients also have myoclonic or absence seizures, and about 10% to 15% have family members with juvenile myoclonic epilepsy or juvenile absence epilepsy. Although the basis for the associ-

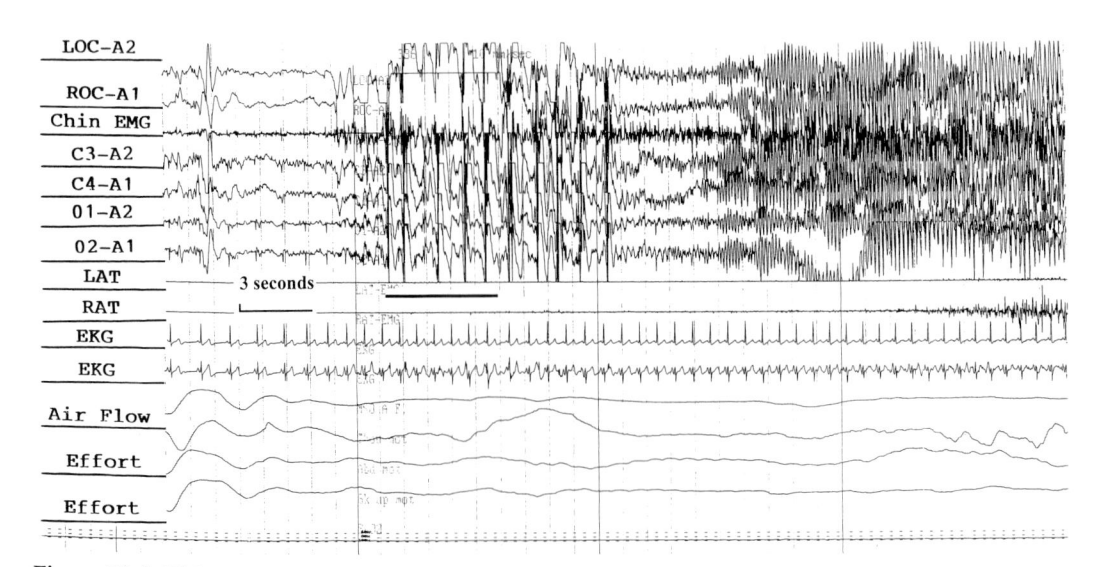

Figure 20–5. This 38-second sample recorded in a 34-year-old man with epilepsy illustrates the initial portion of a generalized seizure that occurred during stage 2 sleep. A series of high-amplitude, multiple spike and wave bursts (initial portion underlined) is followed by diffuse rhythmic activity that evolves from medium-amplitude, 8-Hz activity to high-amplitude, 5-Hz activity. See Figure 20–1 for abbreviations.

ation with awakening is unknown, it is the awakening itself, rather than the circadian phase, that predisposes the patient to seizures.[40] Sleep deprivation and alcohol use also increase the likelihood that a patient with this syndrome will have a seizure.

The morning occurrence is the key to diagnosis, as generalized tonic-clonic seizures do not occur preferentially at this time in most other forms of epilepsy. Neurological examinations and neuroimaging studies are normal. The EEG patterns are variable: irregular 3- to 4-Hz spike-wave discharges may occur during drowsiness, during stage 1–2 sleep, in association with K-complexes and arousals, and occasionally during wakefulness.

Although treatment with valproate, carbamazepine, phenytoin, or phenobarbital (2 to 3 mg/kg) results in complete seizure control in the majority of cases, the relapse rate is high after medication withdrawal, and most patients require lifelong therapy.

Juvenile Myoclonic Epilepsy

About 10% to 15% of epileptic patients have juvenile myoclonic epilepsy, also called *impulsive petit mal* or *juvenile myoclonic epilepsy of Janz*. The disorder, which usually begins in the second or third decade,[40–44] consists of myoclonic seizures characterized by one or more sudden, brief, bilaterally symmetric jerks of the upper extremities and, less commonly, the lower extremities or the entire body.[45,46] In most cases, seizures occur mainly or exclusively within a few minutes of awakening in the morning or after a nap, but they may also occur during nocturnal awakenings.

In some patients, repetitive myoclonic jerks may be followed by a generalized tonic-clonic seizure; in others, isolated generalized tonic-clonic seizures occur. About 15% to 30% of patients also have absence seizures.[35] Sleep deprivation may precipitate seizures, and some patients report that seizures are more frequent when they are drowsy or relaxed. Common precipitants in addition to sleep deprivation include alcohol, stress, and flashing lights.[46,47]

The disorder is familial: A gene on the short arm of chromosome 6 appears to contribute to susceptibility.[42,43] The concordance rate for monozygotic twins is 70% or more, and about 50% of first- and second-degree relatives have epilepsy with myoclonic, absence, or generalized tonic-clonic seizures.[42] In addition, asymptomatic siblings may have epileptiform abnormalities on EEG.

Diagnosis is based on the morning occurrence of bilateral myoclonic jerks without loss of consciousness, associated with a normal neurological examination. Mild myoclonic jerks may be misdiagnosed as hypnic jerks or PLMs, and family members may mistake myoclonic jerks for clumsiness.

An abnormal daytime sleep-wake EEG is usually sufficient for diagnosis, although a recording after sleep deprivation may be required if epileptiform abnormalities are not present on an initial EEG. Paroxysmal, generalized 4- to 6-Hz polyspike-and-wave discharges are typical interictal EEG features that are most prominent upon awakening and at sleep onset; they are infrequent during REM sleep and during the deeper stages of NREM sleep. About 15% to 20% have 3-Hz spike-and-wake or polyspike-and-wave complexes typical of absence epilepsy. During most seizures, the EEG shows 10- to 16-Hz spikes followed by irregular slow waves, but in some patients, absence seizures may be associated with typical 3-Hz spike waves (see below).

Although lifelong therapy is generally required, most patients remain seizure-free or nearly so with valproate, the preferred medication. Patients should be advised to avoid sleep deprivation and alcohol consumption, since they often precipitate seizures.

Childhood Absence Epilepsy

In childhood absence epilepsy, absence (petit mal) seizures cause staring and loss of awareness associated with prominent 3 Hz spike-and-slow-wave patterns on the EEG. In many patients, seizures are more frequent during drowsiness and less frequent during periods of high vigilance. Ethosuximide (20 to 60 mg/kg) or valproate are effective treatments.

Interictal 3 Hz spike-wave discharges are present during all sleep stages, maximally during stage 2 to stage 4 sleep and minimally during stage 1 and REM sleep. During NREM sleep, however, they lose the regular appearance that is characteristic of wakefulness and that is preserved during REM sleep.[12] Although seizures are probably infrequent during sleep in most patients, seizures that do occur usually cannot be detected clinically because motor manifestations are absent or limited to brief clonic movements of the eyelids. In some cases, the unexpected discovery during polysomnography of 3 Hz spike-and-slow-waves brings the disorder to medical attention.

Lennox-Gastaut Syndrome

The Lennox-Gastaut syndrome begins in early childhood, often between ages 3 and 6 years, and is a common cause of intractable seizures in mentally retarded children.[48,49] About two-thirds of cases are caused by brain lesions acquired during the perinatal period or infancy; the remainder are idiopathic.[50] About one-third of cases begin as West's syndrome in infancy and evolve to Lennox-Gastaut syndrome in childhood. In other cases, sudden falls caused by atonic seizures are the initial manifestation, followed by personality and behavioral disturbances, multiple seizure types, and intellectual deterioration.

A 1- to 2.5-Hz ("slow") spike-and-wave pattern, sometimes accompanied by polyspike-and-wave discharges, is a characteristic although not specific EEG feature. During NREM sleep, the discharges increase in frequency and bursts of generalized paroxysmal fast activity may develop.[51] Seizures occur during wakefulness and sleep; during NREM sleep, brief tonic seizures are common, with flexion of the head and trunk, stiffening of the neck and arms, apnea, and sudden flattening of the EEG or bursts of rapid spikes. All-night polygraphic recordings often reveal a much higher than suspected frequency of seizures.

The prognosis is poor, and treatment is unsatisfactory; many patients continue to have frequent seizures despite trials of a variety of anticonvulsants.

PARTIAL EPILEPSIES ASSOCIATED WITH SEIZURES DURING SLEEP

Partial epilepsies are characterized by the occurrence of partial seizures with or without secondary generalization; some patients also have generalized seizures. Many patients have at least some of their seizures during sleep; the likelihood that seizures will occur during sleep is dependent on the location of the focus and on the nature of the epileptic syndrome. For example, in patients with benign childhood epilepsy with centrotemporal spikes (benign rolandic epilepsy), most seizures are nocturnal, and in an inherited form of frontal lobe epilepsy, seizures are exclusively nocturnal (see Frontal Lobe Epilepsy below).

Although most nocturnal partial seizures occur during NREM sleep, some occur during REM sleep, and occasional patients have partial seizures almost exclusively during REM sleep. In one series of 35 nocturnal partial seizures in 32 patients; 21 occurred during NREM sleep, 12 during periods of wakefulness, and 2 during REM sleep.[52] Interictal abnormalities are usually more prominent during NREM sleep than during REM sleep.

Benign Childhood Epilepsy with Centrotemporal Spikes (Benign Rolandic Epilepsy)

With the genetically-based syndrome of benign childhood epilepsy with centrotemporal spikes (benign rolandic epilepsy), one of the most common of the childhood epilepsies, seizures usually begin between 5 and 10 years of age. Although some patients have seizures soon after awakening, most have them only during sleep.[53–55] Intellect and neurological examinations are usually normal.

Most seizures are partial, with motor, sensory, or autonomic manifestations including twitching of the tongue, jaw, face, or throat; unilateral numbness or paresthesia of the tongue, gums, or cheek; hypersalivation; speech arrest; or guttural sounds. Occasionally patients have generalized seizures, usually during sleep, and the initial manifestation of the disorder may be a nocturnal generalized seizure. Seizures without motor manifestations that occur only during sleep may be impossible to diagnose without polygraphic monitoring, and many cases of sleep-related seizures that are exclusively partial probably go undiagnosed. Partial status epilepticus with drooling, facial twitching, and dysarthria is a rare complication.[56]

Diagnosis is based on the age of onset, the absence of neurological or intellectual deficit, the occurrence of partial seizures during sleep, and the presence of a centrotemporal spike focus on an otherwise normal EEG (Fig. 20–6).[55] The EEG spike discharge rate increases in drowsiness and in all stages of NREM sleep, and about one-third of patients have spikes only during sleep. In a child with a normal neurological examination who presents with suspected nocturnal seizures, a diagnosis can be made on the basis of an EEG that shows a normal background and broad, diphasic, high-voltage spikes in the centrotemporal region that occur independently on both sides of the head, in isolation or in clusters, and increase markedly during stage 1–2 sleep.[51] On the other hand, the characteristic EEG abnormality is inherited as an autosomal-dominant trait, and less than 25% of children with the trait have seizures;[57] thus, this finding, if observed during polysomnography performed for other reasons, is usually of little clinical significance. The factors influ-

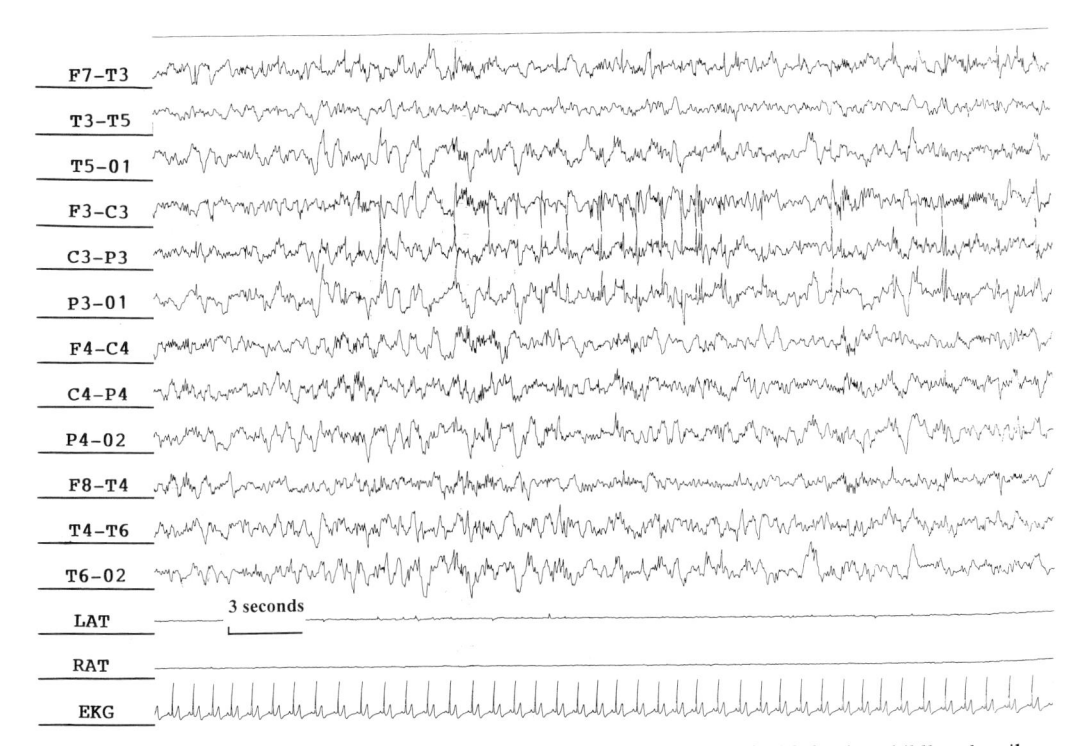

Figure 20–6. Portion of a polysomnogram obtained from an 8-year-old girl with benign childhood epilepsy with centrotemporal spikes (benign rolandic epilepsy). During slow-wave sleep, numerous sharp waves and spikes are apparent over the left central-parietal region (F3–C3, C3–P3, and P3–O1 derivations). See Figures 20–1 and 20–2 for abbreviations.

encing the development of seizures in children with the trait are unknown.

The prognosis is excellent: Seizures almost always stop spontaneously in adolescence, even in children with frequent seizures or frequent or atypical spike discharges.[54] In a series of 168 patients, only three had seizures in adulthood.[58] Seizures, if bothersome or generalized, are usually easy to control with carbamazepine or valproate, which can then be discontinued in adolescence.

Temporal Lobe Epilepsy

Most partial seizures originate in the temporal lobes, apparently because portions of the temporal lobe contain neurons and networks of neurons that more easily generate and sustain the abnormal electrical activity characteristic of epilepsy than other brain regions. Seizures of temporal lobe origin are usually complex partial seizures.

In the majority of patients with complex partial seizures of temporal lobe origin, abnormal epileptogenic tissue involving the hippocampus and other mesiotemporal structures is present in one or both temporal lobes. Less commonly, the epileptic focus is in the neocortex of the temporal lobe. Most patients are neurologically normal apart from subtle memory disturbances.

CLINICAL FEATURES

Complex partial seizures usually begin in childhood or adolescence. Many patients have an aura preceding alteration of consciousness, often characterized by an uncomfortable epigastric sensation and fearful sensations. Some patients also have simple partial seizures during which the aura alone occurs. Alteration of consciousness, which usually lasts 1 to 2 minutes, may be accompanied by a motionless stare, chewing, lip smacking, and gestural automatisms. The seizure is often followed by a period of confusion lasting a few minutes, and aphasia may occur during the postictal period if the seizure originated in the language-dominant temporal lobe. Apart from the initial aura, most patients are amnesic regarding the events of the seizure. Secondary generalization is uncommon.

At least half and possibly as many as 80% of patients with temporal lobe epilepsy have seizures during sleep, mainly NREM sleep. Although most also have daytime seizures, about 5% to 10% have seizures only during sleep.[39] Seizures during sleep, which usually resemble the seizures that the patient has during wakefulness, appear to be more common in the first half of the night (Fig. 20–7).

DIAGNOSIS

Because of the extraordinary range of behaviors that can occur with nocturnal complex partial seizures, descriptions of seizures from parents or bed partners may suggest a variety of other sleep disorders (see Table 20–2). The likelihood that seizures are the cause of unexplained nocturnal spells is greater if the patient has daytime complex partial seizures, particularly if the nocturnal spells resemble daytime seizures. Polysomnography may be required, however, even in patients known to have partial seizures during wakefulness, because behavioral aspects of seizures occurring during sleep may differ from behavior associated with daytime seizures.

If nocturnal complex partial seizures are suspected in a patient who does not have daytime seizures, a daytime EEG should be the initial diagnostic study. The characteristic feature is unilateral or bilateral temporal spikes, usually visualized best from electrodes over the anterior temporal lobe, including F7, T1, and the earlobe. Most patients with temporal lobe epilepsy have such spikes, although they are not always apparent on an initial EEG.[59]

If the diagnosis is uncertain after an EEG is obtained, and the nocturnal events in question are occurring several times per week, VPSG is indicated. In patients with histories strongly suggestive of complex partial seizures and daytime EEGs that are normal or nonspecifically abnormal, interictal epileptiform abnormalities

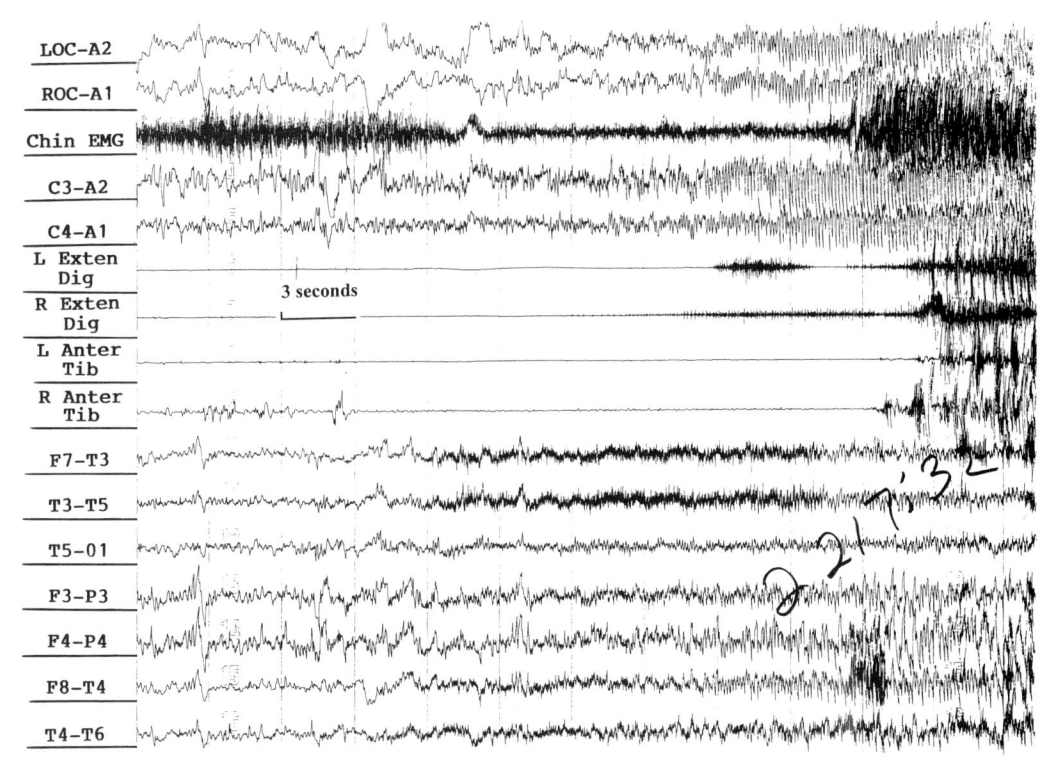

Figure 20–7. Portion of a polysomnogram obtained from a 20-year-old man with a history of suspected nocturnal seizures involving fumbling movements of the hands, occasionally accompanied by getting out of bed and walking. This sample illustrates the initial portion of a complex partial seizure that began during stage 2 sleep. Rhythmic activity that evolves in amplitude and frequency is evident from most EEG channels during the latter third of the sample. Although no definite focal onset is apparent in the sample, prominent slowing was apparent over the right temporal lobe during the postictal period, consistent with a right temporal origin of the seizure. (Exten Dig, extensor digitorum surface EMG; Anter Tib, anterior tibialis surface EMG.) For other abbreviations, see Figures 20–1 and 20–2.

during sleep may be sufficient for diagnosis (Fig. 20–8). If a seizure is recorded, focal ictal patterns are usually evident from scalp electrodes in the form of rhythmic theta activity—sometimes associated with or evolving into spike-and-slow-wave patterns—that varies in frequency and amplitude over the course of the seizure. This activity may be obscured in some instances, however, by artifacts related to muscle activity or movement. Behavioral analysis of video recordings is also helpful, particularly if artifacts obscure the EEG tracings; stereotyped behaviors and postures, similar during several spells, are strongly suggestive of seizures.

Some seizures of temporal lobe origin cannot be identified with surface EEG recordings because they are not associated with scalp EEG changes or because the abnormal EEG activity is obscured by artifact. Thus, the absence of scalp EEG changes during a nocturnal spell does not exclude a diagnosis of temporal lobe epilepsy. Failure to appreciate the limitations of EEG recordings obtained from scalp electrodes may have accounted for descriptions of "paroxysmal arousals" and "episodic nocturnal wanderings" that were not associated with ictal discharges and that responded to antiepileptic medications.[60-62] In such cases, inpatient recordings in specialized epilepsy centers are usually required for definitive diagnosis.

MANAGEMENT

Carbamazepine and phenytoin are the medications most commonly used to

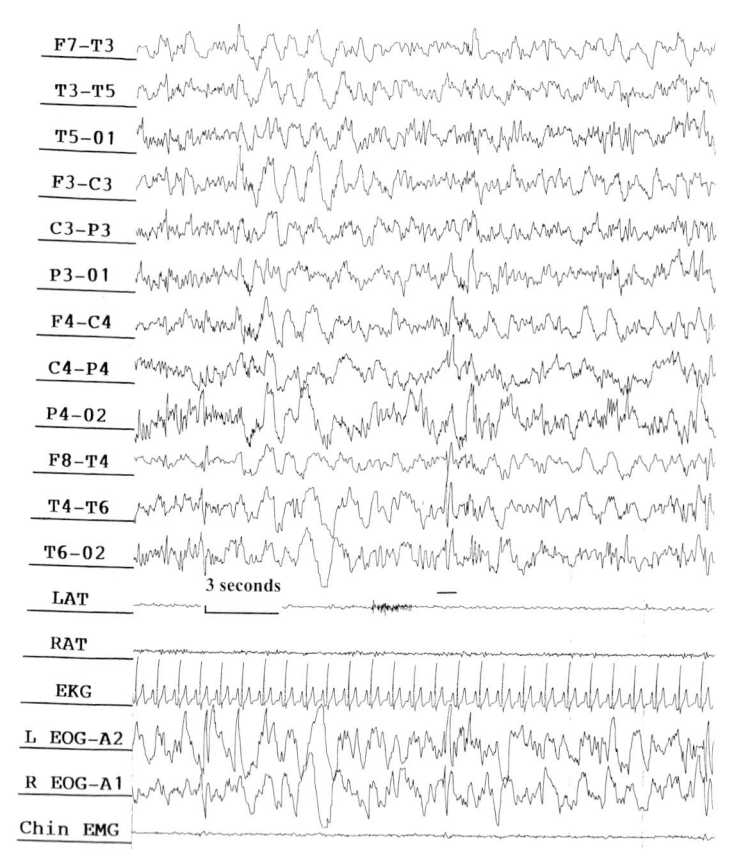

Figure 20-8. This sample, obtained from a 70-year-old woman with four nocturnal seizures in the preceding 6 weeks, illustrates a focal spike (underlined) over the right temporal lobe during slow-wave sleep, a finding characteristic of a partial seizure disorder of right temporal lobe origin. See Figures 20–1 and 20–2 for abbreviations.

treat temporal lobe epilepsy. If these medications are unsuccessful, valproate, gabapentin (10 to 30 mg/kg), or lamotrigine (2 to 15 mg/kg), may be used, but the likelihood of good control is low. In persons with seizures that originate from only one temporal lobe, anterior temporal lobe resection eliminates seizures completely or almost completely in at least 70%.

Frontal Lobe Epilepsy

Frontal lobe epilepsy is less common than temporal lobe epilepsy. For this reason, and because it is difficult to obtain diagnostic EEG recordings in many patients with frontal lobe seizures, the disorder has received less attention than temporal lobe epilepsy. However, it is important for sleep specialists to be knowledgeable about the disorder because seizures are more likely to occur during sleep in patients with frontal lobe epilepsy than in those with temporal lobe epilepsy, and in many patients, the seizures occur exclusively or almost exclusively during sleep.

CLINICAL FEATURES

The frontal lobes have diverse functions; thus, it is not surprising that the clinical manifestations of frontal lobe seizures are varied.[63-66] Head-turning, lateral eye deviation, staring with loss of consciousness, amnesia, forced thinking, stereotyped behavior, floating or déjà vu sensations, drop attacks, blinking, chewing, lip smacking, swallowing, and complex automatisms including running and walking may occur depending on the location of the focus (Table 20–3). Seizures originating from the mesial surface of the frontal lobes are particularly striking; because of the frenetic, bizarre behavior characteristic of such seizures, these patients are often mis-

Table 20–3 **Clinicoanatomic Correlates in Frontal Lobe Epilepsy**

Anatomic origin	Clinical manifestations
Supplementary motor area	Postural; focal tonic (fencing posture); vocalization or speech arrest
Cingulate gyrus	Complex motor gestural automatisms; autonomic disturbance ; changes in mood and affect
Anterior frontopolar	Forced thinking or loss of awareness; head and eye aversion; contraversive movements; axial clonic movements, falls
Orbitofrontal	Motor and gestural automatisms; olfactory hallucinations; autonomic disturbance
Dorsolateral	Tonic; clonic with versive head and eye motion; speech arrest
Opercular	Mastication, salivation, swallowing; speech arrest; epigastric aura; fear, autonomic disturbance; contralateral or ipsilateral clonic movements; sensory symptoms; gustatory hallucinations
Mesial	Shouting, screaming, laughing, unusual breathing patterns, clapping, rubbing hands, snapping fingers, hugging oneself, genital manipulation, pelvic thrusting, pedaling movements, thrashing; nocturnal occurrence; clustering
Motor cortex	Partial motor seizures; speech arrest or vocalization

diagnosed as having psychogenic spells. In one series of 10 patients, 8 had been given an initial diagnosis of hysteria.[63] Behaviors may include shouting, screaming, laughing, or panting; clapping hands, rubbing hands, or snapping fingers; or hugging oneself, manipulating the genitals, thrusting the pelvis, making pedaling movements, or thrashing.

Seizures of frontal lobe origin, which usually last less than 60 seconds, may occur in clusters of up to 40 per night. Other features typical of frontal lobe seizures include the absence of postictal confusion, rapid secondary generalization, prominent motor manifestations, tonic or postural movements, and complex gestural automatisms at onset.

Two examples of patients with nocturnal episodes consistent with frontal lobe seizures are described below.

CASE HISTORY: Case 20–1

A 28-year-old woman had a 4-year history of stereotyped nocturnal spells characterized by brief twitching of all four limbs, shallow breathing with facial twitches, diaphoresis, and a brief period of apnea. She then sat up in bed or stood, appeared fearful, and rubbed her thighs and stomach. Although she was not fully responsive to her husband during the episodes and did not recall the rubbing movements, she described some awareness and could sometimes remember conversations with her husband during the spells.

Spells initially occurred three to four times per year, but then increased to six per night several times per month, and then occurred nightly for stretches of 2 to 3 weeks at a time, awakening her every 15 to 30 minutes. No similar symptoms occurred during wakefulness or during daytime naps. She did not have seizures, sleepwalking, or sleep terrors as a child. Neurological examination and daytime EEG during awake and drowsy states were normal. During two nights of VPSG monitoring, she had more than 20 highly stereotyped episodes—one during REM sleep and the remainder during NREM sleep—lasting 10 to 30 seconds, during which she abruptly awoke and sat up, appeared fearful, and rubbed her thighs. No abnormal EEG activity was apparent during any of the episodes. After treatment with carbamazepine, the spells decreased to three to four per year.

Although no abnormal EEG activity was recorded, the behavioral features of the episodes were so stereotyped that video recordings of the episodes were virtually identical. This extreme stereotypy, combined with the duration of the episodes, the predilection for episodes to occur during NREM sleep, the abrupt arousal, and the response to an antiepileptic medication supported the diagnosis of frontal lobe epilepsy.

CASE HISTORY: Case 20–2

A 24-year-old woman first experienced nocturnal spells at age 15. The spells, which occurred almost every night, were characterized by a sudden arousal accompanied by cursing or screaming. She also had daytime attacks, lasting about 30 seconds, of anxiety and a sense of being at the center of attention. The attacks were associated with hyperventilation, racing heart rate, grunting or singing sounds, running in circles, and inability to respond despite being aware of others around her. Daytime attacks occurred in either clusters of up to 12 per day or not at all for months.

She was treated for panic attacks with alprazolam with some improvement. Later, alprazolam was discontinued and carbamazepine was begun with resolution of daytime episodes; however, she continued to have nightly episodes of sudden arousal, restless movements, and grunting sounds or screams, sometimes relieved if her husband comforted her. Neurological examination, brain magnetic resonance imaging, and EEG were normal.

During two nights of VPSG, she had 12 episodes characterized by abrupt arousal, twisting movements with pelvic thrusting, and singing or yodeling sounds lasting 20 to 45 seconds. The EEG was obscured during several episodes and showed normal arousal patterns during the others. Carbamazepine dose was increased, and a greater proportion of the daily dosage was given at bedtime, with subsequent marked reduction in the frequency of episodes.

DIAGNOSIS

Diagnosis of frontal lobe seizures is difficult because the unusual behaviors may not suggest the possibility of seizures to inexperienced clinicians and because much of the frontal lobe is inaccessible to scalp EEG recordings.[67] On daytime EEG, the interictal EEG in frontal lobe epilepsy often shows no abnormalities or only a mild background asymmetry, particularly in patients with deep foci. In some patients, unilateral or bilateral frontal spikes or sharp waves are apparent (Fig. 20–9). Differential diagnosis may include partial seizures of temporal lobe origin, sleep terrors, nocturnal panic disorder, nocturnal dissociative disorder or other psychogenic conditions, and REM sleep behavior disorder (RBD). In some patients, daytime seizures develop, which supports a diagnosis of epilepsy as the cause of the nocturnal episodes.

VPSG is useful for suspected frontal lobe seizures because the occurrence in many patients of numerous seizures each night permits detailed EEG and behavioral analysis. During seizures, ictal discharges consisting of unilateral or bilateral fast activity, spikes or slow waves, or bilateral high-amplitude sharp waves followed by amplitude suppression may be recorded from frontal or, less commonly, temporal regions.

Unfortunately, myogenic artifacts often obscure the EEG during the episodes, and it may be necessary to record dozens of episodes over the course of several nights before obtaining a recording that reveals a focal ictal discharge. Furthermore, epileptic foci responsible for frontal lobe seizures are often distant from scalp electrodes, and seizures may not be accompanied by changes in the scalp EEG. In such cases, especially in those patients who show little or no response to antiepileptic medications, the determination of whether the episodes are epileptic is difficult, and lengthy evaluations with numerous nights of recording in sleep laboratories and epilepsy units may be required before the cause is determined. For some patients, ictal discharges can be demonstrated only with intracranial recordings.

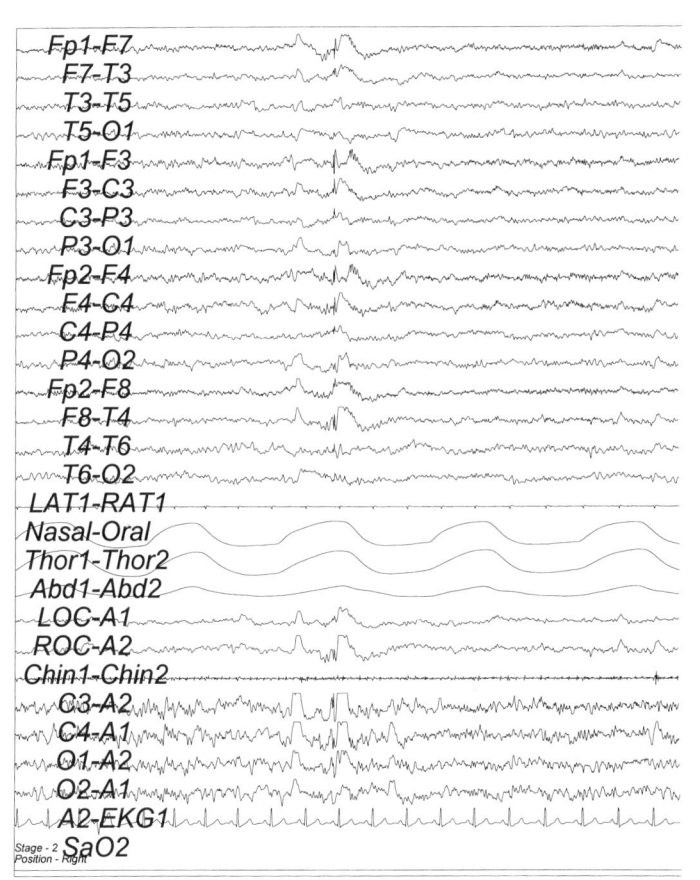

Figure 20–9. Portion of a polysomnogram in a 63-year-old man with a history of repeated episodes of awakening with a "nervous feeling" followed by shaking movements of the arms and legs. A single spike is evident in association with a K-complex. The bianterior distribution of the spike, consistent with frontal lobe epilepsy, is apparent in the extended EEG montage. See Figures 20–1, 20–2, and 20–4 for abbreviations.

In the absence of definite ictal discharges, the abrupt onset from NREM sleep of stereotyped behaviors is the most helpful finding. As in Case 20–1 above, the behaviors during a dozen or more episodes in one night often are virtually identical; this extreme stereotypy does not occur with panic disorder, sleep terrors, RBD, or dissociative episodes. As seizures almost always occur during NREM sleep, RBD can usually be excluded based on the polysomnographic findings. Sleep terrors rarely occur more than three to four times per night, whereas frontal lobe seizures often occur 10 to 40 times per night. Abrupt onset of episodes from sleep are not typical of nocturnal dissociative episodes, which are usually preceded by a few minutes of wakefulness (see Chap. 15).

Other behavioral features also help distinguish frontal lobe seizures from psychogenic spells. Compared to patients with psychogenic seizures, those with frontal lobe seizures are more likely to have had onset of symptoms in childhood, and symptoms in these patients are more likely to occur at night, during sleep, and in the prone position.[68] On the other hand, the occurrence of pelvic thrusting, kicking, pedaling, side-to-side head movements, and rapid postictal recovery does not help to distinguish epileptic from psychogenic events. Secondary gain from the episodes is an important feature because it accompanies many psychogenic spells and generally does not occur with frontal lobe seizures. Daytime psychopathology, almost universal in patients with nocturnal dissociative disorder, is probably no more common in patients with frontal lobe seizures than in the general population.

MANAGEMENT

Pharmacologic management of frontal lobe seizures is similar to that used for temporal lobe seizures, although carbamazepine often seems to produce better results than phenytoin and other antiepileptic medications. Although some patients do well with carbamazepine, others do not respond to any anticonvulsant and have numerous nightly episodes despite a variety of treatments. Compared to treatment for temporal lobe epilepsy, surgical treatment for frontal lobe epilepsy is more difficult because techniques for localization of the epileptic focus are less successful. Spontaneous remissions occur in a few patients, sometimes lasting for years.

Nocturnal Paroxysmal Dystonia

With nocturnal (hypnogenic) paroxysmal dystonia, patients have spells that arise abruptly from sleep. These spells consist of dystonic postures and ballistic, choreoathetoid movements lasting 10 to 45 seconds.[69,70] Episodes most commonly arise during stage 2 sleep, although they may also occur out of stage 3–4 sleep. The patient is usually awake with eyes open during the spells; the movements are usually bilateral but may predominate on one side or may be exclusively unilateral. Patients may have occasional attacks or as many as 30 to 40 episodes per night. Although the usual presenting complaint is nocturnal sleep disruption and violent behavior, some patients may also complain of excessive daytime somnolence.

Although first described as a syndrome without EEG abnormalities, the clinical characteristics closely resemble partial seizures originating in the supplementary motor area, and it now appears that in most patients with this clinical entity, the movements are caused by partial seizures of frontal lobe origin.[71] In rare patients with attacks of long duration, the cause may be a basal ganglia disturbance rather than epilepsy. Diagnosis and management are similar to the approach outlined above for patients with other forms of frontal lobe epilepsy.

Autosomal Dominant Nocturnal Frontal Lobe Epilepsy

In this disorder, reported in families from Australia, Canada, and Britain, brief nocturnal seizures begin in childhood and occur several times per night in NREM sleep through adolescence and adulthood.[72] The gene for the disorder is on chromosome 20 (20q13.2-q13.3); a mutation in the alpha-4 subunit of the neuronal nicotinic acetylcholine receptor appears to be responsible.[73] The daytime EEG may show frontal epileptiform activity, or it may be unremarkable. The seizures usually respond to carbamazepine.

Epilepsy with Continuous Spike Waves During Slow-Wave Sleep

The rare disorder of epilepsy with continuous spike waves during slow-wave sleep (SWS) is notable because seizures are uncommon or absent during wakefulness despite continuous or nearly continuous epileptiform discharges during NREM sleep (Fig. 20–10).[74–76] The cause is unknown; seizures usually begin in late infancy or childhood, either after normal development or in children with hemiparesis, spastic quadriplegia, or ataxia. Myoclonic, generalized tonic-clonic, or atypical absence seizures may occur during wakefulness or sleep, but unlike the Lennox-Gastaut syndrome, atonic seizures are rare. After the onset of the disorder, there is usually developmental delay or regression. Some patients have the Landau-Kleffner syndrome (acquired aphasia syndrome) with seizures, progressive language loss, and inattention to auditory stimuli.

The typical EEG changes appear 1 to 2 years after the first seizure and are associated with behavioral deterioration. Initially, the EEG during wakefulness and sleep may show focal spikes, slow waves, or diffuse spike waves. Later, NREM sleep is associated with abrupt onset of essentially continuous epileptiform activity, accompanied by temporal lobe hypermetabolism,[77] while epileptiform activity remains rare or

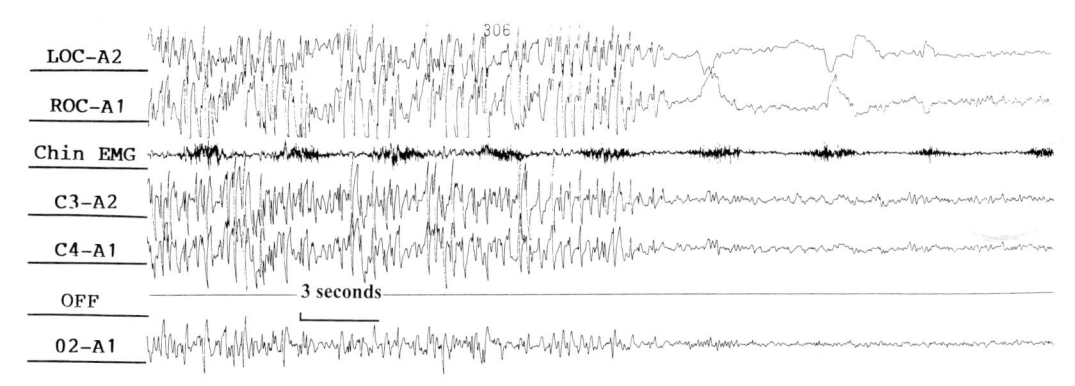

Figure 20–10. Portion of a polysomnogram in a 10-year-old boy with the syndrome of epilepsy with continuous spike waves during slow-wave sleep. The initial portion of the sample illustrates continuous epileptiform activity during NREM sleep. With the onset of REM sleep at about the midpoint of the illustration, epileptiform activity is suppressed. See Figure 20–1 for abbreviations.

absent during wakefulness and REM sleep. Epileptiform activity is enhanced during drowsy wakefulness in some patients.[75]

The behavioral and neurological deterioration in these children is presumably due, at least in part, to the continuous spike waves during SWS, as those with the most epileptiform activity have the most deterioration.[76] The seizures usually stop during the second decade and the continuous epileptiform activity during NREM sleep also attenuates, sometimes accompanied by modest behavioral and neurological improvement.

Although diagnosis requires EEG recordings during sleep, all-night recordings are rarely necessary. Valproate or carbamazepine generally prevent seizures, although they do not eliminate the sleep-related epileptiform activity or the behavioral and neurological deterioration.

SUMMARY

The occurrence in many epileptics of seizures during sleep has been recognized since antiquity. Sleep influences the expression of epileptic disorders in a number of ways: (1) it affects the appearance of epileptiform activity; (2) seizures are more frequent during sleep than during wakefulness in many epileptics; (3) in some epileptic disorders, sleep deprivation influences the likelihood of seizures and of recording epileptiform activity; (4) disorders that cause sleep disturbance may affect seizure frequency; (5) epilepsy can affect sleep continuity; and (6) antiepileptic medications can affect sleep and wakefulness.

Because of the varied clinical manifestations of seizures and the many possible causes of episodic behaviors during sleep, the approach to diagnosis of a suspected seizure during sleep depends on the clinical information. Although the clinical history or home videotapes may be diagnostic for generalized seizures, even in such cases, an EEG recording often helps to determine the type of epileptic syndrome and the approach to management. In patients with suspected partial seizures, the type of EEG study obtained depends on the clinical setting and may include daytime EEG, VPSG, ambulatory EEG monitoring, or long-term inpatient or outpatient video-EEG monitoring.

A substantial proportion of patients with generalized epilepsy with tonic-clonic seizures have seizures exclusively during sleep, almost always during NREM sleep. In the syndrome of epilepsy with grand mal seizures on awakening, seizures develop soon after awakening from nighttime sleep or from naps. In juvenile myoclonic epilepsy, the myoclonic seizures also occur mainly after awakenings. Diagnosis is based on the morning occurrence of bilateral myoclonic jerks without loss of consciousness, associated with a normal neurological examination.

Many patients with partial seizures have at least some of their seizures during sleep. The likelihood that seizures will occur during sleep is dependent on the location of the focus and on the nature of the epileptic syndrome. In the genetically based syndrome of benign childhood epilepsy with centrotemporal spikes, seizures usually begin between 5 and 10 years of age; although some patients have seizures soon after awakening, most have them only during sleep. The prognosis is excellent: Seizures almost always stop spontaneously in adolescence. Among patients with temporal lobe epilepsy, 50% to 80% have seizures during sleep, and about 5% to 10% have seizures only during sleep; in such cases, descriptions of seizures from parents or bed partners may suggest a variety of other sleep disorders. Although frontal lobe epilepsy is less common than temporal lobe epilepsy, seizures are more likely to occur during sleep, and in many patients, the seizures occur exclusively or almost exclusively during sleep. Furthermore, the bizarre manifestations may make diagnosis difficult in some patients. The rare childhood disorder of epilepsy with continuous spike waves during SWS is notable because seizures are uncommon or absent during wakefulness despite continuous or nearly continuous epileptiform discharges during NREM sleep.

REFERENCES

1. Gross, RA: A brief history of epilepsy and its therapy in the western hemisphere. Epilepsy Res 12:65, 1992.
2. Gowers, WR: Epilepsy and other chronic convulsive diseases. London, 1881.
3. Juul-Jensen, P, and Foldspang, A: Natural history of epileptic seizures. Epilepsia 24:297, 1983.
4. Commission on Classification and Terminology of the International League Against Epilepsy: Proposal for revised Classification of Epilepsies and Epileptic Syndromes. Epilepsia 30:389, 1989.
5. Aird, RB, and Gastaut, Y: Occipital and posterior electroencephalographic rhythms. Electroencephalogr Clin Neurophysiol 11:637, 1959.
6. Lothman, EW: Basic mechanisms of seizure expression. Epilepsy Res Suppl 11:9, 1996.
7. Lowenstein, DH: Recent advances related to basic mechanisms of epileptogenesis. Epilepsy Res Suppl 11:45, 1996.
8. Gibbs, EL, and Gibbs, FA: Diagnostic and localizing value of electroencephalographic studies in sleep. Res Publ Assoc Res Nerv Ment Dis 26:366, 1947.
9. Malow, BA, et al: Relationship of interictal epileptiform discharges to sleep depth in partial epilepsy. Electroencephalogr Clin Neurophysiol 102:20, 1997.
10. Autret, A, et al: Influence of waking and sleep stages on the inter-ictal paroxysmal activity in partial epilepsy with complex seizures. Electroencephalogr Clin Neurophysiol 55:406, 1983.
11. Niedermeyer, E, and Rocca, U: The diagnostic significance of sleep electroencephalograms in temporal lobe epilepsy: A comparison of scalp and depth tracings. Eur Neurol 7:119, 1972.
12. Sato, S, et al: The effect of sleep on spike-wave discharges in absence seizures. Neurology 23:1335, 1973.
13. Bazil, CW, and Walczak, TS: Effects of sleep and sleep stage on epileptic and nonepileptic seizures. Epilepsia 38:56, 1997.
14. Pratt, KL, et al: EEG activation of epileptics following sleep deprivation: A prospective study of 114 cases. Electroencephalogr Clin Neurophysiol 24:11, 1968.
15. Papini, M, et al: Alertness and incidence of seizures in patients with Lennox-Gastaut syndrome. Epilepsia 25:161, 1984.
16. Ellingson, RJ, et al: Efficacy of sleep deprivation as an activation procedure in epilepsy patients. J Clin Neurophysiol 1:83, 1984.
17. Degen, R, et al: Sleep EEG with or without sleep deprivation? Does sleep deprivation activate more epileptic activity in patients suffering from different types of epilepsy? Eur Neurol 26:51, 1987.
18. Daly, D: Epilepsy and syncope. In Daly, DD, and Pedley, TA (eds): Current Practice of Clinical Electroencephalography, ed 2. Raven, New York, 1990, pp 269–334.
19. Devinsky, O, et al: Epilepsy and sleep apnea syndrome. Neurology 44:2060, 1994.
20. Vaughn, BV, et al: Improvement of epileptic seizure control with treatment of obstructive sleep apnoea. Seizure 5:73, 1996.
21. Wyler, AR, and Weymuller, EA Jr: Epilepsy complicated by sleep apnea. Ann Neurol 9:403, 1981.
22. Malow, BA, et al: Usefulness of polysomnography in epilepsy patients. Neurology 48:1389, 1997.
23. Peled, R, and Lavie, P: Paroxysmal awakenings from sleep associated with excessive daytime somnolence: A form of nocturnal epilepsy. Neurology 36:95, 1986.
24. Manni, R, et al: Nocturnal partial seizures and arousals/awakenings from sleep: An ambulatory EEG study. Funct Neurol 12:107, 1997.
25. Collaborative Group for Epidemiology of Epilepsy: Adverse reactions to antiepileptic drugs: A multi-center survey of clinical practice. Epilepsia 27:323, 1986.
26. Bonanni, E, et al: A quantitative study of daytime sleepiness induced by carbamazepine and add-on vigabatrin in epileptic patients. Acta Neurol Scand 95:193, 1997.

27. Salinsky, MC, et al: Assessment of drowsiness in epilepsy patients receiving chronic antiepileptic drug therapy. Epilepsia 37:181, 1996.
28. Wolf, P, et al: Influence of therapeutic phenobarbital and phenytoin medication on the polygraphic sleep of patients with epilepsy. Epilepsia 25:467, 1984.
29. Kellaway, P: The incidence, significance, and natural history of spike foci in children. In Henry, CE (ed): Current Clinical Neurophysiology: Update on EEG and Evoked Potentials. Elsevier, North Holland, 1980, pp 151–175.
30. Mizrahi, EM: Avoiding the pitfalls of EEG interpretation in childhood epilepsy. Epilepsia (suppl 1)37:S41, 1996.
31. Westmoreland, B: Benign EEG variants and patterns of uncertain clinical significance. In Daly, DD, and Pedley, TA (eds): Current Practice of Clinical Electroencephalography, ed 2. Raven, New York, 1990, pp 243–252.
32. El-Ad, B, et al: Should sleep EEG record always be performed after sleep deprivation? Electroencephalogr Clin Neurophysiol 90:313, 1994.
33. Kubicki, S, et al: Short-term sleep EEG recordings after partial sleep deprivation as a routine procedure in order to uncover epileptic phenomena: An evaluation of 719 EEG recordings. Epilepsy Res Suppl 2:217, 1991.
34. Carpay, JA, et al: The diagnostic yield of a second EEG after partial sleep deprivation: A prospective study in children with newly diagnosed seizures. Epilepsia 38:595, 1997.
35. Aldrich, MS, and Jahnke, B: Diagnostic value of video-EEG polysomnography. Neurology 41:1060, 1991.
36. Lagerlund, TD, et al: Long-term electroencephalographic monitoring for diagnosis and management of seizures. Mayo Clin Proc 71:1000, 1996.
37. Ebersole, JS: Ambulatory EEG: Telemetered and cassette-recorded. Adv Neurol 46:139, 1987.
38. American Electroencephalographic Society: Guideline twelve: Guidelines for long-term monitoring for epilepsy. J Clin Neurophysiol 11:88, 1994.
39. Billiard, M: Epilepsies and the sleep-wake cycle. In Sterman, MB, et al (eds): Sleep and Epilepsy. Academic Press, New York, 1982, pp 481–494.
40. Janz, D: The grand mal epilepsies and the sleeping-waking cycle. Epilepsia 3:69, 1962.
41. Janz, D: Epilepsy with impulsive petit mal (juvenile myoclonic epilepsy). Acta Neurol Scand 52:449, 1985.
42. Durner, M, et al: Possible association of juvenile myoclonic epilepsy with HLA-DRw6. Epilepsia 33:814, 1992.
43. Greenberg, DA, et al: The genetics of idiopathic generalized epilepsies of adolescent onset: Differences between juvenile myoclonic epilepsy and epilepsy with random grand mal and with awakening grand mal. Neurology 45:942, 1995.
44. Grunewald, RA, et al: Delayed diagnosis of juvenile myoclonic epilepsy. J Neurol Neurosurg Psychiatry 55:497, 1992.
45. Sharpe, C, and Buchanan, N: Juvenile myoclonic epilepsy: Diagnosis, management and outcome. Med J Aust 162:133, 1995.
46. Janz, D: Juvenile myoclonic epilepsy. Cleve Clin J Med 56:S23, 1989.
47. Dreifuss, FE: Juvenile myoclonic epilepsy: Characteristics of a primary generalized epilepsy. Epilepsia 30:S1, 1989.
48. Lennox, WG, and Davis, JP: Clinical correlates of the fast and the slow spike-wave electroencephalogram. Pediatrics 5:626, 1950.
49. Gastaut, H, et al: Childhood epileptic encephalopathy with diffuse slow spike-waves (otherwise known as "petit mal variant") or Lennox syndrome. Epilepsia 7:139, 1966.
50. Markand, ON: Slow spike-wave activity in EEG and associated clinical features: Often called "Lennox" or "Lennox-Gastaut" syndrome. Neurology 27:746, 1977.
51. Drury, I: Epileptiform patterns of children. J Clin Neurophysiol 6:1, 1989.
52. Montplaisir, J, et al: Sleep and epilepsy. In Gotman, J, et al (eds): Long-Term Monitoring in Epilepsy (EEG suppl. no. 37). Elsevier Science, New York, 1985, pp 215–239.
53. Drury, I, and Beydoun, A: Benign partial epilepsy of childhood with monomorphic sharp waves in centrotemporal and other locations. Epilepsia 32:662, 1991.
54. Beydoun, A, et al: Generalized spike-waves, multiple loci, and clinical course in children with EEG features of benign epilepsy of childhood with centrotemporal spikes. Epilepsia 33:1091, 1992.
55. Loiseau, P, and Duche, B: Benign childhood epilepsy with centrotemporal spikes. Cleve Clin J Med 56:S17, 1989.
56. Roulet, E, et al: Prolonged intermittent drooling and oromotor dyspraxia in benign childhood epilepsy with centrotemporal spikes. Epilepsia 30:564, 1989.
57. Heijbel, J, et al: Benign epilepsy of childhood with centrotemporal EEG foci: A genetic study. Epilepsia 16:285, 1975.
58. Loiseau, P, et al: Prognosis of benign childhood epilepsy with centrotemporal spikes: A follow-up study of 168 patients. Epilepsia 29:229, 1988.
59. Williamson, PD, et al: Characteristics of medial temporal lobe epilepsy: II. Interictal and ictal scalp electroencephalography, neuropsychological testing, neuroimaging, surgical results, and pathology. Ann Neurol 34:781, 1993.
60. Pedley, TA, and Guilleminault, C: Episodic nocturnal wanderings responsive to anticonvulsant drug therapy. Ann Neurol 2:30, 1977.
61. Montagna, P, et al: Paroxysmal arousals during sleep. Neurology 40:1063, 1990.
62. Maselli, RA, et al: Episodic nocturnal wanderings in non-epileptic young patients. Sleep 11:156, 1988.
63. Williamson, PD, et al: Complex partial seizures of frontal lobe origin. Ann Neurol 18:497, 1985.
64. Bancaud, J, and Talairach, J: Clinical semiology of frontal lobe seizures. Adv Neurol 57:3, 1992.
65. Munari, C, and Bancaud, J: Electroclinical symptomatology of partial seizures of orbital frontal origin. Adv Neurol 57:257, 1992.
66. Quesney, LF, et al: Seizures from the dorsolateral frontal lobe. Adv Neurol 57:233, 1992.

67. Williamson, PD: Frontal lobe seizures: Problems of diagnosis and classification. Adv Neurol 57:289, 1992.
68. Saygi, S, et al: Frontal lobe partial seizures and psychogenic seizures: Comparison of clinical and ictal characteristics. Neurology 42:1274, 1992.
69. Lugaresi, E, and Cirignotta, F: Hypnogenic paroxysmal dystonia: Epileptic seizure or a new syndrome? Sleep 4:129, 1981.
70. Lugaresi, E, et al: Nocturnal paroxysmal dystonia. J Neurol Neurosurg Psychiatry 49:375, 1986.
71. Meierkord, H, et al: Is nocturnal paroxysmal dystonia a form of frontal lobe epilepsy? Mov Disord 7:38, 1992.
72. Scheffer, IE, et al: Autosomal dominant nocturnal frontal lobe epilepsy: A distinctive clinical disorder. Brain 118:61, 1995.
73. Steinlein, OK, et al: A missense mutation in the neuronal nicotinic acetylcholine receptor alpha 4 subunit is associated with autosomal dominant nocturnal frontal lobe epilepsy. Nat Genet 11:201, 1995.
74. Yasuhara, A, et al: Epilepsy with continuous spike-waves during slow sleep and its treatment. Epilepsia 32:59, 1991.
75. Jayakar, PB, and Seshia, S: Electrical status epilepticus during slow-wave sleep: A review. J Clin Neurophysiol 8:299, 1991.
76. Kobayashi, K, et al: Epilepsy and sleep: With special reference to nonconvulsive status epilepticus with continuous diffuse spike-waves during slow-wave sleep. Brain Dev 22:136, 1990.
77. Rintahaka, PJ, et al: Landau-Kleffner syndrome with continuous spikes and waves during slow-wave sleep. J Child Neurol 10:127, 1995.

INDEX

Acetazolamide, 248, 312
Acetylcholine. See Cholinergic neurons
Actigraphy, 105, 328
Activation-synthesis hypothesis, 85–86
Active Sleep, 15, 71–72, 77–80
Adaptive functions of sleep, 22
Adenosine, 36
Adjustment sleep disorder, 146
Advanced sleep phase syndrome, 195
Akathisia, 175, 178
Akinetic mutism, 16
Alcohol, 300–302
    heartburn and, 315
    insomnia and, 133, 145
    juvenile myoclonic epilepsy and, 361
    nightmares and, 278
    obstructive sleep apnea syndrome and, 216, 302
Alcoholism, 301–302
Alcohol-dependent sleep disorder, 301
Alpha rhythm. See EEG, alpha rhythm
Alpha-delta sleep, 13, 318–320
Altitude insomnia, 241
Alzheimer's disease, 325–329
Ambulatory monitoring. See home monitoring
American Board of Sleep Medicine, 96
American Sleep Disorders Association, 95–96
Amphetamines, 30, 303. See also Stimulants
Anemia, 177–178, 320
Angina, nocturnal, 308–310
Angina, Prinzmetal's, 308
Anorexia nervosa, 299
Anti-epileptic medications, 354
Antihistamines, 136
Anxietas tibiarum, 175
Anxiety disorders, 297–299
Apnea of infancy, idiopathic, 238, 242, 243, 246–247, 249
Apneic threshold, 240
Apneustic center, 39, 48
Apparent life-threatening event, 238, 246, 250–251
Arginine-vasotocin, 36
Arnold-Chiari malformation, 346
Arousal disorders, 261–267
Arthritis, 317–318
Ascending reticular activating system, 6–7, 27–30
Asthma, 312–314. See also Lung diseases
Automatic behavior, 156–157
Autonomic nervous system activity during sleep, 17
Autonomic nervous system disturbances
    fatal familial insomnia and, 340
    Kleine-Levin syndrome and, 343
    Parkinson's disease and, 330

ß-adrenergic receptor blockers, 278
Barbiturates, 127, 134, 303
    nightmares and, 278

Basal forebrain, 6–7, 28, 30–31
Behavioral system for respiratory control, 49, 241–242
Benzodiazepines, 127, 134–136, 146–147. See also Hypnotics
    Alzheimer's disease and, 329
    anxiety disorders and, 298
    arousal disorders and, 267
    central sleep apnea and, 248
    lung diseases and, 314
    nightmares and, 278
    periodic leg movements and, 182–183
    REM sleep behavior disorder and, 274
    restless legs syndrome and, 180
    sleep bruxism and, 280
Benzodiazepine-receptor agonists, 134–136, 146–147
Bereavement, 295, 296
Beta rhythm, 5, 10
Bilevel positive airway pressure, 227, 247, 253, 256
    brainstem lesions and, 346
    lung diseases and, 314
Biological clock, 58–62, 187, 190, 193, 195, 196, 197–198
Bipolar disorder, 290–295, See also Mood disorders
    Kleine-Levin syndrome and, 343
Blindness, non-24-hour sleep-wake syndrome and, 197–198
Botzinger complex, 46–48
Brainstem. 5–7, 27–34, 46–48. See also Midbrain; Pons; Medulla
    compression causing sleep apnea, 317
    lesions, 345–346
    progressive supranuclear palsy and, 333
    multisystem degeneration and, 334
    neoplasms, REM sleep behavior disorder and, 273
    sleep disorders, 338–349
Breathing, 39–55
    age-related changes in, 77–79
    central nervous system controller for, 46–48
    load compensation, 50, 52
    mechanics in infants, 77–79
    metabolic regulation of, 48–49, 52, 239–241
    paradoxical, in infants, 79
    paradoxical, in obstructive sleep apnea syndrome, 209
    periodic, 49, 53, 77–78, 237–238, 240–242
    voluntary control of, 49, 241–242
Bruxism. See Sleep bruxism
Bulimia nervosa, 299

Caffeine, 36, 123, 145, 303
    REM sleep behavior disorder and, 272
    heartburn and, 316
canarc-1 gene, 161

375

Cannabis, 303
Cardiac arrhythmia, 308
  obstructive sleep apnea syndrome and, 217–219
  sudden unexplained nocturnal death and, 284
Cardiovascular function during sleep, 17
Cataplexy, 155, 159–162, 166–167, 171, 345
Central alveolar hypoventilation syndrome, 251–252
Central sleep apnea, 238–249
  brainstem lesions and, 317, 345–346
  caused by upper airway obstruction, 210, 242, 246
  Cheyne-Stokes breathing and, 245, 308–312
  heart disease and, 241, 308–312
  lung diseases, chronic, and, 313–314
  multisystem degeneration and, 334
  post-arousal, 53, 240–241
  premature infants and, 242
  renal failure, chronic, and, 320–321
  sleep-onset, 53, 240–241
Cerebral blood flow and metabolism during sleep, 19
Cerebrovascular disease, REM sleep behavior disorder and, 273
Chemoreceptors, 45–46, 242, 251
Cheyne-Stokes breathing, 49, 77, 226, 237–238, 241, 244–245
  heart disease and, 307–312
  mixed apnea and, 310
Choking, differential diagnosis of nocturnal, 315
Cholinergic neurons, 28, 30, 32–34
  degeneration in Alzheimer's disease, 327
  depression and cholinergic overactivity, 292
  narcolepsy and, 160
Chronic obstructive pulmonary disease. *See* Lung diseases.
Chronobiology, 56–69
Chronobiological disorders, 186–201
  Alzheimer's disease and, 327
  Parkinson's disease and, 330
Circadian distribution of sleep, 57–58
Circadian rhythmicity, physiologic functions with, 57
Circadian temperature rhythm, 62
Circadian timing system, neuroanatomy of, 58–59
Claudication, 178–179
clock gene, 59
Clonazepam, 135, 180, 183, 274
Cocaine, 303
Coma, 16
Confusional arousals, 261–262
Continuous positive airway pressure (CPAP), 225–228, 253
  central sleep apnea and, 247, 248
  heart disease and, 312
  lung diseases and, 314
  neuromuscular disorders and, 256
Cortisol secretion during sleep, 18
Cortistatin, 36
Craniomandibular disorders, 279
Craniopharyngioma, 345
Cystic fibrosis, 313

Day residue, 88
Degenerative neurologic diseases, 325–337
Delayed sleep phase syndrome, 192–195
  depression and, 291
Delirium tremens, 301

Delta sleep. *See* Slow wave sleep
Delta sleep-inducing peptide, 36
  insomnia and, 136
Delta waves, 9, 12–13, 31–32, 74, 75
Dementia, sleep disorders in, 325–329
  REM sleep behavior disorder and, 273
Dental appliances, 229–230
Depression, 290–296. *See also* Mood disorders
  Alzheimer's disease and, 327
  alcoholism and, 302
  delayed sleep phase syndrome and, 193–194
  fibromyalgia and, 319
  impaired nocturnal penile tumescence and, 276
  narcolepsy and, 157, 168
  rheumatologic disorders and, 317
Desmopressin, 283
Diabetes insipidus, 282
Diabetes mellitus
  autonomic neuropathy and nocturnal diarrhea, 316–317
  impaired nocturnal penile tumescence and, 275
  sleep enuresis and, 282
Dialysis, 320–321
Diaphragmatic pacing, 252, 257
Diarrhea, nocturnal, 316–317
Diencephalon, lesions of, 338–349. *See also* Hypothalamus; Thalamus
Dopamine agonists, 179–180, 182–183, 330–332, 334
Dopaminergic neurons, 30, 160–161
  Parkinson's disease and, 330
Dorsal raphe, 29–30, 33
Dorsal respiratory group, 46–48
Dorsolateral small cell reticular group, 33
Dreams, 82–91. *See also* Nightmares
  fatal familial insomnia and, 340–341
  Freud's theories of, 88–89
  REM sleep behavior disorder and, 271–272
Dreamwork, 88
Drowsiness. *See* Sleepiness.
Duchenne muscular dystrophy, 255
Dyssomnias, 96–97
Dysthymic disorder, 295. *See also* Mood disorders
  fibromyalgia and, 319

Eating disorders, 299
Ekbom's syndrome. *See* Restless legs syndrome
Electroencephalogram (EEG), 5–15
  alpha rhythm, 9–10, 74, 76
  alpha-delta sleep, 13, 319–320
  ascending reticular activating system and, 29
  benign epileptiform transients of sleep, 355
  delta brushes, 71, 72
  delta waves, 9, 12–13, 31–32, 74, 75
  epilepsy and, 351–359
  fatal familial insomnia and, 341
  generalized paroxysmal fast activity, 362
  hypnagogic hypersynchrony, 10, 74
  idiopathic recurring stupor and, 344
  infancy and, 72, 74
  inpatient monitoring, 357
  Kleine-Levin syndrome and, 343
  K-complex, 11–12, 74
  nocturnal dissociative disorder and, 283
  portable recordings, 357

polyspike-and-wave, 352
posterior slow waves of youth, 10, 75
rhythmic movement disorder and, 268
sawtooth waves, 14
sharp waves, 352
sleep deprivation and, 353–354
sleep-related seizures and, 354–359
sleep spindles, 12, 31–32, 74
sleep terrors and, 263–267
slow spike-and-wave, 362
spikes, 352, 355, 359, 362–366, 369
spike-and-slow-wave complex, 352–353, 356, 357, 359
tracé alternant, 72, 74
tracé discontinu, 71
unexpected finding during polysomnography, 357–359
vertex sharp waves, 10, 74
video-EEG polysomnography, 265–267, 356–357, 364–365, 368–369
waking, 9–10, 74
Electromyogram, 8–9
Electrooculogram, 8–9, 11
Encephalitis lethargica, 338–339
Endocrine function during sleep, 18
Endozepine-4, 344
Enuresis. See Sleep enuresis
Environmental sleep disorder, 145–146
Epilepsy, 350–374
  benign childhood, with centrotemporal spikes (Benign Rolandic epilepsy), 362–364
  childhood absence, 361–362
  continuous spike-waves during slow wave sleep, 370–371
  electroencephalogram and, 351–359
  frontal lobe, 366–370
  generalized, associated with seizures during drowsiness, sleep, or upon awakening, 359–362
  juvenile myoclonic, 361
  partial, associated with seizures during sleep, 362–370
  temporal lobe, 264–265, 358, 363–366
Epworth sleepiness scale, 116
Evoked potentials, 117
Extended constant routine, 64–65

Factor S, 35
Familial dysautonomia, 251
Fatal familial insomnia, 340–341
Fatigue, 100–101
  lung diseases and, 312
  rheumatoid arthritis and, 317–318
  fibromyalgia and, 319
Fibromyalgia, 318–320
Flumazenil, 344
Food-allergy insomnia, 149
Forbidden zone for sleep, 65
Freud, Sigmund, 6
  dream theories of, 88–89
Functions of sleep, 20–23

γ-aminobutyric acid, 31
Ganglioneuroblastoma, 251

Gasping, differential diagnosis of nocturnal, 315
Gastroesophageal reflux disease, 315–316
  lung diseases and, 313–314
  obstructive sleep apnea syndrome and, 206–207, 315–316
Gastrointestinal diseases, 314–317
Gastrointestinal function during sleep, 18
Generalized anxiety disorder, 297–298
Genioglossus advancement with hyoid suspension, 228–229
Glucose secretion during sleep, 18
Growth hormone secretion during sleep, 18
Growth hormone-releasing hormone, 36

Headbanging. See Rhythmic movement disorder
Heartburn, 315
Heart disease, ischemic, 307–312
  central sleep apnea and, 308–312
  Cheyne-Stokes breathing and, 308–312
  congestive heart failure, 241, 308–312
  obstructive sleep apnea and, 308–312
Hirschsprung's disease, 251
Histaminergic neurons, 30
Home monitoring, 222–223, 250–251
Human leukocyte antigens, 157–159
Hypercapnia, 239–241, 251–257. See also Hypoventilation
  lung diseases and 313–314
  ventilatory response to, 49
Hypersomnia. See also Sleepiness
  idiopathic. See Idiopathic hypersomnia
  Kleine-Levin syndrome and, 342–344
  psychogenic recurrent, 343
  recurrent, 342–344
  thalamic infarcts and, 345
Hypertension
  impaired nocturnal penile tumescence and, 275–276
  obstructive sleep apnea syndrome and, 206–207
Hypnic jerks. See Sleep starts
Hypnogenic paroxysmal dystonia, 370
Hypnagogic and hypnopompic hallucinations, 156, 159, 167, 171
Hypnogram, 13, 107
Hypnotics, 133–136. See also Benzodiazepines
  arthritis and, 318
  jet lag and, 188
  obstructive sleep apnea syndrome and, 216
Hypnotic-dependent sleep disorder, 146–147
Hypomania, 294
Hypopneas, 210–212
Hypothalamus, 6–7, 28–32, 58–65
  astrocytomas of, 345
  encephalitis lethargica and, 339
  Kleine-Levin syndrome and, 343
  Prader-Willi syndrome and, 342
  sleep-onset REM periods and, 163, 345
Hypoventilation. See also Hypercapnia
  brainstem lesions and, 345–346
  central alveolar, 251–252
  neuromuscular disorders and, 253–257
  obesity-hypoventilation syndrome and, 252–253
Hypoxia
  obstructive sleep apnea syndrome and, 217

Hypoxia (*continued*).
  ventilatory response to, 49
Hysteria, 283

Idiopathic hypersomnia, 163–166, 168–170
Idiopathic recurring stupor, 344
Impaired nocturnal penile tumescence, 274–277
Inadequate sleep hygiene, 144–145
Infancy, sleep duration in, 73–74
Inflammatory bowel disease, 316–317
Insomnia, 101, 127–151
  alcohol and, 129, 301–302
  altitude, 241
  anxiety disorders and, 297–299
  behavioral treatment of, 140–142
  bereavement and, 296
  central sleep apnea and, 238
  conditioning and, 132–133, 140–142
  depression and, 291–292
  dysthymic disorder and, 295
  fatal familial, 340–341
  fibromyalgia and, 319
  heart disease and, 308
  hypnotic use and, 133–136
  idiopathic, 142
  medications causing, 145
  midwinter, 198
  obsessive-compulsive disorder and, 299
  older adults and, 76–77
  panic disorder and, 298
  Parkinson's disease and, 329
  progressive supranuclear palsy and, 332
  psychiatric disorders and, 130–131, 288
  psychophysiologic, 139–142
  relaxation therapies for, 141–142
  renal failure, chronic, and, 320
  schizophrenia and, 289
Insufficient sleep syndrome, 120–123
Insulin, 18, 35
Interleukin-1, 19, 35
  sleeping sickness and, 340
Intermittent positive pressure ventilation. *See* Ventilation, positive pressure
Internal desynchronization, 58, 63–64
International Classification of Sleep Disorders, 96–98
Interstitial lung disease, 313
Intestinal disorders, 316–317
Irregular sleep-wake pattern, 195–197
Irritable bowel syndrome, 316–317

Jactatio capitis nocturna. *See* Rhythmic movement disorder
Jet lag, 186–188

Kleine-Levin syndrome, 342–344
Kleitman, Nathaniel, 6
  Mammoth Cave experiments, 65
Kyphoscoliosis, 255

Landau-Kleffner syndrome, 370
Laser-assisted uvulopalatopharyngoplasty, 229
Leigh's syndrome, 251
Lennox-Gastaut syndrome, 362
Levodopa, 179–183
  nightmares and, 278
  Parkinson's disease and, 331–332
Light
  Alzheimer's disease and, 327–328
  delayed sleep phase syndrome and, 193–195
  effects on biological clock function, 59–60
  jet lag and, 188
  phase shifting effects of, 60
  seasonal affective disorder and, 296
  shift work sleep disorder and, 191
Limit setting sleep disorder, 148
Limp man syndrome, 345
Load compensation, 50, 52
Locus ceruleus, 7, 29, 32–33
Long face syndrome, 214
Long sleepers, 122
Lung diseases, 312–314

Maintenance of Wakefulness Test, 106, 118
Major histocompatibility complex, 158
Mania, 294–295
Masking stimuli, 61
Maxillo-mandibular osteotomy, 229
Medical causes of disordered sleep, 307–324
Medroxyprogesterone, 248, 314
Medulla, 5, 33, 45–48. *See also* Brainstem
Melatonin, 61
  Alzheimer's disease and, 328–329
  insomnia and, 136
  jet lag and, 188
  shift work sleep disorder and, 191
Memory formation during sleep, 19–20
Mental retardation
  irregular sleep-wake pattern and, 195–197
  sleep bruxism and, 280
Mesencephalon. *See* Midbrain
Metabolic encephalopathies, sleepiness and, 114–115
Methylphenidate. *See* Stimulants
Midbrain. 5–6, 27–34. *See also* Brainstem
  astrocytomas of, 345
  encephalitis lethargica and, 339
  traumatic injury of, 346
Micro-sleep, 119
Midwinter insomnia, 198
Mixed apneas, 210, 211, 244, 310
Modafinil, 170
Monoamine oxidase inhibitors, 278, 294
Monoamines, 7, 29–30, 160
Mood disorders, 290–296. *See also* Depression
Mood stabilizing agents, 295
Multiple sclerosis, REM sleep behavior disorder and, 273
Multiple Sleep Latency Test, 106, 117–118, 123
  depression and, 292
  Kleine-Levin syndrome and, 343
  narcolepsy and, 167–168
  obstructive sleep apnea syndrome and, 224
  Prader-Willi syndrome and, 342

Multisystem degeneration, 334–335
Muramyl peptides, 35
Muscles, respiratory, 40–45
Myasthenia gravis, 253–257
Myopathies, 253–257
Myotonic dystrophy, 253–257

Naps, 73–76
    narcolepsy and, 170–171
Narcolepsy, 152–174
    brain lesions and, 162–164, 345
    canine, 160–161
    cataplexy, 155, 159–162, 166–167, 171, 345
    cholinergic systems and, 160–161
    driving and, 171
    genetic basis, 157–159
    human leukocyte antigens and, 157–159
    monoamines and, 160–161
    monosymptomatic, 161–162
    MSLT and, 167–169
    naps and, 170–171
    nocturnal sleep disruption with, 156, 171
    pathophysiology, 159–161
    polysomnography and, 167
    psychiatric symptoms and, 157
    sleep-onset REM periods and, 159–160, 167–169
    twin studies of, 158
Nasal continuous positive airway pressure. *See* Continuous positive airway pressure
Neuromuscular diseases, 253–257
Neuropathies, small-fiber, 178–179
Neurons, respiratory, 46–48
Nicotine, 303
Nightmares, 265, 277–279
    Parkinson's disease and, 329, 331
Nocturnal complex motor activity, causes of, 264, 355
Nocturnal delirium. *See* Sundown syndrome
Nocturnal dissociative disorder, 283
Nocturnal eating/drinking syndrome, 148–149
Nocturnal leg cramps, 179, 269–270
Nocturnal myoclonus, 175
Nocturnal paroxysmal dystonia, 370
Nocturnal penile tumescence, 274–277
Nocturnal wandering. *See* Sundown syndrome
Non-24-hour sleep-wake syndrome, 197–198
Noradrenaline. *See* Noradrenergic neurons
Noradrenergic neurons, 29
    Alzheimer's disease and, 327
    thalamic infarcts and, 345
NREM sleep. 6–7, 10–13, 15, 20–22, 29–32. *See also* Slow wave sleep
    seizures and, 352–353
Nucleus
    ambiguus, 46–47
    Kolliker-Fuse, 48
    lateral dorsal tegmental, 30
    magnocellularis, 33
    pedunculopontine tegmental, 30
    raphe, 29–30
    retroambigualis, 46–47
    suprachiasmatic, 58–59, 196. *See also* Biological clock
    tractus solitarius, 31
    tuberomamillary, 31

Obesity, obstructive sleep apnea syndrome and, 206–207, 214
Obesity-hypoventilation syndrome, 203, 252–253
    Prader-Willi syndrome and, 342
Obsessive-compulsive disorder, 297, 299
Obstructive sleep apnea syndrome, 202–236
    accidents, 220
    alcohol and, 216, 302
    Alzheimer's disease and, 327
    bilevel positive airway pressure and, 227
    body position and, 212–213
    brainstem lesions and, 317, 345–346
    cardiac function, 217–219
    central apneas and, 210
    cephalometric radiographs, 224
    complications, 220–222
    continuous positive airway pressure and, 225–228
    dental appliances for, 229
    dysmenorrhea, 220
    end-tidal pCO2 recordings, 223
    gastroesophageal reflux disease and, 206–207, 315
    genetic factors, 213–214
    genioglossus advancement with hyoid suspension for, 229
    heart disease and, 308–312
    hypopneas in, 210–212
    intracranial hemodynamics, 219–220
    intraesophageal pressure recordings, 223–224
    laser-assisted uvulopalatopharyngoplasty for, 229
    maxillo-mandibular osteotomy for, 229
    medications for treatment of, 230
    mixed apneas and, 210, 211
    multisystem degeneration and, 334
    obstructive hypopneas, 210–211
    oxygen for treatment of, 230
    paradoxical breathing during apneas, 209
    Parkinson's disease and, 330
    Prader-Willi syndrome and, 342
    pulmonary arterial pressure, 217
    renal function, 220
    renal failure, chronic, and 320–321
    rheumatoid arthritis and, 317–318
    seizures and, 354
    sleep enuresis and, 282
    surgical treatment of, 228–229
    systemic effects of, 217–221
    tonsillar hypertrophy and, 213
    tonsillectomy with adenoidectomy for, 228
    tracheostomy for, 229
    uvulopalatopharyngoplasty for, 228–229
    vascular morbidity and mortality, 220–221
    weight loss for, 230
Olivopontocerebellar atrophy. *See* Multisystem degeneration
Ondine's curse. *See* Central alveolar hypoventilation syndrome
Ontogeny of sleep, 70–81
Opiates, 178, 180–181, 303
Osteoarthritis, 317–318
Oxygen therapy
    central sleep apnea and, 247

Oxygen therapy (continued).
　heart disease and, 312
　lung diseases and, 314

Painful legs and moving toes syndrome, 179
Panic disorder, 297–298
Paradoxical breathing, 209
Parasomnias, 260–287
Parathyroid hormone secretion during sleep, 18
Parkinson's disease, 329–332
　REM sleep behavior disorder and, 273, 330
Paroxysmal depolarizing shift, 352
Paroxysmal nocturnal dyspnea, 308
Peduncular hallucinosis, 271
Peptic ulcer disease, 316
per gene, 59
Periodic limb movements, 177–178, 181–183
　Alzheimer's disease and, 327
　Parkinson's disease and, 330
　post-traumatic stress disorder and, 298–299
　renal failure, chronic, and, 320–321
　restless legs syndrome and, 177–178
　rheumatologic disorders and, 317–318
Persistent vegetative state, 16
Phylogeny, 3–4
Physiologic changes during sleep, 17–20
Pickwickian syndrome. See Obesity-hypoventilation syndrome
Pneumocardiogram, 246
Pneumotaxic center, 39, 48
Polymyalgia rheumatica, 179
Polysomnography, 105–106
　arousal, 14–15
　central apnea, 210, 245, 246, 248, 309
　generalized seizure, 360
　hypnagogic hypersynchrony, 74
　hypopnea, 212
　mixed apnea, 211, 218, 244, 310, 311
　obstructive apnea, 208, 209, 210, 218, 219, 226, 243
　partial seizure, 358, 365
　periodic leg movements, 177, 182
　REM sleep, 14
　sleep-onset REM period in narcolepsy, 159
　slow wave sleep, 12–13, 75, 319
　spikes and spike-waves, 356, 357, 359, 363, 366, 369
　spike-waves, continuous, during NREM sleep, 371
　stage 1 sleep, 10–11
　stage 2 sleep, 11–12
　unexpected EEG finding during, 357–359
　upper airway resistance syndrome, 213, 223
　wakefulness, 9–10
Pons 6–7, 27–34. See also Brainstem
Pontine-geniculate-occipital spikes, 34
Post-polio syndrome, 255
Post-traumatic hypersomnia, 162, 166
Post-traumatic stress disorder, 278, 297–299
Prader-Willi syndrome, 341–342
Presleep behavior, compulsive, 345
Pre-Botzinger complex, 46–48
Pre-dormital myoclonus. See Sleep starts
Process C, 20, 65–67, 187
Process S, 20–21, 65–67, 187

Progressive supranuclear palsy, 332–334
Prolactin secretion during sleep, 18
Prostaglandins, 35
　sleeping sickness and, 340
Pseudoinsomnia, 142–144
Psychiatric disorders, 131, 138, 288–306
　sleepiness and, 115–116
Psychostimulants. See Stimulants
Pulmonary congestion, central sleep apnea and, 241
Pupillometry, 117

Quiet Sleep, 71–72, 78–80

REM sleep, 14, 21–23, 32–34
　central apneas during, 242
　delirium tremens and, 301
　depression and, 291–293
　deprivation, 120
　dreams and, 83–86
　epilepsy with continuous spike-waves during slow wave sleep and, 371
　epileptiform activity and, 352–353
　fatal familial insomnia and, 341
　genital changes during, 18–19, 274–277
　lung diseases and, 312
　multisystem degeneration and, 334
　narcolepsy and, 159–160, 167–169
　obstructive sleep apnea syndrome and, 212
　parasomnias related to, 270–279
　penile erections during, 274–277
　progressive supranuclear palsy and, 333
　REM sleep behavior disorder and, 271–274
　schizophrenia and, 288–289
　seizures and, 352–353
　short latency to, conditions associated with, 293
　sleep paralysis and, 270
　sleep talking and, 269
REM sleep behavior disorder, 271–274
　Alzheimer's disease and, 327
　alcohol withdrawal and, 301
　fatal familial insomnia and, 341
　Lewy body dementia and, 327
　multisystem degeneration and, 334
　narcolepsy and, 156
　Parkinson's disease and, 330–332
REM sleep-related sinus arrest, 274
Renal failure, chronic, 320–321
Renal function during sleep, 18
Respiration during sleep, 17–18, 50–53. See also Breathing
Respiratory load compensation, 50, 52
Respiratory pause, 77
Restless legs syndrome, 175–181
　renal failure and, 320–321
Reticular formation, 27–31
Retinohypothalamic tract, 58
Rheumatoid arthritis, 317–318
Rheumatologic disorders, 317–320
Rhythmic movement disorder, 267–268
Rhythmie du sommeil. See Rhythmic movement disorder

Sarcoidosis, central nervous system, 345
Schizophrenia, 288–290
Scoring of sleep studies, 9, 106–108
Seasonal affective disorder, 295–296
Sedatives. See Benzodiazepines; Hypnotics
Seizures, 350–372. See also Epilepsy
   absence, 351, 361–362
   atonic, 351, 362
   complex partial, 351, 364–366
   generalized, 351, 359–361
   myoclonic, 351, 361
   partial, 351, 358, 364–370
   simple partial, 351
   tonic, 351, 362
Senescence and sleep, 76–77
Serotonergic neurons, 28–30
   degeneration in Alzheimer's disease, 327
Serotonin reuptake inhibitors, selective, 294
Shift work sleep disorder, 189–192
Short sleepers, 122
Shy-Drager syndrome. See Multisystem degeneration
Sickle cell anemia, 282
Sleep, 3–26
   Active, 15, 71–72, 77–80
   alpha-delta, 13, 318–320
   clinical features of, 15–16
   consciousness and, 15–16
   core, 21–22, 122
   delta. See Slow wave sleep
   drug effects on, 100
   drug-induced, 115
   indeterminate, 71
   insufficient amounts of, 114
   optional, 21–22, 122
   Quiet, 71–72, 78–80
   slow wave. See Slow wave sleep
Sleep attacks, 154–155
Sleep bruxism, 279–280
Sleep deprivation, 118–120
   activating procedure for EEG, 353–354
   depression and, 294
   juvenile myoclonic epilepsy and, 361
Sleep disorders
   classification of, 96–98
   evaluation of patients with, 98–104
Sleep drunkenness, 121, 164
Sleep enuresis, 281–283
Sleep factors, 34–36
Sleep fragmentation, 77, 114
   in Alzheimer's disease, 327
Sleep history, 99–102
Sleep homeostasis, 20
Sleep hygiene, 144–145
   schizophrenia and, 290
Sleep inertia, 121
Sleep jerks. See Sleep starts
Sleep log, 104–105
   Alzheimer's disease and, 328
   non-24-hour sleep-wake syndrome and, 198
Sleep medicine, development of, 95–96
Sleep need, 23, 73–77
Sleep onset, 15, 31
Sleep paralysis, 155–156, 167, 171
   isolated, 270–271
Sleep questionnaires, 104–105

Sleep restriction therapy, 140–142
Sleep spindles, 12, 31–32
Sleep stages, 9–15, 71
Sleep starts, 15, 268–269
Sleep state misperception, 142–144
Sleeptalking, 269
Sleep terrors, 261–267
Sleep-onset myoclonus. See Sleep starts
Sleep-onset REM periods
   causes of, 168, 293
   infancy and, 72
   narcolepsy and, 159–160, 168
   Prader-Willi syndrome and, 342
Sleep-related painful erections, 277
Sleep-onset association disorder, 147–148
Sleep-wake rhythm
   constant environmental conditions and, 62–65
   development of circadian pattern, 72–73
   evolution after birth, 72–77
   non-24 hour sleep-wake schedules and, 65–67
Sleep-wake schedule, 99
Sleep-wake state differentiation, 70–72
Sleep-wake transition disorders, 267–270
Sleepiness, 100–101, 112–118. See also Hypersomnia
   alcohol and, 300
   brainstem lesions and, 345
   body temperature and, 58
   circadian aspects of, 57–58, 112
   differential diagnosis, 166
   idiopathic recurring stupor and, 344
   myotonic dystrophy and, 255
   obstructive sleep apnea and, 205, 216–217
   Prader-Willi syndrome and, 342
   psychiatric disorders and, 115–116, 291–292
   thalamic infarcts and, 345
Sleeping sickness, 339–340
Sleepwalking, 261–267
Slow wave sleep, 9, 12–13, 20, 22, 31–32, 67, 75, 319
   arousal disorders and, 263
   schizophrenia and, 289
Smoking
   impaired nocturnal penile tumescence and, 275
   obstructive sleep apnea syndrome and, 206–207
Snoring, 204, 213, 215
Spinal cord injuries, impaired nocturnal penile tumescence and, 276
Spontaneous internal desynchronization, 63–64
Stanford sleepiness scale, 116–117
Stimulants, 170
   diencephalic lesions and, 345
   encephalitis lethargica and, 339
   Kleine-Levin syndrome and, 343
   Prader-Willi syndrome and, 342
Stimulus control therapy, 140–141
Stress
   insomnia and, 130–132, 146
   nightmares and, 278
Stridor, 334
Stupor, idiopathic recurring. See Idiopathic recurring stupor
Substance-related disorders, 303–304
Sudden infant death syndrome (SIDS), 249–251
Sudden unexplained nocturnal death, 283–284
Sundown syndrome, 327–328
Systemic lupus erythematosus, 317

Temporal isolation facilities, 62–63
Testosterone, 18
    nocturnal penile tumescence and, 275–276
Thalamus, 6–7, 28–32
    fatal familial insomnia and, 341
    infarcts of, 345
Thermal regulation during sleep, 18
tim gene, 59
Thyroid stimulating hormone, secretion during sleep, 18
Tonsillar hypertrophy, obstructive sleep apnea syndrome and, 213–214
Tonsillectomy with adenoidectomy, 228
Top-of-the-basilar syndrome, 345
Tracheostomy, 229, 253, 256–257
Tricyclic antidepressants, 135–136
    Alzheimer's disease and, 329
    depression and, 293–294
    fibromyalgia and, 320
    nightmares and, 279
    Parkinson's disease and, 332
    REM sleep behavior disorder and, 272
    sleep terrors and, 267
Trypanosoma brucei gambiense, 340
Trypanosoma brucei rhodesiense, 340
Tsetse fly, 340

Ultradian rhythms, 67
Ultra-short sleep-wake schedule, 65, 67
Upper airway, 41–46
    obstructive sleep apnea syndrome and, 207–215

Upper airway resistance syndrome, 211, 213, 223–224
Uvulopalatopharyngoplasty, 228–229

Vasoactive intestinal peptide, 35–36
Ventilation. *See also* Breathing; Bilevel positive airway pressure; Continuous positive airway pressure
    brainstem lesions and, 345–346
    changes during sleep, 53
    positive pressure, 252, 256, 314
    negative pressure, 252, 256
Ventilatory pump, 40–41
Ventilatory response to hypercapnia, 48–49, 239–241
Ventral respiratory group, 46–48
Ventral tegmental area, 28, 30, 330
Vesper's curse, 179
Video-EEG polysomnography, 265–267, 356–357, 364–365, 368–369

Wakefulness stimulus, 50, 52, 239–240
Waking dyskinesias, 176
Whipple's disease, central nervous system, 345
Wilkinson Auditory Vigilance Test, 117

Zeitgebers, 61